P9-CKV-866

Pandora's Poison

Pandora's Poison

Chlorine, Health, and a New Environmental Strategy

Joe Thornton

MIT Press
Cambridge, Massachusetts
London, England

This book was printed on paper bleached without chlorine gas, chlorine dioxide, or any other chlorine-based bleaching agent.

This book was set in Sabon by Graphic Composition, Inc. and was printed and bound in the United States of America.

Library of Congress Cataloging-in-Publication Data.

Thornton, Joe.

Pandora's poison : chlorine, health, and a new environmental strategy / Joe Thornton.

p. cm.

Includes bibliographical references and index.

ISBN 0-262-20124-0 (hc : alk. paper)

1. Organochlorine compounds—Toxicology. 2. Organochlorine compounds—Environmental aspects. 3. Environmental policy. I. Title.

RA1242.C436 T48 2000

615.9′51—dc21

99-057011

Contents

Preface

This book presents a new way to solve one of the most pressing environmental problems of our time: the worldwide accumulation of toxic pollutants. It focuses on one very large group of chemicals—chlorinated organic substances, called organochlorines for short—that dominate all lists of global contaminants and environmental health hazards. Organochlorines are produced when chlorine is used in the chemical, paper, water treatment, and other industries, and they include a variety of familiar and obscure plastics, pesticides, solvents, refrigerants, and accidental byproducts. Organochlorines are not the only chemicals worthy of concern, but they are an obvious priority, and the proper course of action to deal with this group of substances illustrates how we can best address the rest of the pollutants that confront us.

In the 1970s, industrialized countries adopted a framework for assessing and regulating toxic chemicals that remains in force today. This approach, which I call the Risk Paradigm, attempts to manage individual pollutants using scientific and engineering tools, including risk assessment, toxicological testing, epidemiological investigations, pollution control devices, and waste disposal technologies. The effort to manage organochlorines in this way has failed resoundingly because regulations focused on severe local health risks cannot effectively address the subtle, long-term, global hazards posed by these substances. In part I of this book, I describe the problem that has emerged as a result of this failure. Since the chemical industry began producing organochlorines in large quantities around World War II, a witch's brew of toxic, persistent pollutants has come to blanket the entire planet, from suburban backyards to the deep oceans, from our own bodies to snow at the North Pole. Even at very low doses, these compounds interfere with fundamental biological processes and cause a stunning array of health effects. Levels once thought to be safe have been repeatedly found to cause subtle forms of health damage not previously investigated or detected. Organochlorines that have accumulated in our bodies are passed to the next generation across the placenta and through mother's milk, so the developing child, exquisitely sensitive to chemical injury, is at particular risk. Seen through the established lens of risk assessment, this may not be a particular problem; from a new perspective that focuses on multigenerational, population-wide erosion of our ability to have healthy children, fight off disease, and function to our full biological and intellectual potential, however, the evidence establishes cause for great concern. Organochlorines pose a long-term threat to the health of humans and wildlife, and a growing body of evidence indicates that the damage has already begun to occur.

That is the problem, but this book is focused on a solution. In part II, I describe the problem's cause—the production, use, and disposal of chlorine and organochlorines in modern industry—in a way that leads directly to a new paradigm for environmental policy. The characteristics of organochlorines—their sheer number, the ways they are produced, their chemical and toxicological properties, the overwhelming lack of data on the hazards most of them pose, and the economic dynamics of the industry that produces them—cannot be addressed or understood by the current approach to environmental policy and science. An adequate solution to organochlorine contamination requires a major shift in how chemicals are assessed and regulated. The new approach must take account of the complexity of natural systems and the impacts of chemical pollution upon them, the limitations of science to predict and diagnose those effects, and the inability of technological fixes like pollution control and disposal devices to prevent the continuing buildup of these compounds around the world.

In the book's final part, I describe this new approach, which I call the Ecological Paradigm, and the specific actions it prescribes to protect health and the environment from organochlorines. In contrast to the Risk Paradigm's narrow, mechanistic focus, the new model takes a holistic approach to the assessment of environmental hazards, focusing on the complex and unpredictable nature of natural systems and the subtle, long-term, and large-scale damage that human activity can do to them. Because scientific tools cannot adequately understand the hazards of pollutants on a chemical-by-chemical basis—and technological tools cannot effectively manage them this way—the Ecological Paradigm shifts the focus to broad classes of hazardous substances and the technologies that produce them. To address the hazards these chemicals and processes pose, the new approach relies on the commonsense principle of Clean Production: instead of trying to manage individual pollutants, we should always use the cleanest available production technologies to fulfill society's needs without using and producing toxic chemicals. In chapter 9, I show in detail how the Ecological Paradigm can be applied to the family of chlorine-based products and processes in a practical program of implementing safer, chlorine-free technologies, which I describe in chapter 10 for the major uses of chlorine.

This book uses scientifically-derived information to make an argument about policy. It is an open attempt to persuade readers that society should pursue a specific set of policies in a democratic framework. It is a work of synthesis, analysis, and advocacy, not a traditional scientific review that cites all the evidence on a narrow subject and evaluates it from a position of apparent objectivity and distance. My subject is broad and interdisciplinary, embracing information from industrial and environmental chemistry, toxicology, molecular biology, epidemiology, ecology, wildlife biology, economics, engineering, politics, and ethics. If I were to review every relevant study, this book would be thousands of pages in length. Further, no analyst of policy can be truly objective, because the process of weighing options for social action always filters the findings of science through a set of political and ethical assumptions and values. With that in mind, I have tried to do two things: to make explicit the ethical and political views that undergird my own evaluation of the science and to be as fair as possible in my presentation of the scientific evidence. I cover what I believe to be the most important information relevant to the case I am making and evaluate its strengths and weaknesses, but I do not claim balance or objectivity, because these are neither appropriate nor possible in this kind of effort.

That science and politics cannot be untangled in the sphere of environmental policy is a theme throughout the book. In each chapter, I deconstruct the empirical basis for the scientific claims that guide and justify the current regulatory system, show how politics have shaped this "knowledge," and propose an alternate view that is consistent with the available data and with a politics and ethics that I view as democratic and humane. With this approach, I combine a critical scientific and policy analysis with the insights developed over the past two decades in the sociology of scientific knowledge. To say that science veils and is shaped by hidden political positions runs counter to the dominant model of science as a source of purely objective truths about the natural world. It also raises serious questions about the proper relationship between science and policy—and between experts and the general public—in a democratic society. These issues, which I explore in the final chapter, are central to the Ecological Paradigm, which is designed to wrest decisions about the environment, health, and technology from the hands of experts and elites and place them in their proper sphere of democratic debate and determination.

I have worked to make this book accessible to any literate person, not just specialists in the environmental field. To be persuasive, however, my argument requires documentary support so extensive that it would make the text impenetrably dense. To preserve the book's accessibility along with its rigor, I have included appendixes and extensive notes that follow the main text; the appendixes present data, and the notes give citations to the scientific literature, explain the basis for calculations I present, and provide more detail on some specific issues and studies for which there was not room in the chapters themselves. This format is intended to create two books under one cover: a readable essay on a pressing subject of environmental policy and a documented exploration for those who would pursue the issue further. The introduction provides a brief overview of the book's argument, without examples or references, so that readers of either type have an idea of the book's trajectory before jumping on board.

I could not have written this book without the prior work and continuing support of countless colleagues in the environmental movement and the scientific community. I began to develop this argument and the information that supports it during a long period at Greenpeace, where I benefited enormously from the research of others. I owe particular debts to Pat Costner and Paul Johnston of Greenpeace International's Science Unit, who made pioneering efforts to establish a base of information on organochlorine contamination and its sources. I thank Ruth Stringer, Charlie Cray, Mark Floegel, Thomas Belazzi, Manfred Krautter, Bonnie Rice, and Beverly Thorpe, all of whom made important contributions to my understanding of chlorine chemistry. I also draw on the work of Jeff Howard, who developed a comprehensive database on contamination of the human population, and Susan Cooper, who researched chlorine disinfection and its alternatives.

I thank all the scientists who did the primary research that I cite throughout this book: the epidemiologists, toxicologists, ecologists, and chemists whose work has established our current understanding of the effects of toxic chemicals on natural systems. A number of scientists have also made major contributions in synthesizing this body of information and its implications for public health. I draw extensively on work of this type by Theo Colborn, Paul Connett, Devra Lee Davis, Sam Epstein, Glen Fox, Ken Geiser, Peter Montague, David Ozonoff, Mary O'Brien, Arnold Schecter, Sandra Steingraber, and Tom Webster: their work has informed my thinking and led me to a wealth of information of which I would otherwise have been unaware. I owe special thanks to Jack Valentine and Ross Hume Hall, formerly of the International Joint Commission's Science Advisory Board, who were the first to make a scientific and moral argument for addressing organochlorines as a class. Finally, there is Barry Commoner, who understood these issues thirty years before the rest of us and remains a pioneering thinker to this day.

Whose ideas about politics are truly original? The positions I advocate in this book emerged from years of working with a team of people who set for themselves the modest goal of helping to bring about a just and healthy society. I had the privilege of discussing specific issues of chlorine chemistry and broader ones of environmental politics, often in strange places at odd hours, with a group of brilliant individuals, including Jack Weinberg, Rick Hind, Charlie Cray, Joan D'Argo, Jay Palter, Jeff Howard, Scott Brown, Gail Martin, Beth Newman, Niaz Dorry, Margie Kelly, Brian Hunt, Kenny Bruno, Lynn Thorp, Dave Rapaport, Bill Walsh, and Lisa Finaldi. I am also indebted to communities across the world that are tirelessly confronting pollution sources, fighting battles at the front lines of the global effort to protect health and the environment. I am particularly grateful to the people of East Liverpool and Fairborn, Ohio, and of numerous communities around the Great Lakes, all of whom have been a rich source of wisdom, perspective, and inspiration.

I thank those who reviewed drafts and made suggestions that strengthened the book in innumerable ways: David Cleverly, Paul Connett, Charlie Cray, Devra Davis, Dave Derosa, Ken Geiser, Michael Gilbertson, Jeff Howard, Kelly Moore, and David Ozonoff. Lincoln Abbey drew figure 6.4; I appreciate his talent and generosity almost as much as his friendship.

Columbia University's Department of Biological Sciences and the Center for Environmental Conservation—particularly Darcy Kelley, Rob DeSalle, Don Melnick, and Bob Pollack—gave me the support and liberty to pursue this rather unorthodox project. Kathleen Kehoe, Neil Silvera, and Joy Wang of Columbia's libraries tracked down what would have been, without their help, an unmanageable quantity of scientific literature. I am grateful to Jackie Hunt Christiansen, Charlie Cray, Rick Hind, and Dave Derosa for generously providing literature from their libraries.

My greatest thanks go to my family, Margie Kelly and our son, Harry. As anyone who has ever lived with a writer knows, a book in progress is like an unwelcome guest, intruding constantly on rituals and relationships, demanding endless time and energy. Margie not only suffered this invasion with patience and good humor but commented on drafts, discussed conceptual problems, and offered a wealth of support and love when my spirits flagged. Without her help, I could never have completed the project and might have lost my mind trying. Harry, who is almost exactly the same age as this book, was endless inspiration, just by becoming himself.

Introduction

The Argument

People in the age of science and technology live in the conviction that they can improve their lives because they are able to grasp and exploit the complexity of nature and the general laws of its functioning. Yet it is precisely these laws which, in the end, tragically catch up with them and get the better of them. People thought they could explain and conquer nature—yet the outcome is that they destroyed it and disinherited themselves from it.

—Vaclav Havel, "Politics and Conscience"[1]

In the next month, my wife, Margie, and I will have our first child. Our house is still quiet, for a few more weeks at least, as we wait—endlessly, it seems—for the baby's arrival. Under the surface, however, a host of questions roils: about the delivery, the baby's health, how his personality will develop, what life he will lead, the kind of world he will inhabit. We are ready for some answers, but the baby stubbornly holds out, unwilling to provide them on anyone's schedule but his own.

There are some things I already know about the baby and his history, however, that I might prefer never to have learned. I know that my semen contains scores of pollutants that may have damaged the DNA I contributed to the baby. I know that Margie, over the course of her life, has accumulated hundreds of industrial compounds in her tissues, and these substances have crossed the placenta and entered the baby's bloodstream. I know that these chemicals are flushed out of the body by breast-feeding, so the baby will get even higher doses after he is born. And I know of an emerging body of evidence that exposure to trace amounts of these compounds early in life can cause a range of subtle and severe problems, from cancer to reduced IQ, from infertility to a compromised immune system.

Some kinds of permanent damage may not become manifest until a child reaches adulthood.

Our baby is no different in this way from any other in today's world. Both Margie and I grew up in pleasant suburbs, reasonably far from major pollution sources. Neither of us works in chemical-intensive industries. So how did we get so contaminated? Through the food supply, the air, and the water we all share. We live on a planet that has become the repository for the products, emissions, and wastes of modern industry. Since about 1940, the production of synthetic organic chemicals in the United States has grown more than thirty-fold.[2] Over 70,000 industrial chemicals are now synthesized and sold on the market, some in amounts over a billion pounds per year.[3] Many are resistant to natural degradation processes, so they gradually accumulate in the environment and are distributed across the globe on currents of wind and water. As a result, a cocktail of hundreds or thousands of man-made chemicals can now be found absolutely anywhere on the planet, from the deep oceans to the North Pole, from the Mississippi River to our own bloodstreams.

The result is a dizzying array of environmental problems that have filled the news since the 1960s: DDT and the decline of bald eagles; toxic waste at Love Canal; cancer among Vietnam veterans exposed to Agent Orange; chloroflourocarbons and the ozone hole; PCBs in polar bear tissue; herbicides in groundwater throughout the Midwest; dioxin in fish downstream from pulp and paper mills. For many people, the hazards seem overwhelming in number, complexity, and the technical expertise necessary to understand them. Solutions seem even less accessible; the apparently sophisticated environmental laws of industrialized countries, with their byzantine and costly regulations, have failed to halt the tide of contamination.

From another perspective, however, the situation is far simpler than it first seems. The litany of problems listed above—and hundreds of less infamous but just as serious hazards—all involve chemicals of a single class, called organochlorines because they are organic (carbon-based) chemicals that contain one or more chlorine atoms. Not all pollution is due to organochlorines; some metals and nonchlorinated synthetic organic chemicals that do not contain chlorine also pose public health threats. But organochlorines dominate virtually all official and unofficial lists of hazardous pollutants in the environment, wildlife, and human tissues.

These pollutants all arise from a single root cause: the industrial production, use, and disposal of chlorine gas and chemicals derived from it—a family of processes called chlorine chemistry. Chlorine chemistry begins at a handful of large chemical facilities, where an extremely powerful electric current is passed through a solution of salt water. In chemical terms, salt is sodium chloride, a stable natural compound that circulates constantly through the ecosystem and our bodies, never combining with the organic matter of which we are made. Industry's electrical energy transforms salt's stable chloride ions into molecules of chlorine gas, a heavy, violently reactive, greenish gas that does not occur in nature. About three-quarters of the chlorine is used within the chemical industry as a feedstock for the production of over 11,000 organochlorines,[4] including plastics, pesticides, solvents, and chemical intermediates, virtually all of which are also foreign to nature. The remaining chlorine is sold to other industries for direct use—to pulp and paper mills as a bleach, for instance, or to sewage plants as a disinfectant.

For well-understood reasons, the chemistry of the chlorine atom gives chlorine gas and organochlorines useful properties, but these same qualities create enormous environmental problems. First, chlorine gas is highly reactive, combining quickly and randomly with whatever organic matter it encounters, so it is an effective bleach, disinfectant, and feedstock for synthesizing chemicals. Whenever chlorine is used, however, this same quality means that a diverse stew of hundreds or thousands of organochlorine by-products is formed incidentally.

Second, chlorination radically affects the chemical stability of organic chemicals, usually increasing it but sometimes decreasing it. Stable organochlorines are useful as plastics, refrigerants, and other applications in which long life is a virtue. Organochlorines that are stable in their intended use, however, are also persistent in the environment, resisting natural degradation processes for long periods of time—centuries, in some cases—so they gradually build to higher and higher concentrations in air, water, and sediments. When chlorination decreases a chemical's stability, on the other hand, it makes it more reactive, so some organochlorines make useful intermediates for synthesis processes in the chemical industry. Reactive organochlorines, however, are much more likely than their non-chlorinated precursors to be converted into highly toxic and cancer-causing forms in the body. The chemical effect of chlorination is therefore to increase, in one way or another, the hazard that a chemical poses.

The third effect of chlorination sounds innocuous but creates a terrible problem. Adding chlorine atoms invariably increases the ability of organic chemicals to dissolve in oils, so organochlorines make excellent solvents for industrial processes, like equipment cleaning and surface-coating operations, that involve oil-based materials. Once oil-soluble organochlorines are released into the environment, however, they accumulate in the fatty tissues of living things, a process called bioaccumulation. Bioaccumulative compounds gravitate from the ambient environment into the food web, magnifying in concentration as they move upward from tiny organisms to large predators. By the time they get to the top of the food web—the tissues of people, eagles, polar bears, and other species—some organochlorines reach concentrations many millions of times greater than their levels in the ambient environment.

Finally, chlorination virtually always increases toxicity. This effect occurs because modulating the persistence, reactivity, and oil solubility of a chemical changes its interactions with proteins and fats inside the body in a way that can disrupt the natural processes of physiology and development. These qualities make organochlorines effective pesticides, antibiotics, and pharmaceuticals. But organochlorines not intended to be poisonous tend to be toxic too; once in the environment, the properties that made them useful for killing unwanted organisms also injure humans and wildlife.

If chlorine chemistry were practiced on a minor scale, it might not present a major problem. Industry first produced chlorine gas around the turn of the century, but its use was rather limited until after World War II, when the chlorine industry began to grow at a breakneck pace. Today the world chemical industry produces an astonishing 40 million tons of chlorine annually, most of which is directed into the generation of organochlorine products and by-products. These substances enter the environment in a variety of ways. Some organochlorines, such as pesticides and the by-products of paper bleaching, are dispersed into the environment directly, while others—polyvinyl chloride (PVC) plastic products, for instance, or the complex hazardous wastes generated during its manufacture—enter the ecosystem indirectly through incinerators or landfills.

Because of their persistence and bioaccumulation, organochlorines now contaminate absolutely every inch of the planet. Even in remote polar regions, thousands of miles from any industrial source, a diverse cocktail of organochlorines can be found in the tissues of whales, seals, and polar bears. They contaminate our bodies too. Hundreds of toxic organochlorines are now present in the fat, mother's milk, blood, semen, and breath of the general human population — people subject to no unusual exposures in their workplace or communities. Airborne organochlorines have even drifted to the stratosphere, where the chlorine they contain reacts with the ozone layer, breaking down the shield that filters out the sun's powerful ultraviolet rays. After just six decades of large-scale chlorine chemistry, we can now say that every person and animal on earth is exposed to a complex stew of toxic organochlorines, from the moment of conception—even before, since the developing sperm and egg encounter these poisons too—until the closure of death.

As early as the 1950s, it was clear that a few organochlorines were extremely toxic, causing cancer and disrupting the body's organ systems at low doses. Only in the last decade, however, has the true scope of the problem begun to emerge. Several hundred compounds have now been tested, and virtually all organochlorines examined to date cause one or more of a wide variety of adverse effects on essential biological processes, including development, reproduction, brain function, and immunity. Some organochlorines cause these effects at extraordinarily low doses—in parts per trillion concentrations, a ratio equivalent to one drop in a train of railroad tank cars ten miles long.[5] Further, molecular biology has made possible the study of mechanisms of toxicity, revealing that organochlorines disrupt biological processes at the most fundamental levels. Some are potent mutagens, undermining the integrity of the genetic messages in our DNA, while others block communication between cells or interfere with the control of gene expression, turning genes on and off at inappropriate times and altering the natural course of development and physiology. A large number of organochlorines have been found to mimic or otherwise interfere with the body's natural hormones, the potent chemical signals by which multicellular organisms regulate their development and coordinate the unified function of their parts.

The implications of universal exposure to compounds that can have effects of this sort are profound. Every species on earth, including humans, is now exposed to organochlorines that can reduce sperm counts, disrupt female reproductive cycles, cause endometriosis, induce spontaneous abortion, alter sexual behavior, cause birth defects, impair the development and function of the brain, reduce cognitive ability, interfere with the controlled development and growth of body tissues, cause cancer, and compromise immunity. If we stopped all further pollution today, these compounds would remain in the environment, the food web, our tissues and those of future generations for centuries.

Contamination by persistent organochlorines thus poses a long-term, global hazard to human health and the environment. The scale and severity of the threat is rivaled only by the hazards associated with climate change, nuclear technologies, and the reduction of biological diversity (itself caused in part by chemical pollution). Even more sobering, a growing body of evidence suggests that global toxic pollution is already contributing to a slow, worldwide erosion of the health of humans and other species. By the time stratospheric ozone levels return to normal later in the next century, ultraviolet radiation is expected to have afflicted millions of people with skin cancer, blindness, and immune suppression. People and many wildlife species are routinely exposed to some organochlorines in amounts that are near, equal to, or greater than the doses that cause adverse effects in laboratory animals. There is little doubt that organochlorines in the food web are responsible for major die-offs and population declines in a variety of wildlife, from marine mammals to a host of fish and bird species in the Great Lakes, due to severe reproductive, developmental, and immunological impairment. Humans exposed to organochlorines in the workplace or by accident manifest similar symptoms, and a growing body of epidemiological evidence suggests that the background organochlorine exposures to which the general population is subject may be linked to the incidence of many kinds of cancer, immune suppression, infertility, and developmental problems like birth defects, low birthweight, and an impaired ability to learn. Exposure to organochlorines may thus be an important factor in the increases in many of these diseases and conditions that have occurred worldwide in the past several decades.

The hazards that organochlorines pose are fundamentally different from the health and environmental risks that current models of environmental regulation were designed to address, so they cannot be understood using the tools and concepts of the current system. These tools and concepts constitute a paradigm,[6] a total way of seeing the world, a lens that determines how we collect and interpret data, draw conclusions from them, and determine what kind of response, if any, is appropriate. Today's environmental policies embody an approach, which I call the Risk Paradigm, that attempts to manage pollution by permitting chemical production, use, and release, as long as discharges do not exceed a quantitative standard of "acceptable" contamination. This approach assumes that ecosystems have an "assimilative capacity" to absorb and degrade pollutants without harm. It also assumes that organisms can accommodate some degree of chemical exposure with no or negligible adverse effects, so long as the exposure is below the "threshold" at which toxic effects become significant.

The Risk Paradigm puts these assumptions into operation with the pollutant discharge permit, a license to pollute that sets maximum legal release rates of individual chemicals from individual facilities. Many other forms of chemical regulation, including pesticide registrations and occupational exposure limits, are also based on "acceptable" exposures. Regulators determine the permissible amount of single pollutants with a technique called quantitative risk assessment,[7] which works backward from the acceptable exposure level to calculate the maximum release rate that will ensure that this level is not exceeded. Industries comply with these limits by installing pollution control devices—such as filters, scrubbers, and evaporators—that capture pollutants at the end of the smokestack or discharge pipe and move them to a different place. In rare cases, the Risk Paradigm has taken more restrictive action like banning a chemical, but only when the evidence from epidemiological or ecological studies is overwhelming that a specific substance has caused severe health and environmental damage.

I might have called this model the acceptable discharge paradigm, or the pollution control paradigm, or the technocratic paradigm; all these names refer to essential elements of today's regulatory system. But calling it the Risk Paradigm gets at the heart of the current approach more clearly than any of these other terms. For one thing, risk assessment is this system's primary tool for assessing chemicals and setting acceptable discharges. More subtly, "risks" by definition are quantifiable probabilities of things that either do or do not happen; the word neatly captures the fact that the Risk Paradigm considers only those kinds of health damage that can be expressed in this narrow, numerical way, like cancer or birth defects, while excluding impacts—such as immune suppression, altered behavior, or reduced fertility—that are difficult to quantify and may affect every individual in a population to some degree. "Risk" also says something about the system's faith in the scope and reliability of scientific knowledge: risks can be reliably quantified, as the current framework presumes, only if we thoroughly understand how ecosystems and organisms are organized and how they may respond to human-induced interventions. The word also evokes the Risk Paradigm's reductionist view of the link between causes and effects in nature: risks by definition are created by specific activities or substances; synergy, feedback, unpredictable cascades of effects, and temporal changes in the sensitivity of an organism or ecosystem play no role in this approach. Finally, "risk" says something about the politics of the current system: in common usage, risks are voluntary things that people *take* in expectation of some benefit—when we bet on the stock market, for instance, or board an airplane, or eat fatty foods. People can reduce the risks they take, but there is no way to eliminate them entirely, since the only way to live a risk-free life is to do nothing at all. These connotations resonate in the Risk Paradigm's assumption that some "acceptable" amount of chemically-induced risks must inevitably be taken in the course of economic production.

This book is a case study of the failure of dominant models of environmental science and policy, and it argues for a fundamentally new approach. The Risk Paradigm is utterly ill suited to addressing the long-term, global health threat that organochlorines pose. Its inadequacy begins with the very concept of acceptable discharges: chemicals that persist in the ambient environment or in the bodies of living organisms build to higher and higher concentrations over time, so acceptable discharges ultimately reach unacceptable levels. Further, recent research in toxicology and biology shows that for many effects—including cancer, developmental impairment, immune suppression, and some kinds of birth defects and neurotoxicity—there is no clear threshold of toxicity; any exposure, no matter how small, appears to contribute to the incidence or severity of disease and functional impairment. Moreover, pollution control devices merely shift pollutants from one place or environmental medium to another; they may reduce local pollution, but they do nothing to prevent global contamination.

Most important, the Risk Paradigm's focus on individual chemicals and individual dischargers offers no way to address the total pollution burden now accumulating in the environment. Thousands of individual facilities, each discharging the "acceptable" amount of thousands of different substances, together produce a cumulative global impact; the current system, focused only on the local parts, is and always will be blind to this problem of the whole. Synthetic chemicals are always produced in complex mixtures, and it is these mixtures, not neat packages of isolated chemicals, that cause health and ecological injury. Real organisms are simultaneously exposed to thousands of chemicals that interact in additive, inhibitory or synergistic ways, so an evaluation of the toxicity of a substance in isolation does not accurately predict the hazard it poses in the context of a myriad of other chemicals. Nor can epidemiology and ecology retrospectively link injury to individual substances; the tools available to these sciences can seldom untangle the complex webs of real-world cause and effect, and health damage is caused by exposure to complex chemical mixtures, which also interact with other causes of disease, like radiation and smoking. We can never fully comprehend environmental injury—or take adequate action to prevent more of it— through a lens that sees only singular substances acting in isolation.

The global hazard that organochlorines pose demands a new model for environmental policy. This approach, which I call the Ecological Paradigm,[8] focuses not on managing pollution but on preventing it. The new framework is founded on the view that ecosystems and organisms—and society too—are extraordinarily complex and dynamic systems in which innumerable parts are connected in webs of interdependency, multiple causality, and feedback loops, all of which change over time. The Ecological Paradigm seeks to protect these complex systems from both extreme local risks and the kinds of large-scale, long-term, subtle forms of damage that organochlorines and other chemicals can cause.

First and foremost, the Ecological Paradigm recognizes the limits of science: toxicology, epidemiology, and ecology provide important clues about nature but can never completely predict or diagnose the impacts of individual chemicals on natural systems. The implications for policy are obvious: since science leaves so much unknown, we cannot afford to make risky bets on its predictions or wait to protect health and the environment until we know for certain that some substance or technological practice has caused injury. Instead, we should avoid practices that have the potential to cause severe damage, even in the absence of scientific proof of harm. This rule, called the *precautionary principle*, is common sense: it says that we should err on the side of caution when the potential impacts of a mistake are serious, widespread, irreversible, and incompletely understood, as they are with the hazards of global toxic contamination.

In exhorting us to take early steps to prevent health and environmental damage, the precautionary principle says nothing about what kind of action is appropriate. To guide policy in practice, the Ecological Paradigm needs several additional principles. The first is a new standard for pollution regulations: called Zero Discharge, this rule would eliminate rather than permit the release of synthetic substances that are persistent or bioaccumulative and thus accumulate over time in the environment and our bodies. The second is a new technological approach for achieving environmental goals: Clean Production emphasizes front-end solutions, particularly the redesign of products and processes to eliminate the use and generation of toxic chemicals, before they need to be managed. Third is a new way of evaluating chemicals: Reverse Onus shifts the default state of environmental regulations from permission to restriction; the burden of proof, which now rests with society to prove that a chemical will cause harm, is shifted to those who want to produce or use a novel chemical. These parties must demonstrate in advance that their actions are not likely to pose a significant hazard. Chemicals already in commerce that do not meet this criterion should be phased out in favor of safer alternatives.

In the Risk Paradigm, a lack of data is misconstrued as evidence of safety, so untested chemicals are allowed to be used without restriction. Since the vast majority of chemicals have not been subject to toxicity testing, ignorance becomes the dominant factor in environmental decisions, and a generally laissez-faire system is the result. In contrast, the default state of the Ecological Paradigm is to avoid the use of chemicals that may harm health and the environment: we do not wait for proof of harm but always strive to reduce the use of substances that we have reason to believe may damage the environment. The Ecological Paradigm thus amounts to a program of continued reductions in the production and use of all synthetic substances, with priority given to chemical classes that are known to persist, or bioaccumulate, or cause severe or fundamental disruptions of biological processes.

By reversing the onus in environmental regulation, the Ecological Paradigm simply applies the standard that society now uses for pharmaceuticals—demonstrate safety and necessity before a drug is licensed for introduction into patients' bodies—to chemicals that will enter our bodies through the environment. Reversing the burden of proof would also set straight the twisted ethics of the current system, in which we mistakenly grant chemicals the presumption of innocence—a right that was created for people—while humans and other species are subject to a large-scale, multigenerational experiment of exposure to untested and potentially toxic chemicals.

In the case of organochlorines, reversing the burden of proof means that we address organochlorines as a class, presuming that chlorine-based products and processes are hazardous unless demonstrated otherwise. Industry and some analysts in government and academia have called this approach radical and unprecedented, but in fact it is neither. In making public policy decisions, we always choose, consciously or unconsciously, the appropriate level of intervention. Society does not try to address insect infestations by targeting individual bugs or traffic problems by regulating individual cars. In these cases, society has decided that it is more effective to focus on the systemic causes of problems rather than their manifestations at the level of individual entities, which are too numerous and uncontrollable to be micromanaged.

The same is true of organochlorines. Of the 11,000 organochlorines in commerce, only a small fraction have been subject to the most basic toxicity testing, and the full range of toxic effects is known for absolutely none. There are thousands more organochlorines formed as accidental by-products; the majority of these have not been chemically identified, so we do not even know their names, not to mention their toxicity and environmental behavior. Establishing chemical-specific regulations for every organochlorine would impose an impossible scientific and administrative burden on society, requiring centuries of study and administration before action could be taken to address a pressing health problem. Further, organochlorines are never created in isolation but are always formed in complex mixtures of products and by-products, so there is no practical way to control them one by one. In fact, all chlorine-based products and processes, at some point in their life cycle, result in the production and release of the most dangerous organochlorines, including dioxins and related compounds. Thus, even the least dangerous organochlorines, at some point in their life cycle, result in the incidental production of the most dangerous ones. If we want to restrict only the most extremely hazardous organochlorines, we must still address the full range of chlorine-based products and processes.

On the other hand, all organochlorines share a single root cause: industry's practice of chlorine chemistry. By applying Reverse Onus and Clean Production to the class of chlorine-based substances and technologies, we focus not on the thousands of individual organochlorines but on the much smaller number of processes that produce them. In this way, the Ecological Paradigm represents a shift from the micromanagement style that targets isolated chemicals to a macromanagement approach, which tackles classes of hazardous substances and technologies. Already, many pollutants—PCBs, lead compounds, and CFCs, for example—have been regulated as groups because they share hazardous characteristics or are produced by common sources. In fact, restrictions on assemblages of chemicals like these represent the most successful pollution policies since the 1970s. To apply this approach to the larger class of organochlorines is a significant extension of current practice, but it does not come out of the blue.

The chemical industry has objected that treating chlorine-based technologies as a class requires an unscientific, unsupported judgment that all organochlorines have the same properties, precluding any effort to evaluate individual chemicals specifically. First, I should be clear that the my argument does *not* assume that all organochlorines are equally hazardous. Each substance has unique properties, and the presence of chlorine does not in itself determine how toxic, persistent, or bioaccumulative a compound is. Instead, the effect of chlorination is to amplify the hazardous qualities of organic chemicals, sometimes to an extraordinary extent; organochlorines are virtually always orders of magnitude more toxic and bioaccumulative, and often much more persistent, than their chlorine-free precursors. For the purposes of public policy, then, there is a sound basis to treat chlorine chemistry as an environmentally dangerous activity that should be avoided whenever possible.

Second, the Ecological Paradigm does not preclude specific investigations of individual chemicals and processes; it merely changes the role of these evaluations in environmental policy. Both the Risk Paradigm and the approach I propose begin with a presumption—that the class of synthetic chemicals is safe or dangerous, respectively—and then evaluate specific substances that may represent exceptions. In structure, neither approach is any more scientific than the other. But consider the data: virtually all organochlorines that have been studied so far have been found to cause one or more adverse effects, so it is hardly likely that the rest will turn out to be safe. Presuming organochlorines hazardous thus better satisfies one of the most important criteria by which scientific theories are judged: that they maximize the explanatory power of the data and minimize ad hoc hypotheses—statements concocted after the fact to maintain a theory in the light of data that contradict it, such as, "Most organochlorines are safe; it's just a coincidence that almost all the ones we have so far examined are dangerous."

In practice, applying the Ecological Paradigm to organochlorines requires a simple but far-reaching program: the gradual phaseout of the production and use of chlorine and organochlorines and the phase-in of safer, chlorine-free alternatives. Organochlorines are now used in a wide variety of industrial applications, so this process, called a chlorine sunset, must be implemented with care. Sunsetting does not mean an immediate ban on chlorine, all its uses, and all its end products. Rather, it means a carefully planned process of technological conversion, a transformation of our industrial infrastructure that will take several decades. Sunsetting will require society to set priorities and time lines, make exceptions when necessary, evaluate substitutes, and take steps to minimize and address any economic dislocation that the program causes.

This proposal recognizes not only the shared properties of organochlorines but also their common source. Once chlorine gas is produced, a myriad of chemicals incompatible with the biological processes of complex organisms and the global ecosystem are formed and released to the environment, despite the best efforts of scientists and engineers to control them. Chlorine chemistry is like nuclear technologies: just as human intervention in the structure of matter unavoidably produces both desired and undesirable radioactive materials, the synthesis of chlorine gas inevitably results in the formation of toxic, persistent, and bioaccumulative products and by-products. Like the splitting of the atom, chlorine chemistry is an inherently dangerous technology of great power that interferes with the processes of nature at a fundamental level. Humans can harness this power but never completely control it. Chlorine chemistry is a technology we can and should choose to forgo.

I fully recognize the magnitude of a chlorine sunset and do not take lightly the technological and economic implications of a transition away from chlorine-based technologies. There are a few chlorine uses, such as some pharmaceuticals and some kinds of water disinfection, for which alternatives have not been developed or will take a long time to implement. For applications like these that serve compelling social needs, chlorine should continue to be used until substitutes are developed. But for the vast majority of chlorine uses, safer alternatives are now available, and technological innovation improves them with each passing year. Some of these processes are less expensive than organochlorines, and some are more costly; some substitute less toxic chlorine-free chemicals for organochlorines, and others rely on skilled labor or traditional materials. It will take time and money to convert to safer substitutes, but a well-planned transition will not impose an undue economic burden on society. In fact, experience suggests that less toxic processes are generally more efficient, create more jobs, and impose fewer externalized costs on society, such as cleaning up contaminated sites and treating people with chemically-induced disease. We should thus see a chlorine sunset as an investment in a healthier and more sustainable economy.

The Ecological Paradigm implies considerable technological and policy change, but it is neither absolutist nor all-encompassing. Precautionary action requires a prima facie case, a sound reason to believe that a practice may cause serious or irreversible environmental damage, before measures are taken to anticipate and prevent harm. This book provides such a case—and more—for organochlorines, demonstrating by example how action on other classes of chemicals might be judged necessary or not. Further, once precautionary action is called for, a chlorine sunset would allow exceptions and a balancing of social interests if no alternatives are available for technologies that serve compelling human needs.

My point is not to elbow aside environmental concerns other than chlorine chemistry. Organochlorines are by no means the only pollutants we should be worried about,[9] and there are other ways we could organize our thinking about pollution issues. For instance, we might be concerned about chemicals that cause specific health effects, like carcinogenic substances or those that disrupt the body's hormones. Classifying chemicals by the problems they create, however, does not point to workable solutions, because pollutants that cause one kind of health effect are typically of many different types and come from diverse and unrelated sources. A policy to address endocrine disrupters, for example, logically begins with a program to identify all the substances that interfere with hormone action and then formulates a separate strategy to deal with each one. This strategy immediately bogs down in all the difficulties of the chemical-by-chemical approach. In contrast, a policy that addresses chemical classes focuses on the sources of chemical contamination, organizing the diverse hazards of synthetic chemicals in a way that leads directly to preventive action.

The fact is, organochlorines account for the majority of known endocrine disrupters; a large portion of identified carcinogens; a great number of chemicals that damage the nervous, endocrine, reproductive and immune systems; and virtually all of the world's persistent organic pollutants. The United Nations recently began negotiating an international agreement to address global contamination by persistent organic pollutants, and all twelve substances slated for immediate action are organochlorines. Organochlorines also dominate national lists of priority water pollutants, contaminants found at hazardous waste sites, and chemicals that contaminate the Great Lakes. If we want to prevent any of the major types of chemically-induced health hazards, sunsetting chlorine chemistry is an obvious priority. We can also see the specific policy I advocate as a case study: a chlorine phase out exemplifies the nature of action and assessment in an Ecological Paradigm, which can and should be extended to other classes of hazardous substances and processes.

As public and scientific concern about organochlorines began to grow in the early 1990s, the chlorine industry mustered its resources and began an ambitious public relations and lobbying counteroffensive. The effort was spearheaded by the Chlorine Chemistry Council (CCC), an arm of the Chemical Manufacturers' Association, along with the Chlorine Institute—a trade group of major producers and users of chlorine and organochlorines—and the Vinyl Institute, the association of companies that manufacture PVC plastic and its feedstocks. As of late 1994, the chlorine industry was spending about $130 million per year on its efforts to protect its products and reputation in the public arena.[10] The result has been a contentious debate over the hazards of organochlorines and the economic impacts of phasing them out. Throughout this book, I address the arguments of the CCC and its allies in detail, for two reasons. First, these statements represent the most comprehensive set of arguments that have been made against the position I take, so I see it as my responsibility to take them on directly. Second, one of my central concerns—on which I focus exclusively in the final chapter—is the relationship between science and policy. The words of chlorine's protectors are highly enlightening on this subject, providing insight into the ways that science, politics, and authority have been mixed in important and problematic ways in the debate over chlorine.

Confronted with calls for a chlorine phaseout, for example, the chemical industry and some scientists and government representatives have responded with the remonstrance that environmental policy must be based on "sound science." This sounds quite reasonable, though rather patronizing, if it means we should be well informed and rigorous in the use of scientific knowledge. But it turns out that "sound science" does not mean a way of asking questions but serves as short-hand for a specific answer: the continued use of chemical-by-chemical, risk-derived discharge limits. Any other approach, it is implied, is based on bad science or—even worse—emotion, fear, or some other suspect motive.

I address that charge in two ways. First, I present the information on organochlorines and assess whether it supports or refutes the assumptions of the two models for environmental policy. It is in fact the Risk Paradigm that turns out to be at odds with current scientific knowledge about chlorine chemistry, its products, and their effects on health and the environment. The Ecological Paradigm, on the other hand, was designed specifically to address the picture that ecology, biology, chemistry, and toxicology have painted of the structure and dynamics of organisms and ecosystems, the kinds of damage that synthetic chemicals can cause, and the ways that toxic substances are produced in industrial processes.

Second, we should not be misled that "sound science" requires society to put decisions about environmental policy in the hands of scientists. Decisions about pollution always encompass questions that science cannot answer. How much health or environmental damage is acceptable? How should health threats be weighed against the benefits of a technology? What alternative processes and materials are available that might prevent the pollution altogether? These are social questions that force us explicitly to consider ethics, values, and politics, as well as science. If we restrict the policy process to science alone, these issues are taken as settled, and the public, most of whom feel unable to evaluate scientific information themselves, is excluded from what should be a democratically determined decision. One of the most troubling elements of the Risk Paradigm is the way it limits debate to highly technical issues, like the quantification of toxicological thresholds and cancer risks, obscuring political and moral questions and protecting them from democratic discussion.

We can probe the idea of "sound science" even further. This book's analysis of the scientific claims deployed in the Risk Paradigm shows that political questions must be confronted not only in determining appropriate policies but also in assessing hazards themselves. Scientific knowledge that affects policy is always built on prior judgments about political and ethical issues. What kinds of health damage are worth assessing? Which species do we want to protect? What standard of proof should be satisfied before a conclusion is drawn? What constitutes a "negligible" risk? The science on which the Risk Paradigm is based presupposes answers to these questions that I believe most people would find unacceptable. In this way, not only the deployment but even the content of "sound science" turns out to involve judgments that should be exposed to democratic debate, not veiled behind a mask of scientific authority. The Ecological Paradigm acknowledges that science is a way of knowing about the world that is at once empirical and rooted in a political and social perspective. This new framework makes explicit the political questions that precede and pervade the creation of science in environmental decision making, and it provides means to settle them in a democratic context.

In the current system, the ethical issues may be hidden, but they are ripe. The technologies in use today are not inevitable facts of life but the products of conscious decisions. The world is contaminated with toxic chemicals because people and corporations have chosen to produce these substances, use them, and discharge them into the environment. We are exposed because society, by affirmation or omission, has given them leave to do so. Toxic pollution is by nature an ethical issue that involves the real and potential impacts of one person's actions on the well-being of others. Organochlorine pollution, global in scope and affecting many future generations, is a moral and political issue of great magnitude. The technologies used to meet society's needs and desires now have the potential to inflict far-reaching damage on the health and well-being of every citizen of the world. Choosing which technologies to adopt is thus a properly social decision that should be made explicitly and democratically. While the current system refuses to intervene in decisions about materials and production processes—insisting, at most, on the installation of tacked-on control devices—the Ecological Paradigm gives society the tools to begin a program of democratic technological development that meets its needs without sacrificing the health of future generations.

We have enough evidence now to come to this conclusion: the products and by-products of chlorine chemistry pose serious threats to the integrity of health and the ecosystem, and our current policies do not provide an adequate remedy. As the German government's Council of Environmental Advisors concluded in 1991, "The dynamic growth of chlorine chemistry during the 50s and 60s represents a decisive mistake in twentieth century industrial development, which would not have occurred had our present knowledge as to environmental damage and health risks due to chlorine chemistry then been available."[11] Today we have the knowledge we lacked then, and we must now act on it. Our failure to prevent the development of dangerous industrial practices in the past does not justify their continued use today.

The production of chlorine gas from salt sets the stage for the purposeful and accidental production of a vast number of novel chemicals that disrupt natural systems at their most fundamental level. The practice of chlorine chemistry has unleashed a host of unintended chemical and ecological consequences that our most sophisticated technologies are not capable of preventing. Chlorine chemistry is a Pandora's box, opened less than 100 years ago and still spewing its demons into the environment. While governments, cheered on by those who benefit from the open box, try to chase down each and every tiny demon that escapes, we miss the simplest and most obvious solution: close the lid.

I

The Problem: A Global Health Hazard

1

Organochlorines Around the World

That I am of the earth my feet know perfectly, and my blood is part of the sea.
—D. H. Lawrence, *Apocalypse*[1]

To most people, toxic pollution is a local problem, summoning images of leaking landfills, smokestacks looming over neighborhoods, and discharge pipes pouring effluent into contaminated streams. This local view of toxic hot spots serves as the foundation for virtually all environmental policy in force today. In the United States and Europe, most chemical regulations use risk assessments to determine the permissible discharges from individual polluting facilities, based on exposures that are calculated to be "acceptable" in nearby communities or ecosystems. These local hot spots are quite real. To confirm their existence, one need only drive past Times Beach, Missouri, a tiny village evacuated over a decade ago due to the purposeful spreading of dioxin-contaminated waste oil for "dust control" on its dirt roads. Or sail down the Mississippi River's Chemical Corridor—called Cancer Alley by the area's environmental activists—where over 100 large chemical plants line the river's banks between Baton Rouge and New Orleans. Or visit northeastern New Jersey, where middle- and working-class neighborhoods are nestled among hazardous waste landfills, incinerators and dozens of chemical factories.

Terrible as they are, these toxic sacrifice zones only hint at the full scale of the problem. Thousands of the same poisons found in these communities are slowly accumulating across the entire globe, and the potential impacts extend to every living thing on the planet. If Love Canal taught us a lesson, it should be this: pollutants do not stay where we put them. The thousands of toxic substances released during the last half-century by thousands of industrial facilities have gradually spread out across the earth, creating a global blanket of contamination. In this chapter, I review the worldwide contamination of the environment, the food web, and our own bodies with a host of organochlorines. I discuss why this problem is the inevitable result of certain aspects of chlorine chemistry and show how the current system of environmental regulations has helped contribute to its development. A hazard of global scope cannot be treated as a collection of bounded, local problems; it demands solutions of global perspective and international cooperation.

Two aspects of chlorine chemistry, which I discuss in detail in chapter 5, lie at the heart of this issue. First, because many organochlorines resist natural degradation processes, even very dilute discharges tend to build up in the environment over time. Chemicals accumulate in the environment if their rate of input into the environment is greater than the rate at which they break down. Organochlorines are released in immense quantities—the chlorine industry makes about 40 million tons of chlorine gas every year, most of it directed into chlorinated organic chemicals—and many break down very slowly. Some, such as 2,3,7,8-tetrachlorodibenzo-p-dioxin (TCDD, or simply "dioxin"), do not break down to any appreciable degree in the environment; virtually all the TCDD released into the environment will remain there, in one place or another, more or less indefinitely. Many other organochlorines are persistent but not so eternal. These can be degraded, but very slowly, with environmental half-lives in the years or decades. As large quantities of these chemicals are released into the environment year after year, their concentrations gradually increase. Other organochlorines can be degraded in weeks or months, but in most cases the resulting substances are other organochlorines, which often are more persistent, and sometimes more toxic, than the original compound. An organochlorine is fully broken down only when all chlorine atoms in the molecule have been removed and returned to their original ionic form in some type of salt. The rate of this process varies from one compound to another, but for many organochlorines it is quite slow—far slower than the rate at which they are now entering the environment. As a result, the total environmental burden of organochlorines has grown steadily over the years.

Second, many organochlorines are more soluble in oils and fats than they are in water. These organochlorines bioaccumulate, migrating from the environment into the fatty tissues of living organisms. Taken together, predator organisms at the top of any food web have less total mass but much higher fat content than those at the lower levels. For this reason, as organochlorines move up the food chain—from the water into plankton, from plankton to invertebrates, from invertebrates to small fish, from small fish to large fish, from large fish to marine mammals, for instance—their concentration is multiplied at each step, a process called biomagnification. Thus, species high on the food chain, such as humans, serve as living reservoirs where these contaminants accumulate in higher and higher concentrations (figure 1.1). In the North Pacific Ocean, for instance, striped dolphins carry levels of PCBs and DDT in their bodies that are 13 million and 37 million times higher, respectively, than their concentrations in the water (figure 1.2).[2] In this way, pollutants present in the ambient environment in minuscule concentrations—in the range of a few parts per trillion or even

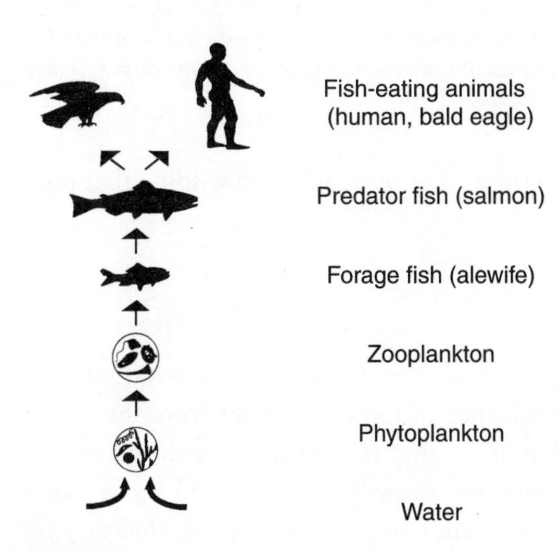

Figure 1.1
Levels in a simplified food chain of a freshwater lake. (Source: After Allan 1991.)

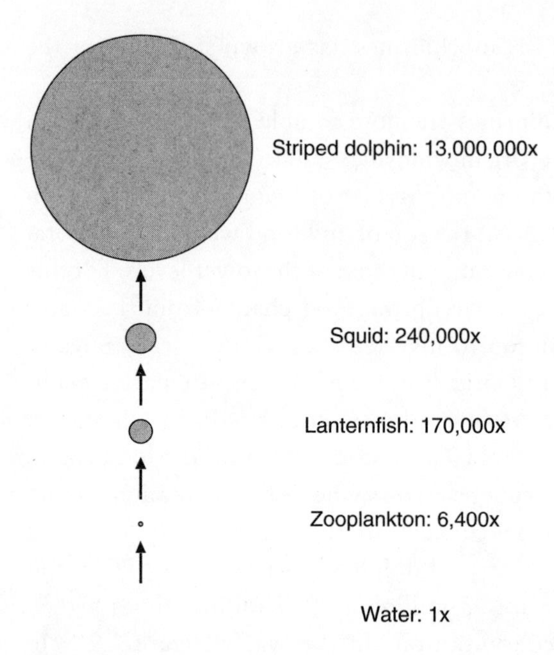

Figure 1.2
Biomagnification of PCBs in a North Pacific food chain. The area of each circle is proportional to the concentration of PCBs in each species. The circle that represents the concentration in water, with a diameter of 3/10,000 of an inch, is too small to see. For each species, the bioconcentration factor—the ratio of the PCB concentration in that species' tissues to the concentration in ocean water—is also given. (Source: Tatsukawa and Tanabe 1990.)

quadrillion—reach levels in the bodies of people and wildlife that pose a real health threat.

In the Air

The earth has mechanisms to keep its atmosphere clean; most natural substances degrade under the influence of ultraviolet light from the sun (a process called photodegradation), or they react with other chemicals naturally present in the air, or they are washed out in rain and snow. Many organochlorines released into the air, however, are so stable that they resist breakdown by sunlight or chemical reaction. For instance, the volatile ozone-depleting compounds carbon tetrachloride, trichloroethane, and

Table 1.1
Persistence in the atmosphere of some volatile organochlorines

Substance	Atmospheric half-life (yrs)
Chloromethane	1.0
Dichloromethane	0.4
Chloroform	0.4
Carbon tetrachloride	24.3
Trichlorofluoromethane	31.2
Dichlorodifluoromethane	69.3
Bromochlorodifluororomethane	11.1
Dichloroethane	0.4
Trichloroethane	3.3
Dichlorotetrafluoroethane	152.5
Trichlorotrifluoroethane	58.9
Chloropentafluoroethane	381.2
Tetrachloroethylene	0.3
HCFC-22 (Dichlorofluoromethane)	8.3
HCFC-123 (Dichlorotrifluoroethane)	1.2
HCFC-124 (Chlorotetrafluoroethane)	4.8
HCFC-141b (Dichlorofluoroethane)	6.2
HCFC-142b (Chlorodifluoroethane)	13.2
HCFC-225ca (Dichloropentafluoropropane)	1.9
HCFC-225-cb (Dichloropentafluoropropane)	5.5

Note: Half-lives are calculated from reported atmospheric lifetimes, the time required for a substance to degrade to 37 percent of its original concentration.
Source: Makhijani and Gurney (1995), Montzka et al. (1999).

chlorofluorocarbons have atmospheric half-lives ranging from 3.3 to 380 years (table 1.1). (A half-life is the amount of time required for a substance to degrade to half its original concentration.) Even a moderate half-life can mean very long-term contamination. If all emissions of these compounds were to stop immediately, it would take 28 to 2,500 years for the current burden of these chemicals in the air to degrade to 1 percent of their current levels. Less volatile organochlorines can also be very persistent: the atmospheric half-lives of hexachlorobenzene, chlordane, and hexachlorocylohexane have been measured in the Great Lakes region at 7 to 13 years; these pollutants would require 50 to 100 years to decay to 1 percent of current values.[3]

Some organochlorines are more readily broken down in the air, but they do not necessarily degrade into harmless substances. For instance, the pesticide mirex can be degraded by ultraviolet light within a few weeks, but the breakdown product, another organochlorine called photomirex, is more persistent and at least as toxic as mirex itself.[4] The dry cleaning solvent tetrachloroethylene (also called perchloroethylene, or "perc") is not persistent at all; it is transformed over a few months into a variety of other organochlorines, including carbon tetrachloride, which has a half-life of 24 years.[5] Some organochlorines can even be transformed into dioxins in the air; when trichlorophenol, a common industrial organochlorine, is exposed to light, for instance, it is transformed into the extraordinarily toxic and persistent TCDD.[6]

With their long lifetimes, organochlorines travel the globe on wind currents and have become truly global atmospheric contaminants. For instance, PCBs—about 1.2 million tons of which were produced worldwide from 1929 to 1992, primarily for use as an insulating fluid in electrical equipment[7]—have been detected in air samples taken over the open ocean in the vicinity of Bermuda, the Grand Banks, Newfoundland, and the Gulf of Mexico and at remote sites in the North, Western, and South Pacific.[8] Rain and snow samples from Nova Scotia, Prince Edward Island, and New Brunswick between 1980 and 1989 contained detectable concentrations of the pesticides lindane, heptachlor, chlordane, endosulfan, aldrin, and DDT, along with PCBs and a variety of chlorinated benzenes, a group of industrial organochlorines used primarily as chemical feedstocks.[9] And at least seven kinds of chlorinated anisoles—breakdown products of the chlorophenols—have been detected at sites over the remote Atlantic Ocean.[10]

Organochlorine pesticides still in widespread use are also important air pollutants. Fog samples collected in Maryland and California contain the popular chlorinated pesticides atrazine, alachlor, and chlorpyrifos at levels many times higher than those found in the ambient air.[11] And the U.S. Geological Survey (USGS) has found that rain samples collected throughout the Midwest and eastern United States—in farmland, cities, and remote areas like Lake Superior's Isle Royale—consistently contain these pesticides, along with DDT, lindane, mirex, and toxaphene. The concentrations are considerable: USGS's data suggest that rain deposits about

140,000 kilograms (308,000 pounds) of atrazine and 82,000 kilograms (110,000 pounds) of alachlor annually onto land and surface waters in the area studied. Airborne deposition of atrazine into the Great Lakes has become so significant that there is now about 36,000 kilograms of the pesticide in Lake Superior water. Noting that only 1 percent of the atrazine in the water degrades each year, the authors concluded, "Accumulation of atrazine in Lake Superior will continue until the annual internal losses of atrazine equal the annual inputs"—which will occur only when atrazine concentrations in rain decline to a small fraction of their current levels.[12]

Chlorinated solvents and refrigerants have also become globally ubiquitous atmospheric contaminants. Air samples taken over the remote Western Pacific, thousands of miles from any industrial source, contain carbon tetrachloride, difluorodichloromethane, and 1,1,1-trichloroethane.[13] Anywhere in the Northern Hemisphere, in fact, the air contains trichloroethane at a "baseline" concentration of about 200 parts per trillion; levels may be 40 or more times higher in industrialized areas, but even in isolated areas of Alaska, the air, rain, and snow are contaminated with this popular industrial degreasing agent.[14] Some chlorinated solvents, such as trichloroethylene and perc, are less persistent in the air, but they degrade into the highly toxic chlorinated acetic acids, which are also widespread contaminants. A survey of scores of rainwater samples taken throughout Austria, for instance, found that trichloroacetic acid, which is toxic to both animals and plants, was consistently present, even in samples taken at stations high in the Alps.[15]

Chlorinated pollutants in air and precipitation are a particular danger to forests. Trees breathe: leaves and needles constantly exchange gases and vapor with the atmosphere. When the atmosphere contains organochlorines, these substances are likely to enter and accumulate in plant tissues. In European forests, Norway spruce needles have been found to concentrate the solvent perchloroethylene to very high levels.[16] Trichloroacetic acid can now be found in conifer needles throughout the forests of Central and Northern Europe at concentrations 3,000 to 10,000 times greater than those in ambient air; the most severe contamination is in northern Finland, far from the major sources of organochlorine solvents. [17] Trichloroacetic acid is extremely toxic to plants, reducing growth, impairing development of the cuticle (the "skin" that protects the plant and regulates its interaction with the environment), and causing chlorosis, a disease condition in which green parts turn yellow and the plant eventually dies.[18] Popular herbicides like atrazine, alachlor, and 2,4-D are specifically designed to harm plants, disrupting photosynthesis and growth of unwanted weeds. When they are distributed throughout the environment, however, these chemicals have the capacity to harm the many other kinds of plants exposed to them. As the health of forests has declined throughout Europe, organochlorine accumulation is suspected as a major cause.[19]

Trees' tendency to accumulate pollutants is bad news for forests but quite useful to environmental chemists. Air samples are difficult to analyze because contaminants are frequently present at concentrations below the limit at which analytical techniques are able to detect them. But tree bark absorbs chemicals from the air and retains them for a long time, so it is a useful indicator of long-term, low-level air pollution. In research reported in 1995, chemists at the University of Indiana analyzed 200 tree bark samples from 90 sites around the world; they discovered that 18 of the 22 organochlorines they tested for—including hexachlorobenzene, DDT, chlordane, heptachlor epoxide, lindane and other hexachlorocyclohexanes, endosulfan, and pentachloroanisole—were "ubiquitous on a global scale." These pollutants were present in tree bark in low but measurable concentrations in some of the world's most remote regions: the rain forests of Venezuela, Ecuador and Belize, "the distant islands of Guam, Bermuda, Tasmania, and the Marshall Islands," and isolated areas of Uganda, Ghana, and Togo. Very high concentrations were measured in the United States, Brazil, Europe, Russia, India, the Middle East, Japan, Taiwan, South Korea, and Australia. These results suggest that organochlorine contamination is most severe in regions where these compounds have been heavily used, but the problem extends across the world's entire surface, from the Cape of Good Hope to northern Siberia.[20]

There is one major exception to the general pattern that areas with the greatest production and use of organochlorines also have the most severe contamination, due to a phenomenon called the global distillation effect. Many persistent organochlorines are semivolatile, which means that they evaporate at the temperatures characteristic of equatorial, tropical, and warmer temperate regions, but they condense at the colder temperatures found in temperate areas during winter months and in polar regions year around. These substances volatilize from the places they are discharged and ride atmospheric currents that blow from the equator to the poles; once they get to colder regions, they are deposited by rain and snow on land and sea, where they hunker down to stay.[21] Because of the distillation effect, for instance, the bulk of the organochlorine pesticides and other persistent pollutants that enter the waters of the Great Lakes come from as far away as the southeastern United States and Latin America.[22]

The polar regions thus become the ultimate global sinks for persistent organochlorines. The most remote parts of the arctic, many thousands of miles from any industrial pollution source, are now thoroughly contaminated with organochlorine pollutants. PCBs and the pesticides hexachlorocyclohexane, endosulfan, dieldrin, chlordane, and heptachlor have been found in snow from the far Northwest Territories, for example.[23] Air samples taken at icy Ellesmere Island—the northernmost land in Canada, at latitude greater than 80 degrees north—contain all these compounds, along with chlorobenzenes, toxaphene, and DDT.[24] Even organochlorines formed accidentally in chemical plants, incinerators, and pulp mills that process chlorine and organochlorines—such as chloroanisoles, chloroveratroles, and octachlorostyrene—are present in measurable quantities in the Canadian arctic.[25]

Finally, organochlorines are ubiquitous in the upper atmosphere. In the stratosphere, a layer of ozone molecules prevents harmful ultraviolet (UV) rays from reaching the surface of the earth. A number of chlorinated solvents and refrigerants are so persistent that they migrate from the lower atmosphere to the stratosphere, where the chlorine they contain catalyzes the breakdown of ozone molecules (O_3) into common oxygen gas (O_2), which does not filter out UV radiation. In the process of destroying ozone, chlorine atoms are not themselves changed; a chlorine atom can catalyze the breakdown of one ozone molecule, and then another, and then another. A single reactive chlorine atom in the stratosphere can destroy about 100,000 stratospheric ozone molecules before it is inactivated.[26] By 1994, an average of 6 to 12 percent of the stratospheric ozone layer over the temperate latitudes of North America, Europe, and Asia had been depleted;[27] in 1993, when particles from the eruption of Mount Pinatubo accelerated the rate of destruction, 17 to 20 percent of the ozone layer over

Canada and Siberia was missing.[28] In the stratosphere over Antarctica, an immense ozone "hole"—an area more than 15 million square kilometers in size in which some 60 to 90 percent of stratospheric ozone has been destroyed—opens up every winter when cold temperatures and stratospheric ice crystals cause the chlorine to destroy ozone molecules even faster.[29]

In the Water

Many organochlorines are persistent in water. For instance, chloroform and perchloroethylene have half-lives in pure water of 1,850 and 990 million years, respectively[30] (see table 1.2). Dioxins (short for polychlorinated dibenzodioxins, or PCDDs)* also resist aquatic breakdown, remaining undegraded for decades or longer; the U.S. Environmental Protection Agency (EPA) now assumes that dioxins do not break down at all in aquatic ecosystems but are instead absorbed onto sediments or into the food chain, where they may persist for centuries.[31]

Some organochlorines are more easily broken down in water by a process of hydrolysis, in which chemical bonds are broken by reaction with water molecules. Many readily hydrolyzable organochlorines are volatile, however, so they escape from surface waters before they can be degraded to a noticeable degree. Trichloroethane, for instance, can be hydrolyzed with a half-life of about a year, but it evaporates quickly into the atmosphere, reaching half its original concentration in a river in just a few hours.[32] Many other organochlorines escape hydrolysis because they are more soluble in organic matter than in water; these attach themselves to sediments or enter food webs—media in which they may be quite persistent. The half-life of hexachlorobutadiene (HCBD), a by-product of the manufacture of many organochlorines, is only a few weeks in pure water;

*Throughout this book, I use the terms *chlorinated dioxins* and *dioxins* to refer to the polychlorinated dibenzo-p-dioxins, a group of 75 different molecules composed of two benzene rings, linked together by two oxygen atoms, with one to eight atoms of chlorine bonded to the benzene rings. The properties of each type of chlorinated dioxin, called a congener, are determined by the number and placement of chlorine atoms on the molecule. Following common usage, I use the singular form *dioxin* to refer specifically to 2,3,7,8-TCDD, the most toxic congener.

Table 1.2
Half-lives of some organochlorines in pure water

Substance	Estimated half-life (years)
Chloroform	1850
Carbon tetrachloride	40
1,1-dichloroethane	61
1,2-dichloroethane	72
1,1,2-trichloroethane	139
1,1,1-trichloroethane	1
1,1,2,2-tetrachloroethane	<1
1,1,1,2-tetrachloroethane	47
Pentachloroethane	<1
Hexachloroethane	1,800,000,000
1,3-dichloropropane	2
2,2-dichloropropane	<1
1,1,2,3-tetrachloropropane	<1
1,1,2,3,3-pentachloropropane	<1
1,1-dichloroethylene	120,000,000
1,2-dichloroethylene	21,000,000,000
Trichloroethylene	1,300,000
Perchloroethylene	990,000,000

Note: Half-lives are calculated based on the simultaneous occurrence of base-catalyzed and neutral hydrolysis in pure water, assuming neutral pH, based on experimentally determined rate constants.
Source: Jeffers et al. (1989).

in real aquatic ecosystems, however, HCBD is quite long-lived, because it quickly accumulates in fish tissues, where it is highly persistent.[33]

When organochlorines can be broken down in water, the products are often more hazardous than the original pollutants, just as was the case in the atmosphere. For example, pulp mills that use chlorine-based bleaches discharge large quantities of chlorolignins, chlorinated versions of complex, heavy molecules called lignins that are natural components of tree fibers. Chlorolignins are relatively nontoxic and short-lived, but they break down into organochlorines of lower molecular weight, such as chlorinated phenols, guaiacols, and catechols, which are more toxic and bioaccumulative than the original substances; these compounds may then break down still further into chloroveratroles, which are even more toxic and long-lived.[34]

As a result of this behavior, organochlorines are now widespread in the waters of the world. In Austria, 609 of 613 samples of surface water samples from 141 sites in both industrialized and remote areas contained organochlorines, generally in the parts per billion range.[35] In the Great Lakes, 362 synthetic chemicals have been "unequivocally identified" in the water, sediments, and food chain, and 168 of these are organochlorines (appendix B). The list includes the most infamous organochlorines, but it also contains a full spectrum of less familiar organochlorines, from simple solvents to a host of complex industrial chemicals and by-products. A glance at this list argues against the claim that our problems can be reduced to a handful of organochlorines that have already been severely restricted.

Organochlorine contamination of water is truly global. PCBs have been found in samples of ocean water from the Sargasso Sea, the North Sea, the Western North Pacific, the South Pacific, the Indian, and the Antarctic oceans.[36] Polychlorinated terphenyls are widespread in the water, sediment, and biota of Northern Europe, including the Rhine and the North Sea, and in the vicinity of the Falkland Islands.[37] And hexachlorocyclohexane pesticides, including lindane, are present in water samples taken from the open Atlantic Ocean off Bermuda.[38]

Covered with water, snow, and ice, polar regions serve as global sinks for waterborne organochlorines. Organochlorines travel to the Arctic not only by air but also on ocean currents and rivers. Once they have arrived, these compounds accumulate, because the water of the Arctic Ocean tends to stay put, with an average residence time of 12 to 300 years, depending on the depth. Thus, PCBs, hexachlorobenzene, and a variety of chlorinated pesticides can be found in falling snow, snowpack, glacial ice, and seawater at sites throughout the arctic.[39]

Because of the very cold temperatures, pollutants that might be degraded in a few months at temperate latitudes can last for centuries near the poles. Atrazine, for instance, has a half-life in water of 60 days at room temperature but does not degrade at all when the temperature drops to 6 degrees Celsius (43 degrees Fahrenheit). For this reason, a host of chlorinated pesticides expected to be nonpersistent—atrazine, chlorpyrifos, endosulfan, chlorothalonil, metolachlor, and terbufos—can now be measured in arctic air, water, and fog.[40]

When organochlorines enter aquatic ecosystems, they do not uniformly mix into the water, causing two additional problems. First, organochlorines that are not water soluble attach to particles in the water that then settle into sediments, reaching high concentrations that may be taken up by bottom-feeding organisms or mobilized when sediments are disturbed by dredging. Second, many organochlorines concentrate in the microlayer, an extremely thin and productive layer of oils and other organic matter that floats on the ocean's surface. Concentrations of pollutants, including many organochlorines, are up to 10,000 times greater in the microlayer than in the water just below.[41] The microlayer serves as home to a tremendous diversity of life, including communities of bacteria, algae, fungi, single-celled organisms, and the eggs and larvae of fish and shellfish. Contamination of the microlayer can disrupt the productivity of marine ecosystems and represents a starting point for the bioaccumulation of organochlorines into the food web.

In the Food Web

Many organochlorines love fat: fish fat, beef fat, butter fat, human fat. They love it so much that they migrate out of the environment and take up residence in the fatty tissues of our bodies, never to let go. This attraction to fats, called *bioconcentration,* results in the gradual buildup of pollutants in the tissues of living things, called *bioaccumulation.* TCDD accumulates in the tissues of fish to levels 159,000 times greater than that in the water in which the fish swims (a figure known as a *bioconcentration factor*).[42] Bioconcentration factors for other dioxins, polychlorinated dibenzofurans ("furans" for short), PCBs, hexachlorobenzene, and octachlorostyrene are all 10,000 or more. Many chlorinated pesticides, including DDT, chlordane, and mirex, have bioconcentration factors in the same range.[43]

Bioconcentration is a characteristic of all sorts of organochlorines. Organic chemicals can be divided into two large groups based on their chemical structures. One, the aromatic compounds, includes any substance that contains one or more benzene rings—six carbon atoms bonded together by single and double bonds in the shape of a regular hexagon (see figure 1.3). Benzene rings are both stable and oil soluble, so aromatic organochlorines tend to be extraordinarily long-lived and bioaccumulative. The group of chlorinated aromatics includes DDT, toxaphene, pentachlorophenol, mirex, and dioxins. It also includes hundreds or thousands of other substances that are "dioxin-like," so-called because they, like the dioxins, are composed of two or more benzene rings with chlorine atoms attached in various places around the rings and thus have similar environmental behavior and toxicological properties to the dioxins.[44] The dioxin-like compounds include the polychlorinated dibenzofurans, biphenyls (PCBs), naphthalenes, diphenyl ethers, dibenzothiophenes, and biphenylenes, all of which have been found in human tissues or aquatic ecosystems in various regions of the world.[45]

Because so many chlorinated aromatics are infamous bioaccumulators, one might think that these are the only pollutants that build up in fat. But many organochlorines in the other group—the chlorinated aliphatic compounds—are also bioaccumulative. Aliphatic compounds by definition do not contain benzene rings but are instead made up of chain structures—one or more carbon atoms linked together in linear shapes, some straight lines, some branched (figure 1.4). The word *aliphatic* comes from the Greek word for fat, and fats and oils themselves are long chains of carbon atoms. As a result, it is not surprising that many aliphatic organochlorines, even those with simple, short chain structures, are also highly bioaccumulative. For instance, hexachlorobutadiene is a simple chain of four carbons and six chlorine atoms, and it has a bioaccumulation factor of up to 17,000.[46] Even the dry cleaning solvent perc, with just two carbons and four chlorine atoms, is very oil soluble. Because it is so volatile, however, perc is more likely to enter the food chain from the air than the water. Once emitted into the atmosphere, it migrates directly into the human food supply. Butter sold from grocery stores near dry cleaning establishments, for example, has been found to contain extremely high concentrations of perc, in some cases greater than 1000 parts per billion—hundreds of times higher than in butter sold in stores not located near dry cleaners.[47]

The most severe contamination of wildlife species occurs in enclosed and semienclosed bodies of water like the Great Lakes, Lake Baikal, the Seto Sea near Japan, and the Mediterranean, Baltic, and Wadden seas in

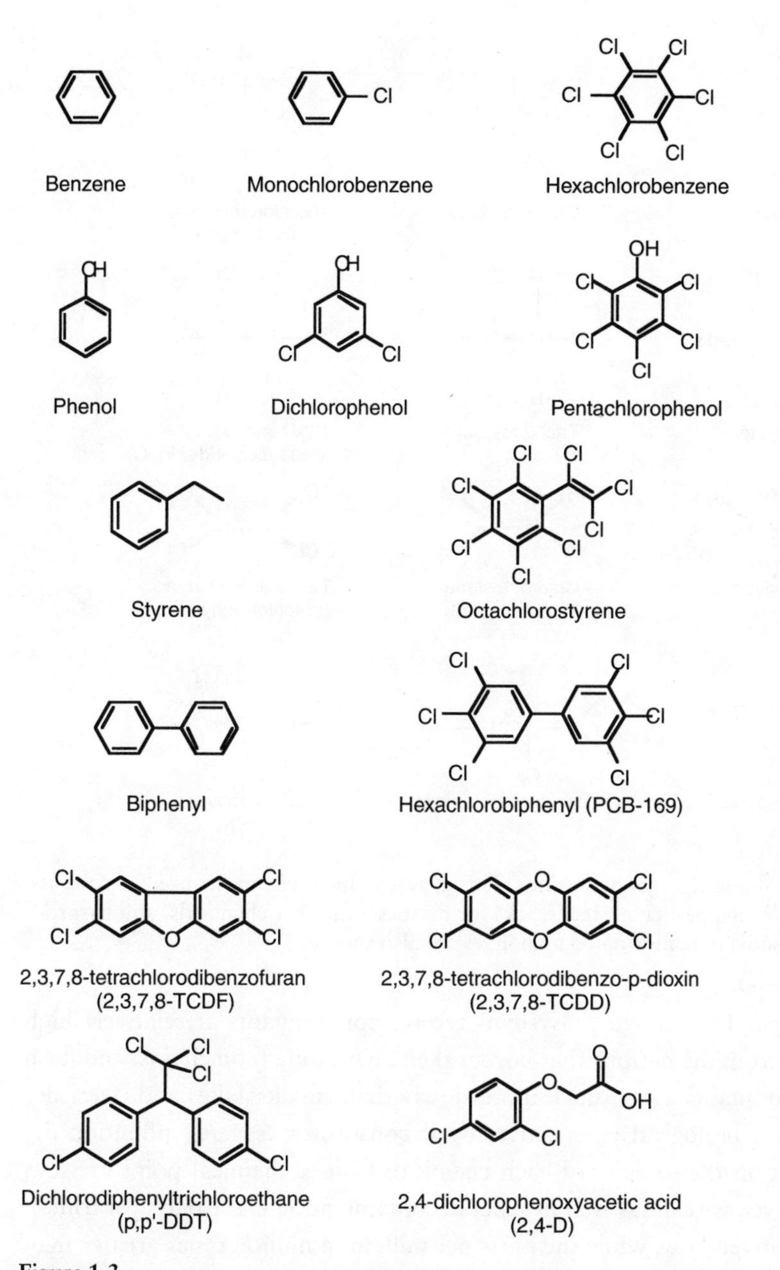

Figure 1.3
Some aromatic organochlorines. All members of this group include one or more benzene rings and one or more atoms of chlorine. Some nonchlorinated analogues are also shown.

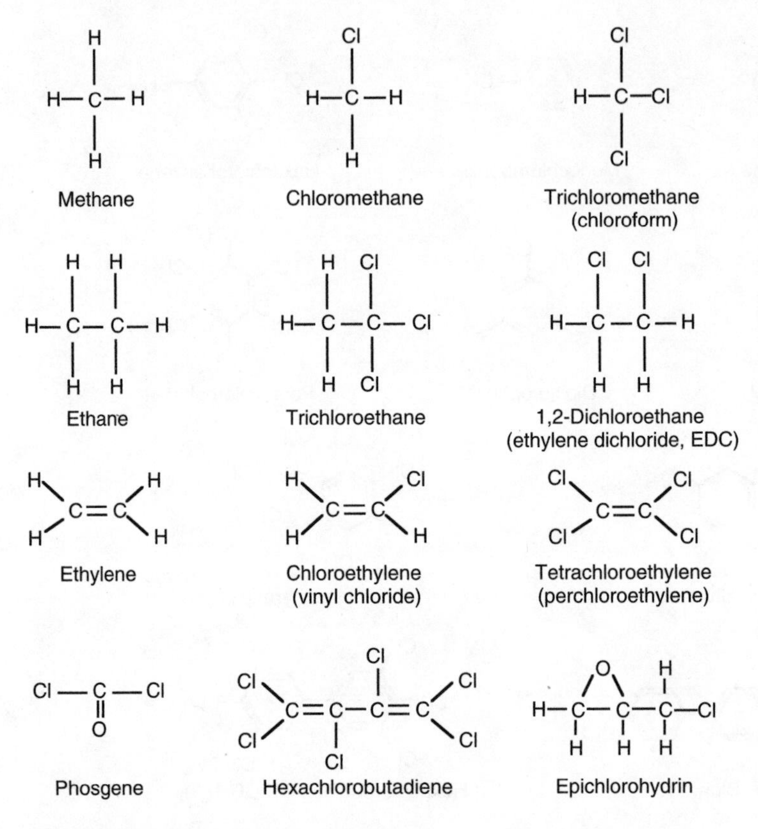

Figure 1.4
Some aliphatic organochlorines. This group includes chemicals used as solvents, refrigerants, pesticides, feedstocks for plastics and other chemicals, and by-products. Some nonchlorinated analogues are also shown.

Europe. These huge ecosystems receive contaminants at relatively high rates from the nations that border them, have long food chains, and flush contaminants away much more slowly than smaller lakes and rivers do. From a biological perspective, what constitutes "severe" pollution depends on the toxicity of each chemical; from a chemical point of view, however, we can say that in general, contamination in the parts per trillion is relatively low, while the parts per million (a million times greater than parts per trillion) range is quite high for organisms with no direct exposure to a chemical. In the Gulf of St. Lawrence in eastern Canada, beluga whales, which feed primarily on eels migrating from the Great Lakes, are contaminated with parts per million levels of chlorobenzenes, PCBs, and a host of organochlorine pesticides.[48] Seals from Lake Baikal carry PCBs and DDTs in their tissues at levels up to 160 parts per million, and a blue-white dolphin stranded on the Mediterranean coast of France had PCB levels over 800 parts per million—more than ten times the level that would make its body legally hazardous waste in the United States.[49]

Bioaccumulation is also a problem in the open ocean, where marine mammals are at the top of a very long food chain. In the 1970s, scientists first found PCBs, DDT, chlordane, and dieldrin in the parts per million range—millions of times greater than their ambient concentrations—in the blubber of at least four whale and five dolphin species. In one Pacific dolphin, DDT was measured at the extraordinary level of more than 1,000 parts per million—greater than 0.1 *percent*.[50] Since that time, the scope of the problem has only grown. Seals from Alaska and the west coast of Sweden have been found to carry extremely high levels of PCBs, DDE, PCB-methylsulfones, toxaphene, chlordanes, dioxins, and furans in their fat.[51] Even orcas—the beautiful black and white "killer whales"—from the open ocean of the Northwest Pacific contain PCBs at levels up to 400 parts per million, along with 35 congeners of dibenzofurans at concentrations thought to be high enough to cause health effects.[52]

Bioaccumulation and the global distillation effect work together to make organochlorine contamination an immense problem in polar regions. As early as 1976, studies conducted in various parts of the arctic found PCBs and a variety of organochlorine pesticides in the bodies of seals, walruses, beluga whales, porpoises and polar bears in various parts of the Arctic.[53] Since then, more pesticides have been added to the list, along with chlorobenzenes, dioxins, and furans,[54] and these compounds have been found in dozens of additional species, including caribou, wolf, and arctic fox.[55] According to a 1990 study that examined scores of animals from the Canadian arctic, every seal sample and all but one polar bear sample contained TCDD, and animals from the latitudes closest to the North Pole carried even higher levels than those from the subarctic Hudson Bay region.[56]

Because polar bears are truly at the top of the food chain, subsisting primarily on marine mammals, they carry some of the world's highest levels of organochlorines in their tissues, with concentrations of PCBs and hexachlorobenzene 6 to 17 times those measured in arctic seals.[57] Contamination of polar bear tissues with dioxins and PCBs is so severe, in fact, that the bears' body burdens exceed by a substantial margin the levels that are known to cause reproductive failure, immune suppression, and altered brain development in other kinds of mammals.[58]

In Our Bodies

Ecologically speaking, humans are part of the environment, though we often forget it. A contaminated ecosystem means contaminated people. If the air we breathe, the water we drink, and the food we eat are contaminated with organochlorines, then we too inevitably become polluted. A host of studies show that organochlorines have invaded not just the far corners of the globe but also the most intimate corners of our own bodies.

People are exposed to organochlorines in drinking water, air, and food. For bioaccumulative compounds like the dioxins, pesticides, and PCBs, the food supply is the major source of contamination. According to EPA, over 90 percent of the average American's dioxin exposures comes through the diet. Animal foods—meat, dairy, fish, and eggs—contain the highest concentrations of dioxin, but smaller exposures also occur through grains, fruits, and vegetables and their oils.[59] For the more volatile or less bioaccumulative organochlorines, air and water play more important roles. For instance, inhalation is an important route of exposure to chlorinated solvents,[60] and drinking water throughout the midwestern United States is contaminated with the herbicides atrazine and alachlor.[61] The myriad by-products formed in the disinfection of drinking water with chlorine also find their way to people mainly through tapwater—by drinking, absorption through the skin, or inhalation during a hot shower or bath.[62] Most people in industrialized nations also experience direct exposures to organochlorines in consumer products, lawn and garden chemicals, dry-cleaned clothing, and the workplace.

Like unwelcome guests, many organochlorines take up residence in our bodies and refuse to leave. Because humans occupy a position at the top of the food chain, our own bodies have become the ultimate dumps for the products of the chemical economy. Analyses of human fat, mother's milk, blood, breath, semen, and urine demonstrate that absolutely everyone—

not just people living near major pollution sources—now carries a "body burden" of toxic organochlorines in his or her tissues. At least 190 organochlorines have been identified in the tissues and fluids of the general populations of the United States and Canada (see appendix A). As one might expect, the list includes the most infamous organochlorines: dioxins, PCBs, DDT, mirex, and so on. Like the ambient environment, however, our bodies also contain a representative sample of the full range of organochlorines, from simple chlorinated solvents and refrigerants to little-known industrial chemicals and by-products.

Although we are all contaminated, some of us have it worse than others. For some people, perfectly innocent habits result in higher-than-average exposures. Sport fishermen in the Great Lakes, for instance, have body burdens of dioxins, furans, and PCBs more than double those of other people in the same region.[63] In others, the problem is simply location. Inuits in the Arctic, like their neighbor the polar bear, eat a diet rich in fish and seal blubber, and their bodies contain some of the highest levels of PCBs and organochlorine pesticides in the world.[64] The breast milk of Inuit women, for example, contains PCBs, dioxins, and organochlorine pesticides at levels two to ten times higher than in women from southern Quebec.[65]

A Legacy for Our Children

There are no predators that routinely eat humans, so we might think that the story of organochlorine exposure stops with people. But there is one class of living things higher on the food chain than human adults: our children. Organochlorines are passed from one generation to the next in extraordinary concentrations through mother's milk and across the placenta.[66]

Once organochlorines have taken up residence in a man's fatty tissues, he has no way to get rid of them. These chemicals will be with him more or less until death, so his organochlorine body burden generally increases slowly throughout his life.[67] A woman, however, has one way to reduce her body burden: she can flush contaminants out of her fat into her breast milk. Sixteen pesticides, 18 dioxins and furans, 65 PCBs, 11 other aromatic organochlorines, and 19 solvents and other chlorinated aliphatics have been identified in the breast milk of the general U.S. and Canadian population (see appendix A). The mother's loss is her child's gain, however: the nursing infant is exposed to these compounds during critical periods of development, when sensitivity to chemical disruption is particularly high.

Just as pollutant concentrations are magnified with each step up the food chain, a breast-fed baby is subject to doses much higher than the mother ever was. An average nursing infant receives doses of dioxins and PCBs that are 20 and 50 times higher than those of the average adult, respectively.[68] In a year of nursing, a typical baby receives 4 to 12 percent of his or her entire lifetime dioxin exposure and accumulates a higher body burden than the average 70-year-old man.[69] In fact, the nursing child's dioxin dose exceeds by a factor of 6 to 10,000 the standard of every government in the world for an adult's "acceptable daily intake."[70] According to one analysis, an infant's dioxin dose in a single year poses a lifetime cancer risk of 187 per million—187 times greater than the government's usual standard for an "acceptable" risk.[71] In the Arctic, where human milk levels are much higher than at temperate latitudes, the problem is compounded by the fact that Inuit women nurse their children for an average of 47 weeks.[72] An Inuit infant's daily doses of dioxins, PCBs, toxaphene, and hexachlorobenzene are 2 to 48 times the Canadian government's "tolerable daily intake" for adults.[73]

Milk contamination is an even greater problem for some other mammals. The milk of dolphins, for instance, is much richer in fat than human milk, so it contains PCBs, dioxins and pesticides at levels many times higher than the already alarming concentrations found in the animals' blubber. Mature female marine mammals have organochlorine body burdens only about half as high as those of males, because they transfer to their young as much as 60 percent of the pollutants they have accumulated in their own tissues.[74] A typical polar bear cub has an organochlorine body burden double that of its mother.[75]

Babies' chemical exposures begin well before nursing begins. One function of the placenta is to serve as a barrier that keeps harmful substances away from the embryo or fetus but fat-soluble organochlorines can cross the placenta. Dioxins and PCBs, for example, have been found in samples of fetal organs, placentas, and umbilical cord blood from the general U.S.

population.[76] As a result, the child is exposed to highly toxic organochlorines during the most exquisitely sensitive periods of development. PCB levels in umbilical cord blood from a substantial fraction of Inuit newborns, for example, are higher than the Canadian government's "level of concern" for adults.[77] In fact, exposure to pollutants begins even before conception, because the gametes—the egg and sperm that will unite to form a single-celled zygote, which will then divide many times to produce the child—are themselves bathed in the organochlorines present in the man's or woman's body.

Because we are so efficient at passing our body burdens to our children, future generations will be contaminated for centuries, even in the best-case scenario. If we stopped all new discharges today and could somehow clean up absolutely all the organochlorines already in the global environment, it would still take six generations—until our great-great-great-great-grandchildren, or about 150 years—until PCBs would no longer be detectable in the bodies of our offspring.[78]

Priority Pollutants and Mystery Chemicals

Why do we keep seeing the same group of pollutants—a dozen or so pesticides, PCBs, and dioxin—on all the lists of pollutants in the environment, the food chain, and in human tissues? Not because they are the only chemicals present. Indeed, the identified organochlorines represent just the tip of the iceberg, because hundreds more pollutants appear to be present but have not yet been chemically identified. In human tissues, for instance, analytical studies routinely identify just a small fraction of the contaminants that are known to be present.[79] EPA scientists estimate that the fatty tissues of the U.S. general population contain at least 700 additional contaminants that have not been chemically characterized,[80] and some of these are sure to be organochlorines. In gull eggs from Lake Ontario, specifically identified substances account for just 25 to 75 percent of the total organically-bound chlorine; in fish tissues from the Baltic Sea, they represent just 5 percent of the total.[81]

Why have so few of the chemicals present in the environment, the food web, and the human population been identified? The first problem is technological: the ability of chemical analyses to characterize contaminants is limited in a number of ways. Compounds can be identified only if they are present in concentrations above a detection limit (usually in the low parts per trillion). Substances present in very low quantities will not be detected, but if there are hundreds or thousands of them, together they may make up the bulk of the total chemical burden. Moreover, analyses can identify only compounds that can be matched against an existing database of chemicals. Substances that are not yet in the database go uncharacterized. Thus, novel or exotic compounds, like industrial by-products, environmental breakdown products, and the metabolic products formed when the body's own chemical processes act on foreign chemicals, will remain unidentified in even the most rigorous analysis.

The second reason for the narrow focus on a small number of compounds is social or political. Inquiry in the environmental field, as in many other scientific disciplines, is driven by a kind of institutional momentum, focusing where existing information or government order has already established a reason for concern. Negative results are of much less professional and practical value than positive results, so scientists tend to look for information where they are likely to find it, not where they have a slim chance of finding something new and interesting but a much greater chance of finding nothing at all. They also look for information in which people and institutions—government and funding agencies, in particular—are interested. That means that chemicals that have already been well-studied and regulated get more and more scientific attention, while few resources are directed toward pollutants about which little is known. For this reason, the scope of existing data does not imply limited contamination but is instead an artifact of the circumscribed reach of existing scientific knowledge and government priorities.

After PCBs, DDT, and other pesticides were phased out in the 1970s, for example, a huge amount of scientific literature began to be generated on their occurrence and effects. In the 1980s, as the toxicity of the dioxins and furans became clear, scientists sought these pollutants too, and found them wherever they looked. But the lack of data on the occurrence of thousands of other organochlorines does not mean that these compounds are not widespread. In 1992, for instance, a team of scientists sought a chemical called tris(4-chlorophenyl)methanol, an exotic industrial organochlorine that had not been the subject of much previous scientific or regulatory attention, in the tissues of birds and marine mammals from a wide variety of locations around the world; they found it at high levels—up to the parts per million—in bird eggs from the Great Lakes, California, and British Columbia, and from whales, seals, and sea lions in California, Australia, arctic Canada, and even Antarctica.[82]

If other obscure organochlorines were sought with the same effort, many more would most certainly turn up. This situation has two major implications for policy. First, we do not know the boundaries of global organochlorine contamination. The number of ubiquitous chlorinated pollutants is likely to be in the hundreds or thousands; whatever the number, the problem is without question a diverse one that is not limited to the handful of well-studied "bad-actor" chemicals. Second, the unidentified chemicals might pose considerable health hazards, but the Risk Paradigm takes no account of them. In a risk assessment, chemicals that have not been specifically identified are treated as if they do not exist: they are assumed to pose a risk of zero. The assumption that unidentified organochlorines are safe is unlikely to be true, since many previously uncharacterized substances that have been identified have turned out to be quite hazardous.[83] When it comes to organochlorines, it is not wise to bet that what we don't know won't hurt us.

Trends in Organochlorine Contamination

The chlorine industry has argued that action to restrict organochlorine production and use is unnecessary, because contamination of the environment is getting better. In its brochure on chlorine chemistry, for example, Dow Chemical responds to evidence of wildlife health damage in the Great Lakes by pointing out that "the amount of pollutants present in the Great Lakes has dropped significantly over the last several years."[84] The Chlorine Chemistry Council and the Vinyl Institute argue that "dioxin in the environment has been falling," so restrictions on PVC plastic are unnecessary.[85]

Is organochlorine contamination actually declining, and do trends indicate that the problem has been adequately addressed? In fact, very little information is available on trends in environmental organochlorine contamination, but the data we do have reinforce the need for—and effectiveness of—international action to phase out persistent chemicals.

The most important point is that contamination trends have been studied for only a handful of organochlorines: ozone-depleting substances, a few pesticides, PCBs, and dioxins, all of which have been subject to major restrictions of one type or another. Absolutely no information is available on trends in the vast majority of chlorinated pollutants, and we have no data on changes in the total organochlorine burden in the environment. With global production of chlorine and organochlorines increasing every year, however, there is no reason to believe that organochlorine contamination as a whole is declining.

The data that are available on specific organochlorines do suggest a pattern, however, that is probably relevant for other pollutants. After production of a synthetic chemical begins, environmental levels rise, until they reach a steady-state at which the rate of "removal"—through degradation or accumulation in sediments, soils, and vegetation—matches the rate of release. (If the rate of input outstrips the environment's maximum rate of removal, then concentrations will increase indefinitely.) If inputs are reduced, levels will slowly decline but then plateau at a new steady state. The more persistent the substance, the slower the decline to the new steady state. The lower the final input level, the lower the ultimate steady state. The steady-state concentration reaches zero sometime after inputs are completely eliminated.[86]

Trends in atmospheric levels of ozone-depleting substances illustrate this pattern. Following their introduction in the 1930s, concentrations of organochlorine solvents and refrigerants increased steadily for decades and continued to do so as long as environmental inputs continued. For instance, measurements at remote sites show that CFCs, trichloroethane, and carbon tetrachloride all increased substantially between 1975 and 1985.[87] Following the adoption in the late 1980s of the Montreal Protocol—the first international agreement on ozone-depleting substances—the production and use of major organochlorine ozone depleters dropped significantly. But these compounds and the hydrochlorofluorocarbons (HCFCs) that replaced them are so persistent that atmospheric concentrations of chlorine and bromine are not expected to begin to decline until the year 2005, and it will take decades before they return to pre-1980 levels.[88]

PCBs tell a similar story, but there is more information, mostly from dated sediment cores. A core is a vertical sample of sediments taken from the bottom of a lake or sea; because particulate matter in a body of water gradually settles to the bottom and builds up in layers over the years, the core contains a record of what kinds of particle-bound pollutants were present in the water at various times in the past. PCB manufacture was initiated in 1929, and it is in sediments from this time that PCBs are first reliably detected. Levels increase steadily from the 1940s to the 1960s, reaching a maximum between the late 1960s and the mid-1970s. PCBs were banned in the 1970s, and deposition in sediment then began to fall slowly. By the 1990s, PCBs in annual sediments had declined by about half, but the decline then slowed to a virtual stop.[89] In Great Lakes wildlife, PCB levels also fell substantially in the late 1970s and then reached a plateau, stabilizing in the 1990s at a level still considered unsafe.[90]

Dioxin trends look similar. Studies of soils in England and sediment layers from the Baltic Sea, the Great Lakes, and lakes in Finland, Germany, and Switzerland all show that dioxin levels were very low or nonexistent before the twentieth century. They began to rise slowly around the turn of the century and then increased rapidly from 1940 to 1970, the period during which the chlorine industry expanded most rapidly.[91] During the 1970s, many governments restricted the use of leaded gasoline (which contains chlorinated additives and was thus an important dioxin source) and major applications of some dioxin-contaminated pesticides, including 2,4,5-T and pentachlorophenol. In the same period, the U.S. Clean Air Act and similar legislation in other nations required a wide range of industrial facilities (such as incinerators, steel mills, and chemical plants) to install particulate-reducing pollution control devices, which are likely to have reduced dioxin emissions to the air. Following those actions, airborne dioxin levels—as measured by dioxin accumulation in plant foliage[92]—declined by up to 80 percent from the late 1970s to the early 1990s. As one might expect, dioxin levels in the milk of cows, which eat foliage, subsequently declined, falling by about 25 percent from 1990 to 1994.[93] Deposition of dioxin into sediment also declined in most places, but more slowly and to a less marked extent, because dioxin is so persistent in the environment.[94]

But falling deposition rates do not necessarily mean a lower burden of pollutants in the environment. Sediment layers provide a reasonably reliable record of the quantity of a substance that settled to the bottom of a body of water in any year, which roughly indicates the amount that entered the water in that year. The annual flux of persistent compounds, however, is not directly related to the total environmental burden; if the rate at which one puts marbles into a jar declines from 100 per year to 50, the total number of marbles in the jar will still continue to increase. For a record of the total amount of dioxin that has accumulated in the environment over time, soils are better than sediments, because pollutants from the recent and the distant past stay near the top of the soil rather than being buried in annual layers. British scientists have found that dioxin levels in soil, unlike those in sediments and foliage, continued to increase without interruption right through the 1980s and into the early 1990s. They concluded, "PCDD/Fs are persistent in soils, such that declines in atmospheric deposition may not result in a decline in the U.K. PCDD/F burden for some time. It may be that even with the anticipated declines in the primary emissions of PCDD/Fs over the next decade, the rate of deposition may still exceed the rate of loss from soils."[95]

Human and wildlife tissues reflect the same pattern, with a delay because of the persistence of these compounds in our bodies. Dioxin levels in several species of wildlife in the Great Lakes declined during the late 1970s, 1980s, and early 1990s.[96] By 1993, however, levels of dioxins in the eggs of Great Lakes trout had stopped falling, reaching, in the words of one group of researchers, "a steady state or a very slow decline."[97] No human tissue analyses are available from the early decades of this century, but dioxin levels in people from the United States appear to have increased steadily during the 1960s; following the regulatory actions of the 1970s, there then appears to have been a moderate decrease in dioxin levels during the 1980s.[98] (There is some difficulty in interpreting these data, according to EPA, "It is not known whether these declines were due to improvements in the analytical methods or actual reduction in body burden levels.")[99] In the 1990s, dioxin levels in the milk of European women declined by 20 to 40 pecent.[100]

This is good news, but not good enough. According to Swedish scientists, the declines are history, not a continuing trend. We should not expect levels to fall further unless we take further action to restrict dioxin sources: "During the last twenty years an overall decrease in the levels [of dioxin in human tissues] is recorded. The major part of this decrease dates back to the late 1970s and the early 1980s. The situation of today seems to be quite constant and resembles what has been found for PCB. Analyses of human breast milk show a similar trend."[101]

One limitation of all these studies is that they document declines on a local or regional basis; lower levels in German milk or British soil do not mean that the total planetary burden of any pollutant is falling. Contaminant levels in foliage, soils, sediments, wildlife, and even human tissues from industrialized countries are determined in large part by emissions from sources in the same general area. If releases in a region are reduced, then environmental levels may decline not because accumulated contaminants are breaking down but because they are being transported to more distant areas by distillation or diffusion.

To understand global trends in organochlorine accumulation, we need to look to the Arctic and the deep oceans. Here, there is little evidence for substantial declines, with the exception of hexachlorocyclohexane, which declined in arctic air during the 1980s as its use as a pesticide fell out of favor in many industrialized nations. In contrast, airborne deposition of PCBs on Ellesmere Island in arctic Canada did not change substantially from 1963 to 1993, and levels in the 1990s are actually higher than they were in the 1980s.[102] At other sites in the Canadian arctic, PCB measurements in snowfall increased substantially from 1986 to 1990.[103] In sediments of most arctic rivers and lakes, PCBs and chlorobenzenes have held steady or even increased in the 1990s.[104]

In wildlife, trends are no rosier. Levels of PCBs and DDT in North Atlantic marine mammals and arctic ringed seals declined during the 1970s, but they did not change significantly in dolphins from the western North Pacific from 1978 to 1986.[105] In the 1980s and 1990s, there has been no substantial change in DDT and PCB levels in arctic seals, beluga, narwhal and sea birds.[106] In polar bears from the Canadian arctic, DDT levels declined by a meager 20 percent from 1969 to 1984, but PCBs, chlorobenzenes, hexachlorocyclohexanes, chlordane and dieldrin all increased by a factor of two to four.[107] As a 1995 review of global trend data summarized, "Residues of organochlorines in Arctic waters may even be increasing. . . . Generally, it appears that the open-ocean environment has exhibited no declining tendency in organochlorine concentrations. . . . Marine pollution by organochlorines may not decline unless strict regulations are imposed on their use throughout the world."[108]

All of this information suggests a pattern with clear implications for policy. Government action to restrict the industrial production and use of specific substances has reduced emissions to the environment of these compounds. On local and regional scales, contamination of the environment and the tissues of living organisms has fallen in response, with the speed of the decline varying among different kinds of sampled material. But these pollutants are so persistent that as long as releases continue somewhere, the global environmental burden of these compounds declines slowly or not at all. If we allow releases to continue at a reduced rate, concentrations will stop declining when a new steady state is reached. To avoid the accumulation of persistent toxic chemicals in the global environment, we need to stop environmental releases altogether.

From a Local to a Global View

Long-term, large-scale contamination has developed not because our governments have no regulations but because they have the wrong kind. Indeed, many organochlorines are subject to pollution control rules, but the reality of global contamination stands in stark contrast to the local focus of the Risk Paradigm. When a facility is granted a permit to discharge pollutants into the air or water (or when the regulations that provide the guidance for these permits are formulated), the risk assessments on which these decisions are based consider only local and immediate exposures through one or a few exposure routes. The central construct of a risk assessment is a "most exposed individual" (MEI)—a hypothetical person who suffers very high local exposures. In the case of an incinerator permit, for example, the MEI stands at the edge of the facility grounds, breathing deeply, 24 hours a day for 20 years or more. The risk assessor uses data and assumptions about the height of the smokestack, the direction and speed of the wind, the amount of air a person breathes, and the toxicity of each pollutant to "back-calculate" a maximum permissible emission rate. Releases below this limit will presumably subject the MEI to exposures that do not exceed an acceptable daily intake—a dose at which cancer risks are "negligible" (usually no more than 1 to 10 per million), and other systemic health effects are not expected to occur.

By this method, any quantity of the pollutant not ingested by the MEI simply disappears. In fact, this untracked fraction represents the vast majority of the emissions, since no MEI takes into his or her lungs all the chemicals released from the smokestack. In the Risk Paradigm, whatever quantity of a pollutant is taken up by wind currents and transported long distances is of no consequence. Further, it is irrelevant that there may be hundreds of other facilities emitting the same chemical, resulting in a cumulative global burden of this substance. If a facility builds a higher smokestack or a longer discharge pipe, ensuring that emissions will be diluted in the air or water, then risks to the MEI will be reduced, and larger quantities of pollutants can be permitted. In this way, risk-based discharge permits allow—and even encourage—global pollution.

A controversial risk assessment, prepared for the Waste Technologies Industries (WTI) hazardous waste incinerator in East Liverpool, Ohio, provides a compelling example of the Risk Paradigm's failure to address long-term accumulation and exposure. When completed, WTI will be the largest hazardous waste incinerator in the world. It is located on the banks of the Ohio River, surrounded by towns and farmland. Four hundred feet away, on a bluff above the facility that places them almost level with the top of the incinerator's stack, are a group of homes and an elementary school. EPA approved WTI's operating permit based on a risk assessment that examined only inhalation exposures and concluded that cancer risks would not exceed EPA's acceptable limit. Local citizens and the environmental group Greenpeace challenged the permit in federal court, arguing that the risk assessment failed to consider exposures to dioxin and other pollutants that would build up in the soil, water, and local food supply.

In February 1993, while the case was in its early stages, an internal EPA memo discussing the agency's legal strategy was leaked to the plaintiffs. Written by Assistant Administrator Richard Guimond, the senior U.S. Public Health Service officer at EPA, to agency chief Carol Browner, the memo admitted that exposure routes ignored in the WTI risk assessment could result in clearly unacceptable health risks.[109] According to Guimond, EPA's Office of Research and Development (ORD) had secretly prepared another preliminary risk assessment, called a "screening analysis," which predicted that dioxin from the incinerator would accumulate in cattle raised at a nearby farm and pose cancer risks to a frequent beef consumer a thousand times greater than those from inhalation alone. There actually turned out to be at least one farm that met the scenario Guimond described: Joy Allison, one of the plaintiffs in the suit against EPA and WTI, raised cattle on her farm near the incinerator site and consumed large amounts of home-grown beef. The obvious similarity between the real Joy Allison and EPA's "hypothetical" most exposed individual lent a macabre quality to EPA's discussion of the potential hazards of the facility.[110]

In the memo, Guimond was straightforward about the implications of ORD's findings for risk assessment as a general practice. Because accumulation in the food-chain always results in exposures many times greater than those by inhalation alone, the plaintiffs' argument applied to many other dioxin sources, threatening a plethora of currently permitted polluting activities. If food chain contamination were considered at all these other facilities, risks would presumably exceed acceptable levels, throwing into question the permissibility of a wide range of technological practices, including virtually any form of incineration. Guimond wrote:

> A preliminary assessment done by ORD does show that risks from beef and milk consumption can be 1000 times higher than risks from inhalation. . . . There are very significant implications associated with adopting risk assessment procedures based on indirect exposure routes for air emission sources. . . . The analyses of the WTI situation show that many air emission sources could be affected if EPA were to adopt the indirect exposure analysis procedures in assessing exposure risks. . . . My staff met with ORD on Friday, January 22 to discuss the analytical steps that must be undertaken in preparation for the upcoming District Court hearing. In addition, my staff is reviewing all of the risk assessment efforts related to WTI and the ORD report on dioxin exposures to ensure that any differences can be explained.

Guimond did not recommend that EPA's findings of potentially high food chain risks be released to the court or to the public, nor did he recommend that the agency do anything to address these threats by revoking or modifying the permit of the incinerator or any other dioxin-releasing facility. Instead, Guimond detailed steps his office was taking to *avoid* acknowledging the problem and to ensure that any inconsistencies could be justified before the court.

When the plaintiffs presented the leaked memo, the court forced EPA to release the ORD risk assessment, over the objection of the agency's lawyers that the document was privileged and should remain secret. Remarkably, ORD's analysis showed that the health threat was even more severe than Guimond had made it out to be: dioxin exposures due to consumption of beef raised near the incinerator could be up to 40,000 times greater than those through inhalation alone. A single year of incinerator operation was projected to pose cancer risks as high as 42 per million—well over EPA's standard for an entire lifetime. If the facility operated for 30 years, cancer risks from dioxin in beef would be on the order of 1,200 per million.[111]

If local risks from one incinerator are this great, consider the global implications of the cumulative emissions from all the facilities like this that EPA and other regulatory agencies permit to operate simultaneously, year after year. There is no accepted method to evaluate the large-scale impacts of global contamination or the impact of individual facilities in the context of the total worldwide exposure burden. By operating under the illusion that pollutants can be traced from the smokestack to a single individual on whom their effects can be reliably predicted, risk assessment inherently precludes a consideration of the global effects of thousands of facilities, each of which emits thousands of toxic substances into the environment.

In the end, the citizens won their case in court, and the judge ordered EPA to stop WTI from operating its incinerator unless a full food chain risk assessment concluded that risks were acceptable.[112] An appeals court overturned the judge's order on matters of jurisdiction, without contesting the facts of the case. WTI and EPA conducted a new risk assessment, which again predicted that the health risks would be "acceptable." EPA awarded WTI its final permit, and the incinerator opened for business in 1994.

The findings reviewed in this chapter provide the first two points on which the Risk Paradigm conflicts with what we know about organochlorines. First, experience shows that there can be no "acceptable discharge" level of persistent toxic pollutants, which accumulate in the environment over time. Even dilute discharges, like the tiny quantities of dioxin released from individual facilities, eventually accumulate to unacceptable levels. For this reason, the system of approving pollutant discharges within calculated limits has failed resoundingly to prevent the long-term buildup of persistent organochlorines. The only successful policies have been those that took an alternative approach: measures that have phased out the production and use of persistent toxic chemicals have reduced the burden of these contaminants in the regional environment.[113] These measures embody the Ecological Paradigm's principle of Zero Discharge, which says that releases of persistent substances must not be permitted in any quantity, a standard that requires these chemicals to be neither produced nor used.

The Risk Paradigm's second flaw is its focus on managing pollution at the local scale, which has allowed regional and global contamination to occur. Controls applied to individual sources of pollution are designed to prevent high local exposures and severe contamination of the local ecosystems. When thousands of such sources are permitted to release pollutants into the environment, however, their emissions are dispersed across regional and ultimately global environments, contributing to a total burden of contamination that the local view never even considered. The Risk Paradigm's focus on each of the trees prevents it from ever seeing the forest of global pollution.

As a result, the countless regulations implemented under the Risk Paradigm have resoundingly failed to prevent contamination of such important ecosystems as the Great Lakes, the Baltic, the Mediterranean, the deep oceans, the Arctic—in short, the entire world. Even the handful of successful bans on chemicals have been implemented at national rather than international scales, so they have created only regional gains. DDT, toxaphene, heptachlor, and other pesticides have been restricted in the United States, Canada, and Western Europe, but they continue to be used in dozens of developing countries; some are even produced in the wealthy nations where they are illegal and shipped to poor countries for use. For instance, over 1 million tons per year of lindane and other hexachlorocyclohexanes, restricted in most industrialized nations, continue to be used on a global basis.[114] In this way, the burden of regional contamination has gradually shifted to developing countries, and global pollution has gone completely unaddressed.

To solve this problem, the Ecological Paradigm requires the sources of pollution to be restricted on an international basis. Pollution knows no borders, and the use of these substances—in the tropics, far inland, anywhere at all—contributes to contamination of regional and global ecosystems. A problem of worldwide scope requires truly international solutions.

There is some cause for optimism on this point. In 1996, largely at the behest of nations with lands in the Far North, the United Nations began work on a global, legally binding instrument to address persistent organic pollutants (POPs) in the earth's environment. This process, much like the one that led to the Montreal Protocol on ozone-depleting substances, represents the first international action to stop the accumulation of persistent toxic substances in the global environment. As conceived at the time of this writing, the treaty is not nearly adequate to its task because it addresses only 12 "priority" pollutants, all of them organochlorines (table 1.3).[115] Even if the POPs treaty applies only to these 12 pollutants, however, it can be a great step forward if it eliminates the production of these substances rather than merely reducing their emissions to specific environmental media. A global agreement to phase out even a handful of persistent toxic substances could provide a framework within which the international community could begin to address the 69,988 other commercial chemicals not included in the initial treaty.

Table 1.3
Persistent organic pollutants addressed by UNEP's International POPs Agreement

Persistent organic pollutants
Aldrin
DDT
Dieldrin
Endrin
Chlordane
Heptachlor
Hexachlorobenzene
Mirex
Toxaphene
Polychlorinated biphenyls (PCBs)
Polychlorinated dibenzo-p-dioxins (PCDDs)
Polychlorinated dibenzo-furans (PCDFs)

Source: UNEP (1997).

2

First-Class Poisons: The Toxicology of Organochlorines

The only solid piece of scientific truth about which I feel totally confident is that we are profoundly ignorant about nature. . . . It is this sudden confrontation with the depth and scope of ignorance that represents the most significant contribution of twentieth-century science to the human intellect.
— Lewis Thomas, *The Medusa and the Snail*[1]

Bodies are resilient things. There are poisons in nature, but organisms have ways of coping with them. Their bodies repair injuries, fight off infections, degrade and excrete foreign compounds, and maintain a steady internal state despite fluctuations in the environment. They have senses that perceive the world and brains that coordinate and direct behaviors, so they can avoid dangers to which their bodies cannot adapt. They even protect the genetic legacy they will pass on to future generations by segregating the germ line—the source of egg and sperm—from the body's other cells. At each generation, life starts anew, unfolding an entirely new body from the protected information in a single fertilized egg.

Suppose, on the other hand, that there were substances that could disrupt the very processes that make organisms resilient in the first place. Imagine that these chemicals could infiltrate the germ line and damage the genetic material we give to future generations. Suppose they could disrupt the body's intrinsic controls over the growth and differentiation of tissues, wreaking havoc with the dance of development and producing cancer in adults. Imagine that they could compromise the body's ability to defend itself against infectious disease and interfere with the hormones that coordinate its ability to join with another and reproduce a new organism. Picture substances that could disrupt the signals by which the body maintains a steady state despite changes in the internal and external environment. Suppose they could wreak havoc with the chemical process by which the brain develops, sends signals, and generates behaviors. Chemicals that could do all this would pose a profound and new kind of threat to the health of our otherwise resilient bodies.

In fact, this litany precisely describes what the organochlorines that have infiltrated the far corners of the globe and the intimate corners of our own bodies can do. In the next three chapters, I argue that the organochlorine cocktail to which absolutely everyone is now exposed poses a long-term threat to the health of people and wildlife. This chapter I devote to the toxicology of organochlorines, focusing on two major points. First, I show why organochlorines, despite their individual differences, can and should be treated as a generally hazardous class of substances: incorporating chlorine atoms in organic compounds has profound effects on the magnitude and nature of their toxicity, radically increasing the biological activity of most chemicals and in many cases creating new hazards that the chlorine-free parent compound did not cause. Second, I show that the basic toxicological assumption of the Risk Paradigm—that science can reliably establish safe exposure levels for individual chemicals—is unrealistic and unsupportable. A close analysis of both the science and the politics of this claim shows that "safe" exposures do not exist in the real world or cannot be discovered with confidence. An Ecological Paradigm for policy would shift the focus of regulation from approving individual chemical discharges to preventing exposure to the entire hazardous class of organochlorines.

Effects of Organochlorines

The health damage that organochlorines can cause has been well established in studies of animals and humans, but the existing data only hint at the full scale of the problem. Only a fraction of organochlorines have undergone basic toxicity testing to reveal obvious acute and chronic effects, and only a handful of compounds have undergone the detailed testing necessary to reveal subtle, long-term health impacts.[2]

According to the International Agency for Research on Cancer (IARC), a division of the World Health Organization, there is limited or sufficient evidence of carcinogencity for over 100 organochlorines or groups of organochlorines, comprising over a thousand individual compounds[3] (table 2.1). The majority are mutagens that damage DNA and initiate cancer, but others are promoters or other kinds of modulators that increase the chance that a latent cancer cell will develop into a malignant tumor. The list includes a broad sample of the universe of organochlorines, from simple low-molecular-weight compounds like dichloromethane, vinyl chloride, epichlorohydrin, and tetrachloroethylene, to more complex aromatic organochlorines like hexachlorobenzene, pentachlorophenol, and the pesticides heptachlor, 2,4-D and 2,4,5-T.

A large number of organochlorines are neurotoxins, causing permanent or temporary damage to the function of the brain and other components of the nervous system.[4] Relatively high doses of organochlorine insecticides, for instance, cause tremor, slurred speech, euphoria and excitement, nervousness and irritability, depression and anxiety, mental confusion, and memory disorders.[5] Chlorinated solvents, intermediates, and the plastic feedstock vinyl chloride dissolve in the membranes of neurons and produce a marked suppression of the brain's ability to transmit signals along nerve pathways.[6] In very low doses, PCBs affect brain chemistry and produce long-lasting behavioral changes and deficits in the ability to think and learn, particularly when exposure occurs before birth or early in life.[7]

Many organochlorines are also reproductive toxicants, causing infertility or other reproductive problems in a number of ways. First, a very large number of organochlorines—ranging from the chlorinated solvents to the chloroketones, a large group of by-products formed in water chlorination and pulp and paper bleaching—are mutagenic:[8] they damage DNA, the genetic material that constitutes the basic hereditary information of all organisms. When mutagens alter DNA in in the germ cells that give rise to egg or sperm they can cause infertility, make a fertilized egg or embryo inviable, or produce birth defects, disease, or compromised function if the offspring survives.[9] Other organochlorines, like the pesticides dibromochloropropane and chlordecone, the chemical feedstock carbon tetrachloride, and two chemicals used in the manufacture of plastics, chlorobutadiene and epichlorohydrin, are toxic to the testis, killing male germ cells or otherwise preventing the production of viable sperm.[10] Still others, such as dioxin, PCBs, and the pesticides endosulfan and atrazine, disrupt

Table 2.1
International Agency for Research on Cancer classification of organochlorine carcinogens

Known human carcinogens
Chlornaphazine [N,*N*-bis (2-chloroethyl)2-naphthylamine]
Bis(chloromethyl)ether
Chloromethyl methyl ether
Chlorambucil
Semustine [2-chloroethyl)-3-(4-methylcyclohexyl)1-nitrosourea]
Mustard gas [Bis-2-chloroethyl sulfide]
2,3,7,8-Tetrachlorodibenzo-p-dioxin [TCDD, dioxin]
Vinyl chloride [Chloroethylene]

Probable human carcinogens
Bis(chloroethyl)nitrosourea
Chloramphenicol
2-Chloroethyl-3-cyclohexyl-1-nitrosourea
p-Chloro-o-toluidine
Chlorozotocin
Dimethylcarbamoyl chloride
Epichlorohydrin
Methylene-bis-2-chloroaniline [MOCA]
Nitrogen mustard [Bis(2-chloroethyl)-methylamine]
Polychlorinated biphenyls [PCBs]
Procarbazine hydrochloride
Tetrachloroethylene
Trichloroethylene
1,2,3-Trichloropropane

Possible human carcinogens
Benzal chloride
Benzotrichloride
Benzyl chloride
Bromodichloromethane
Carbon tetrachloride
Chlordane
Chlordecone [Kepone]
Chlorendic acid [Hexachloroendo-methylenetetrahydrophthalic acid]
Chlorinated paraffins (avg. length C12, Chlorine>60%)
p-Chloroaniline
Chloroform
1-Chloro-2-methylpropene
Chlorophenols
4-Chloro-ortho-phenylenediamine
DDT
1,2-Dibromo-3-chloropropane
p-Dichlorobenzene
3,3′-Dichlorobenzidene
3,3′-Dichloro-4-4′-diaminodiphyenyl ether
1,2-Dichloroethane [Ethylene dichloride]
Dichloromethane
2,4-D [2,4-Dichlorophenoxyacetic acid and salts]
1,3-Dichloropropene
Dichlorvos
Heptachlor
Hexachlorobenzene
Hexachlorocyclohexanes
Mirex
Nitrofen [2′,4′-Dichloro-4-nitro-biphenyl ether]
Nitrogen mustard-N-oxide [2-Chloro-N-(2-chloroethyl)-N-methylethanamine-N-oxide]
Pentachlorophenol
Phenazopyridine hydrochloride
Phenoxybenzamine hydrochloride
Tris(2-chloroethyl)amine [Trichlormethine]
2,4,5-T [2,4,5-Trichlorophenoxy acetic acids and salts]
Toxaphene [Polychlorinated camphenes]
Dry cleaning (occupational exposures)

Table 2.1 (continued)

Limited evidence of carcinogenicity in animals; inadequate or no data in humans
Aldrin
Atrazine
Benzyl chloride
Bis(2-chloroethyl)ether
Bis(2-chloro-1-methylethyl)ether
Captan [N-(trichloromethyl)thio-4-cyclohexene-1,2-dicarboximide]
Chloral hydrate [1,1,1-Trichloro-2,2-dihydroxyethane]
Chlordane
Chloroacetonitrile
Chlorobenzilate
Chlorodibromomethane
Chlorodifluoromethane
Chloroethane
3-Chloro-2-methylpropene
Chlorothalonil
Clofibrate [alpha-p-chloro-phenoxyisobutyryl ethyl ester]
Dichloroacetic acid
1,1-Dichloroethylene [vinylidene chloride]
1,2-Dichloropropane
2,6-Dichloro-1,4-benzenediamine
Dicofol [2,2,2-Trichloro-1,1-bis (p-chlorophenyl)ethanol]
Dieldrin
Heptachlor epoxide
Hexachlorobutadiene
Hexachloroethane
Monuron [3-(4-Chlorophenyl)-1,1-dimethylurea]
Pentachloroethane
Pentachloronitrobenzene
Picloram [4-Aminotrichloropicolinic acid]
Sodium trichloroacetate
1,1,1,2-tetrachloroethane
Tetrachlorovinphos
Trichloroacetic acid
1,1,2-Trichloroethane

Note: Known human carcinogens are agents with sufficient evidence of carcinogeneicity in humans; probable human carcinogens are agents with limited evidence in humans and sufficient evidence in laboratory animals; possible human carcinogens are agents with limited evidence in humans and limited evidence in laboratory animals. Synomyms are shown in brackets.

Sources: Known, probable, and possible human carcinogens from IARC (1997) (groups 1A, 2A, and 2B, respectively). Agents with limited evidence in laboratory animals and inadequate evidence in humans from HSDB (1997).

the body's natural control over the reproductive system by mimicking or blocking the activity of the steroid hormones that regulate reproductive function and behavior or by changing the quantities of hormones that circulate in the body. Organochlorines that act through these mechanisms have been shown to reduce sperm counts, prevent proper estrous or menstrual cycling, increase or reduce the size of the uterus, and cause structural changes in the reproductive tract of both males and females.[11] When pregnant animals are exposed to organochlorines, development of the offspring can be disrupted in a way that causes spontaneous abortion (as is the case for chlordecone, among other pesticides) or birth defects (as occurs following exposure to such organochlorines as vinyl chloride, aldrin, dieldrin, and endrin and the herbicides 2,4-D and 2,4,5-T).[12]

A variety of organochlorines disrupt the development and function of the immune system. For instance, dioxins, PCBs, the solvent trichloroethylene, pentachlorophenol, 2,4-dichlorophenol, and a large number of organochlorine pesticides undermine the body's defenses against bacteria, viruses, and protozoan parasites by suppressing cell-mediated or antibody-mediated immunity.[13] Several organochlorines—including vinyl chloride, PCBs, and chlorinated solvents—are also associated with the development of autoimmunity, in which the immune system attacks the body's own tissues as foreign.[14] Autoimmune diseases include lupus, rheumatoid arthritis, and systemic sclerosis; evidence from a number of studies suggests that chlorinated solvents, vinyl chloride, organochlorine pesticides, dioxins, and other hormone-disrupting pollutants may be associated with an increased risk of these diseases.[15]

A Toxic Class

A wide variety of organochlorines, comprising many types of chemical structures, are highly toxic, but they are not equally or identically so. Each organochlorine produces a unique set of effects that is determined by the solubility of the compound in fat or water, its half-life in the body, the metabolites that are produced when the compound interacts with the body's enzymes, and—above all—the shape of the molecule, which determines the chemical's ability to interact with receptors, enzymes, and other proteins. Further, the toxicological potency of organochlorines varies considerably. For instance, 2,3,7,8-TCDD is considered the most toxic synthetic carcinogen ever tested,[16] but trichloroethylene is much less potent.[17] A very few organochlorines are not particularly toxic at all. For instance, the refrigerant dichlorodifluoromethane is known to produce liver damage and neurotoxicity only at doses in the 1,000 to 5,000 parts per million range, a level that will never be encountered in the general environment.[18] (Of course, this same compound is a major cause of ozone depletion, so it contributes to cancer, immune suppression, and blindness in the global population in a different way.)

Despite the occasional exception, organochlorines are almost universally toxic, in many cases extraordinarily so. The American Public Health Association (APHA), the largest and most important organization of public health scientists and practitioners in the United States, concluded in a consensus resolution in 1993, "Virtually all organochlorines that have been studied exhibit at least one of a range of serious toxic effects, such as endocrine dysfunction, developmental impairment, birth defects, reproductive dysfunction and infertility, immunosuppression and cancer, often at extremely low doses, and many chlorinated organic compounds . . . are recognized as significant workplace hazards."[19]

Among the first to express concern about organochlorines was the Science Advisory Board of the International Joint Commission, which in 1989 recognized the pattern of high toxicity among the class of organochlorines.[20] A thorough exploration of the toxicity of organochlorines did not appear until 1994, when the eminent German toxicologist Dietrich Henschler conducted the first extensive analysis of the mechanisms and toxicological potency of organochlorines. Henschler is anything but a radical environmentalist: since 1965, he has been professor and chair of the department of toxicology at the University of Wurzburg, and he also leads the German government's advisory commission on occupational exposure and is the coauthor of the leading toxicology textbook in German. His study of organochlorines, published as a book-length monograph in German and abridged in English translation in the chemical journal *Angewandt Chemie,*[21] was "suggested and promoted" by the Society of the German Chemical Industry, so it certainly did not set out to show that chlorine chemistry is intrinsically problematic. Nevertheless, its findings strongly support that conclusion.

Henschler examined the impacts that a wide range of organochlorines have on human health, investigated their potency, and explored the mechanisms by which they exert these effects. By comparing organochlorines to each other and to their nonchlorinated analogues, Henschler sought to understand how chlorination affects the toxicity of organic chemicals. He reviewed the toxicity of several hundred organochlorines, including the chlorinated methanes, ethanes, ethylenes, acetylenes, propanes, propenes, butanes, butadienes, benzenes, phenols, paraffins, dioxins, furans, biphenyls, and insecticides. Although he could not include all known organochlorines in his examination, Henschler did manage to analyze a huge number of compounds, representing all the major product groups and a wide range of chemical structures. He wrote, "A series of representative homologous and analogous groups of chlorinated compounds has been evaluated, because they are considered to throw light on structure-activity relationships. . . . Although the present investigations has not included some possibly important groups, conclusions which apply to chlorinated organic compounds in general may be drawn."

From this very large data set, Henschler came to four major conclusions about the toxicity of organochlorines as a class of substances. First, "The introduction of chlorine into organic compounds is almost always associated with an increase in the toxic potential. Only rarely does chlorination produce no increase or even a decrease in effects. This observation applies for all kinds of toxic effect (acute, subchronic and chronic toxicity, reproductive toxicity, mutagenicity, and carcinogenicity)."

In most cases, the nonchlorinated compounds from which organochlorines are derived—methane, ethane, ethylene, and so on—are very weakly toxic, if they are poisonous at all. The chlorinated substances, in contrast, cause neurotoxicity, liver and kidney damage, cancer, damage to DNA, and other effects, generally in concentrations thousands of times lower than those at which their nonchlorinated analogues become toxic. For instance, airborne methane is essentially nontoxic, causing adverse effects only in the parts per hundred range; the chlorinated methanes, in contrast, cause kidney, liver, and reproductive toxicity in the low parts per million. Of the many groups Henschler studied, the only exception to this rule was the chlorobenzenes, which are more potent than benzene in causing acute and chronic toxicity to the liver and other organs but are weaker carcinogens, presumably because they are less subject to conversion into carcinogenic form by the body's enzymes.

Henschler's second finding was that chlorination usually produces entirely new toxic effects. He wrote, "The introduction of chlorine into the molecule frequently gives the substance new properties. New acute and subchronic effects are seen mainly in the parenchymatous organs (especially liver and kidneys, less often in the spleen), circulatory system, and central nervous system." Most of these effects are completely absent from the nonchlorinated compounds but appear when organic molecules are transformed into organochlorines.

His third finding focused on the carcinogenicity of organochlorines: "With the introduction of chlorine into the molecule, most of the organic substances discussed here become mutagenic and/or carcinogenic. . . . A considerable portion of the chlorinated organic compounds which have been studied has carcinogenic activity." None of the chlorine-free compounds examined, with the exception of benzene, are carcinogenic, but *all* the chloromethanes, chloroethylenes, chlorinated propenes, and chlorobutadienes, all but one of the chloroethanes, and many members of the other groups studied are known or suspected carcinogens. An even greater number may turn out to be carcinogens, but "the proportion cannot be determined exactly because chlorinated organic compounds have not been tested systematically."

Finally, "There is also a tendency to increase in toxicity with the number of chlorine substituents in a molecule; this can be seen as a general rule but a less stringent one than that for the effects of introducing chlorine in the first place." For example, among the chlorinated ethanes, toxicity to the liver increases by a factor of five to ten with each chlorine atom added to the molecule; the workplace exposure limit for hexachloroethane is thus 1,000 times lower than that of monochloroethane. The rule is less strict and more complicated than the others, because in many molecules putting the chlorine atoms in different places will result in higher or lower toxicity; for instance, adding a third chlorine atom to 1,1-dichloroethane to produce 1,1,2-trichloroethane increases toxicity tenfold, but adding it in a different position to produce 1,1,1-trichloroethane has little effect. Despite this additional "placement factor," however, the rule of increasing toxicity with increasing chlorination usually holds true.

Henschler's analysis thus makes clear that organochlorines are a hazardous class of substances. There is something inherently dangerous about chlorinating organic molecules, the essential act of chlorine chemistry. All organochlorines are by no means the same, and chlorine alone does not determine the toxicity of a chemical; rather, chlorine is one factor among several that determine the precise nature and potency of a compound's toxic effects. Chlorine *modulates* chemical toxicity, virtually always increasing it, often by a very great degree.

Why Organochlorines Are Toxic

The APHA's conclusion that organochlorines are a hazardous class of compounds is based on abundant but circumstantial evidence: the fact that virtually all organochlorines that have been tested cause one or more severe health effects suggests very strongly but does not prove that the class is generally problematic. Even Henschler's very rigorous analysis is circumstantial: it draws a general conclusion from a broad base of specific examples. This way of drawing conclusions, called induction, is a perfectly reasonable mode of analysis, and it is the way that people arrive at most of their commonsense knowledge. But there is a potential logical flaw in this kind of reasoning: we cannot absolutely rule out the possibility that the many and various organochlorines that have been tested do not represent the total set of all organochlorines, the rest of which will somehow turn out to be harmless.

Farfetched as this objection may be, it does highlight the fact that any generalization from experience can never be strictly proved, because there is always the chance that new examples will come up to contradict the generalization. The APHA's judgment that the class of organochlorines is generally hazardous would be greatly strengthened if it could be supported not only by induction from examples but also by deduction from accepted scientific theories. In deductive reasoning, general principles are applied to specific examples: if the principles are true, then individual judgments based on them must also be true.[22] If I can show, based on the fundamental laws of chemistry and biology, *why* organochlorines are so hazardous, then there will be a theoretical basis as well as an empirical one for our judgment about the class of organochlorines.

There are two reasons that organochlorines turn out to be dangerous, and both arise from the essential nature of the chlorine atom. Chlorination changes the behavior of organic molecules in two critical ways: it makes organochlorines far more fat soluble than organic chemicals that do not contain chlorine, and it changes their chemical stability, making some organochlorines more persistent in the body and others more reactive. Both properties tend to increase toxicity.

In chapter 1, we saw that the fat solubility increases the tendency for a substance to bioaccumulate; it also tends to make it more toxic. As I show in detail in chapter 5, one factor that determines the solubility of a chemical in fats is its size: the larger it is, the more it disrupts the attraction of water molecules for each other, and the greater the tendency for the substance to be excluded from water rather than to dissolve in it. Chlorine is a very large atom—several times larger than an atom of hydrogen, carbon, or oxygen—so chlorination significantly increases the size of organic compounds and as a result almost invariably increases their solubility in fats and oils.[23] The increase in fat solubility—also called lipophilicity ("love of fat") or hydrophobicity ("fear of water")—increases with each chlorine atom added to an organic molecule, and each additional chlorine atom has a proportionally greater effect on bioaccumulation.

Chlorination's effect on solubility increases the toxicity of organochlorines in a number of ways. For one thing, it makes them more likely to interfere with the activity of enzymes, the proteins that carry out virtually all of a cell's essential biochemical functions. Enzymes work by attaching to one or more molecules of chemical substrates—the proteins, sugars, or other substances that are the ingredients in the reaction—and activating specific and useful chemical reactions, a process called catalysis. On many enzymes, the "active site" to which the substrate binds is hydrophobic: the more fat soluble a pollutant is, the greater the chance that it will bind to an enzyme. Once attached to the active site, the foreign substance may exclude the enzyme's natural substrate and prevent the enzyme from doing its proper job. Or, more insidiously, the foreign compound can itself serve as an unintended substrate for the enzyme, which converts it into a new compound, called a metabolite; in the case of organochlorines the metabolite is often far more toxic than the original substance. For instance, human metabolism can transform both DDT and PCBs into new compounds that mimic the activity of estrogen with much greater potency than the original substances.[24] Metabolism is also responsible for the carcinogenicity of many chlorinated solvents, which produce tumors only after they have been activated by the enzymes to which they have a strong affinity. Henschler explains, "The introduction of chlorine substituents makes organic compounds more lipophilic, facilitating interactions with hydrophobic sites, for example, in enzymes, and promoting enzymatic biotransformation in general. . . . Therefore it is to be expected that in spite of their frequently increased chemical stability, chlorinated organic compounds are converted enzymatically more readily than their chlorine-free analogues."

The fat solubility of organochlorines also increases their affinity for other proteins called receptors, like the estrogen receptor or the aryl hydrocarbon (Ah) receptor, which binds dioxin-like compounds. The job of receptors is to mediate the signals sent by the body's natural hydrophobic hormones and vitamins. Each of these substances binds to a specific receptor within target cells; the receptor then binds to specific "response elements" in the cell's DNA and stimulates or represses the activity of nearby genes.[25] The part of these receptors to which the hormone attaches is called the ligand-binding pocket, and it is always hydrophobic—a necessary way to be, because the natural hormones are hydrophobic, too.[26] The problem is that some man-made chemicals also happen to bind to receptors, activating them inappropriately or blocking them from responding to the body's own hormones. All else being equal, the more hydrophobic a chemical is, the stronger its affinity for these receptors will be, which explains why the majority of the pesticides and industrial chemicals that are known to bind to steroid hormone receptors are organochlorines.[27] In fact, the remarkable hydrophobicity of some organochlorines may explain almost entirely their ability to bind to hormone receptors.[28]

The final way that the fat solubility of organochlorines amplifies their toxicity is by increasing their tendency to dissolve in cell membranes, where they can disrupt cell function or make their way to extremely sensitive tissues.[29] All cells are bounded by fatty membranes, which separate the inside of the cell from the surrounding environment and, along with proteins embedded in the membrane, mediate the passage of substances and information into and out of the cell. The more fat soluble a compound is, the more likely it is to dissolve in cell membranes—particularly in the tissues of the bone marrow and nervous system, which are especially fat rich—and disrupt their normal function. Many organochlorines thus accumulate to high concentrations in these tissues, reaching levels that can prevent nerve cells from transmitting signals properly or cause damage to the blood- and immune-cell-forming tissues of the bone marrow. Moreover, several very sensitive organs and tissues are protected by barriers that keep out potentially hazardous substances in the blood; these include the placenta, the blood-brain barrier, and the blood-testis barrier. But these protective boundaries are unable to block the passage of fat-soluble compounds. The more hydrophobic a chemical is, the more easily it crosses these obstacles, reaching sensitive tissues in higher concentrations than a less fat-soluble chemical would.

The second major reason that organochlorines tend to be toxic is that chlorination affects the reactivity of organic chemicals, often stabilizing them but sometimes destabilizing them, depending on the structure of the compound. Either way, the change in reactivity generally increases toxicity. As I discuss in detail in chapter 5, chlorine atoms are extremely electronegative: they exert a very strong pull on the electrons of nearby atoms—far more powerful than the hydrogen atoms they normally replace in organic molecules. The result is that adding chlorine atoms to an organic compound changes the chemistry of the molecule, often in a way that increases its stability. This property makes many organochlorines relatively resistant to the chemical, physical, and biological processes that degrade other organic chemicals.[30]

This phenomenon explains why many organochlorines are extremely persistent in the environment. As the APHA concluded, "Chlorinated organic chemicals . . . comprise the majority of identified persistent xenobiotic substances."[31] Many are also persistent in the bodies of organisms, because the body's primary defense against foreign compounds is to metabolize them in a way that makes them easier to excrete. The increased stability of many organochlorines makes them more resistant to the body's metabolic processes, so they are retained in the body for longer, may accumulate to higher and higher concentrations over time, and will cause more severe toxic effects for a longer period of time than if they were more easily metabolized.

This is a particular problem for organochlorines that bind to hormone receptors. The biological potency of a hormone is a function of two things: its affinity for its receptor and the amount of time it is present in the target cell. Even a relatively weak hormone that is not cleared from the cell will eventually result in the maximum possible biological response; a chemical's impact at the physiological level depends primarily on *how long* the biochemical response is sustained.[32] For instance, a series of classic studies have shown that estradiol and estrone, two natural forms of estrogen, are equally potent inducers of short-term responses in the uterus; because estrone is metabolized much more quickly, however, it fails to stimulate the sustained growth of the uterus that characterizes exposure to estradiol.[33]

If natural hormones are to be effective signals for fine-tuned control of physiological processes, the body must be able to change their levels rapidly, which means they must be quickly degradable. Despite being more stable than estrone, estradiol is cleared from the blood in just 30 minutes,[34] and even synthetic oral contraceptives are 97 percent eliminated in 24 hours.[35] In contrast, organochlorines like dioxin, PCBs, and DDT remain in the body for decades. This persistence translates into a sustained biological response. One study found that treating cells with estradiol caused the number of occupied estrogen receptors in the cell nucleus to increase and then return to normal after 12 hours, but when the cells were treated with the pesticide chlordecone, receptor levels were still elevated to the maximal level when the experiment ended 48 hours later.[36] The extraordinary persistence of dioxin may explain why it is 10,000 to 20,000 times more effective at altering the expression of target genes than nonchlorinated compounds that bind to the same receptor.[37]

There are important exceptions to the generalization that organochlorines tend to resist chemical transformation. For instance, adding a chlorine atom to an organic molecule in which the backbone carbons are held together by a single bond increases the reactivity of the molecule by destabilizing that carbon-carbon bond. But as with most organochlorines that are broken down in the environment, the products of this reaction are simply other organochlorines, which may be more persistent and toxic than the original compound.

If the reactivity of some organochlorines is increased, one might think they would be less toxic than their nonchlorinated analogues. In fact, when chlorination increases the reactivity of a substance—as it does in many lower-molecular-weight organochlorines like solvents and intermediates—the toxicity often increases. This effect occurs because these organochlorines are more likely to be converted by enzymes into highly toxic, reactive metabolites that can then proceed to damage DNA or other essential molecules in cells. "Thus it may be deduced," Henschler writes, "that the introduction of chlorine substituents into organic compounds generally leads to increased chemical and biological reactivity and so to increased toxicity."

This may seem a paradoxical effect: chlorination makes some compounds more persistent, increasing their toxicity, but it makes others less persistent, also increasing their toxicity. This "damned-either-way" phenomenon makes sense if we view it from an evolutionary perspective. The body's detoxification enzymes evolved to break down potentially hazardous organic compounds in plants and other organisms we might eat, but not to handle organochlorines, which have never been naturally abundant in our food. Enzymes optimized by evolution to degrade one kind of molecule may produce harmful metabolites when they encounter novel substances. As a result, chlorine chemistry creates a biological catch-22: reactive organochlorines will probably be converted into toxic metabolites, while the stable ones are often already very toxic and will remain so for a very long time.

We thus have theory to add to the experience of the general toxicity of organochlorines. It is not coincidence that organochlorines dominate lists of priority pollutants around the world or that virtually every organochlorine that has been tested has turned out to be quite hazardous. Nor is it coincidence that organochlorines are almost universally more toxic and more carcinogenic than their chlorine-free counterparts, or that this pattern has been found to apply to virtually every subgroup within the larger class of organochlorines. The fundamental act of the chlorine industry—the chlorination of organic compounds—is biologically and ecologically problematic for reasons that cannot be divorced from our elementary understanding of the processes of nature.

Science in Control

This kind of analysis—an exploration of the hazardous characteristics of a class of chemicals—would never be undertaken in the Risk Paradigm, which is firmly committed to assessment and regulation on a substance-by-substance basis. Virtually all decisions about toxic chemicals in industrialized nations—including discharge permits, pesticide registrations, and occupational exposure limits—are now made following the risk model, in which chemicals are assessed individually to establish a permissible exposure level, environmental concentration, or discharge rate for each one. These limits are usually derived from a quantitative risk assessment, a method of extrapolating from toxicological tests on laboratory animals to specify an exposure level to each compound that purportedly poses no hazard or a negligible one. As long as the proposed technology results in exposures below this level, then it is permitted. If it exceeds this level, addition of another pollution control device can usually bring releases below the "safe" level. This view holds that if local exposures to pollution sources are kept below the threshold of unacceptable risk, there is certainly no reason to be concerned about the much lower background contamination levels to which the general public is exposed.

The chlorine industry and some scientists have fervently defended the Risk Paradigm, deploying its assumptions as a way of dismissing concern about organochlorines as a class. The industry summed up its position in a project that the Chlorine Chemistry Council, with support from other chlorine industry organizations in the United States and Europe, commissioned from the consulting firm CanTox, Inc. The product, a book-length "interpretive review"[38] of the hazards of organochlorines and a shorter article,[39] put forth several "scientific principles" that must govern any evaluation of the hazard posed by chemicals. In fact, these are assumptions, not principles, because they are asserted rather than demonstrated, but they are worth examining anyway, because they represent the intellectual basis of the current regulatory regime. According to CanTox, the first principle is that compounds must be assessed individually, because each compound has unique properties: "The fate and biological activity of a compound are determined by the chemical properties of the compound. . . . These differences among chlorinated organic compounds . . . preclude the generalization that all organic chemicals containing organochlorine behave similarly in the environment." The next principle asserts that all compounds have thresholds: "Compounds do not show adverse effects below certain threshold concentrations, and the magnitude of response is related to dose; inherent metabolic processes allow organisms to accommodate low doses of chlorinated organic chemicals."[40] (The final principle, to which we turn in the next chapter, is that epidemiological evidence is legitimate only if it documents an association between health effects in populations and exposure to a specific compound.)

The Society of Toxicologists (SOT) reiterated and cited CanTox's argument in a 1995 position paper that opposed the idea of treating organochlorines as a class.[41] The SOT is the leading professional organization of American toxicologists; several of the members of the group that wrote the paper were consultants who work frequently for the chlorine industry, and the remainder were academics. According to the SOT, "A truism that has endured for about 500 years is that essentially every chemical, either alone or in combination with other chemicals—in sufficient doses—is capable of producing an adverse effect. In more familiar terms, the dose makes the poison. . . . The SOT takes the position that the most responsible and scientifically sound approach is to assess the toxicity of agents on a chemical by chemical basis, rather than target one class of chemicals (e.g., chlorine-containing compounds) for study and elimination."

Does the Risk Paradigm really provide an adequate way of addressing the hazards of organochlorines? For chemical-specific exposure limits to be a valid and effective means of protecting public health, two assumptions that are directly related to CanTox's "principles" must hold. First, there must be a dose of each pollutant that does not pose a health threat. According to this view, exposures high enough to overwhelm the body's finite capacity to detoxify chemicals and to repair damage to its cells and tissues produce toxic effects, but low doses produce no toxicity at all. In short, the dose makes the poison. The second assumption is that toxicologists can predict the hazard of complex chemical mixtures based on tests of each compound in isolation; a dose of a substance considered safe in isolation must also be safe in combination with other chemical exposures. The whole of exposure must simply be the sum of its parts. The two assumptions are related, because thresholds—which are derived from tests on single chemicals—represent truly safe levels only if the impacts of the substance in isolation are the same as they are in the context of real-world mixtures.

Thresholds are so important to the Risk Paradigm—they are the basis for both approving chemical discharges and dismissing concern that low-level exposures pose a hazard—that they are worth exploring in detail. First, we need to understand how scientists establish thresholds. For all health effects other than cancer, the goal is to determine an acceptable daily intake (ADI, sometimes called a "reference dose," RfD) for humans; wildlife are seldom considered explicitly but are assumed to be protected by any standard that protects humans.[42] To derive an ADI, toxicologists subject laboratory animals—generally mice or rats—to the substance in question at several dose levels and examine the animals for a number of possible health effects. Most studies begin with acute toxicity testing, which seeks the dose level that kills an animal or produces other severe effects within two weeks. Many also involve chronic or "subchronic" toxicity testing, in which the highest dose that does not kill the animal is applied, along with a number of lower doses, for a somewhat extended period of time; the length of the exposure—typically three months to two years—is still much shorter than the seventy-year doses that characterize human exposure in the real world. These tests generally seek one or more forms of obvious "clinical" health damage, such as gross tissue injury, paralysis, tremors, weight loss, cancer, fetal death, and structural birth defects.[43]

Whatever impacts are examined, the useful product of these studies is the lowest observed adverse effect level (LOAEL) or the no observed adverse effect level (NOAEL). The LOAEL or NOAEL is assumed to represent a toxicity threshold, since by definition no adverse effects were observed below this dose. Regulators then divide the LOAEL or NOAEL by a safety factor ranging from 10 to 1,000—to account for differences between rodents and humans, differences in individual susceptibility, and other complications—to arrive at the ADI or RfD.[44]

For chemically-induced cancer, scientists and regulators cannot use this method, because biologists generally agree that exposure to any quantity of a carcinogenic substance—even a single molecule—poses some risk of causing the disease. Thus, a different model is used to establish acceptable exposures for substances that cause cancer. In the typical carcinogenicity assay, toxicologists expose laboratory rodents to the compound at a number of high but nonlethal levels for two years, after which the frequency of cancer in each dose group is recorded. These results provide a basis for calculating the relationship between dose and risk of cancer, assuming that cancer risk is linearly related to the dose. The potency of the substance in causing cancer is the slope of the line that relates cancer incidence to exposure. (There is some controversy about this "linearized multistage" model of carcinogenesis, but the assumption of linearity remains the dominant model in the field of cancer research and environmental regulation.) Regulators can then use the potency to calculate the cancer risk expected from any level of exposure. Because no dose poses zero cancer risk, there is no truly safe dose. Instead, regulators choose a risk level they consider "acceptable" or "negligible"—typically one cancer per 100,000 exposed persons, although the standard has varied from one in 10,000 to one in 1 million, depending on the circumstance. The "acceptable" risk standard allows permissible exposure levels to be calculated for cancer, just as the threshold model allowed them to be established for other effects.

Fact or Artifact of Science's Limits?

Are the safe levels established by toxicologists and risk assessors really safe? Can we be confident that the "acceptable" exposures to which we are all subject are not damaging our health? Are these "safe levels" really scientific at all? It should be obvious that the "acceptable" risk level for carcinogens is a political judgment, not a scientific one, which involves a rather arbitrary choice of some amount of increased cancer risk that is considered unworthy of prevention. When a large population is exposed to a cancer risk of, say, one in 100,000, then at least one or a few people are likely to get the disease. To the people unlucky enough to win the cancer lottery, there is nothing negligible or acceptable about the risk to which they were subjected.

In contrast, chronic toxicity thresholds, emerging as they do from laboratory studies, may seem to be real characteristics of chemicals and organisms. Can we be sure that thresholds are facts rather than artifacts, which Webster defines as "a product of artificial character (as in a scientific test) due to extraneous (as human) agency"?[45] If we look carefully at no-effect levels, it becomes clear that they are artifacts, reflections of the limits, the conventions, and the politics of toxicology.

One way that thresholds are artifacts is that they reflect the methodological limits of toxicological analysis. For us to be confident that a so-called safe level is a real aspect of nature, a testing regimen must be able to detect any and all health effects that might occur at a given dose level, but this is never the case, for six reasons.

First, the NOAEL on which a threshold is based is not actually the dose at which no adverse effects occurred; it is merely the dose at which no effects were *observed*. This is not a mere semantic difference, because the studies from which NOAELs are derived generally examine only a small number of possible impacts, and these are not the most sensitive impacts that can occur. Chronic toxicity tests focus on fairly obvious, clinical-type health effects. Subclinical effects—reduced fertility, compromised immune systems, and reduced intelligence, for example—are not observed not because they have not occurred but because they are seldom sought. To find a threshold for one effect and then assert that it is a safe level for all impacts makes the sweeping assumption that all biological processes display the same relationship between dose and response, which is patently untrue.

Second, even if a NOAEL were a true no-effect level, we would know only that this dose is safe in the species tested, not in humans or other species exposed to the chemical in the real world. The standards derived from thresholds are supposed to protect not just human health but ecosystems and the organisms that live within them, too. The toxicity studies from which thresholds emerge, however, typically involve one or two species very closely related to each other, usually rodents. Rats and mice are fairly good toxicological surrogates for humans, but a NOAEL found in a rodent is a rather tenuous basis for the judgment that no effects will be seen in dolphins, or bald eagles, or trout, or frogs, or butterflies, or krill, or daphnia, or fir trees, or any other distant species of ecological or cultural importance. How does one even begin to evaluate the possibility of eggshell thinning, or mortality at the tadpole stage, or impaired photosynthesis, by testing mammals?

Third, a no-effect level in one species does not even mean that no damage will occur to that species once the chemical is dispersed into the environment, because individuals and populations can be injured by ecologically-mediated effects. A pollutant that has no direct impacts on individuals of one species, for instance, can still decimate populations of that species if it has negative effects on other kinds of animals or plants that play key roles in its survival (as food supply, pollinator, or predator on a competitor, for example). Selective toxicity to one or a few species can affect an ecosystem's structure and change its stability, reducing species diversity, diminishing the survival of certain predator species, or leading to the uncontrolled outbreak of pests. A number of organochlorines have been shown to be toxic to plankton, plants, insects, and other organisms at the base of ecological food webs, with serious implications for all the species that depend on these organisms for survival.[46] Clearly, levels that protect other species are not always safe for these kinds of organisms: airborne concentrations of chlorinated solvents several orders of magnitude lower than the level thought to be safe for humans and other mammals have been found to cause selective toxicity to various kinds of soil invertebrates that play basic and essential roles in forest ecosystems.[47] In the Great Lakes, organochlorine discharges have been linked to severe local alterations in species composition, disrupted food web dynamics, and decimated populations of bottom-dwelling invertebrates.[48]

The fourth problem with thresholds involves the timing of exposure and observation. A no-effect level represents a dose that caused no observable effects only during the specific stage of life when the animals were exposed and then studied. Most testing protocols expose adult rodents to the chemical in question for a fixed period during adulthood, and the animals are assessed immediately afterward. Effects that appear only at an advanced age—like neurological impairment during senescence—may not be observed, because they are not sought at the right time. Further, the animal is seldom exposed during the most sensitive periods of development. The standard carcinogenicity protocol, for instance, involves exposing adult animals to a chemical for two years, at which time they are killed and autopsied.[49] But breast tissue, for example, goes through its period of greatest development and sensitivity to carcinogens well before this time; in utero exposure of breast tissues to some kinds of carcinogens—especially those that interfere with the action of steroid hormones—predisposes the tissue to developing cancer after the individual reaches puberty.[50] Standard tests calculate a negligible risk level without ever seeing this far more sensitive but time-limited effect.

Fifth, the NOAEL is often simply the level at which toxicological tools are not sensitive enough to detect or establish the existence of injury, not a dose at which no effects occur. Subtle subclinical impacts, when they are investigated at all, are far more difficult to measure accurately than frank clinical effects. Determining whether a pup is born dead or with a grossly cleft palate is quite clear-cut; detecting subtle immune suppression or cognitive deficits equivalent to, say, a five-point reduction in human IQ is much more difficult. The tools for measuring any subclinical effect are of limited precision. At some point, they become unable to detect small differences between normal and altered states or to distinguish these differences from natural variation or measurement error. The NOAEL is often the point not at which no effects occur but at which the investigator's tools are no longer sensitive enough to measure them.

Finally, there is statistical power. If subtle differences are observed between exposed and unexposed groups in an experiment, the investigator must show that the difference did not arise by chance. In order to consider a result "statistically significant," most scientists insist on at least 95 percent confidence that the difference was not a mere random fluctuation. The ability to make this determination depends on two things: the magnitude of the difference between the two groups and the size of the groups. In both cases, bigger is better. The more subtle the difference, the more animals need to be evaluated. Most studies, however, are of inadequate power to detect moderate impacts, particularly when there is also natural variability among individuals, as there is for functions like fertility, immunity, and intelligence. EPA guidelines for reproductive toxicity testing, for instance, require tests on two rodent species with at least 20 animals each, a number so small that only the most extreme differences between exposed and unexposed groups would be statistically significant. Even a sample population as large as 200 animals is far too small to establish a statistically significant 10 percent reduction in fertility—a huge effect when we are considering universal exposure to a substance.[51] As a result, many studies find no statistically significant effect, even though the impacts may be real and of considerable biological and social significance.

These six considerations make clear that in many cases, thresholds are likely to be artifacts of the limited scope and sensitivity of testing protocols in toxicology. It is possible that sometimes thresholds are real aspects of nature, but we can never distinguish a threshold that is a fact from one that is an artifact. This means that apparently scientific assurances of the safety of a certain level of pollution have no demonstrable basis in anything that we have reliably observed about reality.

Shrinking Thresholds

If thresholds are not facts of nature but artifacts of the limits of scientific tools, we should expect them to be progressively lowered as scientists gather new data. This is precisely what history shows. For example, as Henschler has shown in another review,[52] the occupational exposure limits for vinyl chloride, ethylene dichloride, and six chlorinated solvents were all progressively reduced from the 1960s to the 1980s. The reductions were substantial, with the new "safe levels" ranging from one-half to one-tenth the original limits. Henschler is quite clear about the reasons for this decline: "Nobody will be surprised by this development because it simply reflects scientific progress in identifying changes in physiological function of pathological significance at the tissue, cellular, and even subcellular levels. In other words: medical progress in diagnosing borderline states of damage or disease has yielded new and more accurate instruments for doctors and toxicologists to determine threshold levels of toxic action more sensitively and precisely."

A particularly telling example of the progressive lowering of thresholds is found in the history of lead regulations.[53] In the 1920s, severe lead poisoning became a major problem among workers involved in the production of leaded gasoline and other industries. Toxicologists and industrial hygienists then established a threshold for lead exposure—80 micrograms of lead per deciliter of blood—based on examination of men with and without severe lead poisoning: as long as a person accumulated no more than this amount there was no health hazard. The much lower levels of lead found in the blood of the general population were considered "natural," and they were assumed to be far too low to cause health effects. This was a reasonable conclusion, because no clinical symptoms had been seen below 80 micrograms per deciliter.

Agreement about the threshold lasted until the 1960s, when Australian physicians showed a connection between lead levels in children and lead in household paint, and geochemists established that the lead levels found in the blood of the general population of industrialized countries were many times higher than in people from less developed nations and in samples from ancient times. These scientists and others began to question whether these low but clearly unnatural doses might cause more subtle effects than those associated with high exposures. For the next two decades, heated controversy over the toxicity of low-level lead exposure engulfed the field, pitting geologists, pediatricians, and environmentalists on one side against the lead and gasoline industry on the other. Toxicologists and industrial hygienists largely allied themselves with the industry, reasserting that background lead levels were natural and posed no hazard.[54] In published rebuttals to those concerned about lower levels, members of these disciplines questioned the qualifications of anyone not trained in toxicology to evaluate the relevant evidence.[55] In 1968 an international group of elite toxicologists published a statement declaring that blood levels below 80 micrograms per deciliter were "acceptable"; symptoms of lead poisoning in patients with blood lead below this level, the statement argued, "are not attributable to lead."[56]

During the 1970s and 1980s, as the controversy continued, new laboratory and epidemiological data began to roll in, showing that lead levels lower than 80 micrograms per deciliter do cause damage to human health. The impacts were more subtle than those associated with lead poisoning, but they were no less frightening, including deficits in the cognitive development of children, leading to reduced intelligence and serious behavioral problems. In the late 1970s the "safe level" was reduced from 80 to 60 micrograms per deciliter. In the early 1980s, the "safe level" was reduced to 40 micrograms. Then it was reduced further to 35 micrograms. Then it was reduced yet again to 30 micrograms. In 1984 the threshold was revised to 25 micrograms. In the 1990s the safe level for children was reduced to 10 micrograms.[57] Today it is understood from detailed research on the epidemiology and biological mechanisms of lead toxicity that exposure to less than 10 micrograms per deciliter also impairs cognitive development, and there is most likely no threshold at all.[58]

An Artifact of Reductionism: The Impacts of Mixtures

For a threshold dose to be a fact of nature—a level that is truly safe in the real world—it must be safe outside the laboratory. As we saw in chapter 1, we are exposed to a constantly changing mixture of hundreds of organochlorines in the air we breathe, the water we drink, the food we eat. Nevertheless, virtually all regulations and risk assessments are focused on single chemicals, as are 95 percent of published toxicology studies.[59] Less than 0.25 percent of studies have evaluated the effects of mixtures of more than two chemicals.[60]

For assessments based on substance-by-substance testing to be valid in the real world, chemicals must exert the same effects in mixtures as they do in isolation. It is clear, however, that chemicals interact to modulate each other's toxicity in surprising and extreme ways. Sometimes the effect of a mixture is just additive, the sum of its component parts. In other cases, chemicals can inhibit each other's toxicity, making the effect of the mixture less than the sum of its parts. But by far the most common result, according to one review of scores of studies on interactions among binary mixtures of organochlorines, is synergy—a multiplicative or exponential effect by which the whole is greater than the sum of its parts.[61] A classic example is the research of scientists in EPA's Superfund program, who concocted a mixture of 25 common groundwater pollutants—12 organochlorines, 7 other organic compounds, and 6 metals—all at levels well below the dose that produces any measurable effects in isolation. Although single-chemical tests predicted the mixture would be safe, rats exposed to it experienced significant immunosuppression and increased susceptibility to several kinds of infectious diseases.[62]

Synergy usually occurs when two compounds affect a single outcome through two different pathways. For instance, combined exposure to a mutagen that initiates cancer and a promoter that increases the rate of replication of the new cancer cells will have a multiplicative effect on the risk of cancer.[63] Synergy can also occur if two pollutants interfere with different aspects of hormone function, one by blocking the binding of one hormone to its receptor, for instance, and the other by reducing circulating levels of that hormone. As one EPA scientist summed up, "We have not begun to understand what the potential adverse effects are of being exposed continuously to complex mixtures of chemicals with varying abilities to affect multiple signaling pathways, both singly and interactively."[64]

There are at least 1,000 reports in the scientific literature of nonadditive toxicological interactions among chemicals.[65] Sometimes the effect is quite extreme. For example, one set of studies found that exposure to a usually nontoxic dose of the pesticide chlordecone increased the death rate 67-fold among rats exposed to an "otherwise inconsequential" dose of carbon tetrachloride.[66] Another study by Dutch researchers examined the effects of exposure to dioxin and non-dioxin-like PCBs on levels of porphyrin, a marker of disturbed liver function. When dioxin and PCBs were applied simultaneously at doses that in isolation caused no effect, porphyrin levels shot up by 150 times. When a dose of PCB that caused just a 1.5-fold increase in isolation was combined with a no-effect dose of dioxin, porphyrin levels increased by 650 times.[67]

Organochlorines interact toxicologically not only with each other but also with other nonchlorinated compounds in the environment. The pesticide fenarimol, for instance, greatly increases the ability of trichloroethylene to cause genetic mutations,[68] and hexanone increases the toxicity of dibromochloropropane to the testis.[69] Conversely, both trichloropropene oxide and dioxins amplify the carcinogenicity of certain polyaromatic hydrocarbons, the primary carcinogens in cigarette smoke and automobile exhaust.[70] Organochlorines can even increase the potency of nonchemical hazards: exposure to EPA's low-level mixture of 25 common groundwater pollutants substantially amplifies radiation's toxic effects on the bone marrow of rats.[71] Ubiquitous modern hazards like cigarette smoke, air pollution, and x-rays thus become even more dangerous in an organochlorine-contaminated world than they would otherwise be.

We thus have a second way that thresholds and acceptable exposures are revealed to be artifacts rather than true facts of nature. A safe level in the laboratory is not likely to be safe in the context of all the other organochlorines, other chemicals, and other hazards that characterize life in the real world. It is the total chemical burden—the complete set of environmental hazards of all types, in fact—that poses a threat to human health; single-chemical studies, conducted completely out of context, offer little insight into the hazard these mixtures actually pose. Thresholds thus emerge not as facts of nature but as artifacts of the simplified and reduced environment of the laboratory, in which animals can be protected from all influences except for exposure to one controlled chemical at a time.

I do not mean that single-chemical studies have no value. With their carefully simplified exposure regimens, these experiments are indispensable for clarifying the mechanism by which individual compounds evoke their toxic effects. But value for science is not the same as value for policy.

Single-chemical studies provide important basic knowledge, but they are not useful for predicting hazards or setting exposure standards.

If isolated single-chemical assessments are inadequate for predicting the effects of real-world exposure, why can't we simply redesign tests on individual chemicals to focus on the effects that might take place against a background of mixtures? This way, policies could still be based on scientific predictions and would still be specific to individual compounds. The problem is that evaluating the role of each individual compound and its interactions with other substances in a mixture would require a multifactorial design that examines all possible combinations of the chemicals. Such studies are astronomically demanding of time and resources. According to scientists at the National Toxicology Program, an abbreviated single-species, 13-week toxicity evaluation of all the interactions in a mixture of just 25 chemicals would require over 33 million experiments at a cost of about $3 trillion.[72] A similar study of all 11,000 organochlorines in commerce would require an unimaginable 10^{3311} experiments—a number far greater than the number of atoms in the universe.[73]

Studying the effect of chemical mixtures, of course, does not really require probing all possible combinations of substances. It is much easier to pluck a real-world mixture of chemicals from some corner of the environment and compare the effects it causes to mixtures from some less contaminated place. For example, several studies have compared laboratory rats fed fish from the Great Lakes to those fed farm-raised or Pacific Ocean fish, and they have found that the chemical mixtures in the Great Lakes fish cause adverse effects on health.[74] But because these studies do not indicate *which* compounds in the mixture are causing the effect, their results become irrelevant and inadmissible when decisions are required to focus on single chemicals. The problem is not that we lack evidence that the low-level chemical mixtures present in the environment can cause health damage. Rather, the problem is that the regulatory system is blind to data that cannot be reduced to a simple cause-effect, single-chemical model.

An Artifact of Simplification: Diverse Mechanisms of Toxicity

The third way in which thresholds are artifacts rather than facts is that they appear only when we radically simplify our view of how organisms work. Traditional toxicological theory assumes that the primary mechanism of toxicity is killing of cells or other extreme and direct impacts on the ability of organs to carry out their proper functions. As CanTox correctly points out, the body can replace some quantity of cells and replenish damaged tissues; as long as chemically-induced injury does not overwhelm this ability, no significant health effects are expected.[75] This model turns out to be extremely limited, however. Evidence gathered since the mid-1970s reveals a remarkable variety of additional ways that organochlorines can cause toxicity at the molecular level, and these kinds of effects are not compatible with a threshold model of assimilable or reparable damage to the body.

Mutations

The first such mechanism to appear, direct damage to DNA, emerged from molecular cancer research in the 1970s. The nucleus of every cell in the body contains very long helical molecules of DNA that represent the hereditary "program" for all the body's functions. These molecules can be thought of as long strings of smaller regions, called genes, each of which carries a coded recipe for one protein or other substance that carries out the basic processes of the cell. A wide range of organochlorines can cause chemical damage to DNA, mutating the code and impairing whatever functions the product of the altered gene controls. When a mutation occurs in genes that control the division and differentiation of cells, it can cause the cell that carries it to grow into a tumor. When a mutation occurs in a germ cell, it compromises the biological information passed on to future generations. A mutation in a liver cell, for instance, will meet its end when the person carrying it dies, but a mutation in the cells that give rise to sperm or egg will be passed on to the child produced from that gamete, and she will pass it on to her children, and so on, forever. Exposure to mutagenic chemicals can thus gradually increase the load of inherited mutations in the population, increasing the burden of genetically-caused infertility, birth defects, disease and functional impairment with each generation.

Biologists generally agree that there is no threshold for chemically-induced genetic mutation: the risk of mutation is linearly related to the concentration and duration of exposure to the chemical. This mechanism explains why the threshold assumption does not apply to cancer and why any quantity of a carcinogenic chemical poses some risk of cancer. According to the current model of carcinogenesis, the development of cancer is a multistage process that begins with one or more mutations in a single cell; this cell then divides and multiplies, growing into a tumor, and goes through additional genetic changes that lead to the development of an aggressive malignancy. Even one molecule of a mutagenic agent can initiate the process of cancer development; higher doses have a greater probability of causing cancer, but there is no threshold below which a mutagenic chemical is truly safe. Thus, very low exposures to mutagens are expected to increase to some degree the incidence of cancer and genetically-mediated reproductive and development impacts.[76]

Endocrine Disruption

Another mechanism by which tiny quantities of organochlorines can cause severe health effects, without producting traditional threshold-style damage to tissues, is by disrupting the body's hormones. Hormones are the chemical signals by which the body regulates many of its most fundamental processes, including reproduction, behavior, immunity, metabolism, and the response to stress. Hormones control the growth and differentiation of certain tissues, so they also play a role in many kinds of cancer.[77] Hormones are essential to proper development of the brain and reproductive tract; tiny quantities during critical periods when these systems are developing cause a "programming" of biological functions that lasts through the organism's entire life.[78]

Hormones are produced by specialized endocrine organs—such as the gonads, the thyroid gland, and the adrenal gland—and they travel through the bloodstream to the tissues they regulate, like brain, breast, uterus, liver, and so on. Cells in these target tissues contain specialized proteins called receptors, each of which uniquely recognizes one hormone and binds to it. The hormone switches the receptor from an inactive to an active state, which enables it to attach to specific "hormone response elements" in the cell's DNA, where it activates or represses the expression of nearby genes. For example, androgens (testosterone and its metabolite dihydrotestosterone) produced in the testis at puberty can bind to the androgen receptor in distant muscle cells, causing those cells to produce greater quantities and different types of muscle proteins, uniquely changing the musculature of the adult male.[79] Because hormones are such powerful signals, the endocrine system must be tightly and delicately regulated. Tiny amounts of hormone produce cascades of coordinated effects throughout the body, but they are quickly degraded; in addition, feedback loops shut down the synthesis and accelerate the degradation of hormones and their receptors when their levels are high, preventing unwanted or runaway biological reactions.

A variety of organochlorines interfere with the endocrine system in a number of ways (fig. 2.1, table 2.2). Some organochlorines mimic hormones, binding directly to hormone receptors, including the estrogen receptor, inappropriately causing them to activate hormone-responsive genes. Others bind to hormone receptors but prevent them from activating gene expression; several pesticides are known to have an antagonistic effect of this sort on the androgen receptor. A number of other organochlorines bind and block the progesterone receptor, proper activity of which is necessary for the maintenance of pregnancy, proper menstrual cycling, normal female reproductive behavior, breast development, control of uterine growth, and other aspects of female reproductive function.[80] A few—including di- and trichloroacetate, metabolites of the solvents trichloroethylene and perchloroethylene, which are seldom thought of as hormone disrupters—bind to and activate another class of receptors called peroxisome proliferator-activated receptors (PPARs), which are involved in the development of liver cancer and diabetes, as well as the differentiation of the adipose cells that produce the body's fat.[81]

Other organochlorines disrupt the endocrine system by changing the quantity of natural hormone present in the body. They accomplish this by affecting the rate at which the hormone is synthesized or excreted. Exposure to the widely used herbicide atrazine, for example, causes the testes of alligators to produce very high levels of the aromatase enzyme, which converts testosterone to estradiol; the levels of aromatase in the testes of exposed males, in fact, are as high as those normally found in ovaries.[82]

The emerging understanding of endocrine mechanisms in molecular biology and toxicology supports the idea that exposure to tiny quantities of foreign compounds can disrupt vital biological processes, often without a threshold. Hormones generally circulate in the blood in the parts per tril-

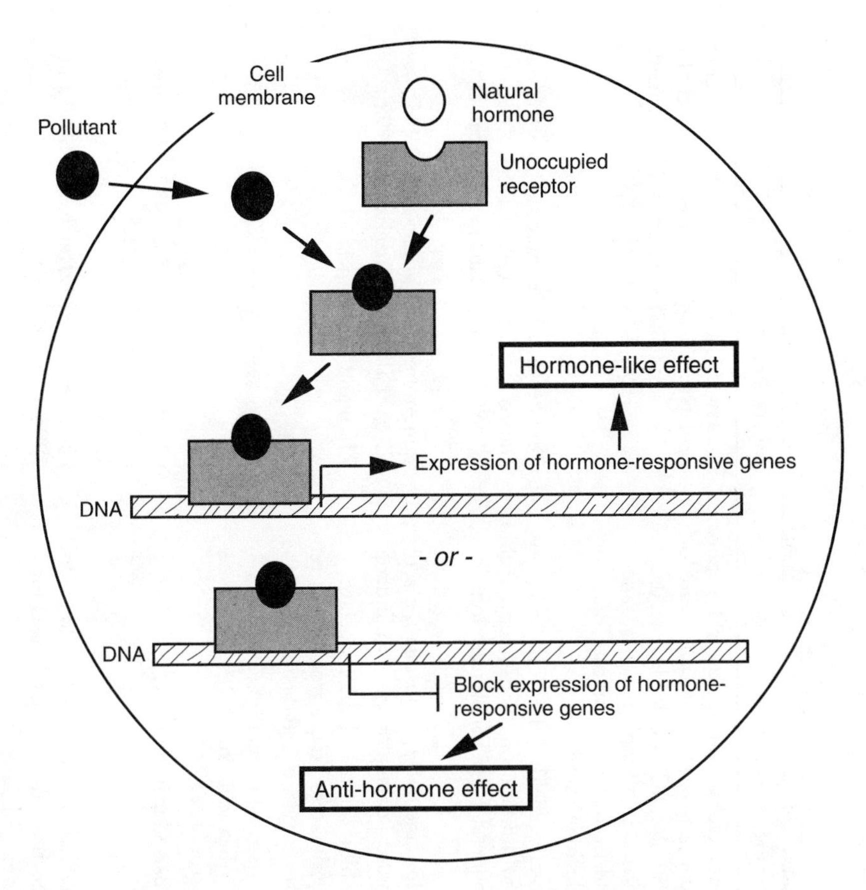

Figure 2.1
Endocrine disruption at the molecular level. Some organochlorines cross cell membranes and bind directly to intracellular receptor proteins for natural hormones. In some cases, the pollutant mimics the action of the hormone, causing the receptor to increase expression of target genes. The result is a hormone-like effect (antagonism), even in the absence of the natural hormone. In other cases, the pollutant blocks the action of the hormone, preventing expression of target genes, causing an antihormonal effect (antagonism). There are many other mechanisms of endocrine disruption as well.

Table 2.2
Effects of organochlorines on hormones, neurotransmitters, and growth factors

Effect	Substance causing[a]	References[b]
Bind and activate or inactivate estrogen receptor (ER agonists and antagonists)	Methoxychlor, DDT, pentachlorophenol, lindane, atrazine, cyanazine, alachlor, nonachlor, PCBs, chlordecone, dicofol, endosulfan, toxaphene, chlordane/dieldrin/toxaphene mixture, alachlor/dieldrin mixture	Danzo 1997, Tezak et al. 1992, Vonier et al. 1996, Arnold et al. 1997, Klotz et al. 1996, vom Saal et al. 1995, Nesaretnam et al. 1996, Soto et al. 1995
Bind and inactivate androgen receptor (AR antagonists)	Atrazine, dieldrin, DDT, lindane, vinclozolin, linuron, PCBs, methoxychlor, kepone, delta-hexachlorocyclohexane,	Tezak et al. 1992, Simic et al. 1991, Danzo 1997, Kelce et al. 1994, Kelce et al. 1995, Waller et al. 1996, Cheek et al. 1998
Bind and inactivate progesterone receptor (PR antagonists)	Chlordane, DDT, endosulfan, alachlor, kepone, pentachlorophenol, lindane	Lundholm 1988, Vonier et al. 1996, Klotz et al. 1997, Cheek et al. 1998
Bind and activate peroxisome proliferator-activated receptors (PPAR agonists)	Trichloroacetate, dichloroacetate (metabolites of trichloroethylene, tetrachloroethylene, and dichloroethylene)	Zhou and Waxman 1998
Bind hormone-binding proteins in blood (SHBG and/or ABP)	DDT, atrazine, methoxychlor, delta-hexachlorocyclohexane, pentachlorophenol	Danzo 1997
Bind transport proteins for thyroid hormones and retinoids	PCBs and metabolites	Cheek et al. 1999, Zile 1992, McKinney and Pedersen 1987, Rickenbacker et al. 1986
Accelerate degradation of testosterone	p,p′-DDE, dieldrin, heptachlor, chlordane, toxaphene, PCBs, hexachlorobenzene, lindane, 2,3,7,8-TCDD	Haake et al. 1987, Devito and Birnbaum 1994, Gray et al. 1992

Accelerate degradation of estrogens	2,3,7,8-TCDD	Devito and Birnbaum 1994, Whitlock 1994
Accelerate degradation of progesterone	2,3,7,8-TCDD	Devito and Birnbaun 1994
Accelerate degradation of thyroid hormones	2,3,7,8-TCDD	Whitlock 1994, Devito and Birnbaum 1994, Goldey et al. 1995
Accelerate degradation of retinoids	2,3,7,8-TCDD	Zile 1992, Devito and Birnbaum 1994, Brouwer et al. 1989
Inhibit conversion of thyroid hormone (thyroxine) to most active form (triiodothyronine)	Lindane	Yadav and Singh 1987
Reduce synthesis of androgens (testosterone or dihydrotestosterone)	2,3,7,8-TCDD, atrazine	Peterson et al. 1992, Babic-Gojmerac et al. 1989
Reduce progesterone levels in blood	Hexachlorobenzene, methoxychlor	Foster et al. 1992, Cummings and Gray 1989
Increase corticosterone levels in blood	Chlordane	Cranmer et al. 1984
Reduce melatonin levels	2,3,7,8-TCDD	Pohjanvirta et al. 1996
Reduce insulin synthesis	Trichloroethylene	Goh et al. 1998
Reduce SHBG synthesis	Trichloroethylene	Goh et al. 1998
Increase activity of aromatase (enzyme that converts testosterone to estrogen)	Atrazine	Crain et al. 1997
Accelerate conversion of estrogens to "hyperestrogenic" forms that covalently bind estrogen receptor	2,3,7,8-TCDD, PCBs	Davis et al. 1997, Tritscher et al. 1996
Alter gonadotropin levels in blood	Chlordimeform, 2,3,7,8-TCDD, PCBs	Goldman et al. 1991, Devito and Birnbaum 1994, Whitlock 1994

Table 2.2 (continued)

Effect	Substance causing[a]	References[b]
Increase prolactin levels in blood	Perchloroethylene	Mutti and Smargiassi 1998, Ferroni et al. 1987
Reduce prolactin levels in blood	2,3,7,8-TCDD	Russell et al. 1988
Reduce synthesis of receptors for estrogen, glucocorticoids, insulin, and prolactin	2,3,7,8-TCDD	Devito and Birnbaum 1994, Whitlock 1994, Lu et al. 1996
Increase synthesis of epidermal growth factor and transforming growth factor alpha	2,3,7,8-TCDD	Devito and Birnbaun 1994, Whitlock 1994
Increase synthesis of protein kinase C, ras, c-jun, co-fos and other factors involved in cell proliferation and differentiation	2,3,7,8-TCDD	Bombick et al. 1987
Bind and/or block GABA neuro-transmitter receptor	Dieldrin, lindane, endosulfan	NRC 1991, Silbergeld 1990, Cheek et al. 1998
Decrease synthesis of dopamine, norepinephrine, serotonin	PCBs (non-dioxin-like)	Seegal and Schantz 1994, Seegal et al. 1997
Increase synthesis of dopamine	PCBs (dioxin-like)	Seegal et al. 1997
Increase synthesis and turnover of dopamine	2,3,7,8-TCDD	Russell et al. 1988
Inhibit gap junction intercellular communication	DDT, dieldrin, toxaphene, PCBs (non-dioxin like)	Kang et al. 1997
Interfere with dopamine signalling	Perchloroethylene	Ferroni et al. 1992

[a]In some cases, metabolites of the listed chemical cause the effect shown.
[b]References include reviews. SHBG = steroid hormone binding globulin; ABP = androgen binding protein.

lion range; many hormonally active organochlorines are present in significantly higher quantities. Blood levels of estradiol, for instance, are normally between 30 and 500 parts per trillion (ppt), depending on sex and stage of the menstrual cycle, while total PCB levels in the blood of the general population range from 2,000 to 8,000 ppt.[83]

How can such tiny concentrations exert major physiological effects? First, the affinity of natural hormones and some organochlorines for receptors is extraordinarily great: at an estradiol concentration of only a few parts per trillion, fully half of the available estrogen receptors in a cell are occupied, and dioxin's affinity for its receptor may be even greater. (The affinity of PCBs and many pesticides for the estrogen receptor, on the other hand, is several orders of magnitude lower than these values).[84] Further, the number of receptors per cell is quite large—5,000 to 30,000 for most hormone receptors—so the chance is quite high that hormone molecules, even if they are few in number, will find receptors to which to bind.[85] Once bound to hormone, receptors have a very high affinity for their DNA response elements, so it takes only a few to begin the biochemical cascades that characterize hormone action.[86] Indeed, only 10 to 20 percent of estrogen receptors need to be occupied by estradiol to produce the *maximal* growth response in the uterus,[87] and measurable biological effects occur when only 0.1 percent of hormone receptors are occupied—a total of just 5 to 30 receptor complexes per cell.[88] If so few molecules of hormone can lead to a biological response, it is clear that a threshold, if one exists at all, is so low as to be of little practical importance.

So long as there are available receptors—as there always are at normal levels of hormone—the number of hormone-receptor complexes will be linearly related to the dose of the hormone or hormone-like chemical.[89] At the next step—effects on gene expression—the situation may be more complicated, but it takes only a very small number of receptor complexes to activate the expression of target genes.[90] At the biochemical level, then, there is not likely to be any meaningful "no-effect" level of exposure to hormone-disrupting pollutants. At the physiological level, some clinical endocrine-mediated effects—menstrual irregularities, visible masculinization, and so on—are likely to have thresholds, but other impacts, particularly those that disrupt the sensitive processes of development, will not. Most important, because any physiological effect is the result of a biochemical response integrated over time, persistent pollutants will eventually overwhelm any thresholds that do characterize the dose-response relationship for the body's own quickly degradable hormones.

The most detailed studies of the relationship between dose and response for an endocrine-disrupting chemical have focused on dioxin, and this evidence argues strongly against the assumption of a threshold. For almost a decade, scientists at the National Institute of Environmental Health Sciences (NIEHS) have been attempting to model the relationship between dioxin dose and response using the most sensitive laboratory techniques available. Never has the "safe level" of a chemical been sought so diligently. But NIEHS's research has not found a threshold for any of dioxin's biochemical effects. At the lowest doses tested, dioxin binds to the Ah receptor, induces the expression of target genes, and activates a cascade of intracellular molecular effects. Nor does there appear to be a threshold for dioxin's ability to promote the growth of premalignant liver tumors or to alter circulating levels of thyroid hormones. For all these impacts, the best estimate of dose-response relationships at very low levels of dioxin is that the severity of the impact is roughly proportional to the magnitude of dioxin exposure.[91]

Supporting the view that there is no practical threshold for dioxin toxicity, several studies have discovered that almost absurdly low doses have significant biological effects. For example, when rats are given a single dose of TCDD as low as 64 millionths of a gram per kilogram of body weight on day 15 of pregnancy, the sexual development, behavior, and function of their male offspring are compromised.[92] Dioxin's immunotoxicity has been documented at even lower levels. Doses of TCDD as low as 2.5 parts per quadrillion—equivalent to a mere 10 molecules per cell—*completely* abolish the ability of cultured immune cells to respond to signals to proliferate and mount an immune defense.[93] In whole animals, dioxin produces immunotoxicity at concentrations in the spleen about five times lower than this—on the order of just two molecules per cell.[94] If there is a threshold for dioxin, it is so low as to be absolutely irrelevant for the purposes of environmental policy and health protection.

Finally, it is important to point out that we have just begun to understand the many ways in which hormone-disrupting chemicals may damage health. Despite the enormous biological power of the Ah receptor, its natural function remains unknown.[95] Receptors for estrogen and testosterone are present in the very earliest stages of development, presumably for some reason, but no one knows what they do.[96] There are scores of receptors in the larger family of proteins that includes the steroid receptors, but little is known about the function of many of them or the natural hormones or synthetic chemicals that may bind to them.[97] Moreover, different chemicals that bind to the same receptor may have profoundly different effects on gene activation.[98] Further, recent research indicates that some hormones can cause effects not only through their traditional intracellular receptors but also through a protein in the blood called sex-hormone binding globulin (SHBG), which, when bound to hormone, then binds to an entirely different kind of receptor attached to cell membranes and triggers rapid and extreme biochemical changes in the cell.[99] Some organochlorines have recently been found to bind to SHBG and displace natural hormones from it (table 2.2); this finding suggests an entirely new pathway by which synthetic chemicals may disrupt the endocrine system, but the physiological impacts and relationship between dose and response remain largely unknown. Considering all these unknowns and vast complexities, the view that we can calculate and predict the effects of chemicals on a system we barely understand is shockingly arrogant.

Neurotoxicity

The human brain is composed of about 1 trillion nerve cells, called neurons, that are linked to each other in a web of staggering complexity; each neuron in the brain may be connected to as many as 1,000 other neurons. One neuron can communicate with another by an electric impulse or a chemical signal called a neurotransmitter. The signal causes or prevents the receiving neuron from sending its own signal further through the web. The brain's function, called computation by those seeking a metaphor with machines we understand, is the product of the passage of countless nearly instantaneous signals among neurons in this unimaginably complex network.

Like efforts in molecular endocrinology, research in neurotoxicity undermines the claim that there are safe exposure levels for all organochlorines. Indeed, the National Research Council (NRC) concluded in a 1992 report that the assumption of thresholds for neurotoxicity is "biologically indefensible."[100] Organochlorines disrupt nervous system function and development in a number of ways. First, some organochlorines—particularly those that dissolve in cell membranes or are metabolized to highly reactive intermediates—can kill brain cells. Unlike most other tissues, brain cells generally do not regenerate; new neurons are not produced after an organism passes early childhood, so the Risk Paradigm's assumption that the body has inherent repair mechanisms that allow it to cope with a certain amount of damage clearly does not apply to the brain.[101] While the brain can often compensate for damage in one area by shifting function to cells in another region, this compensation may merely mask damage that will become manifest only at advanced age, when the number of functional neurons declines to a critical level. If chemicals were to cause an increase in the rate of loss of neurons of just 0.1 percent per year, an effect far too small to detect in the laboratory, continuous exposure over just a few decades would result in a major decline in the functional capacity of the brain by middle age.[102]

Second, organochlorines can interfere with chemical and electronic signal transmission in the brain. Several organochlorine pesticides are neurotransmitter antagonists; they bind directly to neurotransmitter receptors, blocking out the natural transmitter and preventing the neuron from receiving signals through the receptor (table 2.2).[103] They may also affect the production of neurotransmitters: PCBs, for instance, interfere with an enzyme necessary for the synthesis of dopamine, a neurotransmitter that plays important roles in learning and memory in the brains of mammals.[104] By reducing the quantity of neurotransmitters or blocking receptors, a very small quantity of a toxic substance can have a meaningful effect on the nerve cell's ability to transmit signals. The NRC has noted that disruption of receptor-mediated signaling in the brain is not expected to display a toxicity threshold.[105]

Organochlorines can also affect the ability of neurons to communicate with each other by electrical signaling. Transmission of nerve impulses occurs when charged ions flow rapidly from one cell to another through channels called gap junctions, which are tunnel-like proteins that cross the membranes of adjoining cells, connecting the cell interiors to each other. The ability of two neurons to communicate electrically depends directly on the number of gap junctions that connect the cells and the quantity of ions they allow to pass through them. A number of organochlorines—namely DDT, dieldrin, toxaphene, and PCBs—have been found to reduce the ability of cells to communicate with each other through gap junctions.[106] Although the transmission of an impulse is an all-or-nothing effect with a clear threshold, small reductions in gap-junction communication could impair the ability of a cell to respond to a weak stimulus from its neighbors, turning a just-adequate signal into no signal at all.

If the brain is unfathomably complex, only the development of the brain is more so. From a ball of undifferentiated cells, an embryo with specialized tissues develops; trillions of neurons of dozens of specific types are born, migrate to the correct locations, and form the appropriate connections to carry out the complex functions of the brain. Chemical signals called trophic factors guide progenitor cells to migrate, divide, and differentiate into specialized neurons during development; connections are formed and lost depending on the extent to which they are used, according to a use-it-or-lose it rule of neurodevelopment. Chemical disruption of any of these processes during critical periods can lead to permanent changes in brain organization and function.[107] Killing just a handful of nerve cells early in development can potentially cause profound effects, since those cells might have been destined to produce millions of progeny cells, the absence of which will alter the pattern of connections among other neighboring neurons. Similarly, compounds that interfere with trophic factor control over the migration of neurons will disrupt the proper development of the brain; it may be by such a mechanism that exposing bird eggs to dioxin—at doses as tiny as 10 parts per trillion in the egg—results in the development of brains in herons and chickens that researchers describe as "grossly asymmetric."[108] Further, since the survival of the connection between two neurons depends on its use, chemicals that disrupt neurotransmitter or electrical signaling can alter the network that develops within the brain. Finally, hormones—particularly those produced in the thyroid—play key roles in neurological development, controlling the production, differentiation, and maturation of neurons. Disruption of thyroid hormone levels during development leads to problems with gross and fine motor coordination, hearing loss, behavioral problems, impaired memory, and learning disabilities. Organochlorines that reduce circulating thyroid hormone levels, like dioxins and PCBs, cause effects in exposed children similar to those caused by hypothyroidism.[109]

Through all these mechanisms, small local changes and subtle biochemical disruptions can propagate into major alterations in the structure and function of the brain. We should thus never assume that the developing brain can safely absorb a certain amount of damage without adverse effect.[110] An international meeting of neurologists and endocrinologists concluded, "The developing brain exhibits specific and often narrow windows during which exposure to endocrine disrupters can produce permanent changes in its structure and function. . . . A variety of chemical challenges in humans and animals early in life can lead to profound and irreversible abnormalities in brain development at exposure levels that do not produce permanent effects in an adult. There may not be definable thresholds for responses to endocrine disrupters."[111]

The endocrine system, the nervous system, the immune system, and embryonic development are all unfathomably complex. Science's understanding now barely scratches its surface; no matter how rapid the progress in biological research, our ignorance will dwarf our knowledge for a very long time. Without a sound grasp of how a biological system operates, we can hardly pretend to predict the many ways that synthetic chemicals can disrupt its delicate works. Only by ignoring the complexity of biological systems can thresholds be theoretically justified. If we broaden and deepen our model of how organisms are organized (and how chemicals can disrupt those systems), the thresholds we might have expected to see suddenly vanish.

An Artifact of Politics

The final way that thresholds and acceptable intakes are artifacts is that they reflect the political interests of powerful individuals and organizations that are in a position to determine what is accepted as scientifically "true." If thresholds cannot be justified empirically or logically on the basis of rigorously conducted studies, why do the Society of Toxicologists and CanTox continue to say that they represent "sound science" or "basic scientific principles"? The answer is that thresholds provide the intellectual justification for a system of environmental decision making that benefits toxicologists as a professional group and the industries that support much of their research.

If chemicals are approved solely on the basis of whether risks and exposures exceed some quantitative threshold, then the person who calculates the risk has an enormous degree of control over the outcome. In this way, the system of pollution permits puts an overwhelming amount of influence in the hands of technical experts, toxicologists above all. It is quite understandable, then, that many toxicologists have argued in favor of thresholds by dismissing anyone who is not a member of their group as lacking qualifications to speak authoritatively on the issue.[112] As the Society of Toxicologists argued in its defense of the Risk Paradigm and its opposition against a precautionary policy on organocchlorines: "The toxicologist is specially trained to examine the nature of the adverse effects of chemical and physical agents on living organisms and the environment. . . . The determination of unacceptability should be based on scientific data that document the adverse effects of exposure and a weighing of the risks vs. benefits of using the chemical in question. Indeed, based upon sound principles of toxicology, rational and effective assessments of the potential toxicity of chemicals, including chlorinated chemicals, are currently taking place and rigid standards exist for registration of new products to which people will be exposed."

The Risk Paradigm also protects the economic interests of businesses that produce and use toxic chemicals. If some level of exposure poses no hazard, then chemicals can be safely managed using pollution permits and control devices. If there is no such level, then low-level pollution has the potential to threaten health and ecosystems on a global basis, and the only way to protect the public is to prohibit the production and use of many or all toxic substances altogether. The status of thresholds as scientific "facts" thus lends much-needed credibility to the Risk Paradigm.

The "truth" of thresholds also obviates the need to discuss other issues that could prove much stickier for those who benefit financially from the production and use of chemicals. If decisions can be made based on a purely quantitative comparison of predicted exposures versus threshold doses, then political questions—how much risk is acceptable? who gets the risk and who gets the benefits? is there a less hazardous way of fulfilling society's needs?—need never come up for debate. By asserting artifacts as facts, the Risk Paradigm reduces complex political issues to a narrow scientific exercise that need not be opened up to public input, protecting toxicologists and corporations whose interests it serves.

This is not to say that everyone who holds that chemicals have thresholds is uttering a bald-faced and self-serving lie (although some, like Can-Tox, are clearly trying to justify a position that benefits those who pay for their work). Toxicology, like any other scientific discipline, is a culture, with traditions and values that are passed on to its initiates. The idea of thresholds has a long history in toxicology, is taught to all toxicologists in training, and serves as the foundation for a large portion of the body of work built up by toxicologists over the decades. Every field has a prism of concepts through which it perceives the world, organizes data, and carries out new investigations. These traditions are seldom questioned, because it is virtually impossible to be a productive and accepted member of the field without believing in its fundamental concepts.

This explanation begs another question, of course: why hasn't the culture of toxicology changed to take account of all the new information on low-level toxicity, synergy, and endocrine and neurological disruption? Having been egregiously and visibly wrong on lead during the 1970s and 1980s, why would many toxicologists maintain a tradition they have had ample time to revise? There are some toxicologists who have questioned the existence of thresholds, but their voices have not come to dominate the field. Why not? Part of the answer is likely to lie in the fact that industry remains the funding source for a great portion of toxicological research. As the history of lead shows, most of those who questioned the safety of low-level lead exposure were funded by public sources, while many of the toxicologists who defended the threshold concept were funded by industry. For many decades, the most important toxicological work on lead came from the Kettering Laboratory at the University of Cincinnati, directed by a toxicologist who was also the chief medical adviser to the Ethyl Corporation, the largest manufacturer of lead compounds for gasoline.[113]

Today the chemical industry maintains its influence over toxicological research and standards. When federal and state agencies set occupational and ambient air pollution levels, they frequently rely on threshold limit values (TLVs) issued by the American Council of Governmental Industrial Hygenists (ACGIH). According to one analysis, ACGIH gave corporate representatives listed as consultants primary responsibility for developing thresholds on over 100 chemicals, including those manufactured by the companies that employed them. Dow's toxicologists were responsible for developing the standards for at least 30 chemicals manufactured by the company. In addition, unpublished, and thus unreviewed, research by corporate laboratories was an important source of information in developing TLVs for over 100 substances.[114] Industry is also a major funder of research by academic toxicologists, so a great many people in the field of toxicology have economic reasons—to go with the political ones we have already discussed— to hold fast to the idea of thresholds.

An Overwhelming Lack of Data

There is one final reason it is not practical to attempt to regulate organochlorines individually based on toxicological predictions: the data to do so simply do not exist. There are 11,000 organochlorines in commerce, plus thousands more produced as accidental by-products. Basic toxicity data on acute and chronic effects are available for only a tiny fraction. For the vast majority of compounds there are no data at all.

In 1984, the National Research Council investigated the extent to which toxicity data were available for chemicals and pesticides in commerce.[115] At that time, NRC found over 48,000 registered industrial chemicals, 3,300 pesticides, 8,600 food additives, and 3,400 cosmetic ingredients on the market in the United States. It then investigated a random sample of this massive universe of compounds and found a shocking deficiency in the extent of toxicity information. Among the pesticides, adequate data for a complete health hazard assessment were available for only 10 percent, and there were no data whatsoever on 38 percent. For the industrial chemicals, the situation was far worse: a complete health hazard assessment was possible for absolutely none of the substances, and there were no data whatsoever on 78 percent of the chemicals in commerce.

Remarkably, the situation is actually getting worse. Amassing chronic and subchronic toxicology data is slow business. Since the early 1970s, the amply funded National Toxicology Program has evaluated about 500 chemicals for carcinogenicity, an average of about 17 per year.[116] Meanwhile, 500 to 1,000 new chemicals are introduced into commerce annually,[117] so our knowledge base is constantly falling further behind the growing number of substances the chemical industry produces. When researchers at the Environmental Defense Fund (EDF) updated the NRC study in 1997, they found that the situation had not improved. Even among the subset of large-volume chemicals that had already been the subject of specific regulatory attention—those expected to have been the most thoroughly studied—70 percent still lacked even minimal chronic toxicity data. No reproductive toxicity tests were available for 53 percent, no neurotoxicity data for 67 percent, and no immunotoxicity tests for a whopping 86 percent.[118]

These shocking results apply only to chemicals in commerce. For substances that are produced as accidental by-products—as thousands of organochlorines are—toxicity data are even more pathetic. As I discuss in chapter 7, the majority of by-products formed in the processes of chlorine chemistry have not even been chemically identified. These compounds certainly cannot be assessed, since we do not even know what they are.

This situation creates fundamental problems for the chemical-by-chemical approach. It would be supremely impractical to try to assess and predict the toxicity of every commercial organochlorine; generating the necessary information would require an astronomical amount of time and money. Centuries would pass before even those in commerce today would be assessed. In the meantime, substances for which there is no data are treated as if they are harmless. Since regulatory action can be taken to restrict a chemical only when data indicate a hazard, lack of evidence is allowed to serve as evidence of safety. This is the situation for the vast majority of chemicals, so the system works primarily on ignorance, not knowledge.

Some advocates of the Risk Paradigm, including CanTox, have argued that the lack of data is not a fundamental barrier to chemical-specific assessment. Even in the absence of laboratory tests, according to this view, scientists can predict a compound's toxicological potency based on its chemical structure, using so-called structure-activity relationships (SARs).[119] Molecular toxicology has come a long way in the past few years, but this certainly overstates the case. If single-chemical toxicology studies are a poor basis for predicting the environmental hazards of chemical mixtures, then predictions—in the absence of any real data—of what those studies might find if they were actually done are certainly inadequate as well. SARs have their place,[120] but they are a notoriously unreliable indicator of how toxic an untested chemical will actually turn out to be. Indeed, SARs have been used to predict the carcinogenicity of untested chemicals, and they have frequently been wrong.[121] They are unlikely to be useful for predicting hormonal effects either: the surprising structural diversity of organochlorine compounds that bind to the estrogen receptor makes clear that no SAR-based analysis would ever have predicted that many of these substances would turn out to be estrogenic.[122] For predicting neurotoxicity, the National Research Council found that SARs are minimally useful and clearly not a sufficient basis for assessment or policy.[123]

From Prediction to Prevention

The information on organochlorine toxicity reviewed in this chapter suggests a consistent if rather macabre picture. Organochlorines interfere with the basic machinery, information systems, and regulatory circuits by which living organisms create and maintain themselves. Our brains, our hormones, even the germ line—the safe-box of hereditary information—are compromised. The injury emerges over a long period of time, as the symptoms of a slow decay of biological function or a single disruption at a critical time, caused by exposure to complex and interacting chemical mixtures with diverse mechanisms of toxicity. Persistent organochlorine pollution poses a threat of subtlety, severity, and scope that is entirely different from the immediate risks that the current regulatory system, with its focus on short-term exposures, clinical impacts, and threshold doses, was designed to address.

We have also seen that the basic toxicological assumptions of the Risk Paradigm are untenable in light of recent findings on the nature and complexity of chemical toxicity. Single-chemical assessments cannot reliably predict the effects of real-world chemical exposures, and for many chemicals and effects, there appears to be no dose, no matter how low, that can be considered safe. The idea that science can precisely manage thousands of chemicals, calculating harmless doses and permitting exposures below these levels, radically overestimates our understanding of chemical toxicology and basic biology.

In place of the risky bets that the Risk Paradigm makes based on toxicology's limited knowledge, the Ecological Paradigm substitutes a much humbler attitude about the capacity of science to predict and understand nature's complexity. Rather than trying to manage chemical exposures, our priority should be preventing them. The default position in the current system is to permit the production, use, and discharge of chemicals; restrictions are applied only to the extent that chemicals can be shown to pose an unacceptable hazard. In contrast, the Ecological Paradigm makes no attempt to calculate "safe" exposures. Instead, it seeks to minimize the use of potentially toxic chemicals, reducing and ultimately eliminating environmental discharges of chemicals that may cause damage to health. This approach requires that the cleanest available alternatives be used for any process; toxic chemicals are to be used only when there is no less hazardous way to meet society's needs for a product or service. Priority is placed on chemicals that can cause severe and irreversible effects at low doses—mutagens, carcinogens, hormone-disrupting chemicals, neurotoxicants, immunotoxicants, and persistent synthetic substances of all sorts; these substances should be avoided to the greatest extent possible. Instead of approving a certain amount of exposure to each chemical as acceptable, a precautionary approach seeks continual progress toward the elimination of all toxic pollution.

How should such a goal be achieved? Should reduction policies be established for each of the 70,000 chemicals in commerce, plus the thousands more by-products formed by accident? This approach would require priorities to be set based on information from chemical-specific toxicology tests. If we continue to focus the decision process on individual compounds using toxicology-based predictions, we will fall into many of the same traps that characterize the Risk Paradigm: the overwhelming lack of data on most chemicals, the specter of subtle effects not sought or observed in the initial tests, and the unpredictable impacts of chemical mixtures. Even a well-intentioned approach to individual chemicals is doomed to be ineffective.

Instead of focusing on single substances, the Ecological Paradigm prioritizes for reduction the most hazardous classes of chemicals and the technologies that produce them. As I showed in the first half of this chapter, chlorinating organic molecules is an intrinsically dangerous activity that invariably makes the chemicals that result more toxic, more bioaccumulative, and often more persistent. Members of the class of organochlorines can thus be regarded a priori as a threat to health and the environment unless specific information is available to acquit them. If our goal is to reduce global toxic pollution, applying the Ecological Paradigm to the class of organochlorines—and the technologies of chlorine chemistry that produce them—is a logical and obvious place to start.

3

The Damage Done: Health Impacts in People and Wildlife

A man of practical action has always to choose between some more or less definite alternatives, since even inaction is a kind of action.

—Karl Popper, *The Logic of Scientific Discovery*[1]

The information in the previous two chapters is in itself enough to justify immediate action to stop further organochlorine pollution. We know that the chlorine industry has contaminated the entire planet with hundreds or thousands of substances that last for a very long time in the environment, the food web, and our bodies. And we know that very low doses of some of these chemicals can cause severe and insidious effects on reproduction, development, behavior, and the immune system. It takes only common sense to recognize that it is not wise to pollute our world and our bodies with exquisitely toxic, long-lived chemicals.

The widespread occurrence and great toxicity of organochlorines suggest that global contamination has the potential to erode, slowly but severely, the health of the human population and that of other species with which we share the planet. Continued discharges raise the specter of rising rates of infectious disease and cancer, infertility, changed behavior, reduced intellectual ability, and the decimation of wildlife species. The severity and irreversibility of these effects are compelling enough to justify action to prevent even the possibility of their realization.

Does the threat of global health damage lie only in the future, or is contamination compromising our well-being right now? A compelling body of evidence suggests that organochlorine pollution has already begun to cause large-scale damage to the health of humans and wildlife. I offer this information with some hesitation, not because there are any questions about its quality but because, in some ways, it is a red herring. We cannot and should not make environmental policy based on health damage that has already occurred. To make this a criterion for policy would require disease, death, and impairment to take place before we could do anything to protect public health. This kind of "body count" method is an unconscionable approach to environmental policy.

On the other hand, it is surely important that organochlorines have already caused considerable damage to health and the environment. The injury done to humans and wildlife starkly illustrates the urgency of the issue, the scope of the potential harm, and the need for rapid action to prevent further damage. The damage also illustrates the Risk Paradigm's utter failure to protect health and the environment.

In this chapter, I review the evidence that links large-scale organochlorine contamination to five kinds of health injury: reproductive impairment in males and in females, damage to development, impaired cognition and behavior, and suppressed immunity. In the next chapter, I take up the link between organochlorines and cancer incidence. For the evidence I will discuss to make sense, however, we need to understand first how epidemiological data are used in the Risk Paradigm and the new way I propose to treat them in the Ecological Paradigm.

The Limits of Epidemiology

Any system for making policy needs an error-correction mechanism, a way to detect unexpected environmental injury and then take steps to prevent further damage. In the Risk Paradigm, this role falls to epidemiology, the science that examines the patterns of disease in populations and infers their causes, and eco-epidemiology, the equivalent discipline for wildlife species. (For simplicity, I will treat the two together under the name *epidemiology.*) If these sciences can establish a causal link between a pollutant and some type of health damage, then more restrictive actions, such as bans or phaseouts of chemicals, can be justified. Without specific evidence of injury, however, discharge limits based on risk assessments are assumed to be protecting health and the environment.[2]

Using the Risk Paradigm's requirement for specific and conclusive evidence, many commentators have concluded that low levels of chemical pollution are not causing widespread health damage. For example, CanTox has argued, "Except for well-documented instances of toxicity related to specific incidences of accidental releases and certain defined occupational exposures, there have been no definite cause-effect relationships established between exposure to environmental concentrations of chlorinated organic chemicals and adverse health consequences in the human population."[3] In 1998 EPA reviewed an extensive body of evidence that associates endocrine-disrupting chemicals with a variety of diseases and conditions in laboratory animals, wildlife, and humans; the agency found that "although the majority of the effects . . . are of concern, whether these observations represent widespread or isolated phenomena and whether these effects can be attributed to a specific endocrine disrupter will require additional research."[4] And the *New York Times,* in a report critical of the hypothesis that endocrine-disrupting chemicals in the environment might be damaging public health, wrote, "But several leading scientists view such positions as premature at best. They say that the case for ridding the world of these chemicals seems fueled more by hyperbole than facts. . . . These scientists say they are not arbitrarily dismissing fears that trace amounts of synthetic chemicals might injure people and wildlife. But, they say, there is a difference between a hypothesis and convincing evidence."[5]

To require conclusive epidemiological evidence before we can judge health to be at risk makes sense only if science can provide such evidence whenever damage is taking place. If it cannot, then a lack of certainty may reflect the limits of scientific tools rather than a real lack of health injury, and there will be no way to know which is the case. In fact, epidemiology is not equipped to fulfill its assigned task of diagnosing all the impacts of pollution and linking them to their causes. With a mixture of thousands of chemicals, no unexposed control group, and subtle health impacts that may take decades or generations to become manifest, the reality of chemical contamination and its potential effects outstrips the tools available to epidemiologists. Both epidemiology and toxicology provide important clues and partial knowledge about health impacts, but neither can discover and describe all the forms of long-term damage that chemical mixtures may cause. For this reason, judgments in the Risk Paradigm that public health is not being damaged by chemical pollution can often be, like the safe thresholds established in the laboratory, artifacts of science's limits.

It is unethical to subject people to the kind of clean, controlled study designs that make unambiguous interpretations possible in the laboratory. Instead, epidemiologists look at patterns of who gets a disease and then infer the causes based on statistical associations between disease prevalence and the presence or absence of putative causes. This does not stop them from drawing conclusions about causality, but it makes the effort much more complicated. In evaluating hypotheses of causality, epidemiologists often guide their thinking with a set of standards called the Hill criteria. Under these criteria, the ideal body of information would include repeated studies to show that a specific cause has powerful and statistically significant effects on a specific disease, with all potential confounding factors measured and ruled out, a consistent dose-response relationship between the degree of exposure and the severity of the effect, and a biological mechanism to make the association plausible (table 3.1). When many of these criteria are fulfilled, epidemiologists are likely to conclude that a causal link exists, as they have for relationships between smoking and lung cancer, and asbestos and mesothelioma. When many of the criteria are not fulfilled, however, epidemiologists are likely to withhold judgment or conclude that no link exists.[6]

The problem is that the Hill criteria represent unrealistic expectations when we are evaluating whether the ubiquitous low-level mixture of environmental pollutants is causing subtle, long-term health damage. The first problem is the lack of a clean control or reference population. An ideal epidemiological study would compare the rate of an unambiguously measured health impact in a group with clearly documented exposures to the rate in an otherwise identical group with no exposure to that substance. But nowhere in the world is there an unexposed population to serve as a basis for comparison, so epidemiology can never directly examine the contribution of background exposures to health problems. As the epidemiologist Geoffrey Rose has pointed out, if everyone in the country had smoked 20 cigarettes a day, then even the best epidemiological studies would have concluded that lung cancer was a genetic disease: "The more widespread is a particular environmental hazard, the less it explains the distribution of cases. The cause that is universally present has no influence at all on the distribution of disease and may be quite unfindable by

Table 3.1
Epidemiological criteria for inferring causation

Criterion	Inference of causalityis strengthened if...	Problems
Strength of association	The cause is associated with a large increase in risk.	The inability to measure exposure and effect accurately often reduces the apparent magnitude of the association. Weak associations may be of public health significance if exposure is widespread.
Probability of causal association	The association is statistically significant (very unlikely to be due to chance).	Significance requires very large study populations and/or relatively large increases in the effect.
Time order	The introduction of the cause precedes its effect; its removal causes the effect to disappear.	There may be a long latency between exposure and effect, including multigenerational impacts.
Specificity of association	Ideally, a specific cause is linked to specific effects; confounders are controlled or adjusted for.	Chemicals cause suites of many effects. Diseases are caused by chemical mixtures, and interactions among chemicals undermine the detection of specific relationships. Information to adjust for all confounders is seldom available.
Consistency of association	Studies by different authors, in different places, at different times, have similar results.	Impacts of a chemical may be modified by other factors that are not present in all other contexts.
Dose-response relationship	Larger effects are seen in groups with higher exposures.	Difficulty measuring exposures, especially those in the distant past, may obscure a dose-response relationship. Variations in modifying factors between groups may obscure a dose-response relationship. There are many possible kinds of dose-response relationships (e.g., linear, threshold, exponential, U-shaped).
Biological plausibility	A biological mechanism to explain the association is known.	Many biological processes are not well understood. Epidemiological evidence often precedes knowledge of mechanisms by many years.

Source: Criteria adapted from Fox 1991.

the traditional methods of clinical impression and case-control and cohort studies, for these all depend on heterogeneity of exposure."[7]

This means that epidemiologists can never directly assess the health impacts of chemical exposure at the levels found in the general environment. Instead, they must make indirect conclusions by comparing more exposed people to less exposed people. If the groups are large enough and the differences in exposure and health status great enough, epidemiologists can often demonstrate that a population with relatively high exposure to some pollutant—workers in a chemical plant, for example, or communities drinking contaminated groundwater—has an increased incidence of some disease. Findings like these, however, cannot be extended automatically to conclude that the general public's health is at risk, because the general public is the reference population in the first place. In its article on endocrine disruption, the *New York Times* raised this very objection: "Even if high levels of pesticides should turn out to raise the risk of breast or prostate cancer [in exposed farmers and workers], it does not necessarily follow that very low levels would also do so."[8]

On one hand, the *Times*'s point is correct: it is logically impossible to demonstrate directly from empirical data that the general population's health has been damaged by background exposures to pollution. On the other hand, this argument leads to a kind of logical paralysis, since it says that we can judge injury to be occurring only when we have a type of evidence that it is admittedly impossible to provide. This problem explains why epidemiology has never conclusively linked background exposures to public health damage and why science will never be able to do so as long as the standard is direct and conclusive evidence.

We need not buy into such contradictory reasoning. With direct demonstration out of the question, a judgment that universal exposure to chemicals is likely to be harming public health can legitimately be based on extrapolation from a finding of injury in more exposed groups to the possibility of effects at the lower doses that everyone experiences. While chemically-induced disease in people subject to relatively high organochlorine doses does not necessarily mean that these chemicals in the environment cause the same impacts in the rest of us, it is hardly unreasonable to think that they do. In fact, this kind of reasoning would be flawed only if there were specific reason to believe that there is a threshold of toxicity somewhere in the gap between the exposure levels of the higher dose group and the general population. As we saw in the previous chapter, however, this does not appear to be the case for many health impacts; many kinds of disease and functional deficits continue to occur, at reduced frequency or severity, as doses decline.

The logical leap from damage in highly exposed groups to the general population can be made smaller, though never completely eliminated, by studying people exposed to moderately or even slightly elevated levels of pollutants. When epidemiologists begin to compare background exposures to slightly greater ones, however, new complications arise. As the difference between exposed and control groups becomes more subtle, a study's statistical and analytical power gets weaker. To establish that a modest increase in health damage in an exposed population is not due to chance, a study requires a very large group of subjects and highly accurate measures of exposure and health damage. But in our case neither the cause nor the effect can be measured very accurately. There is not a single person whose total exposure to synthetic chemicals has ever been characterized. Everyone on earth now eats, drinks, and breathes a constantly changing and poorly characterized soup of organochlorine contaminants, most of them unidentified. Moreover, health effects may not appear for 30 years or more after exposure, and the impacts are often seen not in the exposed individual but in his or her offspring. There is no way to determine with any confidence the magnitude of chemical exposures that took place during critical periods decades in the past.

Accurate measurement of effect is also a problem. Epidemiological analysis works best with diseases that can be unambiguously diagnosed and quantified, like cholera or cancer. Subtle health impacts like cognitive impairment, reduced fertility, altered sexual behavior, and immune suppression are far more difficult to measure accurately. Further, there is no single "normal" state for any of these functions, all of which vary naturally within some range in the population. This natural variability greatly increases the noisiness of the results and reduces a study's power to establish a statistically significant association of exposure with effect.

The ability to detect a cause-effect link is undermined further by the noise that comes from other factors that contribute to the disease. Virtually all diseases and conditions have multiple causes, and the impacts of chemical exposure are modulated by such factors as genetic susceptibility, diet, and exposure to other pollutants like radiation or cigarette smoke. Thus, two persons with equal exposure to the chemical in question may have different exposure to these other factors, obscuring a causal relationship between the chemical and its effect unless it is very powerful. As Rose points out, the influence of a factor that increases the incidence of a disease with multiple causes by 50 percent or less will usually be impossible for epidemiologists to detect.[9] When the entire population is exposed, however, an increase in incidence of just a few percent will affect millions of people.

The final problem pertains to specificity. Under the Hill criteria, epidemiologists ideally want to see that a single cause is related to a single effect, because undifferentiated associations between many causes and one disease (or one cause and many diseases) may reflect some other true cause hiding within the noisy morass of confounding factors. In its study for the Chlorine Chemistry Council, CanTox took this criterion to an extreme, arguing that epidemiological evidence is valid only when it establishes causal links between individual organochlorines and individual health effects. According to CanTox, "scientific principles" require that a causal relationship between pollution and health damage can only be "inferred between effects and exposures to specific chemicals in the environment."[10]

This demand is utterly unrealistic. There is no way to evaluate the health effects of exposure to DDT in contaminated fish, because those fish also contain dioxin, dieldrin, dichlorobenzene, and thousands of other chemicals. Even studies that have linked specific health effects to body burdens of individual chemicals do not demonstrate that the measured chemical alone is responsible for the effect. The body burden of one bioaccumulated contaminant is roughly correlated with that of all the other chemicals that travel along with it. Associations between levels of an individual chemical and a health effect should be read as links to the total exposure burden, not as an indictment of that substance alone.

Most epidemiologists would recognize that it is unrealistic and unnecessary to demand chemical-specific links before judging that exposure to pollutant mixtures has damaged health. But the Risk Paradigm's focus on specific chemical linkages makes all the information that ties health damage to chemical mixtures inadmissible in the realm of policy. When chemicals are regulated on a substance-by-substance basis, evidence of this kind is irrelevant, because links between chemical mixtures and disease tell the Risk Paradigm nothing about which substances are at fault.

We should keep the correlation of organochlorines with each other in mind when we are struck by the relatively small number of organochlorines that have been directly implicated in large-scale health impacts. Just as was the case for the detection of organochlorines in the global environment, epidemiologists and wildlife biologists generally focus on chemicals whose effects have already been clearly demonstrated in the laboratory and in other field studies. Further, funds for research are most ample for compounds that have already been the subject of regulatory action. Thus, virtually all studies investigating a link between large-scale contamination and health impacts have focused on a very small group of chemicals—PCBs, dioxins, hexachlorobenzene, DDT and other organochlorine pesticides. The inertia that results from studying already well-studied chemicals continually reinforces concern about a few substances we already know are a problem and slows the development of knowledge about the role of other contaminants. Undoubtedly the well-studied organochlorines are among the most hazardous. But the lack of information about other members of the class is not a sign of their safety but an artifact of government priorities and the institutional factors that determine the focus of scientific investigation.

Precautionary Inference

By detailing the limits of epidemiology, I do not mean to imply that there is no evidence to corroborate a link between low-level organochlorine exposure and damage to the health of people and wildlife. In fact, there is a substantial body of information to make this case, including evidence from the laboratory, wildlife studies, occupational studies of highly-exposed people, geographic and temporal patterns of disease in the general population, and studies that show a correlation between disease risk and body burden—the concentration in a person's blood, fat, or other tissues—of specific organochlorines.

But for all the reasons already discussed, none of this evidence is conclusive in itself. By the standards of epidemiological inference, the question of a causal link between background exposure and large-scale health damage remains open. For the purposes of policy, however, this is the wrong standard to use. Instead of insisting on conclusive and specific demonstrations of causality, as the Risk Paradigm does before it takes out-of-the-ordinary action, the Ecological Paradigm uses a very different standard for weighing evidence.

Skepticism is a high value in science: scientists are taught to withhold judgment until a large body of convincing evidence corroborates a hypothesis and rules out alternatives. Their priority is to guard against so-called type I errors—the acceptance of a hypothesis that turns out to be false—and to err on the side of making type II errors, in which they reject a hypothesis that later turns out to be true. In the norms of science, it is better to believe nothing at all than to rush to false judgment.

Scientific skepticism has unarguable merits as a method for creating academic knowledge and for avoiding superstition and unfounded beliefs. But policymakers have adopted this standard too, requiring overwhelming evidence that a specific chemical has caused harm to public health before action is taken to restrict its production or use. The virtues of skepticism in the academic world quickly fade in the arena of environmental policy, where a type II error—failure to recognize the evidence of a threat because it is not conclusive—can result in irreversible damage to the health of millions of people.

By insisting on proof to scientific standards, the Risk Paradigm dismisses any knowledge that does not reach this degree of certainty. In this system, a well-founded suspicion with a body of circumstantial evidence to back it up is little better than a notion with no empirical support. To take account of information that does not reach this standard, the Ecological Paradigm uses a more precautionary method to interpret scientific evidence, with priority placed on avoiding false negative judgments about health damage. This approach takes account of many different types of studies and approaches, and it interprets them from a point of view that seeks to integrate their findings into a judgment consistent with all the evidence and with the goal of protecting public health. When many different types of studies all point in the same direction, the Ecological Paradigm will conclude that health is likely to be at risk.

This method, which I call "precautionary inference,"[11] is an extension of the approach that the U.S. National Research Council (NRC) took when it studied the impacts of hazardous wastes on the health of exposed communities. The NRC argued that because the Hill criteria were so difficult to satisfy, it was more realistic and reasonable to infer a causal relationship when the evidence showed three things: that people are exposed to chemicals in hazardous wastes, that health impacts occur among these people, and that similar effects occur in human and/or laboratory animals exposed to the same chemicals elsewhere.[12] Precautionary inference is also related to the "weight of evidence" approach that the International Joint Commission's Science Advisory Board advocated in 1991. The board's goal was to make "judgments within a preventive framework. . . . One assembles all the evidence: adverse effects on wildlife, adverse effects on humans, adverse effects on ecosystems, and a fundamental understanding of how biological systems, such as the reproductive system can be harmed. As in solving a difficult crime, the weight of evidence together builds a basis for judgment. . . . [This] approach requires new types of evidence and new ways of assembling evidence. Above all, it requires a willingness to act on an integrated body of evidence rather than to wait for irrefutable evidence of a cause-effect link."[13]

This does not mean that the Ecological Paradigm tells us to evaluate evidence uncritically; the quality of a study's design and the clarity of its outcomes should always determine how much weight we give to it. The fact is, however, no evidence in this field is ever perfect, so judgments are always made in the presence of uncertainty. Error is always possible; the question is what kind of error we make it our priority to avoid. The central question in the Risk Paradigm is, Does the evidence conclusively demonstrate a causal link between a pollutant and health injury? In the Ecological Paradigm, the key question becomes, What evidence must we ignore to conclude that contamination is *not* damaging health and ecosystems?[14]

From this angle, the information we already have becomes quite compelling. As we will see, a substantial body of information contradicts the view that organochlorine contamination is not damaging health. Considered together, despite its holes and the weakness of some of its individual parts, the information at hand supports a clear preference for the conclusion that organochlorine exposure is damaging health and the environment on a large scale.

In no way does making this judgment for the purpose of policy preclude further scientific investigation to refine or even refute it. But in the meantime, should we ignore all the evidence from the laboratory, wildlife, exposed workers and communities, and studies that link markers of organochlorine exposure to disease risk in the general population? Should we ignore the evidence of large-scale declines in the health of many wildlife species and the increases in similar forms of disease and impairment in the human population during the same period? It is in this spirit of integrating diverse forms of evidence (fig. 3.1) that the Ecological Paradigm evaluates the epidemiological data already at hand.

Health or Disease?

What is health? How do we know when the health of an individual or population has been harmed by chemical exposure? The Risk Paradigm adopts a negative definition of health that classifies only relatively severe, clinically recognized forms of injury as health damage.[15] This view derives from the traditional medical perspective, which defines health as the absence of diagnosed disease. If disease or damage has not been identified by a qualified physician, then a person is healthy. By this definition, subtle biological impacts, like reduced functional capacity or increased susceptibility to disease, are not a form of health damage because they do not reach the clinical severity that defines disease. A healthy condition becomes a negative artifact—the state that lies outside a doctor's ability or willingness to diagnose disease.

The Ecological Paradigm adopts a positive definition of health, which avoids the problem of negative definitions by focusing on the natural capacity of the organism to function in its environment. The great biologist René Dubos has defined health as the ability to adapt to new or changing circumstances; compromised health may become apparent only when new sources of stress are applied and the individual fails to adapt. The World Health Organization has adopted this view, defining health as "a state of complete physical and social well-being and not merely the absence of disease or infirmity." [16] If chemical exposure reduces a woman's fertility or a

	Lab animals high dose	Lab animals low dose	Wildlife high dose	Wildlife low dose	Humans high dose	Humans low dose	Human trends since WWII
↓ sperm count / semen quality	●	●		●	●	●	●
Male developmental defects	●	●	●	●	●		●
Endometriosis		●				●	●
Pregnancy failure	●	●	●	●	●	●	
Menstrual/estrus dysfunction	●	●		●	●		
Sex hormone changes	●	●	●	●	●		
Birth defects	●	●	●	●	●	●	●
Fetal/embryonic toxicity	●	●	●	●	●	●	
Thyroid changes	●	●		●	●	●	
Cognitive/ behavioral deficits	●	●		●	●	●	
Immune suppression	●	●		●	●	●	
Breast cancer	●	●			●	●	●
Immune cancers	●				●	●	●
Bladder/colorectal cancers	●				●	●	●
Testicular cancer	●		●		●		●

Figure 3.1
Types of evidence supporting a link between organochlorine contamination and large-scale health damage. Circles indicate that one or more organochlorines widespread in the environment have been linked to health damage in laboratory animals, wildlife, or humans. Low-dose categories include exposures at or near those that occur in the general environment; high-dose categories include persons subject to occupational and accidental exposures and wildlife in highly contaminated ecosystems. Trend category indicates conditions that have increased in industrialized nations since the 1940s.

child's intellectual ability, has it not damaged their health? Impacts like these, even if they are not severe enough to be recognized by a physician as infertility or mental retardation, can have major effects on an individual's well-being, happiness and quality of life. When they occur in a large number of individuals, they can have profound implications for an entire species or population. For the purposes of the Ecological Paradigm, then, I define health as the *normal biological state in which an organism retains its full functional and adaptive capacity.*

As the ecologist Glen Fox has pointed out,[17] health damage occurs on a continuum. A healthy individual adapts to stress and can adapt further in the face of additional stress. Subtle damage causes impairment—"early and underlying disturbances of the system or precursors of disease," such as reduced growth, impaired reproduction, increased susceptibility to infection, altered behavior, or reduced capacity to tolerate subsequent stress. As impairment becomes more extreme, it reaches a critical point, and the full-blown, recognizable symptoms of disease and organismal breakdown become apparent.

Impairment of function and adaptability at the physiological level are difficult to identify objectively, because they are hard to measure, vary naturally among individuals, and fluctuate in their expression depending on other factors in the environment. There are, however, indicators of biological disruption at the molecular, cellular, and systemic level that are much easier to measure and quantify. These "biomarkers of impairment"—including genetic mutations, changes in the number and types of immune cells in the blood, and altered levels of hormones, vitamins, neurotransmitters, enzymes, receptors, and growth factors—are useful because major impacts on health at the level of the organism always begin with disruption at the biochemical level. Changes in biomarkers represent a disruption of the body's natural regulatory mechanisms, and they indicate the existence of or potential for functional impairment of the organism. Further, impacts at the organismal level are often irreversible, but their biochemical precursors can usually be detected before catastrophe has occurred. Today's biomarkers are tomorrow's disease. If impacts can be detected early in the development of disability and dysfunction, we can recognize large-scale health damage before it is too late. Biomarkers thus play an important role in the evaluation of epidemiological evidence in the Ecological Paradigm.

Focusing on changes at the biochemical level makes particularly good sense when we are seeking to understand the health impacts of low-level chemical exposure. In healthy populations, there is great natural variability in various functions—fertility, intelligence, immunity, and so on. Where an individual falls in this range is determined, in significant part, by biochemical factors that also vary among individuals. For instance, studies of mice show that natural variation in hormone levels during fetal development causes differences in the size and structure of reproductive organs, the age at which puberty and menopause occur, the length of the estrus cycle, and the degree of aggressiveness and other aspects of adult behavior.[18] Similarly, natural variations in neurotransmitter levels are associated with differences in cognitive ability and emotional state in the "normal" population; this is, in fact, the basis for treating mild cases of depression with Prozac and similar drugs.[19] If subtle variation in biochemical variables leads to natural changes in health and functional capacity, then synthetic chemicals that affect these same and other variables—including hormones, neurotransmitters, and other biologically important substances—should be expected to alter a person's health status as well.[20]

Male Reproductive Health

By several measures, the reproductive health of men has declined on an international basis since World War II. First, developmental disorders of the male reproductive tract—notably hypospadias (incomplete closure of the urethra) and cryptorchidism (undescended testes)—appear to have increased substantially in the United States and some but not all European countries.[21] Second, the incidence of testicular cancer, which strikes primarily young men, has risen steadily in Europe, the United States, and Asia since 1960.[22]

Third, there has been an international decline in the quality of human semen. In 1992 Danish researchers statistically analyzed the data from 61 studies conducted in a large number of countries during the twentieth century and concluded that sperm counts in otherwise healthy men had declined by 50 percent from 1938 to 1990.[23] The report sparked controversy and additional research, as critics questioned the statistical methods the authors had used.[24] In the next five years, several studies were published on trends in semen quality in specific areas—including Paris, London, and parts of Scotland and Belgium—which found notable reductions from the 1960s through the 1980s in sperm count and aspects of semen quality, such as the percentage of sperm that were structurally normal or motile (capable of normal swimming behavior).[25] Other studies in Toulouse, Seattle, and New York City, however, did not find a decrease,[26] so whether there had been a general decline in semen quality remained controversial.

The issue moved much closer to resolution in late 1997, when American epidemiologists, at the request of the National Academy of Sciences, reanalyzed the international historical data, using more sophisticated statistical methods to address the objections raised by critics of the Danish study. Their report concluded that there are important geographic differences in sperm counts and trends, presumably reflecting different environmental and demographic circumstances in each location. In some locales, sperm counts have declined, but in others they have held steady or possibly even increased. Overall, however, the data show a real and substantial overall decline in sperm density in the United States and Europe since World War II. The reduction, in fact, has been even steeper than originally thought, and it has not abated in recent decades.[27] Today the average man's sperm count is about half of what it was in 1940.

What are the health implications of declining semen quality? Sperm count is quite variable in the human population: some men have two to four times the sperm density necessary to conceive, while others have barely enough to fertilize an egg. Thus, a 50 percent drop in sperm count throughout the population would have no effect on the fertility of some men, but it would substantially increase the number of men with sperm counts so low that fertilization is difficult or impossible. Poor semen quality is diagnosed as the cause of infertility in 20 to 50 percent of all couples evaluated in U.S. fertility clinics.[28]

The trends in testicular cancer, developmental defects of the male reproductive tract, and declining semen quality together suggest a real decline in male reproductive health and function in the past fifty years. All the conditions that are on the rise can be caused by changes in the hormonal environment during development, so it is plausible that they have all been caused by increasing exposure to endocrine-disrupting chemicals in the environment during the same period. In 1995 the Danish government convened a group of experts on male reproductive health, who concluded, "All of the described changes in male reproductive health appear interrelated and may have a common origin in fetal life or childhood. . . . The growing number of reports demonstrating that common environmental contaminants and natural factors possess estrogenic activity presents the working hypothesis that the adverse trends in male reproductive health may be, at least in part, associated with exposure to estrogenic or other hormonally active (e.g. anti-androgenic) environmental chemicals during fetal and childhood development."[29]

Impacts on Laboratory Animals

Could the general human population's increasing exposure to organochlorines since the 1940s be a cause of these trends? Studies of laboratory animals, wildlife, occupationally exposed workers, and the general population suggest that the answer is probably yes. The first line of evidence comes from laboratory studies, which give unambiguous cause for concern. As we saw in chapter 2, a large number of organochlorines act as antiandrogens, estrogens, or antiestrogens, or they modulate in other ways the action of hormones that control sexual function and development. Other organochlorines are mutagenic or toxic to the testes, damaging sperm or the germ cells that produce them.

The male reproductive system is especially sensitive during its development. When male rats are exposed to PCBs in the milk of their mothers, they are less likely than unexposed males to impregnate a fertile female.[30] And giving a low dose of the antiandrogenic fungicide vinclozolin to pregnant female rats demasculinizes their male offspring; the effects include smaller male reproductive organs, hypospadias, undeveloped prostate glands, vaginal pouches, and reduced or absent sperm production.[31]

What about all the estrogenic pesticides, like methoxychlor, DDT, and endosulfan? Few have been tested for their effects on male reproductive development. But diethylstilbestrol (DES), a nonchlorinated synthetic estrogen, has been very well studied. Because DES and estrogenic pesticides and industrial chemicals all act through a common mechanism mediated through the estrogen receptor, DES provides a model of the impacts we may expect from exposure to other kinds of exogenous estrogens. From the late 1940s to the early 1970s, DES was prescribed to over 5 million women to prevent spontaneous abortion and other complications of pregnancy. DES was later found not to have been effective for its stated purpose, but it did have profound unintended effects on the children of mothers who took it. Studies of rodents and boys show that in utero exposure to DES is associated with hypospadias, small and undescended testes, very small penises, retention of female reproductive structures, lower sperm counts, and reduced fertility. The similarity of these impacts to the effects that have been rising in the general population since the 1950s is striking.[32]

The most detailed studies have focused on dioxin's ability to disrupt reproductive development. When male rats, still in the womb, are exposed to a single very low dose of dioxin on the fifteenth day after conception—a critical period of early sexual development—they appear perfectly normal at birth, without any conditions that suggest disease or major disability. When they reach puberty, however, the levels of testosterone and gonadotropic hormones—the signals that tell the testes to make more testosterone—in their blood are lower, the descent of their testes is delayed, they have smaller prostates and testes, and their sperm counts are considerably lower than those of unexposed males. Their sexual behavior is changed as well. Placed in a cage with a female in heat, the dioxin-exposed males are slow to mount and penetrate, copulate with less intensity, and wait longer between copulations than unexposed males do. When injected with female hormones and put in a cage with a male, the dioxin-treated males act like females waiting to be mounted, arching their backs in a sexually receptive posture.[33]

The rodents in these studies were subjected to dioxin at levels comparable to the general human population's exposure. The dioxin and furan body burden of the average American is 8 to 13 parts per trillion, expressed as TCDD-equivalents (TEQ),* while people exposed in the work-

*The TEQ approach allows the quantity of dioxins, furans, and dioxin-like PCBs (and potentially other related substances) in a mixture to be expressed in terms ot the toxicity of 2,3,7,8-TCDD. To calculate the TEQ value of a mixture, the concentration of each dioxin or furan congener is multiplied by a factor that expresses its toxicity relative to that of TCDD; the TEQ value for the entire mixture equals the sum of the TEQs for all congeners in the mixture.

place or in accidents have body burdens ranging from 100 to 7,000 parts per trillion.[34] In comparison, a dioxin dose that produces a body burden of just 5 parts per trillion in pregnant rats—below the range for the average human—reduces the sperm count of their sons by 25 percent. At a maternal body burden of 13 parts per trillion—still around the human average—puberty is delayed, and penises and ducts in the testes are smaller. At 56 to 64 parts per trillion—higher than average but still in the range of some people—sexual behavior is demasculinized, sperm counts are cut in half, and prostates, testicular structures, and pituitary glands are all smaller than normal.[35] Current background exposures to dioxin thus appear to offer us no margin of safety.

Impacts on Wildlife

Males of wildlife species exposed to high doses of organochlorines have suffered particularly severe reproductive health damage. In 1980 a chemical spill contaminated Lake Apopka, Florida, with unusually high levels of DDT and dicofol; atrazine, DBCP, and ethylene dibromide are also present. In the next three years, the number of juvenile alligators in Lake Apopka decreased radically, apparently due to impaired reproductive function in the population. Louis Guillette, a zoologist at the University of Florida, investigated and found that males from the lake had extremely small penises—so small that they could not effectively reproduce. Their testes contained abnormal germ cells and produced extremely high quantities of estradiol, and their blood contained levels of testosterone as low as those normally found in females. These effects appeared to be due to pesticide exposure, because treatment of eggs in the laboratory with the same concentrations of organochlorines found in wild eggs caused hormonal and morphological feminization of developing males.[36]

Lake Apopka involved relatively high doses of organochlorines. The case that background contamination impairs male reproduction would be stronger if there were evidence that these effects occurred at still lower doses. Indeed there is: studies of wildlife health in large ecosystems with slightly elevated contaminant levels also indicate that large-scale contamination is severe enough to damage male reproduction. The Great Lakes, for instance, serve as a giant sink for airborne contaminants from much of the United States and Mexico; once deposited, contaminants are slow to leave the lakes, which discharge less than 1 percent of their water into the ocean each year. The levels of contaminants in the Great Lakes food chain are thus five to ten times higher than in freshwater ecosystems of North America that have no direct sources of contamination.[37] This difference is large enough to allow a comparison of the health of wildlife in the two settings but not so large that we should assume that the lower levels found elsewhere produce no effects. In Europe, the Baltic and Wadden seas play a similar role.

Male wildlife in these ecosystems has undoubtedly been damaged. In a number of locations in the Great Lakes, pairs of female herring gulls nest together with double the normal number of eggs in their nest, a phenomenon that occurs only when there is an abnormally low number of reproductively capable males in the population. The frequency of female-female pairs and these "supernormal clutches" of eggs in various locations correlates with the severity of contamination by persistent organochlorines. Most telling, over 70 percent of the male chicks in gull eggs collected from the Great Lakes have partially or fully feminized gonads.[38] Near the Channel Islands of California, where contaminant levels are also higher than normal, female-female pairing and supernormal clutches have also been documented; injection of uncontaminated eggs with DDT or methoxychlor at the levels found in the Channel Islands causes male birds to develop ovary-like gonads and oviducts, presumably due to the estrogenic effects of these compounds.[39] Together these studies suggest that the changes in reproductive behavior of females are due to a large-scale reduction in the number of reproductively capable males in these populations, caused at least in part by the developmental effects of persistent organochlorines.[40]

In the Everglades, the critically endangered Florida panther is a terminal predator that feeds on raccoon and other animals, so it has bioaccumulated extremely high levels of organochlorines. Studies of the panther population reveal an extraordinary frequency of reproductive disorders among the male panthers, including very low sperm counts, a high proportion of abnormal sperm, and total sterility in many animals. Cryptorchidism affects a shocking 90 percent of males in the population—up from just 10 percent in the 1970s—and the levels of estradiol in male panthers are at the high levels typical of females.[41] Because there are no

Florida panthers anywhere else in the world, it is not possible to compare this population to a less contaminated group, so the case for a causal link to organochlorines remains circumstantial. Nevertheless, the suite of reproductive effects in the panther is strikingly similar to that seen in laboratory animals exposed in utero to endocrine-disrupting chemicals.

In marine mammals, which have accumulated some of the highest levels of organochlorines in the entire animal kingdom, biomarkers indicate impairment of male reproductive function. Researchers who measured DDE and PCBs in the blood of Dall's porpoises from the open ocean of the Northern Pacific found a strongly inverse correlation between the body burden of these two compounds and the levels of testosterone in the animals' blood. In porpoises with the highest burdens of organochlorines (about 70 percent higher than the group average), testosterone levels were severely reduced—to about one-fourth the average. The "average" animals were affected too: their testosterone levels were about 25 percent lower than the porpoises with the lowest body burdens of DDE and PCBs.[42] No one knows how much lower the hormone levels were in the least contaminated animals, which still carried substantial body burdens of these compounds, than they would have been in an uncontaminated population.

Effects in Humans

Some of the same kinds of impacts have been seen in men exposed to higher-than-normal quantities of organochlorines. One of the earliest studies of male reproductive toxicity found an epidemic of infertility and low or zero sperm density among workers in a production facility for dibromochloropropane (DBCP).[43] More recently, farmers exposed to the common herbicide 2,4-D have been found to have lower sperm counts and greater percentages of dead, deformed, and immotile sperm than an unexposed comparison group. The effects are strong: sperm counts in the exposed groups were reduced by half, while the fractions of dead, immotile, and anomalous sperm were more than double those in unexposed groups.[44] In the U.S. chemical industry, workers exposed to dioxin have experienced a dose-dependent reduction in circulating testosterone and an increase in gonadotropic hormones.[45] In the electronics manufacturing industry, testosterone levels are reduced in the blood of male workers exposed to the solvent trichloroethylene.[46]

Only a handful of studies have directly explored a link between background chemical exposure and impaired reproduction in human males. There are just four studies—all of them involving small numbers of subjects, with various methodological difficulties—that suggest that the background organochlorine body burden is directly related to reproductive health impairment. One study by the New York State Department of Health found that men with higher concentrations of PCBs in their semen had lower sperm counts and lower sperm motility.[47] An Israeli study found that the semen of infertile men from the general Israeli population had higher levels of DDT metabolites, PCBs, and lindane than semen from a matched group of fertile males, a finding that was particularly strong among men with very low sperm count.[48] A study of fertile young men—donors to a sperm bank at Florida State University—found an inverse relationship between a man's sperm count and the concentration of PCBs, pentachlorophenol, trichlorophenol, hexachloronaphthalene, tetrachlorodiphenyl ether, and HCB in his semen.[49] Finally, a small study at an infertility clinic in Germany found that semen from men with higher concentrations of hexachlorobenzene and PCBs had less success fertilizing eggs in an in vitro fertilization procedure than semen from men with lower body burdens.[50]

Organochlorines are almost certainly not the only cause of declining male reproductive health in humans and wildlife. Developing boys are also exposed to endocrine-disrupting chemicals that do not contain chlorine, including pesticides, drugs given to people and livestock, and plastics additives. Nonchemical factors may also affect male reproduction; tight underwear, diet, and changing exercise habits are frequently cited possibilities. The interplay of organochlorines with other causes makes it difficult to ascertain how much of the decline in male reproductive health is due to organochlorine exposure alone. But neither lifestyle factors nor chemicals to which people are directly exposed through consumer products or drugs can explain the epidemics of reproductive dysfunction in wildlife. The only relevant difference between wildlife in the Great Lakes and other affected ecosystems on one hand and those in cleaner "reference" environments is the quantity of bioaccumulated chemicals to which they are exposed. Great Lakes herring gulls, for instance, are not exposed to estrogens given to dairy cattle or chemicals that may leach out of dental sealants; they presumably eat well, get enough exercise, and do not wear tight underwear. The similarity of the effects seen in laboratory animals, wildlife, and men exposed to higher-than-average quantities of organochlorines and those that are changing for the worse in the general population suggests that organochlorines are likely to be an important factor in the downward trends in male reproductive health in humans and other species.

Female Reproductive Health

Ten to 20 percent of couples in the United States suffer impaired fertility, and about half of those cases are due to reproductive problems in the woman.[51] From 1965 to 1982, the incidence of infertility in the United States tripled among couples in which the woman was between the ages of 15 and 24 years—the group normally considered the most fertile in the population; it then held more or less steady at this higher rate through the 1980s.[52] There are few data on trends in specific aspects of female reproductive health and function, but we do know that endometriosis is increasing[53] and that girls are entering puberty at younger and younger ages. Studies in the Netherlands, Japan, and Poland show a decline in the age at first menstruation among girls born between 1940 and 1965.[54] In the United States, a very large nationwide investigation found that between 1950 and 1993, the average age at which girls enter puberty declined by six months to one year. More alarming, the fraction of girls displaying signs of reproductive maturity at very early ages skyrocketed: by 8 years of age, 15 percent of white girls and a remarkable 48 percent of African American girls have now begun breast development and pubic hair growth, an immense change compared to the 1 percent frequency that used to be considered normal for this age.[55]

What is causing the acceleration of female sexual development? Improved nutrition and higher calorie intake could be partly responsible, but the work of endocrinologist Patricia Whitten of Emory University suggests that a more complex modulation of the endocrine system is more likely to be the cause. If the shift in age at puberty were due to nutritional factors, we would expect other aspects of reproduction that are limited by nutrition—the frequency of non identical twin births, for instance—to have changed too. Rather than rising, however, this kind of twinning declined steadily from the 1950s to the 1980s, ruling out improved nutrition as the sole cause of changes in female reproductive function during this period. The simultaneous downward trends in twinning and age at puberty are consistent, however, with disruption of gonadotropic hormones, which play an organizing role in reproduction and regulate both ovulation and secretion of steroid hormones by the gonads at the beginning of puberty. As Whitten points out, in utero or neonatal exposure to endocrine-disrupting chemicals can cause lifelong changes in gonadotropin levels, suggesting that chemical exposures may explain trends that cannot be explained by nutritional changes.[56]

Impacts on Laboratory Animals

If the problem is endocrine disruption rather than nutritional changes, what is interfering with our hormones? Several lines of evidence raise concern that organochlorines have already damaged the reproductive health of women and wildlife. First, laboratory studies have established that organochlorines can interfere with many aspects of female reproduction. For example, adult female rodents exposed to DDT develop a condition of persistent estrus and infertility because fertilized eggs do not implant in the uterus.[57] Injection of the estrogenic pesticide methoxychlor during the first days of pregnancy causes infertility, apparently because this chemical also reduces levels of progesterone, the hormone necessary for the fertilized egg to implant in the uterine lining.[58] And female rhesus monkeys exposed over the course of several years to very low levels of dioxin or PCBs in their diet experience hormonal changes, altered estrous cycling, uterine hemorrhaging, difficulty conceiving, and an increased frequency of stillbirths and spontaneous abortion.[59]

Organochlorine exposure during critical stages of development can permanently injure the reproductive system, but the effects may not be visible until adulthood. Exposure to methoxychlor, DDT, or PCBs in the first days of life causes female rodents to enter puberty precociously and develop various reproductive disorders, including a state of persistent estrus, the growth of hard tissue in the vagina, excessive growth of the uterus, multiple eggs maturing in a single ovary, and early menopause.[60] In utero exposure to a single, very small dose of dioxin can deform the vagina and clitoris at birth, reduce fertility, change hormone levels, and cause delayed ovulation, abnormal estrous cycling, and bleeding during intercourse.[61]

Effects on Wildlife

Strikingly similar effects have occurred in wildlife. In the Wadden Sea, the seal population collapsed between 1950 and 1980. Per Reinjders, a Dutch biologist, began to suspect bioaccumulated organochlorines as a cause when he found that the ability of females to produce healthy pups had declined severely in the Wadden's western section, where levels of PCBs and other organochlorines are particularly high due to input from the river Rhine. To test this possibility, he fed healthy female seals one of two diets—fish from the western Wadden Sea or the same kind of fish from the less contaminated Northeast Atlantic—for two years. The seals fed fish from the Wadden Sea suffered hormonal changes and severely reduced reproductive success, apparently because fertilized eggs failed to implant in the uterus. The concentration of bioaccumulated chemical contaminants in the diet was the only known difference between the two groups, suggesting that persistent organochlorines at ambient levels in the Wadden Sea are indeed the cause of female reproductive failure in the seal population.[62] A similar crash of the seal population occurred in the Baltic Sea during the period of peak contamination; examination of the animals suggested that chemically-induced damage to female reproduction was the cause there as well. [63]

Persistent organochlorines have caused profound effects on the reproductive health of female wildlife in the Great Lakes. For example, the number of mink living along the shores of the Great Lakes declined for several decades after World War II. The cause was unknown until the 1970s, when ranch-raised mink fed Great Lakes fish began to show signs of severe reproductive toxicity. Female fertility plummeted, apparently due to implantation failure, spontaneous abortion, and resorption of the fetus, and a high proportion of the kits that made it to delivery died soon after birth. Laboratory experiments confirmed that mink fed Great Lakes fish developed severe reproductive and developmental problems with greater frequency than control animals fed fish from less contaminated environments. Further studies showed that PCBs and dioxins cause the same suite of effects on mink reproduction, and doses considerably lower than those found in fish from some areas of the Great Lakes were highly toxic to the animals.[64] If Great Lakes fish cause these effects in the laboratory and on the farm, they presumably do the same to wild populations of mink and probably to other closely related species, such as otters.

Organochlorines have also impaired the reproductive success of female birds in the Great Lakes, but in this case the effect is on behavior rather than physiology—specifically, on gulls' attentiveness to protecting their eggs from cold and predators. In the 1970s, researchers found that herring gull reproduction in the Great Lakes was only 5 to 10 percent of the normal rate, and the reduction was due primarily to embryonic death and the disappearance of eggs from the nest. By placing thermometer-equipped false eggs in the nests of a large number of birds, Canadian scientists showed that gulls on the shores of Lake Ontario spent less time brooding their eggs than gulls from less contaminated colonies on the Atlantic coast; the same behavior was later documented in gull colonies along the Lake Michigan shoreline, as well. Eggs from a large number of nests were measured for organochlorine levels, and the severity of the behavioral effect was correlated with the degree of contamination. Furthermore, when ring doves in the laboratory were exposed to a mixture of organochlorines similar to those found in Great Lakes birds, they too spent less time incubating eggs in their nests than an unexposed control group.[65] That low-level exposure to organochlorine can alter a behavior as basic and instinctive as protecting one's eggs is a powerful testimony to the ability of these chemicals to play havoc with animals' bodies and brains.

Impacts on Women

There are very few studies of reproductive impacts in female workers, because most occupational studies have focused on male workers as the study population and cancer as the health effect. Nevertheless, there are enough data to indicate that women are by no means immune to the effects seen in organochlorine-exposed animals. In the dry-cleaning industry, the majority of employees are women, so there are some studies on female reproductive toxicity in this group of workers. Several large studies conducted in Denmark, Finland, and Britain have found that women who work in dry cleaning facilities—particularly those who operate perchloroethylene-based machines—suffer an increased risk of infertility, delayed conception, spontaneous abortion, and hormonal disturbances.[66]

There is more evidence of reproductive toxicity among women with high organochlorine exposures in nonoccupational settings. In the infamous Yu-cheng incident in Taiwan—in which a batch of rice oil was contaminated in 1979 with PCBs, chlorinated dibenzofurans, and chlorinated quaterphenyls and then sold for consumption to at least 2,000 people—exposed women experienced abnormal menstrual cycles and tended to give birth to babies with low birthweight.[67] In a similar accident in Japan, exposed women experienced hormonal changes, an extremely high incidence of irregular menstrual cycles, and an increased frequency of stillbirth and low birthweight.[68] In Vietnam, several epidemiological studies suggest an increased risk of stillbirth and molar pregnancy—the growth of large cysts in the uterus around a dead or entirely absent embryo—among women who lived in areas sprayed with Agent Orange, and the effects persisted for more than a decade after the war's end.[69]

As for the lower levels to which women in the general population are exposed, the data are again sparse, but several preliminary studies are suggestive. A small German study at an in vitro fertilization (IVF) clinic found that levels of PCBs, DDT, hexachlorocyclohexane, dieldrin and hexachlorobenzene were markedly lower in the follicular fluid of women for whom the IVF treatment successfully resulted in pregnancy than in women for whom the treatment failed.[70] In a later series of studies at another German clinic, women who had difficulty conceiving had higher levels of DDT and pentachlorophenol in their blood than women who had successful pregnancies, and levels of hexachlorocylohexane were elevated in women with uterine fibroids and other reproductive problems.[71] In an Italian study, 120 women who miscarried had higher levels of PCBs in their blood than a matched control group of women who had delivered at full term.[72] Two studies of women from the general population of India have found that levels of DDT, aldrin, and hexachlorocyclohexanes were higher in the blood of women experiencing miscarriage and premature birth than in those whose pregnancies proceeded to full term.[73] All of these reports are subject to common epidemiological problems, particularly small study populations that may not reflect the characteristics of the rest of the country and a failure to control for confounding factors; they are suggestive enough, however, to raise concern that low-level organochlorine contamination may be damaging reproductive health among women in the general population.

Postscript: Endometriosis

Endometriosis, a painful condition that is also a major cause of infertility, afflicts some 10 percent of all reproductive-age women in the United States—about 6 million women.[74] It appears to have been on a steady increase in developed countries in recent decades, a trend physicians and epidemiologists are at a loss to explain. In the United States, the number of hysterectomies performed each year to remedy endometriosis rose from 150,000 in 1965 to 400,000 in 1984; this increase cannot be explained by a higher rate of hysterectomy in general, because operations for endometriosis also increased considerably when expressed as a percentage of all hysterectomies.[75]

In endometriosis, tissues from the uterine lining, called the endometrium, grow ectopically—that is, in areas outside the uterus, such as the fallopian tubes, the ovaries, the lining of the pelvic cavity, the vagina, and the bladder. Like normal uterine tissues, ectopic endometrium responds to the hormones of the menstrual cycle, building up tissue each month, then breaking down and causing bleeding. Endometrial cells in the uterus are shed during each period, but the ectopic growths in endometriosis have no escape from the body, so they cause internal bleeding, inflammation, and scarring.[76]

Since the late 1980s, scientists at the University of Wisconsin have been studying the effects of long-term, low-dose-exposure to dioxin on the reproductive and developmental health of rhesus monkeys. Early in the 1990s, the researchers discovered that female rhesus monkeys that had been exposed for several years to just 5 or 25 parts per trillion of TCDD in the diet were experiencing extreme lower abdominal pain and hemorrhaging. Autopsies revealed that several animals had died from severe endometriosis, an effect the researchers had never thought to investigate. The Wisconsin group then called in Sherry Rier, an endometriosis expert at the University of South Florida, to investigate further. Rier's examination of the entire study and control populations revealed an extraordinarily strong, statistically significant, dose-dependent increase in the incidence and severity of endometriosis in the dioxin-exposed monkeys. Even the low-dose group—in which 70 percent of the animals had endometriosis, 42 percent of them severe—had dioxin body burdens just five to nine times greater than those of the average U.S. woman.[77]

The rhesus monkey study sparked quite a bit of follow-up research. To confirm a link between dioxin and endometriosis, researchers at the National Institute of Environmental Health Sciences surgically implanted a very small amount of endometrial tissue outside the uterus of rats and mice. They then examined the growth rate of the ectopic tissue in animals exposed to dioxin and compared it to that in a control group. Nine weeks after surgery, "massive growth" of the endometriotic tissue was observed in dioxin-exposed animals; no growth was observed in control mice. Dioxin also increased the incidence and severity of the growth of fibroid tissue at the endometrial sites.[78] The mechanism by which dioxin promotes ectopic endometrial tissue to grow into the disease state associated with endometriosis is not clear, but disruption of steroid hormones and modulation of the immune system are plausible explanations. Steroid hormones regulate the growth of endometrial tissues, and dioxin can alter the levels of numerous steroid hormones and their receptors. It is also known to activate immune cells to produce specific kinds of signaling molecules that stimulate endometrial growth and are present at high concentrations in the reproductive tracts of women with endometriosis.[79]

Some epidemiological studies on the association of organochlorines with endometriosis in women are underway, but only two have been published, with mixed results. The first, a fairly small study in Canada, did not find a statistically significant difference in the blood levels of total PCBs and several pesticides in a group of women with endometriosis and a control group without the disease; this study's ability to detect a relationship was limited, however, by its small size, its failure to control for breastfeeding and other potential confounding factors, its lack of attention to dioxin—the organochlorine most strongly suspected of contributing to endometriosis—and its failure to examine a relationship between pollutant levels and the severity (rather than just the incidence) of endometriosis.[80] The other published report did address dioxin. This study measured dioxin levels in the blood of 44 Israeli women with endometriosis and compared them to those in an age-matched control group with infertility but no endometriosis. Women with endometriosis were an astonishing eight times more likely to carry detectable levels of dioxin than unaffected women.[81]

Development

The development of a new organism from a fertilized egg, a precise unfolding of cell differentiation and tissue growth controlled by tiny quantities of biological molecules, is the most sensitive of all processes to chemical disruption. The specific form of injury depends on the chemicals involved and the stage at which exposure takes place. When injury occurs very early in development, for example, the developing mass of cells usually fails to implant and is lost altogether. Disruption at slightly later stages can cause the embryo to die and be resorbed into the mother before she even knows she is pregnant. Later disruptions may cause miscarriage, stillbirth, birth defects, low birthweight, or functional impairment of the child after birth.

The majority of pregnancies fail for one reason or another. Of 100 human conceptions, 50 will fail to implant and will be expelled or resorbed. Of the 50 that do implant, 15 to 30 will fail before pregnancy is even recognized, and 15 to 20 more will end in miscarriage or spontaneous abortion before the end of the second trimester. Of the fetuses that do make it to the third trimester, 10 percent will be stillbirths and 10 percent will be born prematurely. Birth defects are recognized in 3 to 5 percent of all births, and more congenital defects are identified as the child grows and develops.[82]

There are few data on trends in prenatal development. Because implantation failure and fetal resorption occur before pregnancy is even diagnosed, these problems are indistinguishable from maternal infertility. General trends in low birthweight and premature delivery would say little about chemical causes, because inadequate nutrition and prenatal health care are the dominant causes of these conditions. Some kinds of birth defects, however, seem to be on the rise. The U.S. Public Health Service monitors the frequency of birth defects at 1,200 hospitals throughout the nation and in a 1990 study reported on trends in 38 types of malformations. From 1979 to 1987, 29 kinds of birth defects increased at a rate of more than 2 percent per year, while only two decreased and seven remained more or less stable. Cleft lip and defects of the heart, eyes, and central nervous system (excluding spina bifida and anencephaly) all rose substantially.[83] The rapid increases in some subtle internal defects are probably due, at least in part, to improved diagnosis, but other kinds of malformations—like cleft lip, contracted or bent limbs, or absence of eye parts—are obvious to anyone and should not be sensitive to more careful surveillance by doctors. The reported increases in these kinds of birth defects almost certainly reflect a real increase in incidence.[84]

Impacts on Laboratory Animals

Laboratory investigations show that organochlorines can cause a wide range of adverse effects on development. Dioxin, for instance, causes prenatal mortality, low birthweight, cleft palate, structural defects of the kidney, and a fatal accumulation of fluid beneath the skin called subcutaneous edema.[85] Developmental effects are among the most sensitive of all impacts of dioxin exposure; in the eggs of lake trout, doses as low as 40 parts per trillion—about the same as environmental exposures in the Great Lakes—cause mortality, edema, hemorrhage, inhibition of circulation, and arrested development of skeletal and soft tissues.[86] The Wisconsin rhesus monkey study found that dietary dioxin exposure of 25 parts per trillion before and during pregnancy—a dose too low to produce overt toxicity in the mother—increases the incidence of fetal death.[87]

The problem is not limited to dioxin, though other organochlorines require higher doses to produce birth defects. The pesticides mirex, methoxychlor, and pentachlorophenol increase the rate of fetal death, reduce birthweight, and cause defects of the heart, eye, brain, and skeleton.[88] Perchloroethylene causes spontaneous abortion,[89] and trichloroethylene and dichloroethylene cause birth defects of the heart.[90] And forcing pregnant rats to inhale vinyl chloride in the low parts per million range increases the incidence of fetal death, reduced birthweight, and malformations of the brain in their offspring.[91]

Impacts on Wildlife

Are the lower concentrations found in the ambient environment capable of interfering with embryonic and fetal development? Here, the evidence from wildlife is particularly compelling. Numerous species of predator birds and fish in the Great Lakes have experienced epidemics of developmental toxicity. In the 1950s and 1960s, bird populations across North America declined precipitously due to eggshell thinning and breakage caused by DDT. After DDT was banned in the early 1970s, eggshell thinning abated, and populations across the country began to recover. In the Great Lakes, eggshell thinning improved too, but bird populations did not fully rebound. Upon close investigation, researchers began to discover more subtle problems in bald eagles, gulls, terns, herons, and cormorants, including embryonic mortality and birth defects. In fact, these effects had been occurring all along but were masked by the more obvious effect of eggshell thinning.[92] Only when field biologists were able to look beyond the severe and immediate impacts caused by one chemical did they see the more subtle effects caused by other pollutants.

These are true epidemics of reproductive and developmental toxicity. The effects are most severe in locations where contamination is highest, but they affect populations throughout the entire Great Lakes ecosystem. Since their discovery in the 1970s, some of the impacts have abated as levels of PCBs and banned pesticides have begun to decline. But they are by no means a problem of the past. In the early 1990s, when the most recently published field studies were conducted, fully half the bird eggs collected around Lakes Michigan, Huron, and Superior contained dead or deformed embryos.[93] Bald eagle productivity in the Great Lakes were still low in the mid-1990s, and birth defects were still prevalent, despite improvements in the health status of populations in locations not on the Great Lakes.[94]

There is little doubt that these effects have been caused by dioxins, furans, PCBs, and other organochlorines in the food chain. The problems are far more severe in populations that subsist on Great Lakes fish than in nearby inland populations that rely on other food sources. Further, the severity of developmental impacts in different colonies around the lakes correlates very strongly with the concentrations of PCBs and dioxins in eggs from each location. When birds are treated in the laboratory with dioxin-like compounds, they exhibit the same problems seen in wild populations: edema, crossed bills, clubbed feed, missing skull parts, deformed spines, hemorrhaging, brain abnormalities, abnormal feathering, and lack of eyes. These impacts occur at concentrations that are orders of magnitude lower than those found in the Great Lakes ecosystem. In the view of most Great Lakes scientists, these data prove a cause-effect link between organochlorine contaminants in the food chain and widespread developmental toxicity in birds, although a single chemical cannot be entirely blamed for the epidemic.[95]

Other Great Lakes species have been affected too.[96] Most striking is the total failure of fish-stocking efforts in Lake Michigan. Each year from 1965 to 1979, an average of 2 million lake trout were planted in the lake, but a viable, self-reproducing population was never established. The problem was apparently due to a fatal syndrome that affected the fish at the "swim-up" stage, just after hatching, characterized by abnormally retained yolk material, uninflated swim bladders, loss of equilibrium, swimming in circles or corkscrew patterns, and death after several days. The frequency of embryonic mortality correlated strongly with PCB levels in both the eggs and adult tissues. Moreover, the effects can be reproduced in the laboratory by exposure to individual organochlorines—PCBs, DDT, and dioxin—at the same levels at which they are found in the Great Lakes. When eggs from Great Lakes fish are incubated in clean water, the effects still occur, suggesting the problem lies in bioaccumulated contaminants passed from generation to generation.[97] A similar syndrome of embryonic and juvenile mortality, also suspected to be caused by organochlorines, has been observed in salmon in the Baltic Sea.[98]

Impacts on Children

There is little doubt that organochlorines have caused the same kinds of effects in human children, at relatively high doses. Women exposed to PCBs in the manufacture of electrical capacitors tend to give birth to babies earlier and at lower weight than unexposed women do.[99] Children whose mothers were exposed to PCBs and contaminants in the Yu-cheng incident are also at increased risk of low birthweight, as well as birth defects of the skin, mucous membranes, fingernails, toenails, and teeth.[100] And several studies have documented increases in fetal death, birth defects, and infant mortality in Vietnamese villages sprayed with Agent Orange.[101]

Chlorinated solvents and by-products have also been linked to developmental toxicity. Several studies have found associations between exposure to organochlorine solvents in the workplace and an increased incidence of miscarriage and certain kinds of birth defects.[102] Among women living in Tucson, Arizona, a threefold increase in the risk of cardiac birth defects has been linked to drinking trichloroethylene-contaminated water.[103] And a study of birth defects among babies born in the Washington-Baltimore area found that mothers with exposure to degreasing solvents had extremely high risks of giving birth to a child with four different kinds of cardiac birth defects.[104]

Pesticides have also caused developmental toxicity in children. The Washington-Baltimore study found that children born to mothers exposed to pesticides had an increased risk of cardiac malformations.[105] In Minnesota, a detailed epidemiological analysis indicates that cardiac, urogenital, and muscular-skeletal birth defects are elevated among the children of farmers who apply pesticides.[106] The babies of women who work in German day care centers furnished with lindane-treated wood paneling tend to be shorter and weigh less at birth than an unexposed comparison group.[107] Finally, two studies have found that living near hazardous waste landfills is associated in a dose-dependent fashion with low birthweight and the risk of congenital malformations.[108]

What about the lower doses to which the general population is subject? Several studies have demonstrated an association between damage to the health of infants and pesticide levels in the environment. In Iowa, where groundwater contamination with herbicides is a widespread problem, communities with more severe contamination also have higher rates of fetal growth retardation.[109] In Minnesota, the rate of birth defects in the general population is highest in counties of high pesticide use, particularly those with high application rates of 2,4-D and other chlorophenoxy herbicides, where the risk of birth defects was increased by 85 percent. Significantly, the effect is most pronounced for infants conceived in the spring, as expected if herbicides were the cause.[110]

Low-level exposure to solvents and chlorinated by-products also appears to damage infant health, with effects similar to those in laboratory animals. A study by the New Jersey Department of Health of over 80,000 births throughout the state examined a possible link between various adverse pregnancy outcomes and exposure to organochlorine contaminants in drinking water. The study found that exposure to total trihalomethanes

(THMs, a group of by-products formed in chlorine disinfection) in drinking water was associated with reduced birthweight, fetal growth retardation, and malformations of the heart, oral cleft, and nervous system. Carbon tetrachloride, trichloroethylene, perchloroethylene, and the plastic feedstock 1,2-dichloroethane were associated with considerable increases in the risk of low birth weight and various kinds of birth defects.[111] In Iowa, babies born to mothers who live in communities with drinking water concentrations of chloroform—one of the predominant THMs—above 10 parts per billion suffer an 80 percent increase in the risk of intrauterine growth retardation.[112] In Italy mothers who drink water disinfected with chlorine-based compounds give birth more frequently to babies with smaller heads, shorter body length, lower birthweight, and neonatal jaundice than women whose water is not chlorine treated.[113] And a very large, well-designed study by the California Department of Health Services found that the risk of spontaneous abortion is 80 percent higher among women who drank five or more glasses of chlorinated tapwater with relatively high THM levels during pregnancy than among women with lower exposures. The association was apparent at THM concentrations well below the maximum legal level in U.S. drinking water.[114]

Organochlorines in the food chain of North America and Europe are also high enough to interfere with fetal development. One long-term study examined the developmental status of babies born to mothers in Michigan who ate a moderate amount of fish—two to three meals per month—from the Great Lakes. The children of fish-eating mothers had higher levels of PCBs in their umbilical cord serum and gave birth about a week early to babies that weighed less (by an average of about one-half pound) and had smaller heads, compared to the children of a control group of mothers from the same communities who did not eat Great Lakes fish.[115]

A series of studies in the Baltic Sea had similar results.[116] Beginning in 1995, researchers at the University of Lund in Sweden reported the results of their research on the role of eating Baltic fish—a major source of exposure to PCBs, dioxins, and other organochlorines in that country—in low birthweight. They compared babies born to the wives of Baltic Sea fishermen to those in two reference groups: women in the general population of the same area, who ate about half as much fish on average, and fishermen's wives from the less polluted west coast, who were similar to the

Baltic wives in every important way—demographics, lifestyle, and diet—except for the amount of pollutants to which they were exposed. The study found that the Baltic mothers gave birth to infants with low birthweight considerably more often than either comparison group. Women who consumed more than four meals per month of fish from the Baltic Sea were twice as likely to give birth to low birthweight babies than those who ate zero to three meals of the fish per month. Further, the risk was strongly related to PCB concentrations in the mother's blood. These studies indicate that persistent organochlorines in the Baltic food chain are already high enough to impair the development of human children.

Finally, hemorrhaging in newborn babies seems to be linked to dioxin and its impacts on vitamin K. A syndrome called late hemorrhagic disease of the newborn (HDN), in which bleeding in the brain begins a month after birth, has become a recognized cause of death and sickness in infants only since the 1980s. The disease appears to be caused by inadequate levels of vitamin K, which is essential for proper blood clotting. Dutch scientists came to suspect that dioxin-like compounds may contribute to HDN, because these chemicals are known to induce enzymes that degrade vitamin K and cause it to be excreted, and hemorrhage is a well-recognized symptom of dioxin-induced developmental toxicity in laboratory animals. They examined dioxin levels in the breast milk of a group of mothers from the general population and the health of their babies. They found that infants with higher dioxin levels were more likely to have hemorrhage inside the skull than those with lower levels. Because none of the mothers had unusually high pollutant levels in their milk, this result indicates a link between dioxin exposure at background levels, either prenatally or via breast milk, and the risk of brain hemorrhage in infants.[117]

Like any other epidemiological study, all of these reports provide circumstantial evidence. They describe broad associations between indicators of exposure and the frequency or severity of health damage, but they do not rule out all confounding factors or establish unequivocal causal links. We cannot draw definitive conclusions solely on their basis, but we should not dismiss them, either. Instead, we should read this evidence in the context of all the other information that underscores the hazards of organochlorine exposure. Organochlorines cause developmental toxicity in laboratory animals, produce the same effects in people exposed to somewhat higher doses, and cause epidemic toxicity in wildlife exposed to ambient concentrations. Should we not then take seriously reports that the degree of organochlorine exposure of the general population is associated with the incidence of similar effects in people?

Thinking and Behaving

Some of the most worrisome information suggests that organochlorines have impaired the brain function of large numbers of people and wildlife. We saw in the previous chapter that a number of organochlorines interfere with the activity of neurotransmitters, which mediate the transmission of signals from one neuron to another. Organochlorines also affect the levels and actions of thyroid and steroid hormones, which control aspects of brain development; too little or too much thyroid hormone during critical periods of development, for instance, results in hyperactivity, learning disabilities, lack of motor coordination, and hearing loss. An emerging body of evidence links organochlorine contamination to the incidence of these and other conditions on a very large scale.

Impacts on Laboratory Animals

Several studies have established that very low doses of organochlorines during development cause long-term cognitive impairment. Laboratory rats exposed in utero and via breast milk to very low levels of PCBs have reduced levels of thyroid hormone in their blood and lower levels of dopamine in their brains; they are hyperactive, perform poorly on tests of short-term memory, and have hearing impairment years after exposure stops.[118] Prenatal exposure to hexachlorobenzene also results in hyperactivity and a tendency to startle easily and intensely.[119]

When rhesus monkeys, which are much more closely related to people than rats are, are exposed to low doses of PCBs in utero and via breast milk, they appear normal at birth. When they are tested at ages 2 to 6 years, however, they perform poorly on cognitive tests that measure short-term memory and learning ability, according to two studies. In one, the monkeys were exposed only aftter birth to a PCB mixture designed to represent the types and quantities of PCBs found in the milk of a typical Canadian woman. In both studies, the PCB concentrations observed in the bodies of the exposed monkeys were in the middle of the range reported in the general human population.[120]

The Wisconsin rhesus monkey studies—the ones that discovered the link between dioxin and endometriosis—have found that low-dose dioxin exposures cause similar forms of transgenerational cognitive impairment. At birth, monkeys born to mothers exposed to just 5 or 25 parts per trillion in the diet appear to be physically normal, but they behave strangely. They are extremely dependent on their mothers, who in turn treat them as if they were blind or otherwise disabled. At several months of age, the dioxin-exposed monkeys are less attentive to visual stimuli, and they perform poorly in several tests of learning, short-term memory, and pattern recognition. As they grow, the dioxin-exposed young are also unusually aggressive in peer groups, initiating violent behavior more often than unexposed monkeys do.[121] These effects occurred at a maternal body burden that was just three to six times greater than the body burden of the average American woman.[122]

Effects on Wildlife

There is no way to test the cognitive ability of wildlife populations, but it is clear that biomarkers of impaired neurological development have been affected in a number of species. For example, seals fed a diet of fish from the Wadden Sea for two years have much lower levels of thyroid hormones in their blood than seals fed fish from the less contaminated Northeast Atlantic.[123] In the Great Lakes, birds and fish also have altered levels of thyroid hormones in their blood, and the severity of the effect in birds correlates with their exposure to PCBs, dioxin, and other organochlorines. The reach of the problem is extraordinary. A large percentage of gulls and virtually every lake trout ever examined in the Great Lakes suffer from goiter, an extreme enlargement of the thyroid gland, which occurs in response to inadequate levels of thyroid hormone in the blood. Furthermore, when laboratory rats are fed Great Lakes fish, their thyroid hormone levels are depressed relative to control rats fed fish from the less contaminated Pacific Ocean. Thyroid impacts are often attributed to PCBs because of the clear link between PCBs and these effects in the laboratory; however, a diet of Great Lakes fish is at least 10 times more effective at lowering thyroid levels than PCBs alone, suggesting that the cause is the complex mixture of compounds in the environment, not just PCBs.[124] Because thyroid hormones are involved not only in neurodevelopment but also in the control of metabolism, growth, bone density, sexual development, and behavior, reduced thyroid levels in wildlife suggest the possibility of many other manifestations of impaired health, few of which have been investigated.

Interference with hormones and brain chemistry can affect not just cognition but also the ability to cope with stress. To investigate the effects of bioaccumulated pollutants on stress behavior, researchers at the State University of New York at Oswego performed psychological tests on rats fed a diet of Great Lakes salmon. Compared to a control group fed Pacific salmon, the rats fed Great Lakes fish were more disturbed by mildly stressful situations, such as being placed in an unfamiliar environment, receiving a weak electric shock before being given a food reward for performing a task, or replacement of a large reward with a smaller one. Normally a rat will be somewhat hesitant after such an event, but it will quickly recover. In each case, the Great Lakes rats took longer to return to their normal behavior after the stressful event, and some never recovered entirely. These results indicate that organochlorines increase the intensity of the animal's stress reaction to mildly unpleasant, unpredictable, or frustrating situations. Even the offspring of the exposed rats displayed the same kind of stressed-out behavior, whether exposure occurred in utero, through mother's milk, or both.[125] Most important, Great Lakes fish accounted for only 30 percent of the affected rats' diet, which means that wild populations are subject to even higher doses. If rats are as good a model as they are usually taken to be, then ambient levels of organochlorines in the Great Lakes food chain are likely to be capable of undermining the ability of other mammalian populations (including humans) to cope with the everyday stress that characterizes life in the wild (and in society).

Impacts on Humans

The chemically-induced effects observed in the laboratory are strikingly similar to the kinds of cognitive and neurological impairment that have been observed in organochlorine-exposed human populations. For example, children of women exposed to PCBs, chlorinated dibenzofurans, and other chemicals in the Yu-cheng incident are hyperactive, and their behavior is disorganized. At 7 years of age, they scored poorly on several tests of cognitive development, with IQ scores reduced by 4 to 7 points compared to unexposed children. Like the PCB-exposed rats, these children also have hearing loss, which indicates neurological damage caused by altered thyroid hormone levels during development.[126]

Several studies of humans suggest that the much lower levels of organochlorines in the food chain are already impairing behavior and cognition in children. Two studies in the Great Lakes examined the intellectual development of children whose mothers ate organochlorine-contaminated fish from the lakes. The first followed the children of the Michigan mothers, which I described above.[127] At ages 5 and 7 months, these infants whose mothers ate a moderate amount of fish from the Great Lakes performed more poorly on visual recognition tasks than babies in the same communities whose mothers did not eat Great Lakes fish. At 4 years of age, the children were tested again, and those born to fish-eating mothers had impaired short-term memory on both verbal and quantitative tests. The severity of the cognitive effects correlated with the concentrations of PCBs in the umbilical cord serum of the children at birth. The difference was not of clinical significance—the children did not have obvious learning disabilities—but it was not minor either; the exposed children experienced a more than 10 percent reduction in visual recognition memory. The children were tested again at age 11, and prenatal PCB exposure was associated with reduced short- and long-term memory, reduced general intellectual ability, impaired reading comprehension, and difficulty maintaining focused and sustained attention. The group that had had the highest PCB levels in their umbilical cord serum averaged 6.2 points lower on a general IQ test than the lower-exposure groups. The results could not be explained by differences in a host of confounding factors, such as demographic and socioeconomic variables, complications in delivery, and drinking or smoking during or after pregnancy. Despite its focus on PCBs as a marker of organochlorine exposure, these studies did not establish PCBs as the sole cause of cognitive impairment in the children. Although PCBs are likely to have contributed to the effect, we should remember that PCB levels correlate with concentrations of other bioaccumulated contaminants in Great Lakes fish, including dioxins, furans, pesticides, and other compounds. The correlation of the cognitive effect with prenatal PCB exposure thus says more about exposure to chemical mixtures than it does about PCBs alone.

A second study replicated the Michigan findings. Between 1991 and 1994, researchers at the State University of New York—the same group that studied the stressed-out rats—examined newborn babies in Oswego County, New York.[128] The women were classified according to the amount of Lake Ontario fish they had consumed: high consumption (more than 40 pounds of fish, or about two half-pound meals per month for three and a half years), low consumption (less than 40 pounds), or none at all. When assessed for neurological status, babies born to mothers in the high-fish group displayed abnormal reflexes, more intense startles and tremors, reduced attention to visual and auditory stimuli, and delays in getting used to repeated mildly unpleasant noises, an effect that was still present when the researchers adjusted for a wide range of potential confounding factors.

In other parts of the world, too, organochlorines have impaired child development. Several studies have used a different strategy, seeking a correlation between the mother's body burden of specific organochlorines and measures of cognitive development in their children. A series of studies of women from the general population in North Carolina found that levels of PCBs and DDE in mother's milk were associated with poor muscle tone and altered reflexes at birth. At ages 6 months and 1 year, children with higher contaminant levels scored poorly in psychomotor tests that measure muscle control and in short-term tests of visual recognition memory. The effect persisted at ages 18 and 24 months, but it was no longer significant in tests at 5 and 10 years of age.[129]

A Dutch study of newborns from the general population revealed similar effects.[130] Babies whose mother's milk had higher levels of dioxins and furans tended to fare poorly in tests of reflexes, posture, balance, and movements that are considered a good predictor of later neurological performance. The same researchers also found that higher dioxin levels in milk were correlated with lower levels of thyroid hormones in the infants' blood at ages 1 week and 3 months, providing a plausible mechanism for the neurological effect. While the Michigan and North Carolina studies established that in utero exposure was far more damaging than exposure through mothers' milk, the changes in thyroid levels in the Dutch study were linked to dioxin levels in milk, even when the researchers adjusted for prenatal exposure. These results suggest that exposure through nursing alone can affect thyroid status and may thus have an impact on neurological development in newborns.

By no means should this study be taken to mean that women should not breast-feed. Breast-feeding in itself has clear immunological, nutritional, and emotional health benefits. There is no denying that nursing is good for babies and mothers. But there is also no doubt that it would be better for babies and mothers to nurse in a world in which human breast milk were not so laden with toxic chemicals. No mother should ever have to worry that she is doing her baby harm by doing the most natural, nurturing thing in the world.

What is the significance of the subtle changes in cognitive ability documented in these studies? The Michigan studies have shown that children from the general population exposed to the highest levels of organochlorines in utero experience a five-to ten-point reduction in IQ score compared to those with lower exposures. No one knows how great a difference there may be between the less exposed children and the uncontaminated children who no longer exist anywhere on earth. In any case, deficits of this magnitude hinder but do not debilitate an individual. When everyone is exposed, however, the impacts take on a new significance. A five-point reduction in IQ throughout the human population would not only make every individual slightly less intelligent, but it would affect the distribution of cognitive ability in the population; the number of people with IQs over 130 would be cut in half, and the number with scores under 70 would double.[131] As an international conference of neurologists, psychologists, and endocrinologists concluded, changes of this sort can have profound social effects:

> We are certain [that] . . . endocrine-disrupting chemicals can undermine neurological and behavioral development and subsequent potential of individuals exposed in the womb or . . . the egg. This loss of potential in humans and wildlife is expressed as behavioral and physical abnormalities. It may be expressed as reduced intellectual capacity and social adaptability, as impaired responsiveness to environmental demands, or in a variety of other functional guises. Widespread loss of this nature can change the character of human societies or destabilize wildlife populations. . . . Because certain PCBs and dioxins are known to impair normal thyroid function, we suspect that they contribute to learning disabilities, including attention deficit hyperactivity disorder and perhaps other neurological abnormalities.[132]

Immunity

There is great variability in resistance to infectious disease within natural populations; some individuals are simply better able to fight off infection than others. Immunity is not an all-or-nothing proposition but a constant war between the body's immune cells and the populations of bacteria, viruses, and parasites that would like to colonize it. Giving a slight advantage to one side can make the difference between no infection and a brief, chronic, or fatal one. The immune system is delicately regulated to ensure that its response is both adequate and appropriate. One on hand, the response to infection must be speedy and powerful; on the other, too much immune activity can lead to excessive expenditure of energy on the production of immune cells, allergic reactions to foreign compounds that are in fact harmless, or an immune attack on the body's own tissues.

No single immune state is considered normal. Organisms exist somewhere on a continuum of immune states, which can vary from time to time and individual to individual with changes in nutritional, genetic, reproductive, and chemical factors. If intrinsic biological factors place the immune defenses of individuals somewhere in a natural range of intensity, it is reasonable to think that low-level exposures to chemicals that can impair the function or development of the immune system may shift individuals in a population toward greater susceptibility to disease.

Immunity is the end result of the activity of several kinds of immune cells. Macrophages engulf and destroy invading parasites. B-cells (so-called because they mature in the bone marrow) make antibodies, a class of proteins that bind specifically to foreign proteins, called antigens, and facilitate their breakdown and excretion. Natural killer cells kill cancerous and virus-infected cells. T-cells, which mature in the thymus, include cytotoxic T-cells, which also kill virus-infected cells; suppressor T-cells, which reduce the immune response; and helper T-cells, which regulate the proliferation of the appropriate B-cells to combat a specific type of infection. There are billions of kinds of B-cells, each specialized to dispose of one kind of antigen; each B-cell is activated by its own specific kind of helper T-cell. The job of T-cells is to recognize what type of infection is present in the body and activate the correct B-cell population to divide and produce antibodies. An inadequate T-cell response leads to immune suppression; a nonspecific T-cell response can cause autoimmunity.

Impacts on Laboratory Animals

Laboratory studies show that organochlorines can interfere with the activity of all these players, often at doses that cause no other obvious effects.[133] The pesticides aldrin, dieldrin, and endrin, for instance, suppress the activity of macrophage cells and increase susceptibility to viral infection. Chlordane, heptachlor, and the solvent trichloroethylene suppress both T-cell and B-cell responses when an animal is challenged with an infectious agent, like bacteria or viruses.[134] DDT weakens the immune defenses of fish and increases the incidence of protozoan and fungal diseases.[135] Inhalation of ethylene dichloride, perchloroethylene, and dichloromethane all increase the susceptibility of mice to fatal infection by airborne bacteria.[136]

Dioxin is the most immunotoxic organochlorine ever studied. Mice given a single, very small dose of TCDD—as low as 10 billionths of a gram per kilogram of body weight, enough to produce a dioxin body burden of just 10 parts per trillion (ppt), about the same as the average American's—are more likely to die after exposure to the influenza virus than unexposed animals are.[137] In adult monkeys, a dioxin dose that produces a body burden of just 6 to 8 ppt—slightly *lower* than that of the average person—reduces the number of helper T-cells in the blood.[138]

The development of the immune system is particularly sensitive to chemical disruption. The lifelong capacity for the immune response is determined during prenatal and early postnatal development in mammals and before and just after hatching in birds.[139] Chemical exposures during this period can result in permanent damage to immunity. For instance, the offspring of mice treated with chlordane in pregnancy show a profound deficit in immune responses when they reach maturity.[140] The offspring of female rats exposed to the wood preservative pentachlorophenol have suppressed B-cell responses when they become adults,[141] and developmental exposure to 2,4-dichlorophenol—a chemical intermediate and by-product of water disinfection with chlorine—results in depressed T-cell immunity.[142] In utero exposure to dioxin results in atrophy of the thymus and subsequently weakened immune defenses.[143] And when mother rhesus monkeys are exposed to very low levels of PCBs, producing a body burden about equivalent to that of the general human population, their offspring's ability to mount a defense against foreign proteins is permanently compromised.[144]

Impacts on Wildlife

Between 1992 and 1994, researchers collected chicks from four colonies of herring gulls and Caspian terns in the Great Lakes and, for comparison, from an inland colony in Manitoba. They found that T-cell immunity was suppressed in Great Lakes birds by up to 45 percent. Moreover, the severity of the suppression was correlated with the degree of contamination by dioxins, furans, PCBs, and DDE in eggs from the same colonies.[145] Corroborating these findings, mice fed Lake Ontario salmon for several months in the controlled environment of the laboratory were found to have much weaker immune responses to foreign proteins than a control group fed less contaminated salmon from the Pacific. The degree of immune suppression again correlated with PCB levels in the animals' livers.[146]

Marine mammals have suffered especially severe immune consequences. In the 1980s, populations of seals in Northwest Europe and Lake Baikal in Siberia plummeted when viral infections caused mass die-offs of many thousands of seals. These animals are top-level predators, exposed to extraordinarily high levels of bioaccumulated organochlorines, including substances that are known to be potent immunotoxicants. To test whether chemically-induced immune suppression might have played a role in the epidemics, Dutch scientists studied the effect of contaminants in fish on the immune systems of seals held in captivity. One group of seals was fed fish from the Baltic Sea, and a control group was fed the same kind of fish from the less polluted Northeast Atlantic. After two and a half years, the seals fed Baltic fish had blood levels of DDT, PCBs, dioxins, and furans three to six times higher than the control group. These animals were less able to mount an immune response against viral proteins, the activity of their antiviral natural killer (NK) cells was reduced, and their T-cells were less able to proliferate in response to various antigens or to interleukin-2, a natural signaling molecule that normally initiates T-cell activity. Furthermore, the severity of the effects on NK and T-cells correlated with the concentration of dioxins and furans in the seals' blubber. Did compromised immune systems due to organochlorine exposure make the wild seals more susceptible to viruses? The body burdens in the study animals were in many cases even lower than in wild seals, and this degree of exposure was high enough to interfere with immune cells that play central roles in the defense against viruses. It is likely, then, that organochlorine-induced immunosuppression played a role in the viral epidemics by increasing both the number of infected animals and the fatality rate among affected seals. Without organochlorine contamination of the Baltic food chain, it is quite possible that the virus, which probably would have been present anyway, would have had a far less devastating effect on the population.[147]

Dolphins are another top predator species that has suffered large-scale immune suppression. Since 1987 several very large die-offs have killed thousands of wild bottlenose dolphins along the Atlantic coast of North America, and in the Gulf of Mexico, and in the Mediterranean Sea. The mass mortality along the Atlantic coast cut the coastal population in half, and it may take 100 years or more for the population to return to its original numbers. The die-offs appear to have been caused by a startling variety of different kinds of infection, which suggests not the emergence of a new disease but a generalized increase in the susceptibility of dolphins to viruses and bacteria. When very high levels of organochlorines were found in affected dolphins, researchers at the University of Maryland decided to investigate what role bioaccumulation may have played in the North American die-offs. They collected blood from individuals in a wild dolphin population in coastal Florida and examined immune responses and organochlorine levels. The results showed a clear correlation between reduced T-cell immunity and increasing levels of PCBs and DDT metabolites; in fact, the levels of these contaminants explained 80 percent or more of the variance in immune response.[148] These results provide a strong preliminary indication that the dolphin die-offs, like those among seals in Europe, may have taken place because organochlorines made the animals more susceptible to fatal viral infection that otherwise would have been of little consequence.

Impacts on People

A number of studies have demonstrated immune suppression and other immune system changes in people exposed to organochlorines at relatively high levels. Several years after the Yu-cheng incident, PCB-exposed adults experienced more severe skin and respiratory infections and suppressed cellular and humoral immunity, and children of exposed mothers had an unusually high incidence of respiratory and ear infections.[149] Infants born to women workers exposed to PCBs in a Japanese capacitor factory have been found to have more colds, gastrointestinal symptoms, and eczema than an unexposed population, and the severity of the effect is correlated with the length of time the children were breast-fed.[150] Decades after they were exposed to dioxin, male workers in a chemical factory in Hamburg, Germany, have suppressed T-cell immunity, [151] and men exposed to dioxin in an accident at another German chemical plant owned by the BASF corporation have higher rates of infectious and parasitic diseases, upper respiratory tract infections, and total illness episodes.[152]

Other organochlorines have also caused immune suppression in people. One Italian study measured immune parameters of farmers before and after they used the popular chlorophenoxy herbicides 2,4-D and MCPA. After exposure, blood counts of natural killer cells and several types of T-cells were depressed, and the immune systems of these men were less able to respond to a stimulus that normally results in proliferation of immune system cells.[153] In Woburn, Massachusetts, where groundwater was contaminated by the solvents trichloroethylene, perchloroethylene, dichloroethylene, trichloroethane, and chloroform, exposed residents were found to have altered levels of T-cells in the blood, and children exposed to the greatest quantities of contaminated drinking water had more frequent pulmonary and urinary infections than less exposed individuals.[154]

All of these studies involved groups with higher-than-normal organochlorine exposures. There are only two sets of studies that have directly investigated the possibility that environmental levels of organochlorines may be causing immune dysfunction in humans. In the Baltic fishermen's study, Swedish researchers found that men from Sweden and Latvia who eat higher than normal quantities of fish from the Baltic have lower blood counts of natural killer cells than people from the same area with low fish consumption rates. The severity of the immunological effect correlates with their blood levels of PCBs, DDT, and dioxins and furans.[155]

The second study focused on the arctic regions of Quebec, where Inuit children suffer infectious diseases at rates 10 to 15 times greater than children in southern Quebec. To test the hypothesis that organochlorines in the arctic food web were causing immune suppression among the Inuit children, researchers compared the health status of breast-fed babies to bottle-fed babies in the same population. Breast-feeding usually protects against infection, but in this case the nursed babies had more ear infections than the bottle-fed children, who also had much lower levels of organochlorines in their blood. Additional work showed that the breast-fed babies had altered T-cell ratios—a marker of immune suppression—and the severity of this effect correlated with both the duration of breast-feeding and the concentration of organochlorines in breast milk. The degree of immunosuppression in these children is so great, in fact, that some are impossible to vaccinate because their immune systems fail to produce antibodies against the vaccines.[156]

What is the significance of all these findings? A group of 18 immunologists and wildlife biologists reviewed the evidence and noted that laboratory investigations and epidemiological studies of humans and wildlife provide a consistent picture of chemically-induced damage to the immune system. Weak immune systems mean that both the frequency and severity of infectious disease are likely to increase. The group concluded, "Disease patterns are thus likely to be affected by immune modulation induced by immunological toxicants. . . . Our judgment is that the potential exists for widespread immunotoxicity in humans and wildlife species because of the worldwide lack of appropriate protective standards."[157]

Precautionary Action

The cardinal principle of public health practice is disease prevention. The science of epidemiology was founded in the 1850s when an epidemic of cholera was ravaging London. A physician named John Snow believed, contrary to the wisdom of the day, that cholera was passed from person to person through contaminated water. To support his view, Snow prepared a detailed map of the location of cholera cases in the city and saw that a large number were clustered near a single drinking water well. At the time, there was no germ theory of infection, so Snow had no plausible mechanism by which water might cause cholera. Further, the clustering of cases around the well did not prove that the well was responsible for the epidemic, because the true cause could have been anything peculiar to the neighborhood. Unable to convince people not to draw water from the well, Snow took action: he removed the pump's handle. Cholera cases began to subside, and the role of drinking water contamination in infectious disease was eventually accepted. Snow's tale has become the creation myth of epidemiology, illustrating how a disease's causes can be discovered through a detailed analysis of its incidence patterns. It also exemplifies the principle that we can take action to prevent disease even in the absence of strict proof that a risk factor is its cause.

In environmental policy, precaution should play a similarly central role. The precautionary principle states that environmental damage must be prevented before it occurs. Practices that may cause harm, whether or not scientists can establish a firm causal link between the practice and environmental injury, should be avoided. When technologies have the potential to cause serious, large-scale, and irreversible damage to health and the environment, we should look to the future and preempt possible injury, not only damage that has already occurred.

The evidence reviewed here is highly suggestive, but it does not demonstrate conclusively that organochlorines are causing large-scale damage to human health, though it leaves little doubt that they have caused severe injury to wildlife and highly exposed workers and communities. What judgment should we make from this body of information? Has the general public's health been injured, or has it not? In the academic world of pure science, it may be possible to suspend judgment until a case that meets strict standards of proof has been established. But the question of global health damage due to chemical pollution is not a purely academic one. Whatever answer we give—yes, no, or don't know—implies some practical course of action. Academic scientists can withhold judgment and wait for more data, but the rest of us must choose between conclusions of danger and no danger, action or inaction.

Despite its limits, the body of evidence is too great to ignore. It offers support to the conclusion that organochlorine contamination is damaging the health of humans and wildlife, and it contradicts the view that human health is not at risk. Such a judgment does not stop scientists from continuing to study the problem; more information is always a good thing for both knowledge and policy, and perhaps it will prove that chemical pollution does not pose a health threat.

In the meantime, we need to make a decision about the proper course of action. Faced with very suggestive information and a life-and-death situation, we, like John Snow, have enough evidence now to take precautionary action to protect public health. Even short of proof, we can take the handle off the chlorine pump.[158]

4

Organochlorines and Cancer

Industrial developments took place with little if any concern for possible health effects. This was due partly to genuine ignorance, especially about long-term adverse effects such as cancer; later, however, adverse health effects were dismissed, either as unavoidable evils or as of too little importance to justify expensive modifications to production procedures.

—Lorenzo Tomatis, director emeritus, International Agency for Research on Cancer[1]

The twentieth century has brought an extraordinary shift in the diseases that end people's lives in industrialized nations. During the first six decades of this century, in one of the true triumphs of modernity, antibiotics, vaccines, and improved sanitation virtually vanquished the infectious diseases that had been the leading causes of death for hundreds of years. People in the industrialized countries now live longer than they used to because they are much less likely to contract and die from such diseases as tuberculosis, cholera, polio, diphtheria, influenza, and pneumonia.

As these plagues declined, however, new problems arose to take their place. Heart disease has become the number one cause of death, and cancer has slowly but insidiously become the number two killer overall and the leading cause of mortality among those aged 35 to 64.[2] In the United States, cancer now kills one in four, up from one in five as recently as the 1950s.[3] Since the early 1970s, with improved diet and reduced tobacco use, mortality from cardiovascular disease has fallen substantially, but cancer has continued to increase.[4] Early in the twenty-first century, cancer—a minor killer a hundred years ago—is expected to become the industrialized world's number one cause of death.[5]

When and where the increases in cancer have taken place gives the first, most general clue to its causes. Geographically, half of all cancer cases occur in the one-fifth of the world's population that lives in industrialized nations.[6] Temporally, the increase in cancer incidence and mortality in all industrialized countries has closely followed the second wave of industrialization—the rise of the chemical and petroleum industries, nuclear technologies, automobiles, and mass electrification—in the twentieth century, not the first wave that took place around the turn of the nineteenth, which centered on the increasing use of iron, steel, coal, and the steam engine. Cancer mortality has increased throughout the entire industrialized world, and it has recently begun to rise in developing countries as they have adopted the technologies and habits of industrialized nations.[7] Clearly something about life in modern industrial society has caused cancer to grow to epidemic proportions.

As I showed in chapter two, over 100 organochlorines cause cancer in experimental animals or humans (see table 2.2). They do so through a variety of mechanisms, including genetic mutations, hormonal changes, immune suppression, and induction of enzymes that make other compounds more carcinogenic. Although there are some differences between the way rodents respond to toxic chemicals and the way humans do, the vast majority of biological processes, including those relevant to cancer, are common to all mammals. Indeed, on an evolutionary scale, primates and rodents are not particularly distant from each other, having diverged from a common ancestor less than 100 million years ago. Evidence of carcinogenicity in rodents is a very strong indication that a substance will probably turn out to cause cancer in humans, and every carcinogen that has been identified first in humans has turned out to do the same in experimental animals.[8]

Consider two facts we have already discussed. First, every person in the modern world is exposed to a myriad of carcinogenic organochlorines. Second, there is no safe dose for these compounds: even the very small quantities to which the general population is exposed have the capacity to alter biological processes in ways that can contribute to cancer. These two facts alone constitute a priori evidence that organochlorines may contribute to the increases in cancer that have taken place since their large-scale introduction into the global environmental began. A growing body of epidemiological evidence, the subject of this chapter, corroborates this hypothesis, suggesting that organochlorine exposure through the air, drinking water, food supply, and workplace make a significant contribution to cancer rates in industrialized society.

Although easier to diagnose than subtle functional impairment, cancer is not much easier to study epidemiologically. Cancer is a complex group of diseases, the development of which begin decades before diagnosis, due to a combination of many causes, including diet, hormones, use of tobacco and alcohol, inherited susceptibility, and exposures to chemicals and radiation. Untangling the contribution of each of these long-past causes is an extraordinarily difficult task. And there are few data on cancer incidence in wildlife, in part because cancer is generally less common among wild species and in part because few wildlife biologists have focused on it. Epidemiology is thus no better equipped to prove or disprove the role of global organochlorine contamination in human cancer incidence than it is to establish conclusive causal links between pollution and functional impairment.

Despite these limitations, an unequivocal body of research shows that workers and farmers exposed to many organochlorine pesticides and industrial chemicals are at increased risk of cancer. There are also a number of studies that show correlations between certain kinds of cancer and markers of organochlorine exposure in the general population, such as the amount of organochlorines in a person's diet and his or her body burden of specific compounds. Combined with the unambiguous results of laboratory studies, this evidence paints a consistent picture that it would be irrational and imprudent to ignore.

I am by no means suggesting that organochlorines are the sole cause of increasing cancer rates. Smoking is undoubtedly the most important cause of cancer of the lung and several other organs. Some cancers, like certain tumors of the liver and soft tissue, can be caused by viruses. Diet and lack of exercise probably increase the risk for a number of types of cancer. Nor are organochlorines the only potentially important causes linked to modern technology: nonchlorinated chemicals like benzene and polynuclear aromatic hydrocarbons (PAHs, which are by-products of automobile and industrial combustion) and radiation from nuclear and medical technologies may also contribute to cancer rates. As the World Health

Organization points out, cancer is a multifactorial disease, and anything that increases the risk of cancer among exposed persons can be considered a cause. That is, organochlorines need not be the sole reason for cancer to be *a* cause of it; organochlorines need only increase the risk of cancer among a significant number of people to be an important factor in the modern cancer epidemic.

A Cancer Epidemic

In 1990 over 8 million people were diagnosed with cancer, and 5 million died of it.[9] In industrialized countries, cancer accounts for over 19 percent of deaths, about three and a half times the proportion it kills in developing countries.[10] In England and Wales, where records have been kept the longest, mortality from cancers of the lung, breast, pancreas, prostate and ovary, as well as leukemia, has steadily increased since the 1920s.[11] Trends are similar in the United States, where data are available since the 1940s.[12] More recent information is available for a much larger number of nations, which shows that total cancer mortality has increased, particularly among men, in virtually all nations since the late 1960s.[13]

Some specific cancers are declining. For instance, stomach cancer mortality has fallen in industrialized nations due to improvements in food storage and preservation. Cervical, colon, and rectal cancers have also decreased substantially in many nations. These declines are more than offset, however, by increases in cancers of other tissues, so there has been an overall increase in cancer mortality. When stomach cancer is excluded, the increase in cancer becomes even more extreme and universal.[14] According to John Bailar, a leading epidemiologist who has reviewed the international data on cancer mortality, "These sets of trends indicate that the worldwide effort to control cancer has failed to attain its primary objective—substantial reduction of the overall cancer death rate—despite some 40 years of intense effort that has been focused mostly on treatment."[15]

Thanks to earlier detection and improved treatment people with some kinds of cancer do live longer now than they used to.[16] This is good news, but it means that trends in mortality actually underestimate the severity of the increase in cancer cases. From 1950 to 1994, U.S. cancer incidence increased even faster than mortality, rising by 54 percent (figure 4.1). The

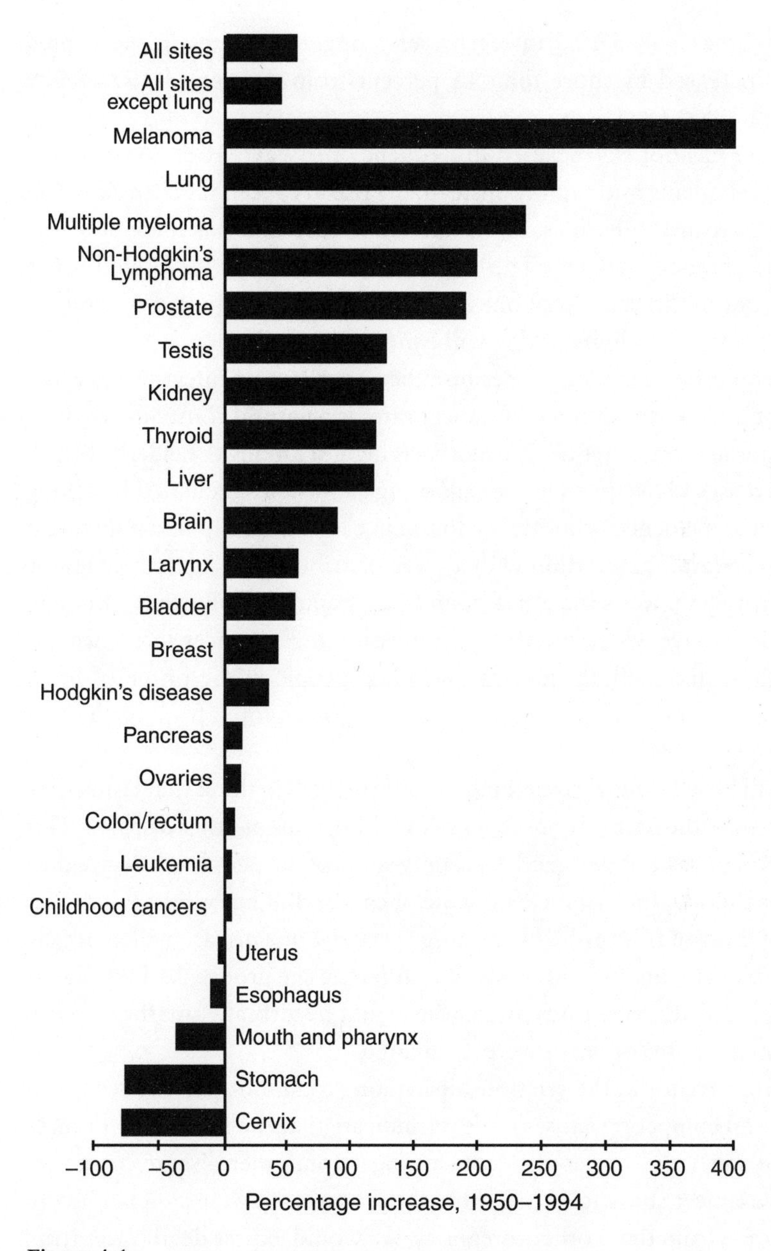

Figure 4.1
Changes in cancer in the United States, 1950-1994. The bars show the age-adjusted percentage change in incidence in each category. (Source: Ries et al., 1997.)

epidemic has not abated in recent years: incidence rates in the United States increased by more than 15 percent from the period 1975–1979 through 1987–1991.[17]

Cancer has not skyrocketed in frequency the way AIDS did, shooting upward suddenly and rapidly. Instead, the rise in cancer has been slow and steady over many decades, an insidious increase that has made it a true scourge of modern society. The lifetime risk of cancer in the United States is now one in three, up from one in four in the 1950s.[18] Sooner or later, 85 million Americans living today will contract cancer.[19]

If people are living longer because they are escaping infectious disease, one might think the increase in cancer rates is a natural consequence of an older population. After all, cancer risk is highest among the elderly. But all the increases I have just mentioned are age adjusted, calculated by taking the cancer incidence or mortality found in each age group and applying it to a "standard" population with a given distribution of ages, a technique that entirely removes the effect of an aging population. Further, although the elderly have experienced the greatest increases, cancer incidence has risen in virtually all age groups, including people in the prime of life.[20] Cancer incidence has even risen substantially in children from birth to 14 years of age.[21]

We all have to die of something, so one might also think that rising cancer rates are the natural consequence of falling rates of other diseases. This may explain part of the trend. Certainly some of the people who once died of tuberculosis, for instance, now die of cancer. But age adjustment takes care of most of this problem, because surviving one disease will generally allow an individual to survive into a different age group. As John Bailar noted, "Age-adjusted rates for cancer would be virtually unaffected even if all other causes of death were eliminated."[22]

Further, trends in the relationships among total mortality, cancer mortality, and competing causes of death indicate that the increases in cancer are not solely due to people "transferring" from other disease categories into the cancer statistics. First, if the cancer category were simply absorbing people from these other categories, we would expect death rates from ill-defined causes and cancer of ill-defined sites to fall. In fact, however, rates in these categories have risen in virtually all industrialized countries during the period of rising cancer mortality.[23]

Second, cancer rates have increased for each group of people born in successive decades, irrespective of the year that the disease emerged. Most epidemiological studies track cancer rates by year of diagnosis or death, so the trends they detect could be due to improved screening, diagnosis, treatment, or changes in other health parameters in society. But two studies, one each in Sweden in the United States, tracked cancer rates based on the decade of birth, an approach that—in the words of one of the reports—reveals the effects of the "unique experiences of a particular generation" rather than role of age and the period when the diagnosis itself was made.[24] The results have been striking. In the United States, men born in the 1940s have twice as much cancer as those born in the 1890s; among women, the increase over the same period is about 50 percent.[25] In Sweden cancer risk at any given age is two to three times higher for those born in the 1950s than for those born in the 1870s (figure 4.2).[26] As the authors of the latter study concluded, "These trends may indicate increasing overall exposure to carcinogenic hazards and predict a continuing rise in the overall cancer incidence."

Some of this increase is due to cigarette smoking. The greatest growth in incidence has been in lung cancer, about 80 percent of which is linked to smoking.[27] But when cancers caused by tobacco use—lung, mouth, larynx, pharynx, and esophagus—are left out of a statistical analysis, incidence of the remaining cancers has still risen substantially. In both the U.S. and Swedish studies, men born in the 1950s have about three times more cancer not related to smoking than men born before the turn of the century. In women, rates have also increased substantially.

Other factors like lack of exercise and dietary intake of fat and cholesterol may also contribute to the increase, but they do not explain all of it. These risk factors are also primary determinants of cardiovascular disease, which began to decline recently. Meanwhile, cancer rates (with the exception of lung cancer in men) continue to increase, indicating that diet and smoking are not the major causes of both diseases. Moreover, studies on the health of farmers, who smoke less and lead more active lives than the general population but who are also exposed to greater quantities of pesticides, solvents, fertilizers, fuels, and exhausts, make the same point. As one might expect from their generally healthy lifestyle, farmers have low rates of cardiovascular disease and lung cancer. If smoking, diet, and

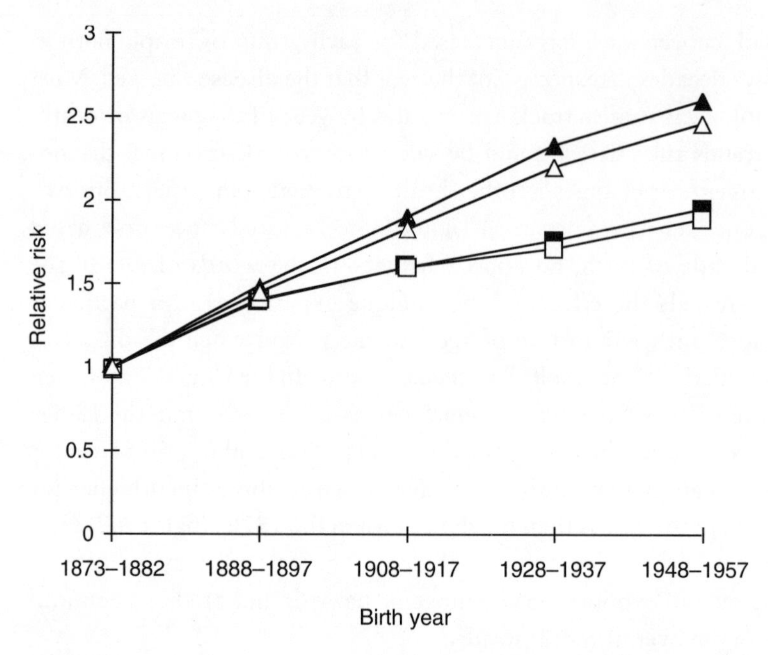

Figure 4.2
Cancer incidence by year of birth in Sweden. Age-adjusted cancer risks relative to the cohort born in the 1870s are shown for women (squares) and men (triangles), for all cancers combined (filled markers) and all except smoking-related cancers (empty markers). Increasing cancer rates in later birth cohorts suggest causes connected to carcinogen exposures rather than changes in diagnostic practices or other causes of death. (Source: Adami et al., 1993.)

exercise were the primary causes of other cancers, too, we would expect their incidence to be low among farmers. In fact, farmers have elevated risks of many of the cancers that are increasing in the general population, including non-Hodgkin's lymphoma, leukemia, multiple myeloma, and cancers of the prostate, brain, and skin.[28]

These patterns suggest that the cancers that are high among farmers and increasing in the rest of the population are unlikely to stem from the dietary and tobacco-related factors that also cause cardiovascular disease, but they may be related to environmental exposure to carcinogenic pollutants. Numerous epidemiological studies, which I discuss later in this chapter, have found specific relationships between farmers' exposure to various organochlorine pesticides and the risk of non-Hodgkin's lymphoma, multiple myeloma, leukemia, and cancers of the prostate, ovary and lung.[29] As one review by three eminent epidemiologists put it, "Could increasing exposures to similar materials in the general population account for the fact that some cancers that are elevated in farmers are also increasing in developed countries? The answers to these questions may provide invaluable indicators of opportunities to prevent cancer."[30]

Skin Cancer

Because the epidemiological data are incomplete, the role of organochlorines in most cancer trends remains controversial. But there is one way that organochlorines are responsible for millions of cancer cases, possibly in the past but certainly in the future, about which there is very little argument. Depletion of the stratospheric ozone layer, caused primarily by organochlorine solvents, refrigerants, and intermediates, is expected to make an immense contribution to the international incidence of skin cancer in the coming century. The impacts will include increases in melanoma, the often deadly cancer of the pigment-producing cells of the skin, caused by severe sun exposure years or decades earlier, and basal and squamous cell carcinomas, the more common and treatable forms of skin cancer, the risk of which is related to cumulative lifetime exposure to ultraviolet radiation from the sun.

We and all other animals that live on land owe our existence to a layer of ozone gas—molecules made up of three oxygen atoms bonded together—that surrounds the globe in the stratosphere at an altitude of 10 to 30 miles. Ozone absorbs solar radiation in the harmful UV range, protecting the earth's surface from wavelengths of light that would otherwise cause severe damage to living organisms, including extreme burning of the skin and eyes, genetic mutations, immune suppression, and skin cancer.

In the 1970s atmospheric scientists discovered that chlorine atoms can break down ozone and warned that the accumulation of long-lived, slowly migrating, volatile organochlorines in the atmosphere could cause catastrophic damage to the ozone. In the 1980s these predictions came true when a hole in the ozone layer—severe thinning to about a third of its former concentration—began to open over Antarctica every winter. Over middle latitudes in the Northern Hemisphere, year-around ozone levels declined by an average of 2 to 4 percent per decade between 1964 and 1994. As more ozone-depleting chemicals reached the stratosphere in the 1980s and 1990s, ozone destruction accelerated further. Since 1979, the average rate of ozone depletion over arctic and temperate latitudes has been twice as high as its average in the period 1964 to 1994, and it has tripled over Antarctica.[31] In 1993—a particularly bad year for the stratosphere, with chlorine levels reaching a new high and the eruption of Mount Pinatubo adding insult to injury—ozone levels at U.S. measuring stations were 13 percent below their pre-1981 levels, and a seasonal hole began to open in the ozone over the Arctic.[32]

With the negotiation of the Montreal Protocol in 1987, the nations of the world agreed to phase out many ozone-depleting compounds. Despite this agreement, however, severe health and ecological damage are expected to occur during the next century due to past and continuing releases of organochlorines into the atmosphere. In the northern midlatitudes, the ozone layer is expected to stabilize at levels 6 to 13 percent lower than natural, depending on the season; in the southern midlatitudes, depletion will be about 11 percent year around. Since 1994, the total stratospheric burden of chemicals that deplete ozone has begun to fall, but the decline is expected to continue only if emissions of other ozone-depleting chemicals, such as the halons and hydochlorofluorocarbons (HCFCs), the primary replacement for CFCs, are also reduced dramatically.[33]

A series of reports by the United Nations Environmental Program (UNEP) has summarized the potential effects of this degree of ozone depletion on human and ecological health. In the northern latitudes, expected ozone losses will allow up to 17 percent more burning rays, up to 32 percent more DNA-damaging rays, and up to 16 percent more cancer-causing rays to reach ground level. This quantity of radiation will cause a steady increase in nonmelanoma skin cancers, resulting in a 25 percent increase in incidence—about 250,000 additional new cancer cases every year. Melanoma, the far more deadly form of skin cancer, is also expected to increase substantially, but the data are inadequate to make a quantitative prediction.[34]

Although ozone depletion will reach its peak in the coming years, it has been taking place for decades; after all, chlorofluorocarbon production began in the 1930s. Ozone depletion may thus already play a role in skin cancer incidence. From 1979 to 1993, the average amount of skin cancer-causing ultraviolet radiation reaching the ground at latitude 45 degrees north—the approximate location of Seattle, Boston, and Milan—increased by 11.1 percent. The increase was 9 percent or more at all latitudes greater than 35 degrees north and south, including virtually all of the United States, Canada, Europe, China, Japan, and New Zealand and much of Australia, Argentina, Chile, and South Africa.[35] There are no data from earlier periods, but there is good reason to believe that ozone depletion began as soon as persistent organochlorines started to reach the stratosphere, decades before 1979.

Meanwhile, the incidence of melanoma has been increasing worldwide, with extremely rapid increases in countries at latitudes where ozone depletion is most severe, including the Nordic countries (6 percent per year) and New Zealand (11 percent per year).[36] In the United States from the 1960s to the 1980s, melanoma incidence quadrupled and nonmelanoma skin cancer tripled. After 1980, melanoma accelerated, tripling again between 1980 and 1995. Some of this rise is probably due to cultural changes that have increased exposure to the sun, like more sunbathing and skimpier clothing.[37] But the increases are common to virtually all nations with light-skinned populations, including many that do not worship the suntan, so it is likely that some of the increase in skin cancer is attributable to ozone depletion.[38]

In addition to cancer, ozone depletion is expected to cause a myriad of other problems that deserve mention. Ultraviolet light damages the cornea and the lens of the eye; ozone depletion is expected to increase the international incidence of cataract, already the most common form of preventable blindness, by over 800,000 cases per year.[39] Ultraviolet light also weakens the immune system, increasing susceptibility to viral, bacterial, fungal, and parasitic diseases. Widespread immune suppression caused by ozone depletion could thus present a profound public health problem, particularly in developing countries, where infectious diseases are already the predominant causes of sickness and death.[40] Ozone depletion is also expected to reduce global food production. UV radiation reduces the productivity of many kinds of plants and marine organisms, including over half of the agricultural crops tested to date; Antarctic ozone depletion has already reduced photosynthetic productivity among phytoplankton by 6 to 12 percent. In a world in which hundreds of millions are already undernourished, reductions in the global and marine food supplies could cause extreme suffering and socioeconomic disruption.[41]

Breast Cancer

"Time trends in the mortality rate of breast cancer indicate an epidemic," according to a report on worldwide breast cancer statistics by epidemiologists at the German federal health office.[42] In 1980, 560,000 women worldwide died of breast cancer;[43] in 2000, more than million women are expected to die of the disease.[44] In virtually all industrialized countries and in all age groups, breast cancer has risen consistently and steadily over the past several decades.[45]

In the United States breast cancer strikes more women than any other type of cancer, accounting for almost one-third of all cancer incidence in females. The lifetime risk of breast cancer is now one in eight, up from one in twenty in 1960. From 1940 to 1982, breast cancer incidence in the United States rose by 1.2 percent annually. In the 1980s it accelerated to an annual increase of 2 percent; later in the decade, the rate increased again to a remarkable 4 percent per year, after which it plateaued. Increased use of mammography may play a role in these recent fluctuations, but the long-term rise in breast cancer incidence appears to be real, not merely an artifact of improved detection.[46]

Hormonal status is by far the most important contributor to a woman's risk of breast cancer. A plethora of studies indicate that increased lifetime exposure to both internal (endogenous) and external (exogenous) estrogens is associated with increased breast cancer risk. For instance, breast cancer risk is elevated in women who experienced early menarche or late menopause, had children late, used oral contraceptives, underwent estrogen replacement therapy, or have high levels of estradiol in their blood.[47] Prenatal estrogen exposure can also cause breast cancer: exposure to DES in the womb predisposes laboratory animals to mammary and reproductive tract cancers when they reach maturity,[48] and girls born to mothers who have high estrogen levels during pregnancy suffer increased breast cancer risk when they become adults.[49] It does not take a powerful estrogen to cause these effects; even very weak estrogenic substances can lead to cancers of the reproductive tract when exposure takes place in utero.[50]

Fat in the diet may play a role in breast cancer risk. Nations with high per capita fat consumption tend to have high breast cancer rates, and a high-fat diet can promote the growth of mammary tumors initiated by other carcinogens in laboratory animals.[51] Further, some epidemiological studies indicate a relationship, albeit a weak one, between fat intake and breast cancer risk.[52] Whether fat itself contributes to breast cancer remains controversial, however. National per capita fat intake is almost always associated with the degree of industrialization and other environmental and cultural changes that come with it; correlations of cancer rates with fat intake may very well indicate an underlying cause other than the fat itself. Most important, no studies of dietary fat and cancer have considered the effect of chemical contaminants in the fat. An apparent connection between fatty diet and cancer could indicate that bioaccumulated chemicals in animal fat—not the fat per se—lead to higher cancer risk.[53]

Inheritance also makes some women more susceptible to breast cancer than others. For instance, having a mother or sister with breast cancer increases a woman's risk by a factor of 1.5 to three.[54] Despite all the attention lavished on several recently discovered "breast cancer genes," however, inherited mutations account for only 2.5 to 10 percent of all breast cancer incidence.[55] Although one might think that heredity would explain the striking differences in breast cancer between countries, it does not. Breast cancer rates among women in Japan, for instance, are only about one-fourth those in American women, but Japanese migrants to the United States adopt the rates of their new country within two to three generations.[56] The bulk of breast cancer is thus associated with environmental or cultural factors of one sort or another, not genetics.

Each of these established causes of breast cancer affects a fairly small portion of the population and is associated with a rather moderate increase in risk. Women with no recognized risk factors at all still suffer an appreciable incidence of breast cancer.[57] Taken together, a history of breast cancer in the family, reproductive factors, and fatty diet explain only about 30 percent of all breast cancer cases. Including socioeconomic status, which is also related to breast cancer risk, brings the attributable fraction to just over 40 percent.[58] Well over half of all breast cancer thus remains unexplained by reference to genetics and lifestyle.

The gradual rise in breast cancer incidence is also inexplicable in terms of shifts in established risk factors. It began long before oral contraceptives and estrogen replacement therapy became available. Changes in the age at which women bear children have been neither large nor consistent enough to account for the steady increase in breast cancer seen in each successive cohort of women since the late nineteenth century.[59] Earlier average age at menarche could explain some of the increase, but these changes are also too small to fully explain the long-standing increases in breast cancer incidence;[60] further, as I showed in the previous chapter, earlier menarche cannot be explained by changes in diet or other recognized factors and may itself have been caused by exposure to endocrine-disrupting chemicals in the environment. Fat consumption has actually decreased in the United States since the early 1950s and therefore cannot account for the breast cancer epidemic.[61]

What then has caused the steady increase in breast cancer over the past five or six decades, along with the majority of new cases today? Exposure to the products and by-products of modern technology may play a important role. Some mutagenic chemicals[62] and ionizing radiation from nuclear weapons and power generation,[63] for instance, are known to increase breast cancer. There is also some preliminary evidence that exposure to electromagnetic fields and electric light at night may be breast cancer risk factors.[64]

Modern exposure to man-made estrogenic compounds, called xenoestrogens, are leading suspects in the breast cancer epidemic.[65] Xenoestrogens include pesticides, industrial chemicals, estrogens given to livestock that end up in dairy and beef products, and pharmaceuticals like DES, oral contraceptives, and estrogen replacement therapy. If lifetime and prenatal exposures to the body's natural estrogens are associated with breast cancer, then it makes sense that unnatural estrogens could also increase cancer risk. Molecular studies support a key role for hormonal changes in increasing breast cancer rates. Two studies have found that the portion of breast cancers in the United States that are estrogen responsive has increased since the 1970s, and the average quantity of estrogen receptors in breast tumors has quadrupled.[66] Thus, while changes in reproductive and dietary behavior do not seem to explain increasing breast cancer rates, externally-induced hormonal changes seem to offer at least a partial explanation. According to the authors of one of these studies, "A change in hormonal events that promote breast development might contribute to both the rising incidence of the disease and a modification of tumor hormonal characteristics."[67]

Among the chemicals that may contribute to breast cancer, organochlorines play a prominent role. At least 23 organochlorines have been specifically shown to cause mammary cancer in laboratory animals (table 4.1),

Table 4.1
Organochlorines that cause mammary cancer in laboratory animals

Pesticides	Source
Aldrin	Fitzhugh et al. 1964
Atrazine	CCRIS 1998
Chlordane	CCRIS 1998
Dieldrin	Fitzhugh et al. 1964
DDT	Scribner et al. 1981
Dichlorvos	CCRIS 1998, Wolff et al. 1996
Industrial and other chemicals	
Benzidine dihydrochloride	CCRIS 1998
Clonitralide [5-chloro-n-(2-chloro-4-nitrophenyl)-2-hydroxybenzamide]	Wolff et al. 1996
3,3′-Dichlorobenzidine	CCRIS 1998
3,3′-Dimethoxybenzidine dihydrochloride	CCRIS 1998, Wolff et al. 1996
3,3′-Dimethylbenzidine dihydrochloride	CCRIS 1998, Wolff et al. 1996
2-Chloro-1-phenylethanone	Wolff et al. 1996
Dibromochloropropane [DBCP]	CCRIS 1998
1,2-Dichloroethane [ethylene dichloride]	Wolff et al. 1996
1,1-Dichloroethylene [vinylidene chloride]	CCRIS 1998
1,2-Dichloropropane	Wolff et al. 1996
Dichloromethane [methylene chloride]	CCRIS 1998, Wolff et al. 1996
Methylene bis-(chloroaniline)	CCRIS 1998
2,3,7,8-TCDD [dioxin]	Brown et al. 1998
o-Toluidine hydrochloride	Wolff et al. 1996
1,2,3-Trichloropropane	Wolff et al. 1996
Uracil mustard [2,6-dihydroxy-5-bis (2-chloroethyl)aminopyrimidine]	CCRIS 1998
Vinyl chloride [chloroethylene]	CCRIS 1998

despite the fact that the standard carcinogenicity assay is poorly designed for detecting mammary carcinogens.[68] The list includes compounds with a wide range of uses, sources, and chemical structures. Some, like DDT, are restricted in many countries, while others, like vinyl chloride and atrazine, are still in widespread use. Some are mutagens that initiate cancers, while others are potent promoters that accelerate the growth of tumors triggered by radiation and other chemicals.[69] Because only a fraction of chemicals in commerce have been tested for cancer of any type (and even fewer for breast cancer in particular), a much large number of organochlorines is likely to turn out to contribute to mammary cancer in the long run.

Organochlorine exposure during development can have particularly profound effects on breast cancer. Vinyl chloride, for instance, causes mammary tumors in adult female rats, and the incidence is even higher in their offspring.[70] Even dioxin, which reduces mammary cancer when given to adult rodents, predisposes the animals to mammary tumors when exposure occurs before birth. In a 1998 study, researchers at the University of Alabama gave pregnant rats a single very small dose (1 millionth of a gram per kilogram of body weight) on day 15 of pregnancy; their daughters experienced a twofold increase in mammary tumors when they reached adulthood.[71]

In humans, several studies have found elevated breast cancer rates among women with high exposures to industrial chemicals, including organochlorines, in their workplace or communities. In some cases, the risk is linked to specific organochlorines. Most notable is vinyl chloride, which was discovered in the early 1970s to have caused extraordinarily high rates of liver cancer, lung cancer, and brain cancer among male workers in the chemical and plastics industries.[72] In 1977 researchers found that women workers exposed to vinyl chloride in 17 PVC plastic fabrication plants in the United States suffered elevated mortality from breast cancer.[73] More recently, atrazine has been implicated in breast cancer in Kentucky, where counties with moderate or high atrazine use and/or groundwater contamination have higher breast cancer incidence than counties with very low atrazine exposure. The increase in risk is modest (15 to 22 percent), but it is statistically significant, and it persists after adjustment for confounding factors like race, income, education, and delayed childbearing. Most important, the magnitude of the risk increases with the severity of atrazine exposure; when a study reveals this kind of dose-response relationship, it greatly increases our confidence that the link is real rather than spurious.[74]

Even dioxin, which can actually inhibit breast cancer in adult laboratory animals,[75] has been linked to breast cancer in humans. Three studies of the same cohort of women workers in an organochlorine pesticide plant in Hamburg, Germany, have found that breast cancer mortality among dioxin-exposed women was two to three times higher than expected; the risk increased with the duration of employment in the facility and was specifically related to work in those processes in which the opportunities for dioxin exposure were greatest.[76]

How should we interpret the contradictory findings in experimental rodents and occupationally-exposed women? One group of scientists has suggested the plausible hypothesis that short- and long-term dioxin exposures may have opposite effects: relatively high doses of dioxin have an antiestrogenic effect in breast tissues and protect against breast cancer in laboratory animals, but longer-term exposure to dioxin—which can suppress the immune system, alter estrogen metabolism to produce carcinogenic compounds, and reduce circulating levels of retinoids, which are known to have a powerful inhibitory effect on breast tumors—may overwhelm the local estrogenic effect, increasing the total risk of breast cancer.[77]

Breast cancer risk has also been linked to chemical mixtures of chlorinated and nonchlorinated pollutants in circumstances where the contribution of individual chemicals cannot be determined. For example, mortality from breast cancer is 63 percent higher among female chemists in the United States than among the general U.S. population, and cancers of the ovary, stomach, pancreas, and immune system are higher too.[78] Studies of occupation and mortality in thousands of blue-collar workers in New Jersey and New York have found that employment in several chemical-intensive industries, including the pharmaceutical, electrical equipment manufacturing, and printing industries, is associated with elevated risks of breast cancer.[79]

Environmental exposures also increase breast cancer risk. A nationwide study by EPA researchers found that counties with waste sites were 6.5

times more likely to have elevated breast cancer rates than countries that did not have such sites.[80] On Long Island, New York, the risk of post-menopausal breast cancer is 60 percent higher in women who have lived within a kilometer of a chemical plant than in those who have not; the more chemical plants in the area of a woman's residence, the higher her breast cancer risk.[81] And a study of breast cancer rates in New Jersey found a correlation between a woman's risk of breast cancer and the density of hazardous waste disposal sites in the area of her home.[82]

Together these studies establish that higher-than-average exposure to synthetic chemicals, both mixtures and specific organochlorines, is associated with breast cancer, but they say little about the lower levels found in the ambient environment. Two other kinds of evidence add to the case that organochlorine contamination may contribute to breast cancer in the general population. First, the health study of Baltic fishermen's wives, discussed in the previous chapter, also looked at cancer mortality. These women constitute a natural experiment in the relationship of low-level organochlorine exposure and cancer, because—as one would expect of fishermen's wives—they consume about twice as much fish from the relatively contaminated Baltic than the rest of the Swedish population does. In a group of about 2,000 women married to Baltic fishermen, mortality from breast cancer was about 30 percent higher than in a matched control group of women from the general population in the same region, and there were no differences between the two groups in other recognized breast cancer risk factors. Even more significantly the Baltic fishermen's wives had breast cancer mortality 68 percent higher than the wives of fishermen from the less polluted Swedish west coast, who were virtually identical in all variables except for the level of organochlorines in the fish they ate.[83]

The second line of evidence supporting a link between organochlorine exposure and the incidence of breast cancer is a statistical relationship, found in a number of studies, between a woman's breast cancer risk and her body burden of specific organochlorines. Interpreting the research on this subject is difficult, because studies with various designs have come to different conclusions. Beginning in the early 1990s, studies of women in Connecticut,[84] New York City,[85] Long Island,[86] and Finland[87] revealed that concentrations of organochlorines in the fat or blood of women with breast cancer are higher—by about 50 percent or so in most studies—than in the bodies of comparison groups of women without breast cancer. A number of organochlorines, including beta-hexachlorocyclohexane, hexachlorobenzene, heptachlor, oxychlordane, hexachlorobenzene, and trans-nonachlor, were elevated in one or more of these studies, but the differences were generally statistically significant only for DDE—the long-lived metabolite of DDT—and PCBs. Women with the highest body burdens of these compounds were at a two- to fourfold higher risk of breast cancer than women with the lowest levels of these compounds.

One explanation for these results might be that cancer causes a woman to lose weight and burn her fat stores, transferring organochlorines from adipose tissue into her blood. In each of these studies, however, samples were taken very early in the development of cancer—or, in the New York City study, months *before* cancer was diagnosed—so higher pollutant levels are not an effect of the disease itself. Another possibility is that organochlorine levels are not a cause of cancer but are correlated with some other factor that is the true cause of increased risk among women with higher body burdens. But these studies, to one degree or another, all took potential confounding factors into account. In the most rigorous—again the one in New York—all major risk factors were adjusted or controlled for, including differences in body mass, age at menarche and first full-term pregnancy, history of lactation, history of breast cancer in the family, and use of tobacco and alcohol.

On the other hand, two large studies have not found a link between a woman's body burden of these compounds and her risk of breast cancer. A 1997 case-control study of women in several European countries did not find a link between DDE levels in fat and breast cancer risk; this study did not, however, adjust for important confounding factors like history of lactation and number of children born.[88] Later that year, researchers at the Harvard Medical School reported on a comparison of PCB and DDE levels in the blood of several hundred women who donated blood in 1989 or 1990 and were subsequently diagnosed with breast cancer.[89] They found no statistically significant relationship between either pollutant and breast cancer risk, even when reproductive risk factors were taken into account.[90] Finally, in 1998, Danish researchers published the results of a long-term study, the largest one to date, of women who had given blood in 1976 and developed breast cancer during the next 17 years. This study found that women with levels of dieldrin, an estrogenic chlorinated insecticide, in the highest 25 percent had more than twice as much breast cancer as women in the lowest quartile. It did not find statistically significant relationships for several other organochlorines, including PCBs and DDT.[91]

We thus have a confusing body of evidence. Five studies, including two large, very well-designed investigations, have linked the body burden of specific organochlorines to breast cancer, but two others—one also of very high quality—have not replicated these findings. Further, even the positive studies do not always find the same organochlorines to be associated with breast cancer risk. How should we interpret these contradictory results? First, we should recognize that not finding a link is not the same as demonstrating that one does not exist. Even the rather large Harvard study could not rule out an increase in risk of 40 percent or less among women with elevated levels of DDE. The negative studies do not confirm the positive ones, but they do not refute them either.

Second, a number of other reports with mixed results suggest several reasons why a link may not always be evident. For instance, a study by Canadian researchers showed that organochlorines increase the risk of some but not all kinds of breast cancer. Only some breast tumors contain the estrogen receptor (ER) and thus proliferate in response to estrogen; the authors reasoned that if estrogenic organochlorines were contributing to breast cancer risk, the effect should be seen only in patients with ER-positive tumors. They obtained samples of fatty tissue from 37 women who were undergoing breast biopsies; 20 women turned out to have early-stage breast cancer, and 17 had benign breast disease. In women with ER-negative tumors, there was no difference between the organochlorine levels in cancer cases and controls. In the fat of women with ER-positive cancers, however, DDE levels were almost three times higher than in women without cancer, and one PCB congener (PCB-99) was about 50 percent higher. (Levels of HCB, oxychlordane, transnonachlor, and some other PCB congeners were also elevated, but the results were not statistically significant.) ER-positive women with the highest DDE levels had a ninefold higher risk of breast cancer than women with low DDE levels.[92] If organochlorines increase breast cancer risk only in some women, a methodology that lumps together women on whom these compounds have no effect with those on whom they do will radically weaken a study's capacity to detect a statistically significant relationship.

Another study suggests that organochlorines may interact with diet or other ethnic factors to modulate breast cancer risk in different ways in different groups of women. In the late 1960s, medical researchers in California collected blood from several thousand women as part of a large health study. More than two decades later, breast cancer researchers found the archived samples and realized their potential to shed light on the relationship between organochlorine body burden and the long-term risk of breast cancer. From the donors, the researchers assembled a study group of 150 women who had developed breast cancer (50 cases each of African American, white, and Asian descent), who were then matched to an equal number of women who did not develop breast cancer. The blood that each had donated was then analyzed for concentrations of DDE and PCBs. Overall, there was no relationship between pollutant levels and breast cancer risk, a result that was widely cited and reported as refuting a link between organochlorines and breast cancer.[93]

It turned out, however, that there was a marked difference among the racial groups. In Asian women, who had the lowest cancer incidence, there was no apparent relationship between breast cancer and organochlorine levels. But when these women were left out of the analysis, there was a strong positive relationship between increasing DDE levels and risk of breast cancer. Among African American women, who had the highest risk of breast cancer, those with the highest DDE concentrations were at an almost fourfold increased risk compared to those with lower pesticide body burdens. Asian women are thought to have a lower breast cancer risk due in part to the antiestrogenic effects of a soy-based diet, which contains natural compounds that may protect against the harmful impacts of both endogenous and xeno-estrogens.[94] If diet and other factors modify the effects of pollutant exposure, combining heterogeneous populations in a single study will obscure positive relationships that may exist in smaller subgroups.

Finally, in early 1998, a large study of women in Buffalo, New York, had mixed results that were particularly interesting.[95] Levels of DDE, PCBs, and mirex were higher in the blood of women with breast cancer than in a matched comparison group of women without the disease, but the increase was not statistically significant. Closer analysis, however, revealed specific kinds of noise in the data that was obscuring the association between organochlorines and cancer. Although the relationship with the level of total PCBs was weak, breast cancer risk was strongly related to concentrations of the subgroup of PCBs that have fewer chlorine atoms on the molecule, a category that includes most of the congeners that produce estrogenic and dioxin-like effects. Use of the catch-all parameter "total PCBs"—on which almost all of the previous studies, both negative and positive, have relied—masked the relationship of breast cancer with a subset of PCBs. Further, when only women who had never lactated were considered, PCBs and mirex levels were strongly elevated in women with breast cancer; breast cancer risk was about three times higher in the one-third of women with the highest PCB concentrations compared to those in the lowest third. Because lactation reduces a woman's body burden and protects against breast cancer, blood levels of organochlorines later in life may not represent total lifetime exposure in women who breast-fed as accurately as they do in women who never nursed their babies. Including women with different breast-feeding experiences thus adds more noise to the analysis and undermines the reliability of a body burden as a marker of long-term exposure to organochlorines.

In summary, there are eight studies with positive or partially positive results and two more with negative results. The Canadian, California, and Buffalo studies, however, do more than just add "votes"; they show just how difficult it can be for epidemiology to untangle the role of organochlorines from that of other causes. In these investigations, the initial finding of no statistically significant relation between body burden and breast cancer risk in the entire study group turned out to be an artifact of a focus on the wrong chemicals as markers or the inclusion of groups heterogeneous for estrogen receptor status, reproductive history, and diet or other factors related to ethnicity. If the inability to document an overall link in these partially positive studies is due to the limits of the epidemiological analysis used rather than a true lack of relationship, the same may be true for the other negative studies, too.

It is possible, of course, that there is no causal link between cancer and the body burden of DDT, PCBs, and other organochlorines; the eight positive or partially positive studies may all be erroneous. This position, however, ignores more evidence than it explains, and it fails to consider the limits of epidemiology, which can account for the results of the two negative studies. To dismiss positive evidence because not all studies corroborate it betrays an excessive confidence in the ability of epidemiology to reveal the causes of disease without ambiguity.

Other Hormonal Cancers

Breast cancer is just one of a number of cancers that are caused or modulated in large part by hormones. Tumors of hormonally-responsive organs, in fact, account for 51 percent of all new cancer cases in women and 36 percent in men in the United States.[96] Of particular interest are cancers of the testis, prostate, and uterus, all of which have international incidence patterns similar to that of breast cancer: high in North America and Europe, low in Japan and developing countries.[97] Since the 1960s the incidence of testicular cancer has risen in numerous industrialized nations by two to four times.[98] Prostate cancer has also increased markedly throughout Europe and North America; although some of the increase is probably due to improved detection of tumors that have not spread, improved detection cannot explain the entire increase.[99] Uterine cancer increased in Europe from 1960 to 1985; in the United States, it rose consistently until 1975 and then fell as the use of estrogen replacement therapy declined and hysterectomy became more prevalent.[100]

The most important identified risk factor for these cancers is exposure during development or adulthood to steroid hormones, particularly estrogens.[101] Prenatal exposure to diethylstilbestrol, for instance, increases the risk in adulthood of testicular, prostate, and uterine cancer;[102] the latter is even increased in the *granddaughters* of mice exposed to DES.[103] Additional exposure to hormones during adulthood further increases the likelihood of tumors in these tissues.[104] Based on these findings, it is quite plausible—a "logical connection" in the words of one group of cancer researchers[105]—that exposure to other synthetic chemicals that disrupt the endocrine system during development or adulthood could have profound impacts on the rates of these hormone-dependent cancers.

There is less evidence on the role of organochlorines in these cancers than there is for breast cancer. In the laboratory, the data on the potential of synthetic chemicals to cause cancers of the testis and prostate are limited, because the most common species of experimental rodents are very poor models for studying these cancers.[106] No organochlorines have been specifically linked to cancers of the prostate, but trichloroethylene and vinyl chloride cause testicular cancer. Ethylene dichloride, atrazine, chloroethane, chlorotrifluoroethane, trichloropropane, and several other organochlorines, including some also linked to breast cancer, cause tumors in the uterus.[107]

Few occupational studies have specifically focused on these cancers either. Testicular cancer strikes mostly very young men, so workplace exposures are likely to be less important than developmental ones. Studies of uterine cancer are also scarce, because most chemical-intensive industries have not had many long-term female employees. The best information comes from studies of farmers. At least ten investigations have shown that farmers suffer statistically significant increases in the risk of prostate cancer, and some of the risk has been specifically linked to organochlorine herbicides.[108] For instance, a case-control study of men with prostate cancer in Canada found that exposure to herbicides—primarily chlorophenoxy herbicides, such as 2,4-D—was associated with a more than twofold increased risk of prostate cancer.[109] In women, atrazine exposure is associated with an almost threefold increase in ovarian cancer risk.[110] Some but not all studies have found elevated rates of testicular cancer, with up to sixfold increases, among farmers.[111]

There is some evidence linking dioxin and Agent Orange to these cancers. Several studies have found that men involved in the manufacture of chlorophenoxy herbicides have elevated risks of prostate cancer,[112] which led the U.S. National Institute of Medicine to conclude that there is "suggestive evidence" linking Agent Orange and its contaminants to prostate cancer.[113] Testicular cancer is twice as high among veterans who served in Vietnam, and thus were potentially exposed to Agent Orange, than among soldiers who served elsewhere during the same period.[114] And U.S. military dogs that served in South Vietnam had nearly twice as much testicular cancer as dogs that served during the same period in other regions and were not exposed to Agent Orange.[115]

There are no studies, positive or negative, that have investigated whether exposure to organochlorines in the ambient environment is associated with these hormonal cancers. We do know, however, that sons born to mothers who later develop breast cancer have a 2.9-fold increased risk of testicular cancer. This finding suggests that at least part of the rise in testicular cancer may be attributable to the same hormonally mediated changes that have produced the modern epidemic of breast cancer.[116] The link between breast and testicular cancers points out the need to avoid viewing trends in various diseases and conditions in isolation. The evidence we have pertains to a suite of diseases and conditions—cancers of the breast and other hormonally responsive organs, falling sperm counts, endometriosis, and abnormal development of the reproductive system—that all have well-established hormonal causes. Although each disease is ultimately unique, we know from experimental studies that disruption of the endocrine system is a primary cause of all of these conditions in laboratory animals and that organochlorines can cause many of them. Epidemiological work confirms the importance of hormonal disruption in all these effects and of pollutant exposures in many. If we take each individual disease in isolation, the evidence leaves large blank spots on the canvas. If we see them as indicators of a general disruption of aspects of health and development that are controlled by hormones, then the body of experimental and epidemiological evidence gives cause for concern.

Immune Cancers

One of the jobs of the immune system is to fight off cancer by detecting and destroying tumor cells. Because the body is always producing millions of new immune cells, however, the immune system itself is prone to developing cancer. When natural controls on the division and differentiation of immune cells are lost, T-cells, B-cells, and their precursors can give rise to tumors. The most common type of immune malignancy is lymphoma—cancer of mature immune cells in the lymph nodes and other tissues of the body. There are two kinds of lymphoma: Hodgkin's disease, characterized by the presence of a special kind of very large malignant cell, and all the rest, which are collectively referred to as non-Hodgkin's lymphomas (NHL). Next in frequency comes leukemia, a cancer of white blood cell precursors, which causes the bloodstream to be inundated with huge numbers of these cells. The least common immune cancer is multiple myeloma, a malignancy of B-cell precursors in the bone marrow, which leads to an extremely painful destruction of bones throughout the body.[117]

Cancers of the immune system account for about 8 percent of all cancer cases in the United States; they are even more important in younger people, making up about one-fourth of all cancer before age 35.[118] As a group, these cancers are increasing internationally at a consistent but moderate rate, but NHL and multiple myeloma are rising very rapidly. From 1945 to 1985, the incidence of NHL increased by 152 percent in the United States. In the 1980s, only skin cancer and prostate cancer rose faster, leading one epidemiologist to call NHL "an emerging epidemic."[119] Myeloma is also increasing steadily, though less rapidly than NHL, with an 11 percent increase in the United States in the 1980s and similar trends in most other nations.[120] The increase in Hodgkin's disease in the United States was more moderate—about 21 percent from 1950 to 1988—and rates in most nations are now relatively stable.[121] In contrast, leukemia rose by only 4 percent during this period and decreased from 1975 to 1991, but the incidence among children continued to increase, making it the most common and one of the fastest-growing cancers among those under age 14.[122]

The causes of these increases are not well understood. Immune suppression is clearly a risk factor for immune cancers, because people who have AIDS or are being treated with immunosuppressive drugs have an extraordinarily high incidence of these malignancies, espcially NHL.[123] Exposure to certain viruses, including the Epstein-Barr virus (EBV), increases the risk of lymphomas, but neither EBV nor the human immunodeficiency virus (HIV) that causes AIDS can explain upward trends in the disease.[124] Atomic bomb survivors and workers at the Hanford nuclear weapons plant in Washington State have an increased risk of multiple myeloma, but the overall role of radiation in this and other immune cancers remains unclear.[125] Oil refinery workers have an increased risk of immune system cancers, suggesting that petrochemicals may also be a risk factor.[126]

Exposure to organochlorines may play an important role. Of the organochlorines that have been tested in laboratory animals, 22 cause leukemia or lymphomas (table 4.2). On the list are a number of chemicals to which millions of people are regularly exposed, including the pesticides atrazine, mirex, and dichlorvos; the solvents perchloroethylene and meth-

Table 4.2
Organochlorines that cause immune cancers in laboratory animals

Lymphomas	Leukemia
Atrazine	Atrazine
Chlorinated paraffins	1,1-Dichloroethylene [vinylidene chloride]
Dichloroacetylene	Dichloromethane
1,2-Dichloroethane	Dichlorvos
2,6-Dichlorobenzonitrile	3,3′-Dimethylbenzidine dihydrochloride
4,4′-Methylenedianiline dihydrochloride	4-Hydroxyaminoquinoline-1-oxide hydrochloride
Nitrogen mustard hydrochloride [di(chloroethyl)methylamine hydrochloride]	Mirex
	Tetrachloroethylene [perchloroethylene]
	2,4,6-Trichlorophenol
Nitrogen mustard N-oxide	Tris(chloroethyl)phosphate
2,4,6-Trichlorophenol	
2,3,7,8-TCDD	
Uracil mustard [5-(di-2-chloroethyl) aminouracil]	

Source: CCRIS 1997. Synonyms are shown in brackets.

ylene chloride; the plastic feedstocks ethylene dichloride and vinylidene chloride; and the ubiquitous environmental pollutants dioxin and chlorinated paraffins. Given the well-documented link between weakened immunity and NHL, we should presume that all the other organochlorines that suppress the immune system—many pesticides, solvents, PCBs, and so on—may also have the capacity to cause lymphomas.

People in a number of occupations with high exposures to organochlorines and other synthetic chemicals are at increased risk of immune cancers. For instance, professional chemists, pharmaceutical workers, and chemical production workers suffer elevated death rates from leukemia and lymphomas.[127] Workers exposed to organochlorines and other chemicals in a Swedish chemical plant are at increased risk of NHL and multiple myeloma,[128] and several studies have found that workers in the pulp and paper industry—a major source of organochlorine by-products as well as other carcinogens—have elevated mortality from leukemia, Hodgkin's disease, and NHL.[129] Overall, exposure to carcinogens in the workplace is thought to cause 4 to 11 percent of all non-Hodgkin's lymphoma, or as many as 5,900 cases each year in the United States alone.[130]

In all of these industries, workers were exposed to mixtures of many kinds of chemicals, so pinpointing the specific agent—if there is one—is impossible. In other studies, the role of one or a few organochlorines has been more clear-cut. For instance, several studies have found that both male and female workers exposed to vinyl chloride in the chemical industry suffer considerable excesses of fatal leukemia and lymphoma.[131] And in chemical industry workers exposed to allyl chloride and epichlorohydrin—carcinogens used in the manufacture of plastics—leukemia mortality is five times higher than expected.[132]

Several studies suggest that perc, which causes leukemia in the laboratory, also causes immune cancer in people. World War II servicemen whose job was to treat uniforms with a perc solution to protect against poison gas attacks are more likely to die of leukemia than men in the same companies who did not handle perc directly. The problem appears to be specifically related to perc, because these men are also more likely to die of leukemia than those who held the same job in other companies that used a water-based solution instead of perc.[133] Women workers exposed to perc for more than one year during maintenance work at an air force base experience elevated risks of NHL and multiple myeloma,[134] and several studies have found increases in lymphomas and leukemia among workers in dry cleaning shops, where perc is the most common solvent.[135] Some other studies have not found a statistically significant association between perc and lymphoma, but all of these were too small to detect a moderate increase.[136] Occupational exposure to perc has also been repeatedly linked to substantial increases in cancers of the esophagus, bladder, kidney, and liver.[137]

Some of the most convincing evidence demonstrates that farmers exposed to organochlorine pesticides suffer excess rates of immune cancers. Several dozen studies have shown that farmers are at increased risk of all four immune malignancies.[138] Organochlorines are not the only possible cause—farmers are exposed to many kinds of potentially carcinogenic chemicals and animal viruses—but several studies have established specific links between immune cancers and individual organochlorines. For instance, studies of farmers in the United States, Canada, and Sweden have found two- to eightfold increases in NHL in farmers who had been exposed to 2,4,5-T and 2,4-D, compared to those who had not used these chemicals.[139] In several of these reports, the risk rose with the degree of herbicide exposure, strengthening the case that the chemicals themselves were the true cause of the increase. Immune cancers are also increased in migrant and seasonal farmworkers, county extension agents, and forestry, sawmill, and leather workers, all of whom tend to be exposed to chlorophenoxy herbicides.[140] Corroborating these findings, workers involved in the manufacture of these herbicides at four German chemical plants have a threefold increase in mortality from NHL,[141] as does a large international cohort of workers in agriculture, forestry, and the chemical industry who were exposed to 2,4,5-T, pentachlorophenol, and dioxin.[142] Based on these kinds of studies, the U.S. National Institute of Medicine concluded in 1993 that there was "sufficient" evidence linking 2,4-D, 2,4,5-T and their contaminants to the risk of NHL and Hodgkin's disease.[143]

Other organochlorine pesticides have also been linked to immune cancers. For instance, one large case-control study of men in Iowa and Minnesota found that leukemia risk was more than twice as high among farmers who reported exposure to DDT, dichlorvos, or methoxychlor than among those who did not use these chemicals.[144] A follow-up of men with NHL in the same states also found increases in farmers who had personally handled DDT, chlordane, dichlorvos, lindane, and toxaphene, and the risks were even higher among those who did not use protective clothing.[145] Studies of farmers in Kansas and in Nebraska have linked atrazine to two- to threefold increases in NHL;[146] in Sweden, the risk of multiple myeloma is elevated among those exposed to DDT and chlorophenoxy herbicides.[147] And a number of studies have linked pesticide exposure to childhood cancers, leukemia in particular. Especially high leukemia risks were found for children exposed during gestation or early childhood to "no-pest strips," in which the active ingredient is usually dichlorvos, a carcinogenic organochlorine insecticide.[148]

Environmental contamination with organochlorines has also been linked to immune cancers in a number of studies. In Seveso, Italy, 15 years after an explosion at a chemical plant released large quantities of dioxin-contaminated chlorophenoxy herbicides, their feedstocks, and dioxin into the local environment, mortality from leukemia, Hodgkin's disease and multiple myeloma is three to six times higher in the contaminated zone than in the general population.[149] And in a Finnish community where drinking water is contaminated with chlorophenols, NHL incidence is also extremely high.[150]

More routine exposures at home may also contribute to immune cancers. 2,4-D is the most common herbicide used on lawns and gardens; 10 percent of Americans use commercial lawn services, and 20 percent apply lawn pesticides themselves.[151] There are no studies of the human health effects of home exposure to chemical herbicides, but researchers at the National Cancer Institute have studied the incidence of immune cancers in dogs, who, like children, are especially likely to be exposed to chemicals when they run and play on the lawn. This case-control study of hundreds of domestic dogs with non-Hodgkin's lymphoma and a comparison group of animals without the disease found that pet dogs whose owners used 2,4-D on their lawns four or more times per year had twice as much NHL as dogs with no 2,4-D exposure, and the magnitude of the risk was directly related to the frequency of herbicide use.[152]

Organochlorine solvents have also been linked to immune cancers. In Woburn, Massachusetts, children exposed to drinking water contaminated with trichloroethylene, perc and other chlorinated solvents suffer a twelvefold increase in leukemia incidence.[153] The New Jersey Department of Health has found that the incidence of leukemia and NHL is substantially elevated in women served by water supplies with elevated levels of perc and trichloroethylene.[154]

In the 1960s and 1970s, thousands of municipalities installed PVC liners in their cement water supply pipes to improve the taste of tapwater. The plastic was applied to the pipe using perc as a solvent, which the authorities expected to evaporate without a trace. By the late 1970s, however, at least six states in New England discovered that substantial quantities of perc remained in the pipes and were slowly leaching into drinking water. Researchers at the Boston University School of Public Health investigated the issue on Cape Cod, Massachusetts, where large amounts of vinyl-lined pipe had been installed. In a 1993 study, they reported that exposure to perc in drinking water was associated with a doubled risk of leukemia. Among people with the highest perc exposures, leukemia was increased eightfold, and bladder cancer was elevated too.[155]

What about low-level organochlorine exposure from the food supply? Three lines of evidence suggest a role in immune cancers. First, although there are no studies of these malignancies in wildlife, there is a host of evidence, discussed in the previous chapter, that wildlife populations are suffering severe and widespread suppression of the immune system, apparently associated with contamination of the food chain by dioxins, PCBs, and other persistent organochlorines. Because immune suppression is known to be a risk factor for lymphomas and myeloma—and because occupational exposure to the same organochlorines has been linked to these cancers in people—this evidence raises the possibility that environmental levels of these chemicals may play a role in the incidence of immune cancers as well.

The study of Baltic fishermen supports this idea. Men who ate larger amounts of fish from the Baltic experienced three times more multiple myeloma than the general population of the same region or a comparison group of fishermen from the less contaminated West Coast. NHL was also elevated among the Baltic fishermen, but not enough to reach statistical significance. Because the concentration of dioxin-like compounds was only about twice as high in the blood of Baltic men than in the other groups, these findings suggest that organochlorine exposure at or near the levels to which the general population is subject can have a detectable impact on immune cancer incidence.[156]

Finally, two studies have sought and found a link between the body burden of organochlorines in the general population and the risk of non-Hodgkin's lymphoma. The first, a small study in Sweden,[157] compared levels of PCBs and other organochlorines in the fat of 27 patients with NHL to 17 matched individuals who were hospitalized for nonmalignant surgical conditions. PCBs were more than 30 percent higher in people with NHL than in the comparison group, and the immunosuppressive pesticide chlordane and its metabolites were about twice as high. Those with PCB levels above the median suffered an almost threefold increase in the risk of NHL compared to those below this level. No one in the study was known to have suffered unusually high exposures to PCBs or any other organochlorines in the workplace or elsewhere.

This study spurred follow-up work by researchers at the National Cancer Institute, who designed a particularly rigorous investigation.[158] Beginning in 1974, researchers had taken blood samples from thousands of people in Maryland and then followed their health status over the next several decades. From this group, the researchers identified 74 people who developed non-Hodgkin's lymphoma, along with a matched comparison group that did not develop cancer. When the blood samples were analyzed, the researchers discovered a strong dose-response relationship between the level of PCBs and the risk of NHL; people in the top quarter of PCB concentrations had 4.5 times more NHL than the group with the lowest levels. The researchers also measured the quantity of Epstein-Barr virus in the blood and found that it too was related to the risk of cancer. Most interesting, PCBs and the virus had a synergistic effect on NHL risk, so that the virus caused the greatest increase in cancer among those with higher levels of PCBs in their blood. This result suggests that organochlorines may interact with other factors that modulate the immune system, viruses in particular, to cause a complex and multiplicative increase in NHL.

Cancer on Tap

One of the great public health triumphs of the twentieth century in industrialized countries has been drinking water sanitation. In modern society, we flush our feces into the same bodies of water from which we draw water to drink; to prevent the spread of infectious disease, we must now disinfect water before drinking it. One cheap and effective way to kill many microorganisms is to treat water with chlorine gas or related chemicals, such as chlorine dioxide, sodium hypochlorite, or chloramine.

Chlorine made its debut in municipal water treatment in 1908 at the New Jersey Water Works in Boonton, New Jersey. It was so effective and met such a pressing public health need that chlorination spread quickly to other towns in the United States, reducing the incidence of typhoid and other waterborne diseases dramatically. At the time no one dreamed that chlorination would pose public health hazards to go with its benefits. Reporting on chlorine's debut in Boonton, the *New York Times* noted that "any municipal water supply can be made as pure as mountain spring water. Chlorination destroys all animal and microbial life, leaving no trace of itself afterwards."[159] Not until 1974 did chemists discover that chlorinated water was not so pure after all: chlorine was reacting with organic matter in the water to produce organochlorines that remained in the water supply all the way to the tap. The first identified by-products were chloroform and other trihalomethanes (THMs), which cause cancer in laboratory animals. A subsequent survey of U.S. water supplies found that THMs were widespread drinking water contaminants throughout the nation and were present in concentrations up to several hundred parts per billion.[160]

In fact, THMs are just the tip of the iceberg. Chlorination produces thousands of organochlorine by-products, several hundred of which have been identified (a subject discussed in detail in chapter 7). A large number of these are carcinogenic or mutagenic in laboratory animals, including chloroform, bromodichloromethane, and the chloroacetic acids.[161] Of particular concern is the now-ubiquitous by-product MX (also known as 3-chloro-4-dichloromethyl-5-hydroxy-2(5H)-furanone). MX is formed in very small amounts but is so toxic that it is responsible for one-third to one-half of all the mutagenic activity in chlorinated drinking water.[162] A 1997 study by Finnish researchers found that small doses of MX in drinking water gave rats cancer of the liver, lungs, pancreas, thyroid, adrenal and mammary glands, as well as lymphoma and leukemia. Even the lowest dose studied was demonstrably carcinogenic.[163]

Today hundreds of millions of people consume small quantities of these compounds in their drinking water daily,[164] and the epidemiological evidence supporting a link to human cancers is nearly conclusive. There have been several dozen such studies in a range of countries, using different types of study design, and the vast majority have found an increase in cancer, particularly of the digestive and urinary organs. According to Kenneth Cantor, an epidemiologist at the National Cancer Institute and one of the leading researchers in the field, "The growing body of toxicological and epidemiological data suggests that risk is likely to be elevated . . . (relative risk of 1.5 to 2 for one or more cancer sites). Rectal and bladder cancers have been identified as those most likely to be associated with long-term consumption of chlorinated surface waters. If excess risks are in the measurable range for these sites, several thousand excess cases each year may be linked to consumption of chlorination by-products from surface water sources in the United States."[165]

The history of epidemiological work on chlorination and cancer is one of steadily improving study designs and repeated findings of elevated cancer risk among those exposed to organochlorine by-products.[166] The first group of studies, conducted in the 1970s, compared THM exposure and cancer mortality among populations in areas with or without chlorinated water. These "ecological" studies found that the risks of bladder, colon, and rectal cancers were associated with consumption of chlorinated surface water and the severity of THM contamination. In the early 1980s, researchers began to conduct case-control studies on groups of individuals who had died of specific kinds of cancer, reconstructing past exposures to chlorination by-products based on the water supply where each person had lived. These studies, which focused on bladder, colon, and rectal cancers, generally confirmed the earlier studies' findings that chlorination increased cancer risk. Later in the 1980s and into the 1990s, epidemiologists began to improve their estimates of people's exposure by conducting personal interviews and acquiring detailed sampling data on the quality of community water supplies. These studies also confirmed the increases in bladder and rectal cancer and raised concerns about other malignancies as well, including cancers of the pancreas and stomach.

None of these studies was perfect: some were small, the excess risks were modest and not always statistically significant, and the validity of the exposure assessment was sometimes uncertain. The results were not absolutely consistent either, in that not all studies found statistically significant increases for cancers of the same tissues. To address these problems, a group of researchers led by Robert Morris of the Medical College of Wisconsin conducted a meta-analysis: they reviewed all studies on drinking water and cancer prior to 1992 and then aggregated and analyzed the data from those that met several criteria of epidemiological quality.[167] The combined results strengthened the impression made by each of the studies considered in isolation. Exposure to chlorination by-products was associated with a statistically significant 15 percent increase in cancers of all organs combined. Bladder and rectal cancers were elevated by 21 and 38 percent, respectively, and the association was even stronger when only the best-conducted studies were considered. Chlorination was also associated with higher risks of breast, colon, esophageal, kidney, liver, and stomach cancer—with increases ranging from 11 to 18 percent—but these results did not reach statistical significance. When bladder, colon, and rectal cancer were excluded from the analysis, however, total cancer risk was still increased by a statistically significant 9 percent, indicating a real increase in other cancers, as well.

Since the meta-analysis, a number of high-quality studies have confirmed the cancer risk associated with water chlorination. In Finland, two very well-designed studies found 10 to 50 percent increases in cancers of the bladder, kidney, rectum, and breast and a 90 percent increase in cancer of the esophagus, among those exposed to mutagenic chlorine by-products.[168] A Canadian study found a 40 percent excess in the risk of bladder cancer in people exposed to chlorinated surface water for 35 or more years, and the risk was even higher among those exposed to THMs at concentrations greater than 50 parts per billion.[169] In Iowa, one large study found that exposure to chlorinated water was associated with elevated risks of bladder and rectal cancers; another found increases in colon cancer, breast cancer, and all cancers combined in postmenopausal women who drank chlorinated water. In both cases, the risk grew with the duration or the level of exposure.[170]

The increases in cancer associated with chlorination are fairly modest, as one would expect given the low levels of contaminants in public drinking water supplies. But with hundreds of millions of people exposed on a daily basis, the implications for public health are profound. Morris's meta-analysis suggested that consumption of chlorinated water is responsible for 9 percent and 15 percent of all bladder and rectal cancer, respectively, in the United States. This proportion translates into about 4,200 cases of bladder cancer and 6,500 cases of rectal cancer every year, or over 100,000 additional cancer cases per decade for just these two organs. Without a doubt, this number is much lower than the number who died from infectious disease before we began to disinfect our water. It is far greater, however, than the number who would die if we would learn to sanitize water without poisoning it with carcinogens, or—better yet—not to contaminate it in the first place.

Dioxin and Cancer

Ironically, the most carcinogenic organochlorine is also the one whose carcinogenicity is most controversial. TCDD (dioxin) is the most potent synthetic carcinogen ever tested in the laboratory. There have been 18 separate assessments of dioxin's carcinogenicity, involving five different species of experimental animals, both sexes, five routes of exposure, and high and low doses over short and long periods of time. In every case, dioxin has caused cancer. Dioxin has caused tumors of nine different types, including those of the lungs, oral and nasal cavity, adrenal glands, liver, soft tissue, skin, lymphomas, and all cancers combined. Although not mutagenic itself, dioxin is a powerful promoter of tumors initiated by other carcinogens. It is also a complete carcinogen, increasing the incidence of tumors in animals treated with dioxin alone, presumably because it increases the rate at which the body converts other chemicals, including its own hormones,[171] to forms that can initiate cancer. Tests at doses as low as 1 trillionth of total body weight per day have failed to provide evidence that there is a level of dioxin that does not cause cancer.[172]

Dioxin's extraordinary carcinogenicity in animals was established in the late 1970s. Until recently, however, it was standard for scientists, regulators, and the media to say that dioxin probably did not cause cancer in people. No one ever offered a compelling explanation *why* or *how* humans were somehow immune to dioxin's great toxicity in other mammals. Nevertheless, years of shoddy science and aggressive public relations, repeated again and again in the media and scientific publications, kept in vogue the "consensus" position that dioxin was not carcinogenic to people.

The U.S. military began spraying Agent Orange in Vietnam in 1962, expanded the practice from 1965 to 1969, and discontinued it in 1970 when scientists discovered that 2,4,5-T caused birth defects in mice. After the war, information gradually emerged that fanned a bonfire of controversy.[173] First, the severity of dioxin contamination in Agent Orange became more widely known to veterans' groups; the news that the government had let its own soldiers be exposed to such an exquisitely toxic substance was highly inflammatory. A few years later, high levels of dioxin were found in the milk of women in southern Vietnam and in the blood of American soldiers who had served there. In 1977, the first experimental work established that dioxin was a powerful carcinogen. Veterans reacted with alarm, believing that Agent Orange exposure had caused the severe health problems that many of them were experiencing. Veterans' groups took political and legal action, putting pressure on Congress and bringing suit against the government and industrial manufacturers of Agent Orange to compensate exposed veterans for health damage.

Those who stood to lose millions to the veterans' litigation deployed their political and scientific resources to deny any evidence of health damage. It was impossible to argue that Agent Orange was not contaminated with dioxin, or that dioxin was not toxic in the laboratory, so the Agent Orange manufacturers—Monsanto and Dow Chemical in particular—argued that there was no evidence that humans suffered the same effects seen in experimental animals. This argument turned out to be very difficult to refute in epidemiological studies of Vietnam veterans, because the lack of specific exposure records made it almost impossible to evaluate the link between health effects and Agent Orange exposure on a soldier-by-soldier basis.

Attention turned instead to a different population of dioxin-exposed men: workers in the chemical factories that had produced 2,4,5-T. Between 1980 and 1984, Monsanto released three epidemiological analyses of workers at its 2,4,5-T factory in Nitro, West Virginia.[174] All concluded that the workers suffered no dioxin-related effects except for chloracne, a painful skin disease. Studies conducted by Dow[175] on its own workers came to similar conclusions, and the government's own Agent Orange project, beset by difficulties in establishing exposure histories, failed to link dioxin to any health impacts in veterans.[176]

These studies became the scientific basis for a consensus that dioxin had never caused any serious health problems in people. The Monsanto studies were particularly influential, playing important roles in legal rulings and government decisions denying compensation to Vietnam veterans. EPA concluded that there was no evidence to support a link between dioxin exposure and human cancer, citing Monsanto's studies as support.[177] Articles and editorials in *the New York Times,* the *Washington Post,* and the *Journal of the American Medical Association* all took the position that, in the words of the *New York Times,* "most effects of dioxins have not materialized among humans."[178] Scientific American carried an article with the following tag line in large print, "Dioxin: concern that this material is harmful to health or the environment may be misplaced. Although it is toxic to certain animals, evidence is lacking that it has any serious long-term effects on human beings."[179] The article was written by Fred Tschirley, a former govenment official who had become head of the Michigan Agriculture/ Business Council, an industry group in the state where Dow's headquarters was located. Citing the Monsanto studies, Tschirley's article claimed that "Investigators are in general agreement that TCDD is less toxic to humans than it is to experimental animals. . . . What the [Environmental Protection] Agency has not done—and might be said to have a responsibility to do—is to try to dispel the public's fear on the basis of the evidence that exposure to low concentrations of TCDD in the environment appears not to have serious chronic effects on human beings."

In Australia, Monsanto and its studies had overwhelming influence on the Royal Commission on the Use and Effects of Chemical Agents on Australian Personnel in Vietnam, which was charged with evaluating the health consequences of Agent Orange exposure and the appropriateness of compensation for Australian veterans. The commission cited the Monsanto studies extensively and concluded there was no evidence that dioxin or 2,4,5-T had harmed human health. Large parts of the Royal Commission's final report were taken verbatim or nearly so from Monsanto's submission to the commission, including critiques of the validity of other studies that did find a relationship between dioxin exposure and cancer. Not only did Monsanto essentially ghost-write parts of the report, but there was no citation, attribution, or other indication of Monsanto's influence on the commission's conclusions.[180] The commission was soundly criticized by independent scientists who had testified before it, but its report became influential within Australia and was widely cited in scientific papers, books, and even an editorial in the *Journal of the American Medical Association.*[181]

Not until 1989 did the consensus that dioxin was more or less safe for people begin to crack apart. First, the Monsanto studies came under scrutiny in private litigation against the company. Pursuing concerns about the studies raised by independent scientists as early as 1986,[182] the plaintiffs reviewed Monsanto's research records in detail and found numerous irregularities in the conduct of the studies. Under examination, Monsanto's chief medical officer conceded that more than a dozen workers with cancer who were known to have been exposed to dioxin had been inexplicably shifted into the unexposed "control" group or excluded from the study altogether, grossly biasing the analysis toward finding no effect.[183] After extensive controversy, media attention, and even a failed libel suit against an environmental journalist for reporting on the evidence of misconduct, Monsanto's studies were largely discredited; today they are seldom cited as legitimate or important scientific findings.

Also in 1989, an independent reanalysis raised serious questions about the validity and conduct of a study that found no health impacts in dioxin-exposed workers at a plant in Germany owned by the chemical company BASF.[184] In response to a claim for compensation filed by BASF employees, the German chemical industry's liability insurance association concluded, based on data supplied by BASF, that there was no evidence of health damage among dioxin-exposed workers and denied the claim. When the workers appealed the decision, a German court appointed Friedemann Rohleder, an independent epidemiologist, to review the studies. Rohleder found that BASF had placed 20 unexposed supervisors in the group of exposed workers, while 23 workers who had chloracne—a hallmark of dioxin exposure—were shifted into the unexposed group. The problems were so numerous and egregious that Rohleder was forced to conclude that BASF had manipulated the data to prevent a relationship between exposure and health problems from becoming apparent. When he corrected only the worst errors, Rohleder found a statistically significant increase in several types of cancer among dioxin-exposed workers.

Less than a year later, the U.S. government's own study of servicemen exposed to Agent Orange, conducted between 1982 and 1987 by the Centers for Disease Control (CDC), also came under scrutiny. After extensive investigation, the U.S. House of Representatives Committee on Government Operations charged that the CDC study embodied "flawed science" and "political manipulation" by the Reagan White House, which "controlled and obstructed" the study to ensure that Agent Orange would not be linked to the veterans' health problems.[185] According to the committee report, the study had been changed from its original format to exclude the soldiers who received the heaviest exposures, and several other aspects of the study appeared to have been manipulated as well. Under pressure from the White House and without scientific justification, CDC had canceled its investigation into a specific relationship between herbicide exposure and health effects in veterans; without that information, no link between Agent Orange and health damage would ever be convincingly demonstrated.

As the scientific basis for the consensus began to crumble, new independent studies appeared, documenting a link between dioxin exposure and cancer. In 1991, the National Institute for Occupational Safety of Health (NIOSH) published an analysis of the health of a combined cohort of over 5,000 dioxin-exposed workers at 12 chemical plants in the United States, including the Monsanto facility.[186] Unlike its industrial predecessors, this study found statistically significant associations between dioxin and cancer. It reported that mortality from all cancers combined was 15 percent higher than expected among all dioxin-exposed workers, and 46 percent higher among those who had been exposed to dioxin for more than one year with a 20-year latency period; in this latter group, deaths from soft tissue sarcoma and lung cancer were 9 and 1.4 times higher than expected, respectively.

Later that year, a team of government and academic researchers in Germany published their study of cancer mortality among almost 1,600 dioxin-exposed workers in the Hamburg pesticide factory where the breast cancer studies I mentioned earlier were conducted.[187] Total cancer mortality in the dioxin-exposed workers turned out to be 39 percent higher than in a comparison cohort of workers in the gas industry with no dioxin exposures; among men with dioxin exposures of the longest duration or the highest intensity, cancer mortality was even higher. As in the NIOSH study, there was a statistically significant increase in lung cancer, and risk of immune cancers was significantly higher as well. After these studies appeared, an editorial in the *New England Journal of Medicine* noted, "The hypothesis that low [dioxin] exposures are entirely safe for humans is distinctly less tenable now than before."[188]

Subsequent studies sealed the case that dioxin causes cancer in humans. Analyses of cancer incidence and mortality in Seveso found that people in the dioxin-exposed zones had experienced increased risks of liver cancer, lymphomas, multiple myeloma, and leukemia; mortality from these immune cancers increased with the length of time a person resided in the area.[189] A follow-up at the Hamburg plant established that total cancer mortality consistently increased with rising dioxin exposures, and statistically significant increases were also found for cancers of the lung, rectum, esophagus, and immune system. Among the group with the highest dioxin exposures, the risk of dying from any kind of cancer was more than three times higher than expected.[190] A study of dioxin-exposed workers from four chlorophenoxy herbicide plants in Germany found statistically significant increases in deaths from all cancers combined, respiratory cancer, and NHL.[191] And two studies by the International Agency for Research on Cancer (IARC) found that exposure to dioxin-contminated herbicides was associated with a 29 percent increase in cancer mortality in men and a twofold increase in women in a large international cohort of chemical and agricultural workers.[192]

These studies established beyond a reasonable doubt that dioxin does increase cancer risk in people. By 1996 an international conference of epidemiologists could report that "almost all participants agreed that TCDD can be considered a human carcinogen,"[193] a remarkable turnaround from the consensus that dominated the field less than a decade earlier. In 1997 the IARC upgraded its assessment of dioxin from "possibly carcinogenic to humans" based on effects in laboratory animals to "carcinogenic to humans" based on epidemiological evidence.[194]

For many researchers, the repeated finding that dioxin increases lung cancer and all cancers combined among dioxin-exposed workers was surprising. Although mice and rats exposed to dioxin develop excess tumors in these categories, human studies had always focused on rarer cancers of the immune system and connective tissues as the most important types of dioxin-induced cancers. The increases in lung and total cancer are modest—about 40 percent when all studies are considered[195]—but the findings are consistent and statistically significant. This pattern suggests that dioxin may act as a general enhancer of cancer in many tissues throughout the body. It may do so by suppressing the immune system, interfering with endocrine control over the growth of tissues and the differentiation of cells, and enhancing the potency of other carcinogens to which people are exposed. The last mechanism is of particular interest because dioxin is known to induce enzymes that catalyze the activation of the body's natural hormones and PAHs—the cancer-causing agents in tobacco smoke and other kinds of air pollution—to their carcinogenic forms.[196] In this way, dioxin makes other pollutants more potent carcinogens than they would otherwise be, raising the possibility that dioxin could increase cancer incidence associated with smoking and other carcinogens. Notably, the risk of lung cancer per cigarette has increased substantially over the past several decades, even while the yield of carcinogenic tar per cigarette has actually decreased; lung cancer among nonsmokers has been inexplicably on the increase, too.[197]

If occupational exposure to dioxin increases cancer risk only moderately, then one might think that the much lower background exposures to which the general public is subject must be virtually harmless. This argument sounds quite reasonable, but it turns out to be wrong, given the extraordinary carcinogenicity of dioxin and the huge number of people exposed. EPA's current official estimate of dioxin's cancer potency, derived from experiments on laboratory animals, suggests that the average person's dioxin exposures pose lifetime cancer risks of 500 to 1,000 per million—hundreds of times greater than the level of risk generally deemed "acceptable."[198] Even if the actual risk were lower than this estimate by a factor of 100, dioxin would still cause a substantial number of cancers every year.

One very controversial aspect of figures like these is the extrapolation of human risk estimates from dioxin's carcinogenicity in rodents. On one hand, dioxin may be more or less potent in humans than in rats and mice; on the other, if dioxin acts by enhancing the ability of other chemicals to cause cancer, its potency may be higher in the real, polluted world than in the clean environment of the laboratory. The Hamburg study of dioxin-exposed chemical workers circumvented the rodents-to-people problem by directly estimating dioxin's carcinogenicity in humans. Using data on the levels of dioxin in the workers' blood to estimate the quantitative relationship between dioxin exposure and cancer risk in humans, this analysis concluded that each additional unit dose of dioxin (one-trillionth of a gram of dioxin per kilogram of body weight per day) is associated with an increase in lifetime cancer risk of 1,000 to 10,000 per million—about an order of magnitude higher than EPA's rodent-based estimate.[199] If we apply this estimate to the general human population in the United States—in which the average daily dose of dioxins and related compounds is about 3 to 6 trillionths of a gram per kilogram (expressed as TCDD- equivalents)—we find that "background" dioxin exposure poses cancer risks ranging from 3,000 to 60,000 per million. That is, the chances of dying from cancer due to dioxin exposure may be as high as 6 *percent*. For those with greater-than-average dioxin exposures—workers in contaminated industries, sport and subsistence fishermen and their families, and people who live near major dioxin sources—the risk is even higher.

These extraordinary and patently unacceptable risk figures imply that dioxin causes from 11,000 to 214,000 cancers every year in the United States alone—1 to 20 percent of total cancer incidence.[200] If the German team's dose-response calculations are accurate, dioxin would turn out to be more than just a serious and long-denied threat to highly exposed workers, soldiers, and communities. Instead, it would constitute a major cause of cancer throughout the industrialized world.

These numbers should be taken with several grains of salt because, like all other risk estimates, they are derived from some data but even more assumptions. They do make a number of unambiguous and important points, however. First, in contrast to an established scientific and social consensus that lasted for more than a decade, dioxin turns out to be no less potent a carcinogen—and possibly an even stronger one—in people than it is in rodents; the scientific community's repeated assurances that humans are not at risk turned out to be baseless. Second, dioxin's occupational hazards are modest enough that it took decades of very careful analyses, large studies, an immense commitment of scientific resources, and many negative or inconclusive results along the way to detect them; this process of gradual illumination illustrates the perils of interpreting a lack of conclusive data as evidence of safety.

Finally, one norm of science is that the results of research are independent of the people or institutions who do the work, but the history of dioxin shows this norm to have been repeatedly violated. Again and again, scientists working for industry and branches of the government with a material interest in dioxin's toxicity have found dioxin more or less safe for humans. Those working for independent academic or government institutions have found the opposite. This dependence of scientific outcome on the interests of the researchers should come as no surprise in the age of corporate-sponsored science. The shocking thing is how willing the wider scientific community, the media, and decision makers were to accept the chemical industry's studies as objective evidence of dioxin's safety in humans. Sick people were denied compensation, polluters absolved of liability for their actions, and dioxin pollution allowed to continue, all based on the naive view that industry's studies of its own workers were valid and trustworthy.

Cancer Is Not a Fact of Life

Having examined the role of individual substances in cancer risk, we can now consider the potential health impacts of the total burden of human exposure to thousands of organochlorines and other synthetic chemicals. If a few well-studied chemicals and exposures—dioxin-like compounds, the by-products of drinking water chlorination, and increased UV radiation—are each associated with many thousands of new cancer cases per year, then the complex mixture of carcinogenic pollutants of which these agents are a part must make an even greater contribution. It is not proven but entirely reasonable to conclude that organochlorines play a role, possibly a very substantial one, in the steadily growing epidemic of cancer.

The first principle of public health practice must be prevention, not proof. We should not delay action until we have quantified the contribution of each potentially carcinogenic organochlorine to cancer incidence. Chemicals that cause cancer, mutations, or endocrine disruption in laboratory animals should be assumed to be hazardous to wildlife and humans as well. Chemicals that have not been tested—organochlorines in particular, given their tendency toward increased carcinogenicity—should be presumed harmful unless demonstrated otherwise.

In most nations, cancer programs focus on finding cures and treatments; comparatively little attention has been paid to prevention, which is far more effective at saving lives.[201] Treatment can extend the lives of many people, but the majority of people who contract cancer will die of it. When cancer programs have pursued prevention, they have generally focused on changing "lifestyle factors" like smoking, diet, alcohol consumption, and inadequate exercise.

If people can be persuaded to change their behavior, these kinds of programs have the potential to reduce cancer incidence considerably. But the emphasis on individual lifestyle choices leaves people powerless to prevent cancer caused by pollution. Once carcinogens have been released into the air, water, and food supply, we cannot avoid being exposed to them;

eating, drinking, and breathing are not matters of choice. Prevention programs focused on lifestyle factors have thus had little to say about pollution as a cause of cancer.

That does not mean that pollution-related cancer is an inescapable fact of modern life. Cancers caused by synthetic chemicals are among the most preventable of all kinds of cancer.[202] Averting them, however, requires action not by the individual but by the larger society. Exposure to synthetic carcinogens occurs because human beings deliberately manufacture these substances and disperse them into the economy and environment. Society can prevent the cancers that synthetic chemicals cause by ceasing to make them. As the dioxin experience shows, those who manufacture carcinogens will not stop doing so voluntarily. To prevent cancer caused by pollution, society must require industry to phase out the products and processes that contaminate the environment and workplace with synthetic chemicals that cause cancer.

Every case of cancer imposes incalculable suffering on its victims and their loved ones. Magnified by millions of cases per year, cancer is one of the great tragedies of our time. Society must do everything it can to reduce the terrible burden of cancer, not limit its actions to those that are politically or economically expedient. Saying "no more" to those who knowingly produce and expose people to carcinogens is a good place to start.

II

The Cause: Industrial Chlorine Chemistry

5

Chlorine in Nature and Industry

So it comes about that salt, sodium chloride, formed from these two violently reactive and poisonous elements, is not only safe to eat, it is essential to life. And the element chlorine, seeking another electron wherever it can be found, . . . is a destroyer of life and a useful bactericide.
—G. E Porter, Nobel laureate in chemistry[1]

"A crusade is ongoing against element 17 in the periodic table, i.e., chlorine, one of the most abundant on earth," writes Albert Fischli, president of the International Union for Pure and Applied Chemistry.[2] "Government could not possibly outlaw the use of one of the most abundant elements in nature," Kip Howlett, head of the industry's Chlorine Chemistry Council, told the American Chemical Society in 1995.[3]

These men are correct on one point: elements cannot be banned. Elements simply are. Nuclear technology aside, humans are in no position to decide whether an element will be allowed to exist. Virtually all the chlorine in nature exists in the familiar form of sodium chloride, commonly known as table salt. There are some 26 quadrillion tons of salt in the oceans, 60 quadrillion tons in the salt beds of the earth's crust, and large deposits of other common salts like potassium and calcium chloride.[4] There is obviously no way that we can ban the element chlorine from the earth.

That is really no problem, however, because chlorine in its natural form is no threat to health. Sodium chloride flows constantly through the ecosystem, our bloodstreams, into and out of our cells, never reacting with the carbon-containing organic matter of which all living things are made. In contrast, the chlorine gas produced by the chemical industry is violently reactive, combining quickly and promiscuously with any organic matter with which it comes in contact to produce the new family of compounds called organochlorines. Both chloride salt and chlorine gas are forms of chlorine, but they are as different in character as Dr. Jekyll and Mr. Hyde. We must thus distinguish between the element chlorine and the set of industrial technologies, brought into large-scale use only in this century, that fundamentally transform chlorine from its natural state into the novel forms of chlorine gas and organochlorines.

Although it would be silly to get anxious about salt, concern rightly focuses on the technologies of chlorine chemistry and the compounds it produces. "Phasing out chlorine" does not mean banning an element; it serves as shorthand for "phasing out industrial chlorine chemistry," or "phasing out the production of chlorine gas and the compounds made from it." This is not a new or strange shorthand. When chlor-alkali manufacturers formed a trade group for lobbying and public relations, it called the organization "The Chlorine Institute," and there are no producers of table salt in the group. When Kip Howlett himself says, "Chlorine is one of the eight fundamental building blocks for modern society," he is talking about his industry's products, not salt.[5]

In this chapter, I discuss the properties and significance of chlorine in its natural and industrial forms. Some authors criticize the case against chlorine chemistry as merely guilt by association: some organochlorines are quite hazardous, so therefore all must be. This would surely be faulty reasoning, but it is not the argument I am making. Rather, adding chlorine to organic matter almost invariably increases one or more hazardous properties—persistence, toxicity, or the tendency to bioaccumulate—and this pattern comes about because of well-understood aspects of the chemistry of the chlorine atom itself. My purpose here is to discuss the properties of organochlorines and explain *why* these chemicals dominate virtually all government lists of priority pollutants. Once we understand that there is something inherently powerful and dangerous about attaching chlorine to organic chemicals, then we can understand why so many organochlorines—not just a few "bad actors" like DDT and PCBs—pose such numerous and severe environmental hazards.

From Salt to Chlorine

To appreciate the chemical transformation the chemical industry wreaks on salt when it produces chlorine gas and organochlorines, we need to understand some basic chemistry. All atoms consist of a nucleus that contains positively-charged protons (and in most cases neutral neutrons, as well), with an equal number of negatively-charged electrons "orbiting" around the nucleus. Each element is characterized by the unique number of protons and electrons its atoms contain. The electrons are organized in orbitals, or shells of fixed sizes. Moving out from the nucleus, the inner level can contain up to two electrons; the next level can contain up to eight; the next also holds as many as eight; and so on. The shells are filled starting at the inner layer, moving outward as each one becomes full. An atom of chlorine has 17 electrons, distributed 2 in the inner layer, 8 in the next, and 7 in the outer layer, where there is room for 1 more. (The elements that require 1 electron to complete their outer shell are called halogens, and all of them—chlorine, fluorine, bromine, and iodine—have similar chemical behavior.) Sodium has 11 electrons: 2 and 8 in the inner layers, but just 1 in the outer level.

Every atom's goal is to reach its most stable chemical state—the state of lowest energy—which occurs when its outer electron layer is completely full. Thus, the chemical behavior of any atom is determined by the number of electrons in its outer shell. An atom of an inert gas—helium, for instance—will not react with other atoms because its outer shell is already full; the atom is satisfied and wants to stay that way. In contrast, chlorine, with 7 electrons in its outer shell, wants to gain an electron from another atom; sodium, with 1 electron in its outer layer, desperately wants to lose that electron, so that the next shell inward, which is complete with its 8 electrons, becomes the outermost level. If chlorine can gain an electron and sodium can lose one, they become stable and unreactive, like an inert gas.

This is precisely what happens when atomic sodium and atomic chlorine are mixed. The two react violently, releasing large amounts of energy as an electron is transferred from sodium to chloride. Each atom is now very stable, and it would take a great amount of energy to rip the extra electron away from the chlorine atom, forcing it back into a reactive state. The transfer of the electron gives the chlorine atom one more electron than it has protons, so it has a net negative charge, while the sodium atom has one fewer electron, giving it a positive charge. Charged atoms are called ions, and the ionic form of chlorine is called chloride. Because oppositely charged particles have a weak attraction for each other, sodium and chloride ions arrange themselves into a lattice to form the crystals found in everyone's salt shaker; when put into water, the crystals dissolve to release the chemically inert ions of sodium and chloride.

The chlorine and sodium atoms now on earth were created during the origin of the solar system by violent nuclear reactions in the sun. These atoms were transformed into ions so quickly and permanently, however, that neither atomic chlorine nor chlorine gas has ever been a significant constituent of the earth's environment. Virtually all of the chlorine in nature thus exists in the form of chloride salts.

The chlorine industry takes this ionic chloride and fundamentally changes its character. Chlorine chemistry begins with the chlor-alkali process, in which an extraordinarily powerful electric current is passed through a solution of sodium chloride. In the special environment of a chlor-alkali "cell"—the chamber in which chlorine is produced—electrical energy forces the chloride atoms out of their stable form and into a new, more reactive chemical state. At one end of the cell, electricity rips electrons away from the chlorine ions, returning chlorine to its unstable atomic form; two of these chlorine atoms then combine with each other to form chlorine gas (Cl_2), an entirely new substance. At the other end of the cell, water molecules are split to yield hydrogen ions (H^+) and hydroxide (OH^-) ions. The electrons from the chloride ions combine with the hydrogen ions to produce hydrogen gas (H_2), and the hydroxide ions combine with the sodium to yield sodium hydroxide (NaOH), also called alkali. The reaction can be represented this way:

$$2NaCl + 2H_20 \Rightarrow Cl_2 + NaOH + H_2$$

At the end of the chlor-alkali process, the sodium is still in an ionic state, but the chlorine is not. In chlorine gas, also called elemental chlorine, the 2 atoms of chlorine now share 2 electrons—an arrangement called a covalent bond—so that each atom now has its outer shell filled, even though there are only 14 outer-shell electrons to fill 16 places. Some covalent bonds are strong, but this one is not: 2 chlorine atoms share electrons with each other very uneasily, so chlorine gas is a very unstable compound.

Chlorine gas is thus highly reactive, particularly toward organic matter. Carbon is the most important element in living things precisely because it is so good at forming stable covalent bonds with other types of atoms, like hydrogen, oxygen, and nitrogen, allowing organisms to create a repertoire of molecules out of a limited number of building blocks. A carbon atom has 4 electrons in its outer shell, so it can form four bonds with other atoms. The possibilities are endless: carbon can form chains and rings with other carbon atoms, and hydrogen (or sometimes oxygen or nitrogen) atoms bond readily with carbon to fill any remaining slots. The resulting organic substances are generally quite stable, and they represent the chemical foundations of all life; DNA, sugars, starches, fats, proteins, hormones, and other natural molecules are all built from a carbon backbone.

With chlorine around, however, carbon's ability to combine with other substances becomes a curse. The chlorine atoms in chlorine gas are less stable when bonded to each other than if each one shared a pair of electrons with a carbon atom. Carbon is happy to oblige. Thus, if an organic substance is available, the two atoms in chlorine gas will dissociate from each other and form new bonds with the carbon atoms, replacing the hydrogen atoms on the original molecule.[6] The original substance is destroyed, and a new chemical—an organochlorine—is produced instead.

Because it destroys the matter of which living things are made, chlorine gas is extremely dangerous. First recognized as an element in the early nineteenth century, chlorine is a heavy, green-colored gas with a powerful odor.[7] If released into the environment, chlorine gas will travel slowly over the ground in a coherent cloud, a phenomenon familiar to World War I soldiers who faced it as a chemical weapon, one of chlorine's first large-scale applications. Also familiar to these men was chlorine's toxicity, which arises from its tendency to combine with and destroy organic matter, like that of the lungs and eyes.

The key to chlorine's behavior is its powerful desire to gain an electron—a property called electronegativity. Chlorine is the third most electronegative element in nature, and chlorine atoms have about a 50 percent stronger pull on electrons than hydrogen atoms do, which explains why they replace hydrogen so readily in organic molecules.[8] But there must be more to the story, because other substances that are hungry for electrons are not so dangerous. An atom of oxygen, for instance, is an even more powerful electron attractor than chlorine; that is why free oxygen radicals in our cells can damage DNA and other molecules, a phenomenon that has given rise to a burgeoning market in "antioxidant" pills. So why then doesn't oxygen spontaneously combine with organic matter the way chlorine does? The answer is that all chemical reactions—even the ones that release energy—require some energy input to get them started by breaking the bonds in the reactants. If this "activation energy" is high, the rate will be slow, because few molecules in a mixture will have enough energy to break apart and get the reaction started. Splitting apart a molecule of chlorine gas to start a reaction is easy: even a mild ray of light can do it. But breaking apart a molecule of oxygen to yield reactive radicals requires more than twice the energy needed to split an unstable molecule of chlorine. Since the rate of a reaction increases exponentially as the activation energy goes down, the same reaction with chlorine molecules will proceed at an astronomically faster rate than it will with oxygen.[9]

Chlorine thus reacts very quickly with any organic matter with which it comes in contact, which makes it a very effective disinfectant and bleach. Due to the properties of the chlorine atom, the organochlorines that result tend to have certain properties—persistence, bioaccumulation, and toxicity—that make them useful for industrial purposes but very dangerous in nature. To those qualities we now turn.

The Persistence of Organochlorines

Many organochlorines resist degradation, for reasons that also arise from the chemical properties of the chlorine atom. As the Canadian chemist Donald Mackay has noted, the electronegativity of the chlorine atom means that "addition of chlorine to organic molecules generally causes them to be less soluble in water, less volatile, less reactive, less flammable, more toxic to organisms, more bioaccumulative in fish and more resistant to breakdown by bacteria. . . . The chlorine atom effectively sucks reactivity from the molecule and renders it more stable to degradation."[10]

There are several reasons that chlorination often has this effect on environmental persistence. First, the carbon-chlorine bond is a fairly strong one, requiring a significant quantity of energy to break, so these bonds tend to resist degradation by light or chemical reactions in the environment. Because the chlorine atom has such a strong pull on electrons (much stronger than the hydrogen it replaced), many organochlorines are resistant to breakdown by other chemicals that would compete with the chlorine to share electrons with a carbon atom. Thus, for instance, polyethylene plastic—a polymer of linked ethylene molecules—is very flammable, which means it reacts rapidly with oxygen when temperatures are high; PVC plastic, in contrast, a polymer of ethylene molecules each containing one chlorine atom, is prized for its fire resistance. It is generally true that the more chlorine atoms added to an organic molecule, the more it will tend to resist degradation. Dichlorobenzene, for instance, is somewhat persistent, and hexachlorobenzene is extremely so.[11]

Second, chlorine can affect the stability of the bonds between the carbon atoms that form the backbone of the compound. In several kinds of organochlorines, chlorine atoms stabilize the carbon backbone in a way that prevents the natural breakdown of the compound. When chlorine is attached to a benzene ring—as in such important organochlorines as dioxins, PCBs, DDT, toxaphene, chlorophenols, and chlorobenzenes—the chlorine atom stabilizes the ring structure, so that the new organochlorine is much less chemically reactive than it would be if hydrogen were attached. The more chlorine atoms that are attached to the ring, the stronger the stabilizing effect.[12] Also, when chlorine is attached to a chain of organic molecules in which the carbon backbone is held together by double bonds, as it is in such important organochlorines as vinyl chloride and perchloroethylene, chlorine's hold on electrons tends to stabilize the neighboring double bond, so the backbone is again protected from the degradation processes that normally attack these molecules in nature.[13] In some other substances, the large chlorine atoms pull electrons outward away from the carbons, creating a shield of electrons at the outside of the molecule that protects the carbon backbone from chemical attack. Organochlorines such as these—generally substances in which chlorine is substituted for most or all of the available hydrogen atoms, such as carbon tetrachloride, 1,1,1-trichloroethane, and trichlorotrifluoroethane—

often have extraordinarily long lifetimes in the environment, but their chlorine-free analogues are easily degraded.

Not all organochlorines resist chemical transformation. For instance, adding one chlorine atom to an organic molecule in which the backbone carbons are held together by a single bond destabilizes the carbon-carbon link and increases the molecule's reactivity. But as with most other organochlorines that are broken down in the environment, the product of this reaction is simply another organochlorine.[14] As we saw in chapter 1, when one organochlorine is transformed into another, the new substance is often more persistent and toxic than the original compound. Degradation is not complete until all the components of a compound are returned to forms that can reenter natural ecological cycles; all the chlorine must be removed and converted to chloride before we can say that an organochlorine has been truly degraded. This process is typically a long and slow one, which explains why some organochlorines have environmental lifetimes estimated in the decades or centuries.

Chemicals in the environment are broken down not just by physical and chemical processes but also by biological ones. Over millions of years, microorganisms have evolved methods to metabolize naturally-occurring organic compounds. But because most organochlorines do not occur naturally in the ambient environment, few organisms have had the opportunity to evolve enzymes that are adapted specifically to break them down. Consequently, the biological transformation of organochlorines into chloride in the environment generally takes place very slowly, if at all.[15] When EPA listed the compounds most resistant to biodegradation in sewage treatment plants, all were chlorinated or brominated compounds. According to the agency, "The extent of halogenation also influences the relative biodegradability of the compounds, i.e., the more halogens in a compound by weight, the less biodegradation will be in evidence."[16]

There are numerous examples of organochlorine biodegradation, most of which involve individual species of microorganisms or fungi that can break down specific organochlorine compounds. But thousands of man-made organochlorines now circulate in the environment, so these substance-specific mechanisms do little to prevent the general buildup of organochlorines. Further, in most cases, biodegradation of an organochlorine usually produces a new organochlorine, which is often more hazardous than the original compound. For instance, the pesticide DDT can be metabolized into DDE,[17] which then persists for decades, bioaccumulates, and is highly toxic in mammals. The chlorophenols discharged by chlorine-bleaching pulp and paper mills are metabolized by aquatic organisms into the more toxic chloroveratroles.[18] Some species—fungi and bacteria in particular—have enzymes that can "dechlorinate" some organochlorines, but most are substrate-specific, so they work on one or a few organochlorines, rather than all of them.[19]

These considerations explain why the chemistry of chlorine so often increases the resistance of organic chemicals to environmental breakdown, making many organochlorines persistent. I am not arguing that all organochlorines are persistent; some, in fact, are quite reactive and are used for further synthesis in the chemical industry for just that reason. But it is true that a vast number of organochlorines are more persistent than their nonchlorinated analogues, often extraordinarily so. These substances are less compatible with the ecosystem's natural cycles and more likely to build up in the global environment than other kinds of organic chemicals, explaining why organochlorines dominate government and scientific lists of persistent organic pollutants.

Out of the Environment and into Our Fat

Organochlorines also tend to be bioaccumulative—far more so than organic chemicals that do not contain chlorine. The tendency to bioaccumulate is determined by the solubility of a substance in fat, and chlorination generally increases the fat solubility of chemicals. As anyone who has ever tried to mix a salad dressing of oil and vinegar knows, fatty materials do not mix with water but instead segregate into their own separate layer. Add to your oil and vinegar a substance that is water soluble, like sugar, and it will dissolve in the vinegar compartment; but add something fat soluble—some other kind of oil, say—and it will eschew the watery layer and join the fatty compartment.

The layers in salad dressing have their analogues in the ecosystem. The world's watery compartment is all around us and within us. Any substance that dissolves in water—salt, for instance—circulates through the bloodstream of animals and out again the same way it circulates through the seas. The quantity of water in the world is unimaginably immense, covering 70 percent of the world's surface. In contrast, virtually all fatty substances in nature are produced by living things, which make fats and retain them in their bodies as cell membranes and as energy reserves. The world's oily compartment is thus primarily inside the bodies of living organisms. (The major exception is petroleum, which is contained in underground deposits formed from the bodies of dead organisms.) If we deposit a fatty substance in a lake, it will be drawn out of the water into the fats contained in the bodies of the creatures that live in the lake. Because the total quantity of matter in the fatty compartment is so minuscule compared to that in the watery compartment, oil-soluble substances will accumulate in the fats of living organisms at concentrations far greater than their levels in the ambient environment.

As we move up the food chain, the problem becomes more severe. When a predator eats another organism, any fat-soluble substances in the prey that the predator does not digest will be retained in its fat. Meal after meal, as an animal eats many times its body weight in the flesh of other creatures, the contaminants in all its prey will be concentrated in the fat of a single predator. The food web is thus a kind of food pyramid, and concentrations of fat-soluble chemicals are magnified with each step upward. In this way, species at the top of the food chain, including people, become living reservoirs for organochlorines that have been dispersed across the planet.

Organochlorines tend to be more bioaccumulative than nonchlorinated organic chemicals, because chlorinating an organic molecule increases its solubility in oil and decreases its solubility in water. This effect occurs because chlorine atoms are extremely large, with a volume about equal to that of a methyl group, which contains a carbon and three hydrogen atoms; thus, adding a single chlorine atom to a molecule of methane will approximately double its volume. The larger the molecular volume of a substance, the more difficult it is to dissolve in water. The reason for this behavior arises from the strong tendency of water molecules to attract each other. Molecules with larger volumes have a more disruptive effect on the interactions among water molecules. If the large molecules associate with each other, however, the surface area of the resulting "glob" is lower than when each molecule is in isolation, and there is less total disruption of the interactions among water molecules. Because this is the favored situation, the large molecules aggregate together, separately from the water molecules. Substances with large molecular volumes thus dissolve poorly in water, and chlorination increases the tendency of organic chemicals to bioaccumulate.[20]

Each chlorine atom added to an organic molecule increases the volume of a molecule further, so an increasing degree of chlorination progressively reduces water solubility. Each new chlorine atom, in fact, increases bioaccumulation by a greater amount than the one that came before.[21] Thus, the first chlorine atom has a modest effect: for the benzenes and phenols, the monochlorinated types are about five times more soluble in fats and oils than the chlorine-free ones. By the time we add a fifth or sixth chlorine atom, each one increases oil solubility by several hundred-fold, so that hexachlorobenzene is over 1,500 times more oil-soluble than benzene, and pentachlorophenol is over 4,500 times more so than phenol (see fig. 5.1, appendix F).

The relationship between chlorination and oil solubility is a general one, applying to virtually all types of organochlorines (fig. 5.1, apendix). Bioaccumulation is thus not limited to a few "bad actor" organochlorines. Indeed, fat-soluble organochlorines include a great number of compounds, representing a wide range of structures and applications, including both aromatic (benzene-based) compounds and aliphatic ones (those with chain structures). This does not mean all organochlorines are *very* bioaccumulative; add a single chlorine atom to a water-soluble compound, like an organic acid, and the resulting organochlorine will have only a modest attraction to fatty tissues. But a vast number of organic chemicals—particularly those used in the chemical industry, which are derived from petroleum—are oil soluble from the start. Adding chlorine atoms to these molecules creates a vast array of bioaccumulative organochlorines.

Two additional issues amplify the bioaccumulation problem with organochlorines. First, the bioaccumulation of an oil-soluble chemical will be moderated if the body can metabolize it to a water-soluble form and excrete it, but organochlorines are more difficult to metabolize and excrete than their nonchlorinated analogues are. Vertebrates have a battery of enzymes they have evolved to partially degrade and excrete

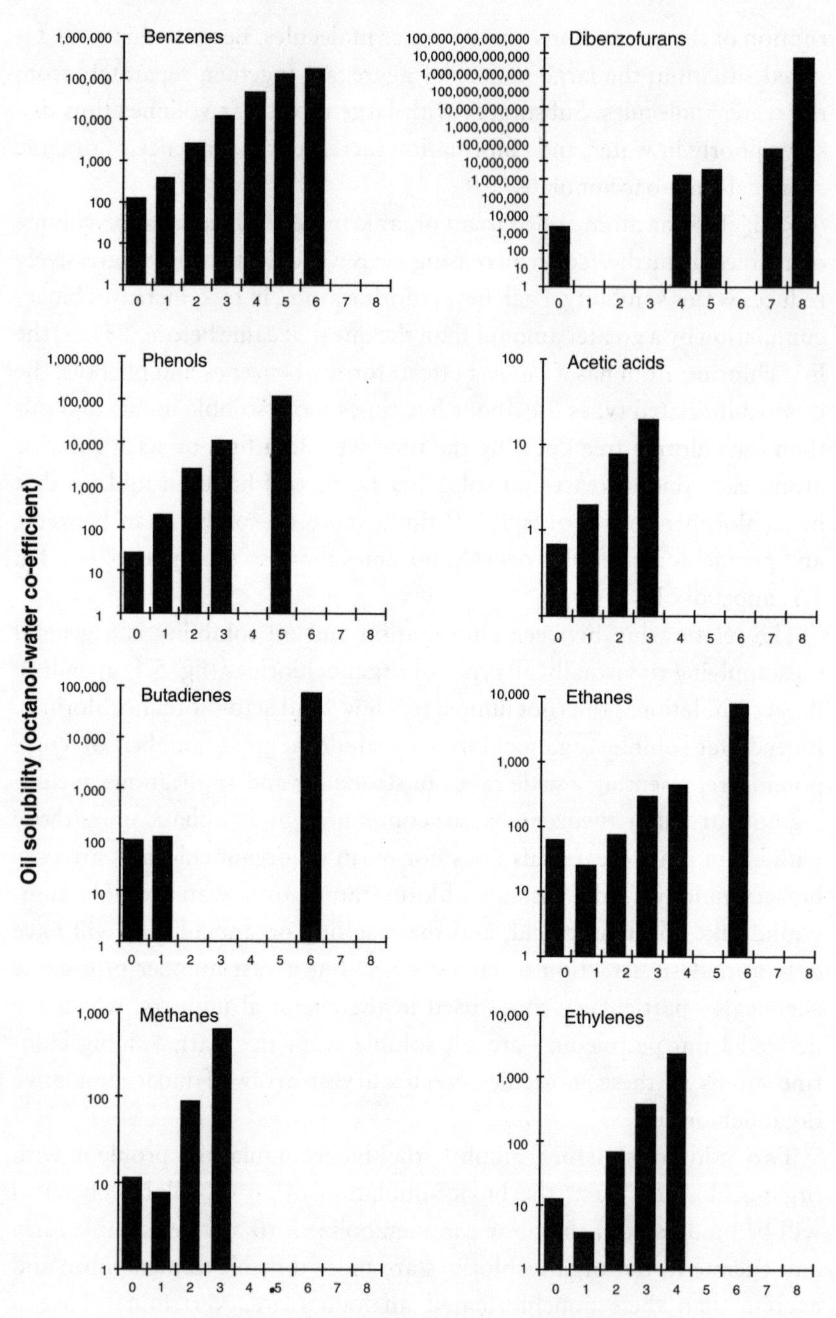
Benzenes
Dibenzofurans
Phenols
Acetic acids
Butadienes
Ethanes
Methanes
Ethylenes
Oil solubility (octanol-water co-efficient)
Number of chlorine atoms on molecule

potentially toxic chemicals they encounter in the environment and the plants they eat; these enzymes are effective against a wide variety of natural organic compounds. But because organochlorines are not naturally abundant in land plants, vertebrates or the ambient environment, humans and most other so-called higher species have not experienced the pressures of natural selection that would lead them to evolve specific mechanisms to break them down. The body does have "all-purpose" detoxification enzymes that break down many different kinds of substances,[22] but they are ineffective against organochlorines that are very stable chemically. For example, the enzymes the body uses to metabolize and excrete a class of organic chemicals called polynuclear aromatic hydrocarbons (PAHs) are induced if the body is exposed to chlorinated PAHs like dioxin, but the enzymes have little or no ability to break dioxin down.[23] Thus, TCDD has a half-life in the human body of 7 to 11 years, and octachlorodibenzofuran has a half-life that EPA estimates as "infinite";[24] the nonchlorinated dioxins and furans, in contrast, do not pose a bioaccumulation problem. The body's general detoxification enzymes can break down some organochlorines into other organochlorines that are more easily excreted, but, as we saw in chapter 2, the new substances are frequently more toxic than the original compound.

The second problem is that persistence and bioaccumulation work together to magnify their hazards. A substance that is long-lived in the environment has more time to migrate into the tissues of living things, and one that bioaccumulates is protected from breakdown by physical and chemical agents in the ambient environment. The double-whammy of persistence and bioaccumulation together begins to explain why organochlorines account for the vast majority of substances on lists of pollutants that have built up in the tissues of wildlife and humans. Longer lived and thus more likely to build up in the environment, more bioaccumulative and thus more likely to accumulate in our bodies, and more dangerous once they are inside us, organochlorines are clearly an extraordinary environmental

Figure 5.1
Effect of chlorine on bioaccumulation of organic chemicals. In all major groups, chlorination increases the tendency of a chemical to dissolve in fats and oils, and each additional chlorine atom has a greater effect. Oil solubility is shown as the octanol-water coefficient, the ratio of a substance's solubility in octanol to its solubility in water. Blank cells indicate no data available. (Source: See appendix F.)

threat. Members of the class of organochlorines tend to be hazardous not because of guilt by association, but because chlorine chemistry *by its nature* gives them properties that threaten health and the ecosystem.

Ironically, the properties that make the products of chlorine chemistry a threat to health and the environment are the same ones that make them useful in industrial applications. The reactivity of chlorine gas makes it an effective bleach, disinfectant, and chemical reactant, but it also results in the creation of a vast number of organochlorine by-products in any chlorine-based reaction. The stability of organochlorines makes them useful as plastics, refrigerants, and pesticides, but it also causes them to persist in the environment. The oil solubility of organochlorines makes them effective solvents for industrial cleaning operations and cutting oils for metallurgical processes, but it also results in bioaccumulation. And the toxicity of organochlorines makes them good pesticides and pharmaceuticals, but it makes them dangerous when they are released into the environment. To argue that there is nothing special about the hazards of organochlorines, one would have to say there is nothing special about their utility, either. Chlorination gives organic chemicals novel powers to perform industrial work; produced on a massive scale, however, these same abilities become powers to destroy. A useful invention becomes an uncontrollable scourge as it enters the environment.

Natural Organochlorines

Although chlorine gas and organochlorines are not naturally abundant, they do occur in nature; it is likely that at least several hundred organochlorines are naturally produced, particularly by algae, fungi, and microorganisms. Thus, although it takes a considerable amount of energy to transform chloride salts into organochlorines, it is not impossible for organisms to do so. What, then, is the extent of natural organochlorine production, and what does it imply about the hazards or safety of man-made organochlorines? Does the existence of natural organochlorines mean there is no reason to be concerned about man-made ones?[25]

It remains unclear how many organochlorines are naturally produced. Some scientists have estimated the number of naturally occurring organochlorines at over a thousand.[26] But we must be skeptical of reports of natural organochlorine production simply based on their occurrence in an organism, without ruling out the possibility that the compound may be an industrially-produced contaminant or a metabolite of one. It is also important to ensure that an organochlorine detected in a sample is not merely an artifact of the analytical procedures carried out in the laboratory, because organochlorines are frequently introduced into samples as part of the process of analyzing them, and they may combine with chemicals in the sample to produce novel chlorinated substances. This, in fact, has turned out to be the case with a number of the organochlorines once reported to occur naturally in plants.[27] With the entire planet now coated with man-made organochlorines—and every laboratory now filled with them—it is very difficult to assert with confidence that finding an organochlorine in an organism implies that it was produced there naturally.[28]

Only through the controlled growth of organisms in the presence of specially labeled radioactive chlorine atoms, called isotopes, is it possible to distinguish internally produced from externally acquired organochlorines. Most studies have not been so rigorous before claiming to have isolated another "natural organochlorine." For instance, Gordon Gribble, a Dartmouth College chemist who has become the most outspoken cataloguer of natural organochlorines, says that an organochlorine occurs naturally in cows.[29] A 1983 study did find a chemical called 3-chloro-9H-carbazole in a sample of bovine urine, but the authors did nothing to verify that the compound had actually been produced by the cow. They did note, however, that it "was obtained only in very small amounts . . . and the origin of this compound is mysterious. . . . The usefulness of urine as a starting material for the isolation of endogenous [compounds] is questionable. It is likely that exogenous compounds are encountered as well."[30] This is a particularly important example, because this is the first and only report of an organochlorine in a mammal; Gribble's repeated citation of the study to support the contention that mammals produce organochlorines, without noting the authors' own suspicion that the chemical is not a natural product, is misleading. There is no telling how many of the other substances listed as "natural organochlorines" may not be truly natural.

Nevertheless, there have been enough well-conducted studies to show that a considerable number of organochlorines are produced in nature, primarily by single-celled microorganisms, fungi, some plants, and marine organisms like algae, sponges, and coral.[31] Absolutely no organochlorines, however, are known to occur naturally in the tissues of humans.[32] In fact, organochlorines appear to be completely foreign to all terrestrial vertebrates—mammals, birds, reptiles, and amphibians—with a single known exception, a rare Ecuadoran tree frog that produces a toxin with a single chlorine atom to deter predators.[33] Every one of the hundreds of organochlorines found in our tissues and those of wildlife is, as far as we know, an industrial contaminant.

Although the number of natural organochlorines is large, only one is known to circulate in the environment in significant quantities. Chloromethane is produced, mostly by algae and fungi, in quantities estimated at 2 to 5 million tons per year.[34] The simplest and one of the least persistent organochlorines, this substance is thought to play a role in the natural regulation of the stratospheric ozone layer.[35] Nature maintains stratospheric ozone at its proper abundance by balancing the creation of ozone (from oxygen emitted into the environment by photosynthetic plants and bacteria) with its destruction by the natural production of chloromethane by marine organisms. If chloromethane were produced at a rate faster than it can be broken down, levels of stratospheric chlorine would rise; ozone would be depleted, allowing dangerous ultraviolet light to reach the earth's surface, and populations of many organisms, including those that produce chloromethane, would be reduced. In this way, the natural production of chloromethane is strictly regulated at the steady-state level required for proper maintenance of the ozone layer.

Other organochlorines are formed in trace quantities. Most natural organochlorines are produced as chemical deterrents against predators or parasites.[36] That is, the vast majority of natural organochlorines are made in very small quantities for internal purposes only and are not released into the environment. For example, the chlorinated terpenes that algae produce to deter other organisms from feeding on them are held in special compartments surrounded by membranes that isolate them from the alga's own internal processes.[37] Within the organism that produces them, organochlorines are generally maintained at very low concentrations, usually in the very low parts per million range.[38] A few organochlorines are made for external purposes but on a very local scale: for instance, ticks synthesize dichlorophenol as a pheromone to attract mates, and the German cockroach makes a chemical called monochloroblattellastanoside-A for the same purpose.[39] But these pheromones are so powerful that the quantities produced are tiny. Each tick produces 5 billionths of a gram of dichlorophenol, of which a little over 2 billionths of a gram is chlorine. This minuscule quantity stands in stark contrast to the 40 million tons of chlorine gas the world chemical industry now produces each year—about 6,600 grams per person, over 3 trillion times the "per-capita" production seen in ticks.[40]

The character of organochlorines that are naturally produced tells us something else about this class of compounds. In general, there are some obvious differences between the kinds of organochlorines found in nature and those that are produced by the chemical industry. The natural ones typically have fewer chlorine atoms and are less persistent and less toxic than their human-produced cousins.[41] Even so, virtually all the natural organochlorines ever discovered are produced by organisms precisely because they are potent disrupters of biological systems. The vast majority serve as natural pesticides, toxins, and antibiotics to deter attack by predators, parasitic fungi, and microbes; they are toxic to a wide variety of animals and microorganisms, including humans. There is a strong correlation, for instance, between the antimicrobial toxicity of material extracted from marine organisms and the material's organohalogen content.[42] These findings confirm the view that organochlorines, even those naturally produced, tend to be toxic and potentially harmful to living things.

A few other natural organochlorines—like the two chlorinated insect pheromones and a plant growth hormone—serve not as toxins but as chemical signals that regulate biological processes, confirming the view that organochlorines tend to be biologically active. A chlorinated growth hormone produced by pea seeds in an early stage of development is a remarkable example. Growth hormones are essential to regulating growth and senescence during any plant's life cycle, but this one appears to function as a "death hormone," in the words of the botanist who discovered it. The substance is not produced by the adult plant but is synthesized by the seed, still in its pod, which releases it into the "body" of the mother plant. The chlorine atom on the molecule makes the hormone many times more powerful than nonchlorinated growth hormones, and it forces the mother plant into early senescence and death, eliminating the mother from competition for sunlight and soil resources with the progeny plant.[43]

In addition to these specific compounds that are made in trace amounts, there is some evidence that chlorine may be incorporated into organic matter in significant quantities in some terrestrial environments. A series of Swedish studies has found that soils from forests and peat bogs contain unexpectedly high concentrations of a parameter called AOX, which measures the total amount of organically-bound halogens in a sample.[44] The halogenated organic matter in soil consists mostly of very large, complex molecules, so it is not amenable to the methods normally used to characterize specific substances. As a consequence, no one knows what chemicals are actually in this AOX, but there are some clues. When the soil is divided up into various chemical fractions, the AOX is highest in the fraction that contains the humic and fulvic acids—compounds derived from the rich, complex humus formed when soil organisms degrade and polymerize lignin, a major component of wood. When the AOX from humic matter is degraded in the laboratory into smaller molecules that can be analyzed, mono- and dichlorinated derivatives of lignin are found, suggesting that chlorine has been incorporated into the humic material itself.[45]

The source and environmental significance of these organochlorines remain unclear, however. First, we do not know where the chlorine in chlorinated humic matter came from. It could be that soil organisms actually turn chloride ions in the soil into an active form that they then incorporate into humic acid; indeed, enzymes that can carry out this process have been identified in soil organisms.[46] But there are also enzymes that can incorporate organochlorine pollutants in the soil into humus, so the chlorinated humic acids could be the result of a natural process acting on industrial organochlorines that are now widespread in the environment.[47] Thus, the AOX in soil could be a natural product, an unnatural one, or a mixture; the available information cannot tell which is the case.

There is one leading candidate as a natural precursor of chlorinated humic acid. The chlorinated compounds found when humus is degraded in the laboratory resemble a group of chemicals called chlorinated anisyl metabolites (CAMs). Some kinds of fungi that degrade fallen wood generate a specific kind of CAM—chlorinated anisyl alcohol—which they use as a chemical intermediate in the production of hydrogen peroxide, which in turn they use to degrade the lignin in the wood. The fungus produces this chlorinated chemical precisely because it is very stable. It needs a compound that is resistant to its own powerful wood-degrading enzymes, and the nonchlorinated version of the chemical breaks down too quickly to be useful. In the reaction in which the fungus produces hydrogen peroxide, CA alcohol is changed into CA aldehyde, which the organism can then recycle back into the alcohol form, and the cycle can begin again. The fungus thus reuses its CAMs rather than having to produce ever more of them. If either compound is released into the environment, it can be broken down, releasing chloride, or it can be incorporated into humic material. Samples of rotting forest matter consistently find that CAM is present, so it is quite possible that this process is an important source of the AOX found in soil.[48]

We do not know the environmental significance of this natural AOX, but it appears to be less worrisome than the kind of organochlorines that industry produces. Chlorinated humic acids produced by incorporation of chloride into humus are more easily degraded than those produced when organochlorines are bonded to humic acid,[49] and a significant portion of the former type are degraded back to chloride, in both the soil and groundwater.[50] Further, it appears that naturally chlorinated humic acids are not toxic. They are of such high molecular weight that they have low bioavailability, meaning that organisms cannot easily ingest or absorb them.[51] CAMs, in particular, are considered poorly bioavailable and nontoxic when they are incorporated into humic acids.[52] As one group of scientists studying the issue noted, their toxicity is so low that chlorinated humic material can serve as a nonhazardous storage medium for organochlorines; this group suggested that incorporation of organochlorines into humus might be a useful means of detoxifying land contaminated by industrial operations.[53]

All these data suggest that soil AOX, to whatever extent it is produced naturally, is part of a natural terrestrial cycle, the same way that chloromethane is part of an atmospheric-marine cycle. Chlorinated humic acid is produced by soil organisms, sequestering potentially toxic organochlorines produced by fungi; eventually the chlorinated humic compounds are degraded and returned to chloride. There is thus a constant incorporation of chloride into organic matter, balanced by its degradation and release back into chloride. A natural cycle clearly limits the accumulation of natural organochlorines, because the ratio of chlorinated to nonchlorinated components in the humus is far lower than it would be if the incorporation of chlorine into organic matter were not limited by some natural constraint, such as the degradability of these compounds and the limited occurrence of the enzymes that mediate their synthesis.[54]

The existence of AOX in soils has led to some wild claims. Some authors have argued that the quantity of naturally produced organochlorines dwarfs the amount produced by industry. They base this claim on studies by Swedish researchers that found more AOX in a peat bog than could be accounted for by measurements of organochlorines in precipitation, which they take as evidence that most of the organochlorines in the soil had been produced in the bog itself.[55] But this conclusion does not necessarily follow from the data. First, it is probable that industrial organochlorines entered the bog through routes the authors did not measure, including fog, dew, and mist, in the vapor phase, attached to airborne particles, and in groundwater. These unaccounted sources of pollution could explain the failure of the researchers' input-output equations to balance, obviating the need to posit an unseen force like massive natural production of organochlorines. Second, the AOX method measures the concentration of all organohalogens—not just organochlorines but also organobromines and organoiodines, which are known to be produced naturally in much greater quantities than organochlorines. An imbalance in AOX says nothing about an imbalance in organically-bound chlorine; nature may simply be producing large quantities of organohalogens of different types. Third, the AOX method overestimates the true amount of organohalogens if the sample contains ionic chlorides, bromides, or iodides, as it will if it contains any living matter, including microorganisms. One study of soil found that AOX results declined by 10 to 80 percent when an additional step that efficiently removes halide salts was added to the procedure.[56] Most soil samples are likely to contain halides, so estimates of the quantity of organically-bound halogen present will often be artificially high. Finally, peat bogs are a very special environment, because they are unusually acidic. This quality slows natural degradation processes—so that either natural or synthetic organochlorines can build up to particularly high levels—and it can also speed up natural halogenation processes. Thus, even if a large quantity of organohalogens is produced in peat bogs, this would say little about the quantity or character of organohalogens produced in the rest of the world's environment, which is chemically very different from a peat bog.

In summary, organochlorines may be formed naturally in peat bogs and forest soils, but the quantitative and environmental significance of their production is not established. Production of organochlorines in soil, if it takes place in large quantities, occurs within the context of a natural system that limits their production, persistence, bioavailability, and toxicity. There is no reason to believe that global accumulation of organochlorines that are bioaccumulative, bioavailable, persistent, or toxic is not entirely a twentieth-century phenomenon. The lessons of global warming, ozone depletion, and eutrophied lakes should have taught us this: the existence of natural cycles does not give humans an excuse to overload those cycles with massive quantities of industrial compounds.

For millions of years, natural feedback processes have limited the amount of organochlorines in the global ecosystem to relatively small quantities. Now the industrial production of 40 million tons of chlorine each year has severely disrupted this balance. The generation and dispersal of persistent synthetic organochlorines have far outstripped nature's ability to break them down. Whether there are 3 or 3,000 naturally produced organochlorines, whether they are produced in quantities measured in micrograms or kilotons, the advent of industrial-scale chlorine chemistry has overwhelmed nature's ability to degrade and cope with these compounds.

Unnatural Dioxin

In 1980 scientists from Dow Chemical—the world's largest chlorine manufacturer and a major producer of Agent Orange, then facing liability suits from Vietnam veterans—advanced the position that dioxin occurs naturally in the environment and has been with us "since the advent of fire." With this theory, which Dow called "Trace Chemistries of Fire," the company argued that dioxins are formed in any combustion process in which natural chloride salts are present, including forest fires, volcanoes, and household stoves. Natural dioxin, according to Dow and its allies, meant that chemical companies could not be blamed for dioxin exposure, and it meant that attempts to control dioxin produced during the life cycle of industrial chemicals would be misplaced and ineffective.[57] If dioxin were produced by all forms of combustion, addressing it would require action directed not at the products of chlorine chemistry but at a much broader base of processes intrinsic to industrialization, including the burning of coal, oil, gasoline, and wood.

Dow's theory still plays a role in the public debate over dioxin. In a 1995 speech, Chlorine Chemistry Council managing director Kip Howlett argued that "dioxin occurs as a natural—though unwanted, by-product of combustion. Automobiles, forest fires, incinerators, wood stoves, and even barbecues can be sources of dioxins and furans."[58] The Vinyl Institute, in turn, commissioned a report which I analyze in detail in chapter 7 that purported to find that dioxin is formed in any and all types of combustion facilities, irrespective of the presence of chlorinated chemical wastes, due to the presence of salt.[59]

Whether burning salt really does produce tiny quantities of dioxin has not yet been settled, because it is impossible to avoid contamination of materials being burned, including fuels and air, by organochlorines, which certainly produce dioxin when incinerated. For example, the measurement of dioxins in air or soot from a forest fire in 1998 does not mean that these substances are natural products, because forests everywhere are now heavily contaminated with organochlorine solvents, pesticides, and other products of industrial chlorine chemistry, which are known to produce dioxin when burned.[60] There are mechanisms that could plausibly account for the formation of dioxins from salt at high temperatures if suitable catalysts are present; the contamination problem, however, means that there are no direct and reliable empirical data to indicate whether these processes actually do produce dioxin in nature.

However that issue is resolved, Dow's assertion that combustion of chloride is a major dioxin source has been proved false. A number of lines of evidence indicate that dioxin contamination of the environment is almost exclusively a result of the manufacture and dispersal of chlorinated organic chemicals, not the natural burning of chloride salts. First, the quantity of dioxin released from various sources indicates that if any dioxin is produced naturally, the quantities are negligible. EPA's dioxin source inventory estimated that more than 99 percent of all dioxin in the United States comes from industrial sources,[61] and a global inventory indicates that no more than 3 percent of dioxin comes from the burning of "biomass."[62] Even these low figures include "unnatural" dioxin, because forests and fields contain pesticides, solvents, and other organochlorines that are applied to them on purpose or accumulate as contaminants, and these produce dioxin when burned.

Second, the data in Dow's original paper indicate that natural and industrial combustion sources that burn chloride are insignificant compared to processes associated with chlorine chemistry. Soil samples from Dow's own chlorine chemical manufacturing and incineration facility in Midland, Michigan, contained dioxin in concentrations thousands of times higher than in urban areas, which are themselves subject to contamination by general industrial sources like power plants, vehicles, and heating boilers. Trace amounts of dioxin were found in combustion residues from automobiles, home fireplaces, and cigarettes, but the dioxin levels found in particulate matter from Dow's own incinerators were orders of magnitude higher.

Third, the dioxin levels found in the preserved tissues of ancient human beings indicate that dioxin was not a significant pollutant before the advent of chlorine chemistry. Several studies have analyzed the dioxin and furan content of mummified and frozen remains of people several hundred to several thousand years old, including individuals from cultures that cooked over indoor fires and were exposed to considerable amounts of combustion emissions. All of these studies have found that dioxin levels (measured as TCDD-equivalents) in ancient tissues were no more than 1 to 2 percent of the amount found in modern humans, and even this could represent contaminants deposited in the samples in modern times, especially during handling and analysis.[63] As EPA noted, "The theory that much of today's body burden could be due to natural sources such as forest fires has been largely discounted by testing of ancient tissues which show levels much lower than those found today."[64]

A shorter-term but more reliable record of dioxin contamination is found in sediment cores from lakes and seas in the Great Lakes and Europe. Every sediment study shows that dioxin levels were extremely low before the twentieth century, despite the fact that natural and industrial combustion processes were abundant in this period. Sediments in Swedish lakes show no measurable dioxin before 1945,[65] and those in the Great Lakes show none before 1920.[66] In the Baltic, dioxins and furans were present in a sediment sample dated to 1882, but the levels were 20 times lower than the peak concentrations in 1978.[67] A study of two lakes in Germany's Black Forest found that sediments from the seventeenth and eighteenth centuries contained very small quantities of dioxins and furans—77 and 34 times lower than the maximum concentrations from this century (figure 5.2). Expressed as TCDD- equivalents, the ratios were even higher: 310 and 90 times greater in modern than prechlorine sediments.[68] In New York's Green Lake, very small quantities of dioxins and furans are present in layers from the late 1800s, but at concentrations 1,500 times lower than those found in the 1960s.[69]

There is some question about whether even these very low concentrations are truly from the era before chlorine chemistry. Sediment layers are almost never perfectly sequential; forces that disturb the sediment can mix newer material downward, contaminating older layers with more recent pollutants. Moreover, samples may become contaminated during preparation and analysis by the dioxin that is now present everywhere in dust and air; indeed, most sediment samples are air-dried for a long period of time before analysis, leaving ample time for modern dioxin to be deposited onto historical materials. Finally, error is always a problem in efforts to date materials, so there is the possibility that a sediment layer thought to be from 1880 is in fact from 1910.[70] One recent study does suggest, however, that there probably were very small quantities of dioxin created before the advent of chlorine chemistry. In this study, scientists found an archived sample of British soil from the 1880s, prepared it as carefully as possible to eliminate all known sources of contamination, and found that it contained low concentrations of dioxins and furans.[71] It is impossible to rule out other sources of contamination of which the researchers were not aware, and this work should be replicated on more than a single sample; nevertheless, these results do suggest that dioxin ex-

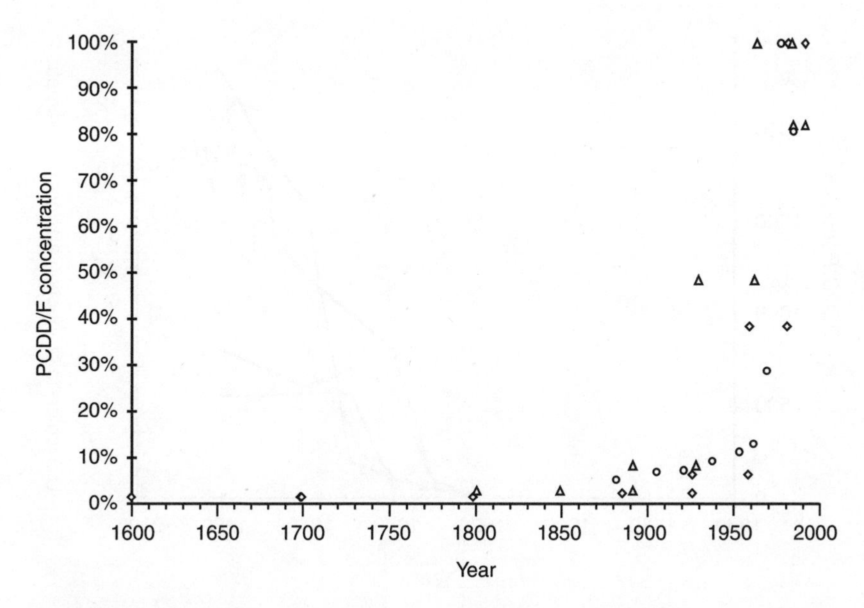

Figure 5.2
Dioxin deposition in European sediments. The vertical axis shows concentrations of total dioxins and furans in sediment cores from the Baltic (circles) and two German lakes—the Wildsee (triangles) and the Herrenweiser See (diamonds)—expressed as a percentage of the highest levels measured in each location. In all locations, levels were extremely low prior to the advent of chlorine chemistry, and they rose rapidly thereafter. (Sources: Juttner et al 1997, Kjeller and Rappe 1995.)

isted before chlorine chemistry, albeit in very small amounts, presumably due to industrial combustion of chloride-containing materials.

Only with the advent of chlorine chemistry did dioxin levels begin to rise. In all samples, dioxin concentrations began to increase slowly in the early decades of this century, and they shot up rapidly beginning around World War II, rising 25-fold or more between 1935 and 1970.[72] This pattern is consistent with the rise of chlorine chemistry, peaking in the 1960s or 1970s and declining somewhat thereafter as restrictions on dioxin-contaminated pesticides and chlorinated gasoline additives went into effect. Dioxin trends do not in any way, however, track the history of combustion, either industrial or natural. Two studies by Ronald Hites and his colleagues at Indiana University reinforce this point. One was conducted on a lake in the United States, the watershed of which suffered a

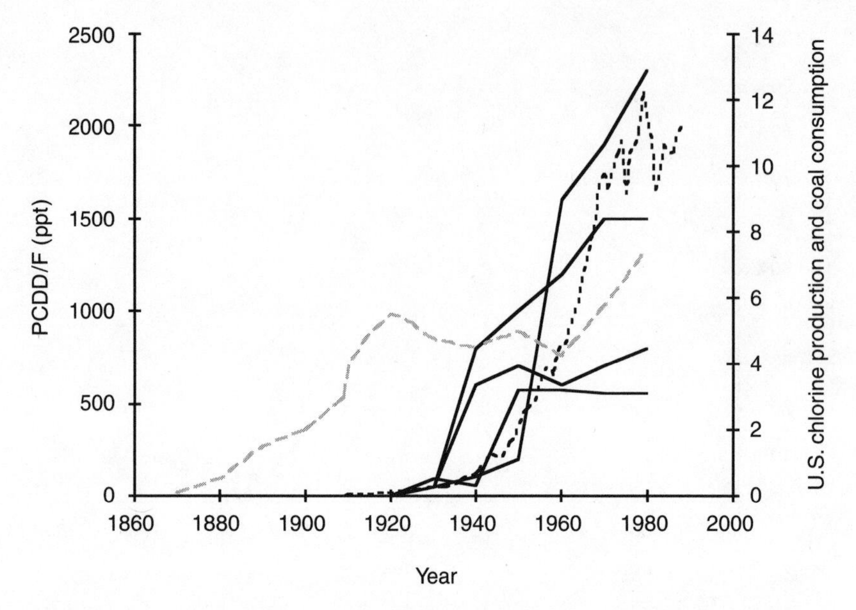

Figure 5.3
Dioxin deposition to Great Lakes sediments. Heavy lines show levels of total dioxins and furans in four sediment cores from Lake Huron; the dotted line shows U.S. chlorine production capacity (in millions of short tons per year); the dashed line shows U.S. coal combustion (in 100 millions of short tons per year). Dioxin levels were low or zero when coal combustion was at its peak, increasing only with the growth of chlorine chemistry. (Sources: Dioxin and coal redrawn from Czuczwa and Hites 1984; chlorine from Chlorine Institute 1991, Leder et al. 1994.)

major forest fire in 1937; dioxin levels in the lake sediment showed no change whatsoever around or after the time of the fire, suggesting that natural combustion is not a significant dioxin source.[73] The second study examined dioxin trends in Great Lakes sediments. This study found that dioxin levels do not follow trends in combustion of coal, which was practiced on a massive scale long before dioxin concentrations began to rise, but they do correspond quite closely to the rise of the chlorine chemical industry (figure 5.3). These results suggest that industrial combustion processes—including coal-fired power plants, steel mills, rail engines, furnaces for heating, and other industries powered by coal, which contains chloride salts—have never been major sources of dioxin, either. The authors of the Great Lakes studies summarized their results so succinctly that they are worth quoting:

There is an abrupt increase in PCDD and PCDF concentrations around 1940. . . . Starting at this time, the production of chlorinated organic compounds such as chlorobenzenes and chlorophenols increased substantially. These compounds are used in a variety of products, including building supplies, herbicides and packaging. Much of these materials eventually become incorporated in solid wastes. The trend for the production of chloro-organic compounds is very similar to the sedimentary PCDD and PCDF profiles. The agreement between these two trends is convincing despite the uncertainties introduced by sediment mixing and the errors inherent in the dating and quantitation techniques. . . . It is clear that the high levels of dioxins and furans found in presently accumulating sediments are not due to the advent of fire.[74]

Together, all these data refute the Trace Chemistries of Fire theory. As Hites and his colleague Louis Brzuzy concluded in a 1996 review, "There is no experimental evidence to support the abundant, natural production of PCDD/F. . . . Several experimental studies specifically rule out major natural sources of PCDD/F. . . . Natural sources are certainly not 'significant' sources of PCDD/F to the environment."[75]

Chemistry Evolving

What are the implications of what we have learned about organochlorine production in nature and industry for our two frameworks for environmental policy? The Risk Paradigm, by allowing the production of vast numbers and quantities of chemicals except when information is available to justify restrictions on specific substances, makes the tacit assumption that most novel synthetic compounds are compatible with natural processes. But we have seen that this assumption is false for organochlorines, virtually all of which have one or more of the ecologically incompatible characteristics of persistence, bioaccumulation, or toxicity. Even more to the point, adding chlorine to organic molecules—the essential action of chlorine chemistry—virtually always makes substances less compatible than they would otherwise have been with natural processes such as degradation, excretion, and various aspects of physiology. This pattern provides sound justification for the Ecological Paradigm's generally restrictive policy on the technology of chlorination and the class of substances it produces.

The ecologically incompatible characteristics of organochlorines begin to explain why they are not major components of the ecosystem or the natural biochemical processes of most organisms. This is an interesting reversal of the familiar argument that chemicals that are foreign to nature are dangerous. The fact that organochlorines are not naturally widespread is not a particularly persuasive rationale for their hazardous qualities; after all, one can imagine chemical technologies specifically designed to synthesize chemicals that are at once novel and degradable, nonbioaccumulative, and nontoxic. The argument works much better in reverse: organochlorines are not naturally abundant because they tend to have hazardous qualities. Chemicals that are incompatible with ecological and physiological processes have not become part of the fabric of life because that fabric was woven by the process of natural selection. The chemicals that living things produce are the result of evolutionary processes that have acted on organisms over the last three and a half billion years. An organism that produces chemicals that it cannot degrade or excrete—or that build up in the environment and poison the other species on which it depends for survival—will ultimately perish in the course of natural selection. The biochemical reactions that do survive are optimized to produce substances that are compatible with ecological and physiological processes.

If we look carefully at biochemistry, there are three striking differences between the way that nature's chemistry has evolved to be and the current practice of industrial chemistry.[76] The first lies in the way that substances are produced. As I detail in the next chapter, whenever chlorine gas is used in an industrial process, hundreds or thousands of by-products are formed by accident. Organisms, in contrast, do not create chemicals at random. Biochemical reactions are catalyzed and controlled by specific enzymes; each of these highly adapted proteins carries out a specific reaction in a coordinated multistep pathway that yields, ultimately, an intended product. Enzymes are not promiscuous; almost all are responsible for transforming a specific reactant into a specific product, which is then transformed by the next enzyme in the series. Enzymes do not set in motion random or uncontrolled reaction schemes; when by-products are formed, they are very limited in number and quantity.

The second difference lies in the fate of the chemicals produced in nature and industry. Industrial chemicals are synthesized, used, and then disposed of or discharged into the environment; reuse and recycling take place to a very limited extent. In nature, products and by-products are always recycled. Biochemical recycling often takes place within an organism, because the waste products of reactions are usually valuable substances that can be cycled back into the same or another reaction, as we saw with fungal production of CA alcohol. When a natural product cannot be reused, it is broken down and excreted, because no organism can tolerate the internal buildup of any kind of chemical. Once excreted, natural waste products turn out to be food for some other organism; they are recycled in the ecosystem rather than the individual. In the simplest and most important example of ecological recycling, animals produce and exhale carbon dioxide as a by-product of respiration, the process by which they extract energy from food. The buildup of carbon dioxide would present a crisis were it not for plants, which consume the carbon dioxide in the process of photosynthesis, releasing oxygen that animals then inhale for use in the respiration process.

The third difference is that natural systems have feedback mechanisms that sense when levels of a substance increase and then respond by reducing its production or hastening its breakdown. Consider the controls that adult mammals have over the levels of steroid hormones in their bodies. If the concentration of a hormone in the blood rises, more hormone receptors are switched into an active state. These receptors then signal the body's genes to do two things: reduce the production of the enzymes that produce fresh hormone and increase the synthesis of the enzymes that degrade it. As a result, hormone levels began to fall, receptor activity gradually declines, and both return to their normal steady state. The same kind of feedback processes take place in the ecosystem, as the carbon dioxide example shows again: if carbon dioxide levels in the natural atmosphere begin to rise, plants prosper, more carbon dioxide is consumed, and atmospheric levels return to normal. As the biophysicist Harold Morowitz has pointed out, material cycles and feedback loops are the essence of natural chemistry and are prerequisites for an "environmental matrix in which biological systems can arise and flourish."[77]

Biochemistry's superefficient, cyclical organization could not be more unlike the one-way flow of materials through industrial chemistry: virgin feedstocks like hydrocarbons and chlorine gas are transformed into huge quantities of diverse products and waste chemicals, which are then dispersed into the environment, without any expectation that these compounds will ever be taken up into natural cycles. Moreover, there is no feedback from nature into the control of these proceses, no natural check on the quantity or types of chemicals that industry produces. The signals that regulate industrial production come from the market—revenues and profits—which are more likely to drive production faster and faster than to keep it within any kind of natural limit.

In the absence of natural constraints, we need to impose our own limits on industrial chemistry, and we should do so by imitating the rules of nature's chemistry. The goal of the Ecological Paradigm is to limit the production of synthetic substances to types and amounts that we are confident can be integrated into the natural materials cycles on which life depends.[78] To accomplish this end, the Ecological Paradigm prohibits the production of substances that resist natural degradation processes and avoids the manufacture and use of novel synthetic compounds whenever possible. Organochlorines, largely foreign to nature and characterized by a pattern of persistence, bioaccumulation, and toxicity, are an obvious priority for putting this perspective into action.

Of course, we could take our ecological conservatism too far. Organisms have learned to cope with new chemicals throughout the course of evolution. In a famous example of evolutionary novelty, plants have spent the last few hundred million years evolving new toxins to repel the insects that would consume them, and the insects have evolved ways to detoxify those poisons.[79] But this story should not make us sanguine about our own ability to cope with the products of industrial chemistry. Evolution proceeds slowly over hundreds or thousands of generations by the selective death of those who cannot cope with a new compound; a happy ending for the species comes after centuries of suffering for countless individuals. Moreover, the natural production of toxins is highly specific, so the response that evolves against it is specific, too. No organism has ever had to adapt to thousands of novel chemicals at the same time. Finally, the most common adaptation in the face of a toxic food source is to find something else to eat. Birds, for instance, learn not to eat monarch butterflies once they have had one nauseating meal. When toxins are not restricted to a single species but contaminate the entire planet and its food web, however, this sensible strategy becomes utterly useless. We have not just poisoned the monarchs, or even all the insects, but everything around us, including ourselves.

6

The Chlorine Business

The hydra throve on its wounds, and none of its hundred heads could be cut off with impunity, without being replaced by two new ones which made its neck stronger than ever.
—Ovid, *Metamorphoses*[1]

The second great task of the ancient hero Hercules was to fight the many-headed hydra. All who had tried to vanquish the hydra had been killed, overwhelmed by the beast's ability to grow two fearsome new heads each time one was cut off. In a fierce battle, Hercules slew the hydra not with brute strength but with strategy: he cauterized the stump of each severed head with a burning torch in order to prevent new heads from sprouting. Finally, he reached the central, immortal trunk from which all the hydra's heads grew; this he severed and sealed with the torch, burying the remains under a giant boulder.

Chlorine chemistry is a diverse and many-headed creature that we can look at in one of two ways. The Risk Paradigm views organochlorines as a collection of thousands of individual and unrelated chemicals to be assessed and controlled one by one. The Ecological Paradigm, in contrast, sees this group of substances as the end products of a group of technologies unified by a single history, a common economic dynamic, and a single root material that flows through them all. All chlorine-based chemicals trace their descent back to the root of chlorine chemistry—the production of chlorine in the chlor-alkali process. From this origin, chlorine moves through this complex into a limited number of feedstocks, which are then used to produce a larger number of primary products, which are directed into the further production of a far larger number of substances, and so on.

Which paradigm's view is more accurate? To formulate an effective remedy to organochlorine contamination, we need to understand something about the chlorine industry's history, how it is organized, and the factors that drive its behavior.

The Rise of Chlorine Chemistry

Chlorine chemistry is a twentieth-century phenomenon, but its general invasion into modern industrial life is restricted to the past sixty years (figure 6.1). During the second half of the nineteenth century, chemical companies produced calcium hypochlorite, an inorganic chlorine-based bleaching powder, on an industrial scale for use in textile production, but they made neither chlorine gas nor organochlorine chemicals. Alkali, which was always in demand by manufacturers of glass, soap, paper, textiles, and other products, was made primarily by processing mined soda ash in high-temperature kilns, a technique that presented problems because of the limited availability of the ash and the presence of impurities in the final product. Just before the turn of the century, German chemical companies introduced a new method—the electrolysis of brine—to make alkali, which also produced chlorine and hydrogen gases as coproducts.

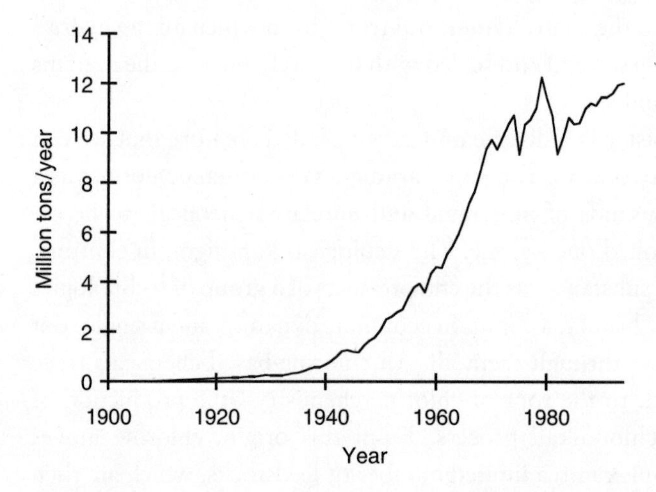

Figure 6.1
U.S. chlorine production capacity. (Sources: Chlorine Institute 1991, Leder et al. 1994.)

This chlor-alkali process was more efficient, required only brine and electricity, and produced a purer alkali product, so it came quickly to dominate the field.[2]

The first commercial chlor-alkali facility in the United States went on line in 1893 in Rumford, Maine, followed by plants owned by the Mathieson Chemical Company in Saltville, Virginia, Dow Chemical in Midland, Michigan, and the Castner Electrolytic Alkali Company and the Niagara Electrochemical Company, both of Niagara Falls, New York. In 1906, Hooker Chemical Company, later bought by Occidental Chemical, opened a very large chlor-alkali plant in Niagara Falls that dwarfed all the existing plants in capacity.[3]

Making chlorine was not the point. Companies invested in the chlor-alkali process in the industry's early days to meet a rapidly growing demand for caustic soda. Since chlorine gas had never existed before, there were few markets for it, which put these new alkali manufacturers in a difficult position. To make a ton of caustic in the chlor-alkali process, one must make about nine-tenths of a ton of chlorine. Chlorine gas is extremely dangerous to store and all but impossible to dispose of, so it must be used or sold quickly. For this reason, chlor-alkali companies can produce only as much caustic as they have viable uses for chlorine.

From the beginning, this new branch of the chemical industry had to develop applications that could serve as sinks for the chlorine formed in the process of alkali production. As one history of the industry put it, "When the electrolytic process appeared, the chlorine was first used in bleaching powder. . . . Generally, markets had to be *created,* as with almost all of the products introduced by the electrochemicals industry."[4] The first new applications emerged in the new century's first decade, when chlorine gas began to be used in the recovery of scrap tin and in water disinfection. None of these new uses grew fast enough, however, to allow the chlor-alkali industry to fully exploit the rapidly rising demand for caustic soda, so the pressure to invent new uses for chlorine did not fade.[5]

In its first three decades, chlor-alkali companies operated on a very small scale. Even in Germany, where the chemical industry was most advanced, only 25,000 tons of chlorine were produced in 1900, a tiny fraction of the 3.5 million tons of chlorine made in that country 85 years later.[6] Development in the United States was even slower; not until 1910 did the

U.S. chemical industry produce 25,000 tons of chlorine, less than 0.2 percent of current annual production.[7] Even at that point, there was little or no synthesis of organochlorines. Virtually all of the chlorine these plants produced was converted on-site into bleaching powder and sold to the textile and paper industries. Just before World War I, German chemists began to use small amounts of chlorine in the synthesis of organic chemicals, mostly in the production of chlorinated textile dyes from coal tar. The strict German patent system prevented other countries from developing independent chemical industries, however, so there was virtually no organochlorine production outside Germany.

With World War I, chlorine found a major new use as a war gas. Elemental chlorine was first deployed in 1915 on the battlefields of Ypres, with horrific consequences. The military industry soon began to make other chlorine-based chemical weapons, such as phosgene and mustard gas (dichlorodiethyl sulfide), which made their debuts during the next two years at Verdun and again at Ypres. Chlorinated chemicals were particularly effective chemical weapons because they were highly toxic and oil soluble, so they could cross cell membranes and destroy the tissues of lungs, eyes, and skin, incapacitating soldiers and causing extreme pain. As both sides developed sophisticated means to deliver war gases in shells, grenades, and other armanents, casualties of chlorinated chemical weapons rose into the tens of thousands. By the end of the war, more than a quarter of all munitions contained chemical agents, and all of these were chlorine based.[8]

The military demand for chlorine was a great boon for the chemical business. In France, a chlor-alkali industry was established for the first time, explicitly to supply chlorine for the manufacture of chemical weapons.[9] In England, the government requested the Castner-Kellner Company to increase its liquid chlorine capacity thirtyfold to keep pace with military demand.[10] In the United States, chlorine production shot up at a rate of 10.3 percent per year in the decade 1914–1923.[11]

After the war, the German patent system was dismantled. The U.S. government seized all German-owned patents and sold them to American companies for the highest bid. With this change, an independent science-based chemical industry began to develop in the United States during the 1920s.[12] Just as important, the chemical industry shifted from coal tar to petroleum as its principal organic feedstock, setting the stage for the development of the modern petrochemical industry and the production of large quantities of synthetic organic chemicals. Chlorine played only a minor role in the early days of the new chemical industry, but after about a decade, new organochlorine products began to appear. In the middle of the 1920s, the first chlorinated solvents for cleaning and degreasing came on the market; although these products provided little revenue, they served as a reservoir for excess chlorine.[13] In 1929, Monsanto introduced PCBs as a nonflammable insulating fluid in capacitors and transformers, as a pump fluid, and as a cutting oil. In the early 1930s, Du Pont brought chlorofluorocarbon (CFC) refrigerants on the market, which it had developed at the request of General Motors.[14] These new sinks allowed chlor-alkali production to continue its strong growth during the 1920s, with an annual increase in production of over 7 percent.

The chlorine industry began to come into its own during World War II. In 1937, the insecticidal properties of DDT were discovered, and the Allies used it on its own troops for mosquito control throughout the war years.[15] Polyvinyl chloride (PVC) plastic, first marketed in 1936, also found a number of applications in the war.[16] But the most important legacy of the war effort for the chlorine industry was the new and permanent alliance of government, industry, and academic scientists. Blessed with immense resources and opportunities for institutional cooperation, the scientific-military-industrial complex allowed chemical engineering to emerge after the war as a major source of new products and technologies.[17] The pace of innovation quickened, and chlorine became an ingredient in many new chemicals, including solvents, pesticides, plastics, and intermediates used in the production of other chemicals. Established uses, like pulp and paper bleaching, also increased as the economy began to expand again. In the 1940s alone, chlorine production tripled, reaching 2 million tons per year by the decade's end.

The good times continued for the chlorine industry during the 1950s, as chemical use in U.S. agriculture and industry kept expanding, and chlorine production increased by 9 percent per year.[18] Strong growth continued in the 1960s, but the first environmental trouble began to appear for the industry with the publication in 1962 of Rachel Carson's *Silent Spring*, which documented extensive harm to wildlife due to chlorinated pesticides.[19] At the same time, chemical engineers were developing more efficient ways to manufacture some compounds that had previously been synthesized using chlorine-based intermediates, so some chlorine markets began to decline for the first time. To pick up the slack and maintain overall growth, the industry began to emphasize the expansion of markets for polyvinyl chloride plastic, seeking opportunities to replace traditional materials—wood, metals, glass, and textiles—with this inexpensive, chlorine-rich plastic. Chlorine's strong growth continued, but at a less astronomical 7.7 percent per year.

In the 1970s, chlor-alkali growth slowed to just 3 percent annually and then went slightly negative in the 1980s as the economy ran out of steam and some important chlorine uses, such as chlorinated additives in leaded gasoline and numerous pesticides, were phased out. A number of chlor-alkali plants closed, and several chemical companies withdrew from chlorine manufacture altogether. Only with rapid increases in PVC consumption was total chlorine production able to stay about flat. This upheaval set the stage for new growth in the 1990s, when U.S. chlorine production began to rise again at 1 to 2 percent annually. Virtually all this growth was due to increased production of PVC plastic. Domestic PVC consumption increased at about the same rate as the gross national product (GNP), and the export of PVC and the feedstocks from which it is made—vinyl chloride (VC) and ethylene dichloride (EDC)—became the fastest growing demand sector for American chlorine industry.

How to Make Chlorine

There are three ways to produce chlorine and alkali: the mercury, diaphragm, and membrane processes. All involve passing electricity through a solution of brine to produce sodium hydroxide and chlorine gas in a fixed ratio of 1.1 to 1. Because chlorine gas and sodium hydroxide react with each other on contact, the key to each process is to separate the chlorine from the alkali immediately in a specially designed electrolytic chamber, called a cell. The three methods differ in the ways that they separate these two materials.

The mercury process, the oldest and most energy intensive of the three, involves two cells connected to each other. In the first cell, salt is split into chlorine gas and sodium at the cell's positive terminal (called the anode);

the sodium forms an amalgam with a layer of liquid mercury, which then flows into another cell, where it reacts with water to form sodium hydroxide and hydrogen gas. Most of the mercury is recycled, but significant quantities are routinely released into the environment through air emissions, water discharges, products, and waste sludges. In this century as a whole, chlor-alkali production has been the largest single source of mercury releases to the environment.[20] As recently as the 1980s, the chlorine industry was second only to fossil fuel combustion as a mercury source in Europe.[21]

Many mercury cell plants have been retired, and controls on existing plants have improved, but chlor-alkali plants remain a major source of mercury pollution. The chlorine industry is the largest mercury consumer in the United States; presumably it is even more important in Europe, where the mercury cell process is even more common.[22] Based on estimates by Euro-Chlor, the trade association of the European chlorine industry, the world chlor-alkali industry *consumed* about 230 tons of mercury in 1994; this is the quantity not recycled but lost from production processes each year. Exactly where the mercury goes remains controversial, but if we use Euro-Chlor's data, about 30 tons were released directly into the air and water, 5 tons remained as a contaminant in the product, more than 150 tons were disposed on land, and 36 tons could not be accounted for (table 6.1).[23] The actual worldwide totals are likely to be even higher, be-

Table 6.1
Mercury releases from the world chlor-alkali industry, 1994

	Grams of mercury per ton of chlorine produced[a]	Metric tons of mercury world total[b]
Consumption	16.6	229.8
Air emissions	1.9	26.3
Discharges to water	0.2	2.8
Contaminants in products	0.4	5.5
Disposed on land	11.4	157.8
Unaccounted for	2.6	36.0

[a]Ayres 1997. These data are based on mass balances prepared by the chlorine industry for facilities in Europe and may not accurately represent global averages.
[b]Based on consumption and release rates in the first column, assuming global chlorine production of 39 million metric tons per year, 35.5 percent of which is produced by the mercury process (Leder et al. 1994).

cause the well-regulated facilities of Europe are not likely to be representative of those in other regions of the world.

Mercury is an extremely toxic, bioaccumulative global pollutant that causes irreversible health damage to wildlife and humans, developing children in particular. The most tragic and infamous example of mercury pollution took place in Minimata, Japan, where the Chisso Chemical company routinely dumped mercury-contaminated waste into the local bay from the 1930s to the 1960s. Fish in Minimata Bay bioaccumulated mercury to levels 40 to 60 times higher than those in nearby ecosystems, and the local community, in whose diet fish played a key role, sufferred very high mercury exposures. In the early 1950s, symptoms of chronic mercury poisoning, including neurological toxicity, paralysis, coma, and death, began to appear in adults, and a horrifying outbreak of severe birth defects and mental retardation occurred in children. Ultimately mercury poisoning killed hundreds and injured over 20,000 people in the area.[24] Chlor-alkali production is not traditionally assumed to have been the source of the Chisso's mercury releases, because the company had been using mercury as a catalyst in fertilizer production since the 1930s. As one history of the event points out, however, Chisso began using the mercury process to make chlorine for PVC plastic in 1952. In 1953, symptoms of mercury poisoning began to appear in the local population, and over the next four years the number of victims correlated with Chisso's growing production volume of vinyl chloride.[25] These facts suggest that mercury releases from the chlor-alkali process played at least some role in the Minimata epidemic.

Mercury cells are now banned in Japan and are gradually being phased out in the United States and Europe, but releases of mercury remain a problem. A significant fraction of existing plants in North America (14 percent) and Western Europe (65 percent) continue to use this technology,[26] and despite improved control of emissions, mercury discharges still occur. In the 1980s, for instance, a major British chlor-alkali facility was found to be discharging up to 100 kilograms per day of mercury into local waterways; more than a decade later, mercury levels in the sediment remained extremely high.[27] In Italy, elevated levels of mercury in air, soil and plant tissues have been found in the vicinity of a mercury-based chlor-alkali plant owned by the Solvay Company.[28] In India, a 1990 study of waterways around a chlorine facility documented severe mercury contamination of fish and sediments.[29]

In the two other methods of chlor-alkali production, there is only one cell, which is divided by a semipermeable barrier that separates the chlorine gas from the sodium. In the diaphragm process, brine enters the cell and is split at the anode, yielding chlorine gas and sodium ions. The ions then flow through an asbestos membrane to the other pole, where they react with water to form sodium hydroxide and hydrogen gas; the chlorine, which cannot pass through the membrane, remains near the anode. The diaphragm method was developed after the mercury process and, as of 1994, accounted for 77 percent of all chlorine production in the United States and 25 percent of that in Europe.[30]

The remaining 7 percent of chlor-alkali production in both regions is based on the membrane process, the most recent of the three methods. The membrane technique is similar to the diaphragm process, except that a synthetic membrane rather than asbestos is used to separate the compartments in which chlorine and caustic are formed. The membrane process uses slightly less energy and yields products of higher purity than the other kinds of cell, but retrofitting a chlorine plant with membranes is very expensive. Because most chlor-alkali plants in the United States and Europe were built decades ago, few plants in these countries use the membrane process, but most new facilities, particularly those in Asia and Latin America, are constructed with it.[31]

Making chlorine requires enormous amounts of energy. Chlor-alkali electrolysis is one of the most energy-intensive industrial processes in the world. The production of 1 ton of chlorine requires about 3,000 kilowatt-hours of electricity, and the global chlor-alkali industry consumes about 117 billion kilowatt hours of electricity each year.[32] This quantity is about 1 percent of the world's total demand for electricity,[33] costs about $5 billion per year,[34] and is equivalent to the annual power production of about 20 medium-sized nuclear power plants.[35] As a major energy consumer, chlorine chemistry contributes considerably to all the environmental problems—global warming, acid rain, air pollution, radioactive contamination, and so on—that are associated with energy production.

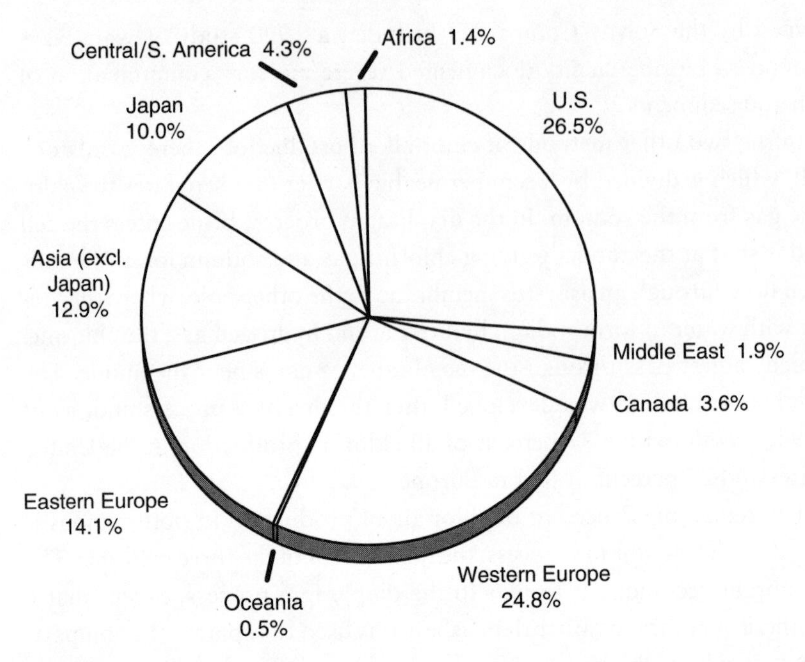

Figure 6.2
World chlor-alkali production capacity by region, 1993. (Source: Leder et al. 1994.)

The Producers

The world chlor-alkali industry produced about 39 million metric tons of chlorine in 1997. About two-thirds of world chlor-alkali capacity is located in the wealthy nations of North America, Western Europe, and Japan, with significant development in Eastern Europe and Asia as well (figure 6.2). Because chlorine markets in the rich nations have become more or less saturated, the industry is rapidly expanding production and marketing efforts in the Pacific Rim, Latin America, and the Middle East.

In the United States, chlorine production is concentrated in a few geographical regions. There are 42 chlor-akali facilities in the United States (table 6.2), but over 70 percent of production capacity is located at just 12 large plants in the Gulf Coast regions of Louisiana and Texas. In the post-

Table 6.2
U.S. chlorine producers, 1994 (capacity in kt/yr)

Company	Location	Capacity (kt/yr)	Production method
Dow Chemical	Freeport, TX	2,250	diaphragm
PPG Industries	Lake Charles, LA	1,241	mercury, diaphragm
Dow Chemical	Plaquemine, LA	1,185	diaphragm
Formosa Plastics	Pt. Comfort, TX	715	membrane
Occidental Chemical	Taft, LA	640	diaphragm, membrane
Occidental Chemical	LaPorte, TX	529	diaphragm
Occidental Chemical	Corpus Christi, TX	480	diaphragm
Olin Corporation	McIntosh, AL	402	membrane
PPG Industries	Natrium, WV	392	mercury, diaphragm
Occidental Chemical	Deer Park, TX	383	mercury, diaphragm
Occidental Chemical	Niagara Falls, NY	323	diaphragm
Occidental Chemical	Convent, LA	307	diaphragm
Vulcan Materials	Geismar, LA	268	diaphragm
Vulcan Materials	Wichita, KS	263	diaphragm, membrane
Olin Corporation	Charleston, TN	260	mercury
Niachlor	Niagara Falls, NY	240	membrane
Occidental Chemical	Tacoma, WA	215	membrane
LaRoche Chemicals	Gramercy, LA	200	diaphragm
Formosa Plastics	Baton Rouge, LA	198	diaphragm
Elf Atochem	Portland, OR	186	diaphragm, membrane
Pioneer Chlor-Alkali	St. Gabriel, LA	176	mercury
Weyerhauser Co.	Longview, WA	158	diaphragm
Occidental Chemical	Muscle Shoals, AL	146	mercury
Occidental Chemical	Delaware City, DE	139	mercury
Geon/BFGoodrich	Calvert City, KY	120	mercury, diaphragm
Pioneer Chlor-Alkali	Henderson, NV	115	diaphragm
Olin Corporation	Augusta, GA	112	mercury
Georgia-Pacific	Bellingham, WA	90	mercury
Miles Inc.	Baytown, TX	90	other
DuPont	Niagara Falls, NY	85	other
HoltraChem	Orrington, ME	80	mercury

Table 6.2 (continued)

Company	Location	Capacity (kt/yr)	Production method
Vulcan Materials	Port Edwards, WI	76	mercury
GE Plastics	Mt. Vernon, IN	55	diaphragm
HoltraChem	Acme, NC	53	mercury
Occidental Chemical	Mobile, AL	45	membrane
Ashta Chemicals	Ashtabula, OH	40	mercury
Cedar Chemical	Vicksburg, MS	40	other
GE Plastics	Burkville, AL	26	diaphragm
Magnesium Corp. of America	Rowley, UT	17	other
Fort Howard	Green Bay, WI	9	diaphragm
Fort Howard	Muskogee, OK	6	membrane
Fort Howard	Rincon, GA	7	membrane

Source: Leder et al. 1994.

war decades, the chemical industry concentrated its expansion in this region, which offered cheap and abundant petrochemicals, water, shipping lanes, and electricity. The rest of the nation's chlorine producers are scattered geographically. Three plants in Niagara Falls, New York, produce about 6 percent of the nation's chlorine, lured by extraordinarily cheap hydroelectric power. The rest of the nation's chlor-alkali plants, mostly small facilities, dot the Pacific Northwest and the Southeast, the remnants of an era when cheap power was available and pulp mills provided a ready commercial market for chlorine and alkali.

The U.S. chlorine industry is dominated by a few major players. The five largest producers—Dow, Occidental, PPG, Olin, and Vulcan—account for over 78 percent of chlorine production capacity (figure 6.3). America's largest chlorine companies are also the world's largest producers. The top ten chlorine manufacturers in the world, in order of capacity, are Dow, Occidental, ICI (Britain), PPG, Solvay (Belgium), Bayer (Germany), Atochem (Netherlands), Formosa (Taiwan), Hoechst (Germany), and Enichem (Italy). Most of the European producers, and Dow as well, are multinational, operating facilities not just in their home countries but in a number of other nations in Europe and Asia.[36]

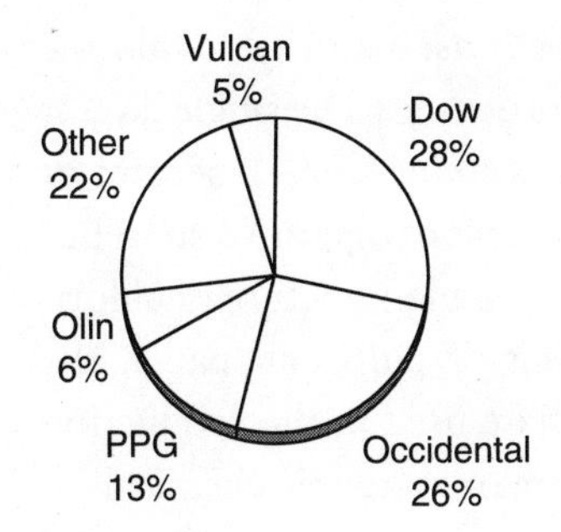

Figure 6.3
U.S. chlorine manufacturers, percentage of total production capacity, 1993. (Source: Leder et al. 1994.)

Most chlorine producers are diversified chemical companies that make a range of chemicals, some chlorine based and some not. These are the old, established firms that over the decades have developed expertise and production lines in many kinds of organic chemical synthesis, including but not limited to products that contain chlorine are or made using chlorinated intermediates. Only a few major producers in the United States—notably Occidental, Vulcan, and Georgia-Gulf—are almost exclusively invested in chlorine chemistry, with product lines dominated by chlorine gas, caustic soda, PVC and its feedstocks, and chlorinated solvents.

The Uses of Chlorine: A Family Tree

The way we look at chlorine chemistry—how we organize our perspective on the uses of chlorine—affects the kinds of policies we can envision to address its environmental impacts. If we simply list all the chemicals made from chlorine, we automatically begin to ask which individual substances on the list are worthy of regulation. There are some 11,000 organochlorines in commerce, plus thousands more formed as accidental by-products in chlorine-based processes. Making policy at this level of complexity is clearly impractical, guaranteeing a shotgun approach to a few of the most infamous chemicals on the list.

There is a better way to organize the uses of chlorine than just listing its products, because the thousands of compounds ultimately produced do not play equal roles in the chlorine industry. The vast majority of chlorine is used in just a handful of product lines and processes. The single largest application, the manufacture of PVC plastic, accounts for 41 percent of chlorine demand in the United States and 38 percent of that in Western Europe. The four largest uses—PVC production, the manufacture of chlorinated solvents and refrigerants, the bleaching of pulp and paper, the synthesis of three kinds of chemical intermediate used in the production of nonchlorinated plastics, and the manufacture of inorganic chemicals—account for almost three-quarters of all chlorine consumption in both the United States (table 6.3) and Western Europe.[37]

In nations that do not have highly specialized chemical industries—essentially, every country but the United States, Japan, and the larger nations of Western Europe—even higher portions of chlorine production go into one or two major applications. In the Middle East and Korea, over 60 percent of chlorine is used in the production of PVC alone.[38] Countries with large forestry industries use proportionally greater amounts the paper industry. In Canada, pulp bleaching and PVC accounted for 77 percent of chlorine demand in 1992.[39]

On one hand, we have a few major uses that consume the majority of the chlorine. On the other, a few areas of chlorine chemistry use just a little chlorine but lead to the synthesis of a disproportionately large number of organochlorine products. For instance, less than 1 percent of all chlorine goes into the production and use of chlorinated benzenes, from which thousands of specialty chemicals, including dyes, deodorizers, and pharmaceuticals, are produced. A little over 1 percent goes into the manufacture of pesticides, which includes hundreds of different compounds, all of them used for related purposes.

These two aspects of chlorine chemistry—the concentration of most of the chlorine in a handful of uses and the proliferation of many products from a few small applications—suggest a more productive way of looking at chlorine chemistry than just listing chemicals. Organochlorines can be classified into coherent product groups that are derived from common feedstocks or are used for similar purposes, such as pesticides, plastics, or pharmaceuticals. We should think of chlorine chemistry not linearly, as a list, but hierarchically, as a tree—like a family tree or evolutionary tree—that organizes the many chlorine-based processes into classes of related activities (figure 6.4).[40]

Table 6.3
U.S. chlorine uses, 1997

Use	Percentage of total	Growth, 1992–1997 (%/year)
Organochlorine products		
PVC	29.5	3.0
Exports of EDC/VCM	11.6	4.0
CFCs, HCFCs, HFCs	4.1	14.1
Vinylidene chloride	1.2	2.2
Trichloroethylene	0.5	0.0
Perchloroethylene	0.3	–8.4
Methyl chloride[a]	0.3	1.0
Methylene chloride (paint stripper)	0.9	–6.4
Chlorobenzenes[b]	1.0	–1.6
Chloroprene	0.9	1.0
Pesticides[c]	1.4	–2.1
Chlorinated paraffins	0.5	0.0
Other[d]	3.0	–5.4
Intermediates		
Propylene oxide[e]	7.8	0.0
Isocyanates[f]	6.7	2.5
Epoxy resins[g]	3.3	2.4
Synthetic glycerin	0.9	–0.9
Polycarbonates[h]	1.1	5.0
Inorganics		
Hydrogen chloride	3.0	0.5
Hypochlorite bleaches	2.1	2.0
Phosphorous chlorides	0.8	4.0
Bromine	0.8	2.4
Ferric chloride	0.4	5.0
Other	3.1	0.0
Direct uses		
Pulp and paper[i]	6.2	–8.3
Water treatment[j]	4.4	–1.3
Titanium dioxide	3.2	2.0
Metals production	0.3	1.7
Other	0.4	–5.1

Source: Leder et al 1994.
[a]Does not include methyl chloride for silicones, which is derived from waste HCl.
[b]Some use as an intermediate to nonchlorinated products.
[c]Includes direct use of chlorine in pesticide production and indirect use in production of intermediates for pesticide synthesis.
[d]Excludes intermediates for pesticides; includes ethyl chloride, ethylenamines, cyanuric chloride, and other organic chemicals.
[e]From propylene chlorohydrin, for polyurethanes and chemicals.
[f]From phosgene, for polyurethane.
[g]From epichlorohydrin.
[h]From phosgene.
[i]Does not include chlorine dioxide used in the pulp industry.
[j]Of this, approximately 80 percent is for wastewater treatment, 20 percent for drinking water.

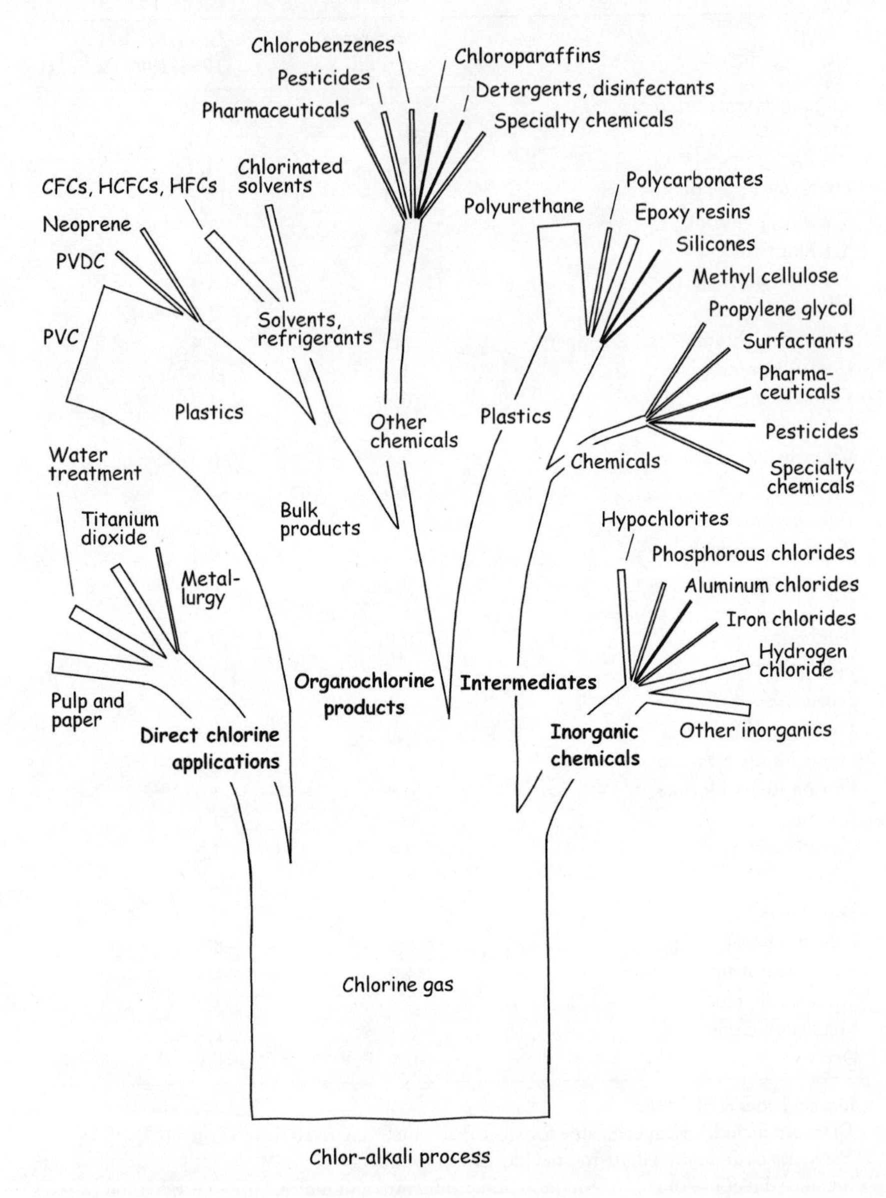

Figure 6.4
Tree of chlorine uses. The width of each branch is proportional to the amount of chlorine that goes into the associated applications in the United States (see table 6.3). Many terminal twigs on this tree can be subdivided further to represent all of the individual substances in each group. At the root of the tree is the production of chlorine gas in the chlor-alkali process.

All branches of chlorine chemistry derive from the trunk—the production of chlorine in the chlor-alkali process. The limbs represent the major feedstocks and product groups; the branches, twigs, and the tips represent finer divisions within these groups, at greater and greater levels of specialization. The hierarchical approach explicitly allows us to choose, for policy purposes, the level of generality on which to focus. For instance, we can assess and regulate each chlorinated pesticide separately, or we can focus on the group of chlorinated pesticides, all of which may be amenable to reduction by similar alternative farming methods. Similarly, we can try to stop the generation of each of the hundreds of organochlorine byproducts formed in the paper industry, or we can address all of these together as downstream consequences of the introduction of chlorine into the pulp bleaching process.

From the trunk of chlorine production, there are four limbs along which chlorine travels. The first represents the sale of chlorine to other industries for direct application, including pulp and paper bleaching, water disinfection, and certain metallurgical processes. This limb accounts for about 14 percent of chlorine use in the United States. The second and by far the largest limb, consuming 56 percent of all U.S. chlorine production, represents use within the chemical industry to make commercial organochlorine products. The third consists of the 20 percent of chlorine directed into chlorinated intermediate chemicals that are used for the synthesis of nonchlorinated chemical products. The final limb, accounting for 10 percent of all the chlorine produced, leads to the manufacture of inorganic chlorides, including hydrogen chloride, hypochlorite bleaches, and other chemicals.

Each limb can be subdivided into branches. From the first limb—the sale of chlorine to other industries for direct application—the bleaching of wood pulp in the manufacture of paper is the largest branch, taking up over percent of total chlorine production. Moderate quantities of chlorine are used to treat industrial and municipal wastewater and to remove titanium from mined ores for the production of titanium dioxide, a common white pigment. Less than 1 percent of chlorine goes into drinking water disinfection, and even smaller portions of chlorine production are used in the metallurgical industries. As I show in the next chapter, these processes involve the reaction of chlorine gas with materials that contain substantial amounts of organic matter—producing large quantities of organochlorine by-products that are discharged into the environment—so this area of chlorine chemistry is an important one from an environmental perspective.

The tree's second limb—the synthesis of organochlorine products—represents the most environmentally important sector for two reasons: it includes the majority of chlorine consumption, and the resulting organochlorines are dispersed into the economy (and ultimately, to a large extent, into the environment). This limb splits into three branches. The first—and by far the largest of all branches on the tree—leads to the manufacture of chlorinated plastics. Virtually all the chlorine in this area goes into the feedstocks for PVC, with very small amounts (less than 1 percent each) going to polyvinylidene chloride (plastic wrap), and polychlorobutadiene (neoprene). The second subbranch leads to the synthesis of solvents and refrigerants. The last branch on this limb leads to a multiplicity of relatively small-volume chlorinated chemicals, including pesticides, industrial lubricants, coolants, plasticizers, and pharmaceuticals.

The third major limb of the tree is the manufacture of chlorinated intermediates for the synthesis of chlorine-free products. Virtually all the chlorine in this limb leads to the synthesis of three plastics. Polyurethane is made by reacting isocyanates (which are synthesized from the organochlorine phosgene) with propylene oxide (synthesized from propylene chlorohydrin), epoxy resins are made from the reaction of epichlorohydrin with bisphenol-A, and polycarbonates are made from the reaction of phosgene with bisphenol-A. Much smaller quantities of chlorine find their way into chlorinated intermediates used to make a variety of industrial specialty chemicals, pesticides, and pharmaceuticals. Although organochlorine intermediates are used and transformed within the chemical industry, significant quantities of organochlorine feedstocks, intermediates, and by-products are released to the environment through emissions and wastes. This limb thus represents an intermediate environmental priority.

The final limb is the manufacture of chlorinated inorganic chemicals. The first branch in this sector is the manufacture of high-purity hydrogen chloride, which is used in the food, electronics, and pharmaceuticals industries. The second is the use of chlorine in sodium and calcium hypochlorites, used as bleaches and swimming pool chemicals. Other inorganic chemicals, including phosphorus, iron, aluminum, and sulfur chlorides, account for very minor fractions of chlorine demand. Unlike organochlorines, inorganic chlorides are not important pollutants, so this branch represents the lowest environmental priority. Organochlorines are formed as by-products when inorganic chlorides are made from chlorine gas, however, so this branch cannot be completely ignored.

Viewing industrial processes as interconnected is central to the Ecological Paradigm. With just four limbs and fewer than a dozen major branches, the chlorine tree represents a much more comprehensible way of viewing chlorine chemistry than making a list of thousands of substances. Organizing chlorine-based technologies hierarchically, according to the flow of chlorine through the tree, has another advantage: it reveals pollution that occurs during prior portions of the life cycle of any chlorine-based product. A product far out on any branch can be made only if its feedstocks—materials on the branches, limb, and trunk leading to the product—have themselves been made, along with any pollutants that occur at these earlier points in its life cycle. For example, PVC plastic is the end-product of several processes, including the manufacture of vinyl chloride monomer from ethylene dichloride (EDC), the synthesis of EDC from ethylene and chlorine, and the production of chlorine gas from salt. The tree allows us to view each product as the result of a series of industrial processes and to consider the enviromental impacts of its full life cycle in the course of formulating environmental policies.

At what level should chlorine chemistry be addressed? Should certain twig tips be pinched off, as the chemical-by-chemical approach would have it? Should some of the branches be chosen as particularly problematic and "pruned," as the Canadian government has suggested? Should the entire tree be cut down? By viewing chlorine chemistry hierarchically, the Ecological Paradigm opens up policy options that cannot even be considered under the current chemical-by-chemical framework.

PVC Heads South

Since the 1960s, PVC has gone from a relatively minor material to the second most important plastic in the world, trailing only polyethylene in production volume. PVC is now by far the largest and fastest growing use of chlorine in the world. In fact, it is the only major chlorine application still growing in the world's wealthy nations, and it is the engine of chlorine growth in developing countries, too.[41] Why has PVC become the dominant use of chlorine? The answer to this question lies in the supply/demand dynamics that have determined the industry's economics since its beginning.

Chemical companies invested in the chlor-alkali process in the industry's early days mainly to take advantage of a growing demand for caustic soda, and markets had to be created rapidly to absorb the by-product chlorine gas. Today the pressure to dump chlorine into a sink product to maintain the chlor-alkali ratio has become more important than ever. Demand for caustic soda continues to grow in its diverse uses (figure 6.5), but chlorine consumption has slackened, because environmental policies have drastically reduced chlorine use in refrigerants, solvents, and pulp bleaching. From 1987 to 1997, the U.S. industry lost over 1.6 million tons of chlorine demand in these sectors—about 15 percent of the total 1987 market for chlorine,[42] and ten chlor-alkali plants that primarily served these

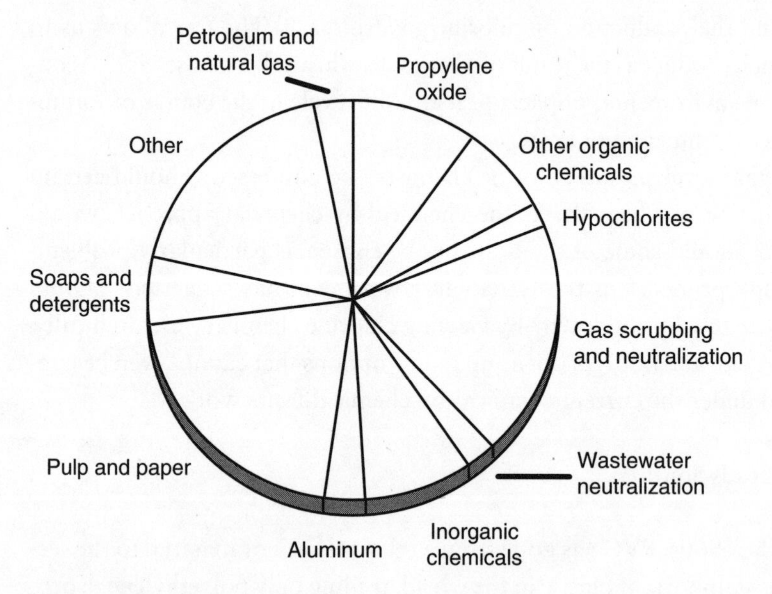

Figure 6.5
Uses of sodium hydroxide (alkali) in the United States, 1997 projections. (Source: Leder et al. 1994.)

markets were forced to close.[43] A similar pattern has unfolded in other industrial nations, where environmental regulations caused chlorine markets to stagnate or shrink (figure 6.6).

This dynamic has created a situation that industry analysts call "the world chlor-alkali imbalance,"[44] which has important economic implications for the chlorine business. When consumption of chlorine cannot keep pace with demand for caustic, the industry's growth slows or stagnates, and it cannot take full advantage of opportunities to sell sodium hydroxide. If the imbalance becomes severe, the industry will begin to lose market share. When caustic supply is short, prices go up; in 1993, for example, solid sodium hydroxide sold for about four times the price of chlorine. At some point, prices go up so much that caustic is no longer less expensive than alternatives, including soda ash, calcium hydroxide, sodium sulfate, and sodium hydroxide made through means other than the chlor-alkali process. When that happens, the chlor-alkali industry begins to lose its alkali customers and the revenue they provide. For these reasons the industry must always keep chlorine consumption growing and avoid shrinking demand at all costs.

As of 1992, chlorine and alkali demand were essentially in balance with each other. But in the 1990s and the first decade of the twenty-first cen-

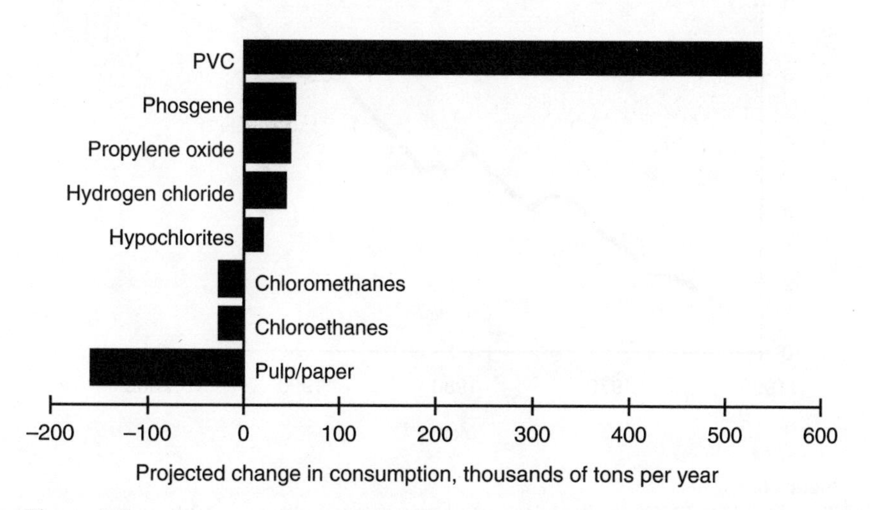

Figure 6.6
Projected change in world chlorine uses, 1995–2005. PVC accounts for virtually all of the expected growth in global chlorine consumption. (Source: Mears 1995.)

tury, world demand for chlorine is expected to grow at only about 0.7 percent per year, while demand for sodium hydroxide is expected to grow at 2 percent annually. By 1997, the worldwide caustic shortage was predicted to be about 2.6 million tons (about 6 percent of total annual consumption).[45] According to an analyst for the chlor-alkali manufacturer Elf-Atochem, "There is a logical progression toward permanent imbalance between caustic supply and demand. Domestic chlorine consumption and chlorinated exports will set operating rates for U.S. chlor-alkali capacity, with the EDC/VCM/PVC chain leading the way."[46]

The industry's strategy to rectify the chlor-alkali imbalance is to expand markets aggressively for PVC and the feedstocks from which it is made, which are already the major global sinks for excess chlorine (figure 6.6). For the past several decades, PVC production and consumption has grown at a remarkable pace (figure 6.7). Recently, however, PVC markets in industrialized nations have neared saturation because vinyl has already replaced so many traditional materials; growth in vinyl in these countries is now no greater than annual increases in gross national product.[47] This

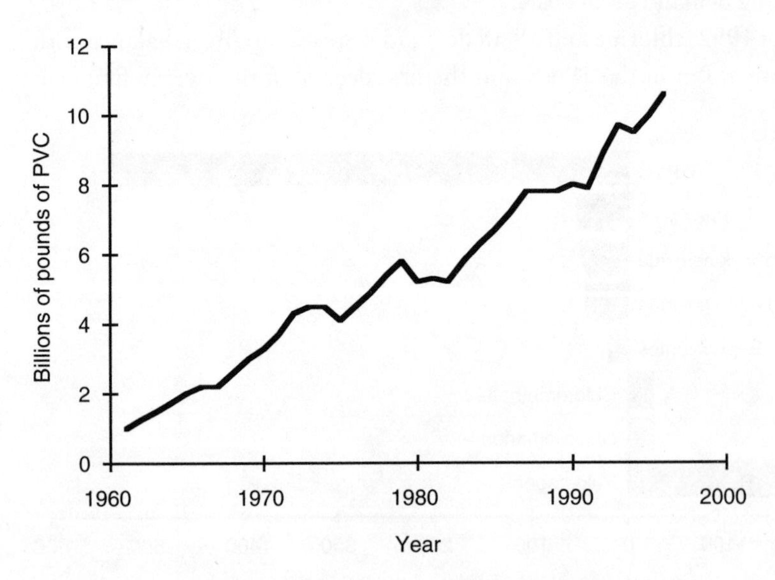

Figure 6.7
Trends in U.S. PVC consumption. (Redrawn from Waltermire 1996.)

rate of growth is not nearly enough to offset the loss of demand in pulp and paper, refrigerants, and solvents, so the industry has focused on expanding exports of PVC and its feedstocks. U.S. net exports of EDC, VCM and PVC now contain about 2 million tons of chlorine per year—over 15 percent of total chlorine production—and were expected to grow by a stunning 14 percent in 1998 alone.[48]

With vinyl markets in the wealthy countries more or less saturated, the exports go primarily to developing nations,[49] particularly those in Latin America and Asia, where PVC consumption is expected to grow at annual rates of 7 percent or more, leading to a doubling of demand each decade (figure 6.8). Why these countries? As an executive of a major Japanese PVC company explained, vinyl is a uniquely marketable product for export because poor countries need to reach only minimal levels of economic and technological development before they can be encouraged to buy plastic, and these nations usually have few environmental regulations:

> Demand for PVC in the high-population developing countries will grow rapidly after their GNP per capita reaches $500 per year. On the other hand, in the world's major industrialized countries where per capita GNP is over $10,000/year, the use

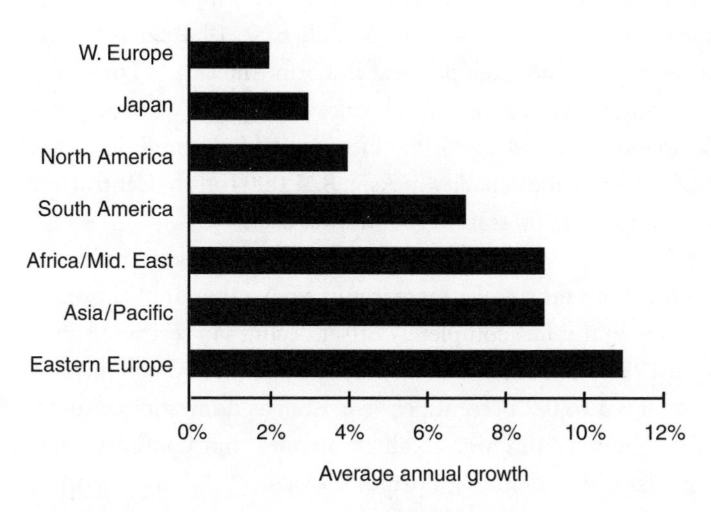

Figure 6.8
Increase in PVC demand by region, 1994–1998. In North America, Western Europe, and Japan, PVC growth is less than or about the same as GNP growth. In other regions, PVC markets are expanding very rapidly. (Source: Waltermire 1996.)

of PVC has come close to its maturity, and the growth rate of PVC may not be as much as the GNP growth rate. The concern over the disposal of waste material is one of the reasons for advanced society to refrain from excessive use of plastics.[50]

The industry has already had success with its strategy of marketing PVC products as replacements for traditional materials and products, particularly in areas where populations are becoming urbanized. The rapid increase in vinyl consumption in developing countries means that despite slow growth in PVC consumption in the wealthy nations, global demand for PVC will rise from 22 million tons per year in 1996 to 28 million tons per year 2000—annual growth of over 6 percent.[51]

PVC plastic itself is shipped abroad, but the fastest growing of all chlorine sectors is export of the feedstocks EDC and VCM. Once they arrive, these chemicals must be processed into PVC, so new vinyl production plants are being set up all over the world, transferring the environmental impacts of production to developing nations. As of 1992, 21 new PVC manufacturing facilities were slated for construction in Asia, Latin America, and the Middle East, compared to one each in Japan and Western Europe, three in Eastern Europe, and none in North America.[52] During the 1990s, PVC production is expected to grow by a stunning 6.5 percent per year in Southeast Asia, 8.1 percent in the Middle East, 14.1 percent in India and Pakistan, and 5.7 percent per year in Latin America.[53] The industry is also expanding production of chlorine, EDC, and VCM in these countries. Between 1992 and 1996, the industry added a million metric tons of new annual chlorine capacity in Asia, 825,000 tons in Central and South America, and 580,000 tons in the Middle East.[54]

The strategy appears to be working. Growth in PVC markets, primarily for export—along with far smaller increases in a few other applications—has been so strong that it has completely offset declines in restricted chlorine uses. From 1987 to 2005, total U.S. chlorine production will increase at a steady rate of 0.5 to 0.7 percent per year, even as domestic consumption shrinks.[55] A number of chlor-alkali companies that concentrate on PVC and its feedstocks—including Formosa, Occidental, Dow, Georgia-Gulf, and Geon (formerly BFGoodrich)—have expanded their production capacity at plants in Louisiana and Texas.[56] An industry analyst summarized the situation in 1995:

The most important structural changes [in the chlorine industry] will be concentration of growth in emerging markets and restructuring in industrialized markets: potential loss of 10–30 percent of current customers in industrialized markets; continued shutdown of inland plants linked to declining uses; three quarters of global demand growth in developing countries; increase in VCM and PVC trade and potential tripling in volume of global EDC trade. It appears unlikely at this point that lost markets will offset growth for PVC and other uses.[57]

One Step Forward, One Step Back

These economic patterns have both political and environmental implications. Since the strategic focus of the chlor-alkali industry is always on total consumption of chlorine and caustic soda, piecemeal regulations that target single chemicals merely shift chlorine from one organochlorine to another. In order not to lose markets for caustic soda, the chlor-alkali industry must always maintain or increase its total market for chlorine. When individual chlorine applications are curtailed, the industry finds new outlets for its chlorine, transferring the chlorine, along with its environmental impacts, to a new application. In this way, the industry can and will always circumvent the Risk Paradigm's chemical-by-chemical regulations. Only a holistic approach to the entire class of substances, like the Ecological Paradigm, can reduce the total environmental burden of organochlorines. By requiring the least toxic available technology to be used, this approach prevents one dangerous substances from being replaced by another of equal or greater hazard.

The expansion of PVC, particularly in developing countries, is a major environmental problem. As I show in the next two chapters, the life cycle of PVC plastic is an immense source of persistent organochlorine pollutants, probably the world's largest. Increasing the production of vinyl—and in turn the world stockpile of PVC that will eventually be disposed of or burned accidentally—promises to erase the environmental gains achieved to date through restrictions on individual organochlorines. Rapid expansion of vinyl production and disposal in the countries of Asia, Latin America, the Middle East, and Africa, where expensive technologies and regulations for pollution control and waste disposal are not as well developed as in the northern nations, is a latent environmental catastrophe.

With environmental regulations on pulp bleaching and ozone depleters, we took steps forward on two serious environmental problems, but we almost immediately took a giant step backward as global PVC production increased to fill the chlorine gap. Because these policies addressed only two specific chlorine applications, the industry was free to transfer the environmental burden associated with chlorine chemistry from one sector to another and from one part of the world to another. We cut two heads off the hydra, and it grew another, bigger and just as fearsome as the ones it lost.

7

Chemistry out of Control

Many a shaft at random sent / Finds mark the archer little meant.
—Sir Walter Scott, "The Lord of the Isles"[1]

Webster defines a tool as "a means to an end." Chlorine chemistry is undoubtedly a tool, one that industry uses to make new kinds of products, solve problems in the manufacture of existing ones, and increase productivity and efficiency. But tools are often means to many ends, some of them unintended and undesirable. The unplanned results of a technology may be minor—like the generation of soapy wastewater by a washing machine—or they may be severe, as in an automobile accident. They can come about as the result of improper use of the tool, as when an electric saw cuts not wood but the flesh of a hand, or they may be inevitable, like the production of radioactive waste by nuclear power plants.

A central question in the debate over chlorine chemistry is how much control humans actually have over this tool and how significant are the unintended consequences of its use. One school of thought argues that industry can control chlorine chemistry with exact scientific knowledge and finely-tuned engineering techniques. "Like steel, chlorine is neutral, neither good nor bad," says Donald Mackay, a professor of chemical engineering at the University of Toronto and a consultant to the paper industry. "It is how we as a society shape it and exploit it that determines the benefits and disbenefits. . . . Molecules can now be designed and built with chlorine as a component to have a desired set of properties. . . . Chlorine is a very powerful and valuable instrument in the chemical orchestra, but it should be played with care."[2]

Lawrence Fischer, a pharmacologist at Michigan State University's Institute for Environmental Toxicology and another prominent critic of efforts to address organochlorines as a class, takes a similar position:

> The biological activities of compounds within the complex mixtures of unknown organochlorine molecules that are byproducts of chlorine chemistry should be of concern to all of us. Within this throwaway mixture may be products with extraordinary biological activity. Such activity might contribute to the cause of chronic disease. Alternatively, it may be useful as a treatment for disease. It is within our technology to gain the needed chemical and biological information about these mixtures and either exploit them for our use or detoxify them, if necessary.[3]

These men suggest that humans can practice chlorine chemistry the way a musician plays his instrument—with such precise command of our actions that the outcome we produce corresponds exactly to our intent. Their words are rather poetic formulations of assumptions that are central to the Risk Paradigm. Current regulations are based on the idea that technological solutions—adding control devices, fine-tuning reaction conditions, improving incinerators, and the like—can effectively minimize the unintended consequences of chemical technologies. This belief requires an even more basic assumption: to control the incidental effects of chlorine chemistry we must first know what they are—that is, we must be able to identify all the by-products that are formed in processes that involve chlorine and organochlorines.

Is this picture of chlorine chemistry accurate? Do we actually have the control over chlorine chemistry that the Risk Paradigm presupposes? Are organochlorines produced in a way that can be addressed by the chemical-by-chemical approach? In this chapter, I survey the major processes in which chlorine and organochlorines are involved, showing that no sector of chlorine chemistry is free of unintended chemical consequences of great complexity and significance. All chlorine-based products and processes give rise at some point to the accidental generation of thousands of organochlorine by-products, impossible to prevent, poorly understood, and extremely hazardous.

The full impact of any chemical product includes releases to the environment not only of the substance itself but also of all feedstocks, by-products, and wastes generated at any time during the entire "life cycle"[4] of the product. The life cycle begins with the extraction or manufacture of starting materials and includes all stages of the synthesis, processing, use, and disposal of the product, plus disposal of all wastes created at any of these stages and environmental transformation of any chemicals released into the ecosystem. The life cycle ends when all of the associated products and wastes have been entirely converted back to chloride ions, a process that can take decades or centuries. Thus, the environmental impact of a single chemical typically involves many different compounds and occurs over a long period of time at many different places. To evaluate the impact of any chlorine-based product or process, we must not only examine the toxicity and environmental behavior of the product itself, as risk assessments do, but also consider the organochlorines that are formed accidentally or deliberately at all points during the chemical's life cycle.

Assessing the full impact of even one application of a single substance is thus extremely complex. It requires attention to a large number of accidental chemical reactions and destinations, beginning before the substance was created and ending only when all its associated products and wastes have been fully broken down. A complete assessment of the life cycle of all major uses of chlorine is beyond the scope of this book; indeed, each major application merits its own book-length investigation. Here, I concentrate on the general pattern of organochlorine formation and its implications for our two paradigms for environmental policy, specifically whether we have the scientific knowledge and technological control to assess and manage chlorine chemistry on the fine scale of individual compounds.

Production of Chlorine

The life cycle of all chlorine-based processes and products begins with the production of chlorine in the chlor-alkali process. If hazardous organochlorines are formed during chlorine production, then the life cycle of any chlorine-derived product or process is implicated in this pollution. The downstream product, by definition, inherits the sins of the feedstock.[5]

From the moment that chlorine gas is formed in the chlor-alkali cell, it will react with any organic matter that is present to form organochlorines. For this reason, manufacturers carefully purify raw materials and equipment surfaces to remove as much organic material as possible. Nevertheless, carbon-containing substances remain present as trace impurities, in plastic materials, or from the graphite electrodes used in some types of chlor-alkali cells.[6] Chlorine combines with these organic contaminants to form persistent organochlorine by-products, such as hexachlorobenzene (HCB) and hexachloroethane (HCE), which are found in the chlorine product itself.[7] Chlorine gas has also been found to be contaminated with PCBs, octachlorostyrene (OCS), and tetrachlorobenzene. The concentrations are rather low, but the quantities are significant: based on the levels found, the chlorine gas that is produced every year carries between 1.6 and 8.2 tons of these highly toxic and persistent by-products into the world's economy.[8]

Even greater quantities of organochlorine contaminants are deposited in the wastes from chlorine production. Swedish researchers have identified very high concentrations of dioxins and furans—up to 650 parts per billion—in the sludges from spent graphite electrodes used in chlor-alkali cells.[9] Severe contamination of fish and sediments with OCS has been documented near eight North American chlorine producers, and large-scale OCS contamination of sediments in Lake Ontario has been traced to disposal of spent chlor-alkali electrodes.[10] Very high levels of polychlorinated dibenzofurans have been found in the blood of Swedish chlor-alkali workers.[11]

Because of this problem, all chlor-alkali plants in North America and many in Europe replaced graphite electrodes with titanium substitutes during the 1970s and 1980s, a move that industry and government assumed eliminated the formation of organochlorine by-products. But recent data indicate that even the most modern chlor-alkali plants produce dioxin-like compounds. With graphite eliminated, traces of organic chemicals are still present, primarily from plastic pipes and valves that release small quantities of their materials into the cell. A 1993 study by Swedish scientists found dioxins and furans in the sludge and plastic piping from a modern chlor-alkali plant with titanium electrodes at levels around 5 parts per trillion (TEQ).[12] Subsequent Swedish research found significant quantities of chlorinated dibenzofurans in the sludge from a chlor-alkali plant with titanium electrodes, apparently due to chlorination of organic compounds in the rubber linings of the cell.[13] In 1997, the UK Environment Agency confirmed that a chlor-alkali plant owned by ICI Chemicals and

Polymers, which replaced its graphite electrodes around 1980, continued to release dioxins in its wastewater.[14]

Uses of Chlorine Gas

When chlorine reacts with organic matter, it produces a broad spectrum of organochlorines. The electronegativity of chlorine atoms and the chemical instability of chlorine gas make it a powerful oxidizing agent that reacts very rapidly with whatever organic matter it encounters. Chlorine atoms attack carbon-containing substances, taking the place of hydrogen atoms on organic molecules; in addition, other bonds are broken and formed, producing a variety of new organic molecules that are then chlorinated. The precise types and quantities of by-products in the mixture depend on the nature of the organic matter present, but certain things are constant. The organochlorine cocktail that results is always very diverse; it always includes extremely toxic, persistent, and bioaccumulative organochlorines; and the vast majority of the substances formed remain unidentified.

Pulp and Paper Bleaching

Wood pulp is composed primarily of cellulose fibers and lignin, a complex group of aromatic compounds linked together in long chains. The goal of pulp bleaching is to remove the lignin, leaving bright-white cellulose fibers behind. For most of this century, chlorine gas has been the bleach of choice because it is a powerful oxidizing agent that causes minimal damage to the cellulose fibers themselves. When chlorine gas makes contact with the organic matter in wood pulp, it oxidizes and chlorinates the lignin, combining with it to form a great variety of organochlorine by-products. Some of the chlorinated lignin is broken up into a diverse group of smaller organochlorines, and some are discharged as organochlorines of high molecular weight.

At least 1,000 organochlorines are thought to form when chlorine gas is used to bleach and delignify pulp.[15] Over 300 such compounds have been identified in chlorine-bleached pulp mill effluents, including a tremendous diversity of chemical structures: chlorinated dioxins, furans, phenols, benzenes, thiophenes, methylsulfones, methanes, ethanes, acids, and many others.[16] Recent research suggests that PCBs, too, are produced in chlorine bleaching of pulp.[17] Many of the organochlorine by-products are known mutagens, carcinogens, and reproductive and developmental toxicants.[18]

Along with to these identified compounds are hundreds of unidentified organochlorines. The identified substances account for only 3 to 10 percent of all the organically-bound chlorine in the effluent; the remaining 90 to 97 percent are mystery compounds that have not been specifically identified or assessed. An estimated 80 percent of the total consists of high-molecular-weight organochlorines that are not amenable to the chromatographic techniques used to identify chemical structures. Unidentified low- and medium-weight organochlorines account for the remaining 10 to 17 percent.[19]

About 10 percent of the chlorine fed to the bleaching process combines with organic matter to form organochlorines, so the quantity of organochlorines formed in this industry has been huge—several million tons each year.[20] These by-products are distributed among the various materials that leave the bleaching process—the wastewater effluent, the air emissions, and the pulp itself. Solids in the wastewater are typically removed by settling or filtering, creating large quantities of sludge that contain very high concentrations of organochlorines.[21] The sludge is then disposed of on land or burned in incinerators; the latter process creates a vast number of new organochlorines as combustion by-products, including additional quantities of dioxins. Even when chlorine dioxide, the "state-of-the-art" method for chlorine bleaching, is used, huge amounts of organochlorines are produced, though the quantities are substantially less than with chlorine gas.[22] When all pathways are considered, the bleaching of pulp with chlorine has been and continues to be one of the world's largest sources of persistent organochlorines into the environment, as I discuss in the next chapter.

Once discharged, organochlorines in the effluent undergo further transformations. The high-molecular-weight material is broken down into smaller chlorophenolic compounds, which are more bioavailable, more toxic, more persistent, and more bioaccumulative than the original compounds.[23] Once organochlorines enter the bodies of fish or other organisms, some are bioaccumulated, and others are transformed by the fish's metabolic processes; a large fraction are stored in fish tissues as a diverse group of chlorinated fatty acids, the biological significance of which is unknown. As a report for the Swedish EPA summarized, "The incorporation of the chlorinated material into the lipid metabolism of the fish opens up a new dimension of interactions involving uptake, distribution, and toxicity. If low molecular weight compounds are taken up and incorporated in naural metabolic pathways, estimates of the toxicity of the parent compounds are of limited value."[24]

Water Treatment

Many sewage treatment plants, industrial facilities, and drinking water systems add chlorine to water to kill micro-organisms that may cause disease. In the process, they address an important biological health hazard but create a new chemical one, because chlorine combines with organic matter in the water to form a diverse array of organochlorines.

The same types of organochlorines form whether the water being chlorinated is drinking water, sewage, or industrial wastewater.[25] Among the by-products are the trihalomethanes, including chloroform and dichlorobromomethane, which have been the subject of intensive research for three reasons: they were the first disinfection by-products to be discovered, they are produced in relatively large amounts, and they are known to be carcinogenic. Significant quantities of trihalomethanes and other volatile organochlorines are also produced in the chlorination of swimming pool water,[26] and greatly elevated concentrations of chloroform have been found in the blood of swimmers shortly after they exercised in chlorinated pools.[27]

These substances are just the tip of the iceberg. More than 70 organochlorines have been identified as chlorination by-products, including a wide range of chlorinated methanes, ethanes, ethylenes, butanes, ketones, acids, phenols, and phenoxyphenols (See appendix C). Among these are many compounds known to be carcinogenic or otherwise toxic; some of them are persistent, as well. The identified compounds represent only a fraction—about 25 to 50 percent—of the total organically-bound chlorine in the water.[28]

Water chlorination even produces dioxin-like compounds. According to Swedish research, drinking water chlorination produces small quantities of chlorinated dibenzofurans. The pattern of congeners resembles that found in other waters treated with chlorine, such as pulp mill effluent, and suggests that PCDFs are formed when natural chlorine-free dibenzofurans in the wastewater are chlorinated.[29] Other researchers have identified chlorinated phenoxyphenols, called "predioxins" because they are easily converted into dioxins, as by-products of chlorination.[30] Yet more dioxins are produced when the organochlorine-rich sludges from water treatment facilities are incinerated.[31]

Metallurgical Processes

We might not expect organochlorines to form when chlorine is used in industrial applications that do not involve organic matter like wood pulp or human wastes. But organic compounds are ubiquitous, and they are present in small amounts in all industrial processes. Thus, organochlorines are formed when chlorine is used in metallurgical processes and in the production of inorganic chemicals. The by-products include dioxins and related compounds, which is not surprising, since many metals are known to play a catalytic role in the formation of these substances.

For instance, chlorine is used to treat ores in the purification of nickel and magnesium, and dioxins, furans, HCB, and octachlorostyrene have been identified as by-products in the releases from facilities carrying out these processes. In the late 1980s, one magnesium factory in Norway was found to be discharging hexachlorobenzene and dioxins at the astonishing rates of 350 kilograms and 500 grams (TEQ) per year, respectively.[32] In the same country, fish caught near a nickel refinery that uses a chlorine-based process carried high concentrations of chlorinated dibenzofurans in their tissues.[33]

A considerable amount of chlorine gas is used in the production of titanium dioxide, the primary ingredient in many white pigments. Surprisingly, there are very few data available on organochlorine formation in this industry. In this process, chlorine gas is used to precipitate titanium (in the form of titanium tetrachloride) from complex ores; the titanium tetrachloride is then oxidized to yield titanium dioxide. Titanium-containing ores contain some quantity of organic matter, so we should expect organochlorines to form when they are treated with chlorine gas. The only data available confirm that titanium tetrachloride contains trace quantities of PCBs,[34] but no one has tried to identify all the by-products formed in this process.

Recently iron and steel plants have been found to be major sources of dioxins and furans, contributing as much as 11 percent of all dioxin emissions from known sources in some inventories (table 7.1). Organochlorines are produced in these processes primarily because chlorinated solvents, cutting oils, and scrap metal containing PVC are often introduced into the furnace.[35] According to one German analysis, "Sintering plants serve for recycling of dusts, scrap, and abrasion from other processes of the metallurgical plant to recover the iron for further use in the blast furnace. But this reasonable waste management method is accompanied by the problem of introducing traces of chlorine and organic compounds responsible for the generation of PCDD/F within these plants."[36]

Production of Inorganic Chlorides

The manufacture of inorganic chlorinated chemicals, including chlorides and bleaches, also results in the formation of organochlorine by-products, again because of the trace presence of organic matter in all processes. Of particular concern are chlorides of iron and copper, because these compounds can catalyze dioxin formation at relatively rapid rates. Tests by researchers at the University of Bayreuth, Germany, have found detectable quantities of dioxins, furans, PCBs, octachlorostyrene, hexachlorobenzene, and tetrachlorobenzene in iron trichloride, aluminum trichloride, copper chloride, and copper dichloride. Concentrations ranged from the low parts per trillion for dioxins and furans in chlorides of copper to the parts per million for hexachlorobenzene in chlorides of iron and aluminum.[37] German and Swedish researchers have even identified low levels of hexachlorobenzene, tetrachlorobenzene, and PCDFs in common household bleach (sodium hypochlorite).[38]

Manufacture of Organochlorines

The majority of chlorine is used within the chemical industry to synthesize organochlorines. When chlorine is combined with hydrocarbon feedstocks to manufacture organochlorine substances, unpredictable and unpreventable by-products are formed along with the intended products. The majority of these by-products have not been characterized, but the identified ones include some of the most persistent, bioaccumulative, and

Table 7.1
Sources of of dioxin air emissions

Source type	Percentage of all identified releases			
	World[a]	United States[b]	United States[c]	Great Lakes[d]
Municipal waste incinerators	37.6	40.1	51.4	20.1
Medical waste incinerators	2.8	17.4	10.3	48.7
Hazardous waste incineration	22.7	5.7	2.6	8.0
Ferrous metals production	11.7	NA	0.8	10.6
Copper smelting	2.6	19.7	2.6	4.1
Forest, brush, straw fires	11.7	7.6	7.7	0
Dioxin-contaminated chemicals	NA	NA	4.7	NA
Cement kilns (no hazardous waste)	10.7	0.6	NA	2.0
Wood combustion	NA	3.3	5.9	1.9
Coal combustion	NA	2.7	NA	2.5
Accidental fires	NA	NA	3.7	NA
Automobile fuels	0.4	1.4	0.4	1.4
Sewage sludge incineration	NA	0.2	1	0.6
Aluminum smelting	NA	0.6	NA	NA
Oil combustion	NA	0.3	0.8	NA
Pulp mill boilers	NA	0.1	0.1	NA
Lead resmelters	NA	0.1	NA	NA

Note: NA = quantitative estimate not available. There are numerous additional dioxin sources for which none of the inventories made a quantitative estimate due to inadequate data. Sources are listed from largest to smallest, by the mean percentage contribution for all four inventories.

[a]Brzuzy and Hites 1996a. Percentage of identified PCDD/F releases (TEQ) to the air. Estimate for hazardous waste incineration includes cement kilns burning hazardous waste; estimate for cement kilns does not. Estimate for forest, brush, and straw fires includes all biomass combustion, including wood.

[b]U.S. EPA 1998. Percentage of all identified releases to air of PCDD/F (TEQ), based on median estimates for 1995. Hazardous waste incineration estimate includes releases from cement kilns that burn hazardous wastes, as well as boilers and industrial furnaces.

[c]Thomas and Spiro 1995. Percentage of identified emissions of total PCDD/F to the air in the United States as of 1989. Municipal waste incinerators include apartment incinerators. Accidental fires include structural fires, PCB fires, and PCP fires.

[d]Cohen et al. 1995. Percentage of identified emissions of PCDD/F (TEQ) to the air that reach the Great Lakes.

toxic organochlorines, such as the dioxins. This problem is not restricted to a few chemical manufacturing processes but appears to be endemic to the entire field of organochlorine synthesis.

There are two reasons for the formation of unintended by-products. The first lies in the intrinsic nature of chemical reactions. The idea behind chemical engineering is to create conditions that generate reactive intermediate substances by mixing the right input materials and providing the proper amount of energy in the form of temperature, pressure, or reactive substances like chlorine gas; catalysts are also added in some cases. Reactive intermediates then recombine with each other to yield the desired product. Once intermediates are generated, however, there is virtually always more than one reaction pathway they can follow, each leading to a different group of stable compounds.

A large-scale chemical reaction is by nature a stochastic process—a large ensemble of random molecular events in which the quantity of each product in the final mix is equal to the probability that it will be formed in any one molecular reaction. In turn, that probability is related to the energy required to make the product; the lower the energy, the more likely it is that any single molecular reaction will make that substance. But no product has a 100 percent probability of being formed. Although there is a single lowest-energy product—a *global* optimum that should ideally be the sole product of the reaction—some of these pathways lead to *local* optima, substances that have higher energy than the global optimum but are quite stable in themselves. Like a mountain meadow surrounded by peaks that prevent access to the lowlands beyond, the local optimum represents a "dead-end"substance that is unlikely to be converted into the preferred substance. The intended chemical is the most likely and the most prevalent product, but it is never the only substance created. When a large number of molecules are present, even relatively improbable substances will be formed in quantities that are not negligible. For example, the synthesis of epichlorohydrin involves side reactions that generate more than 20 chlorinated ethers and diethers—persistent and toxic compounds that have been found in very high concentrations in the sediments downstream from a Dutch facility where epichlorohydrin is produced.[39] Even if reaction conditions could be perfectly controlled, the manufacture of organochlorines would result in the formation of a great variety of unintended by-products.

The other reason for by-product formation is that the real world is not perfect. In reality, input streams are never pure, and reaction conditions are never uniformly optimized. Despite a high degree of control over production conditions including material purity, temperature, and pressure, any synthesis process involves transient variations—deviations from optimal conditions in local regions of space and time. There is always a little bit of benzene left in your ethylene, a cold spot in your reactor, a speck of metal that catalyzes some unintended reaction. Machinery never works perfectly, and minor upset conditions occur quite frequently. As for material purity, the second law of thermodynamics makes clear that feedstocks can never be completely purified, because it would take an infinite amount of energy to remove absolutely all the impurities from any input stream. Chemical processes always receive some quantity of contaminants, which take part in side reactions that create unintended by-products. As a result, the quantity of by-products and wastes formed ends up being significantly greater than the perfect world's inevitable minimum.

Occasionally there are major accidents that cause runaway reactions, explosions, or chemical releases. For instance, in June 1976, a runaway chemical reaction inside a 2,4,5-trichlorophenol manufacturing reactor owned by Hoffman-LaRoche in Seveso, Italy, caused a rapid rise in temperature and pressure. A valve opened to relieve excess pressure, and a cloud of almost 3 tons of organic matter, including at least 600 kilograms of chlorophenols and a large quantity of dioxins, was blown 150 feet into the air. The chemical cloud gradually dispersed over adjacent residential areas, causing severe dioxin contamination of soils, plants, wildlife, and domestic animals; later, dioxin levels were measured in the tissues of people living in the area at concentrations 5 to 20 times above background levels. Thousands of animals died over the next two months, and both severe short-term health problems and long-term increases in cancer and heart disease have been documented in exposed people in Seveso.[40]

Catastrophes like this are rare, but they represent large-scale versions of phenomena that occur regularly to a much milder degree within chemical manufacturing processes. For example, the production of methallyl chloride—an intermediate in the production of sodium methyl sulphonate, which is used to make pesticides and acrylic fibers—requires near perfect conditions. According to one chemical engineering reference: "Proper mixing of the components by the inlet jet is a prerequisite to obtaining a high yield of methallyl chloride: the reaction is so rapid that high local concentrations of chlorine cannot be dissipated, leading instead to the production of more highly chlorinated products."[41] Or consider the chlorination of methane, one of the simplest of all organochlorine manufacturing processes. This process also requires a specific range of conditions: "In the [temperature] region of commercial interest—350 to 550 degrees C—the reaction proceeds very rapidly. If a certain critical temperature is exceeded in the reaction mixture. . . . decomposition of the metastable methane chlorination products occurs. In that event, the chlorination leads to the formation of undesirable by-products, including highly chlorinated or high molecular mass compounds (tetrachloroethylene, hexachloroethane, etc.)."[42]

The quantities of unintended chemicals can be surprisingly large. In some cases, the wastes go into the products themselves. By-products account for almost 20 percent of commercial DDT preparations, and about one-sixth of this is unidentified compounds.[43] Technical grade dibromochloropropane contains up to 10 percent impurities, including epichlorohydrin and allyl chloride.[44]

In other cases, the organochlorines go into hazardous wastes or are discharged directly into the environment. When chlorinated methanes and ethylenes are synthesized by a process called chlorinolysis, 1 to 7 percent of the total yield consists of highly persistent and bioaccumulative organochlorine by-products.[45] In the reactions by which vinyl chloride is synthesized, 10 percent of the yield consists of organic by-products,[46] including the very hazardous HCB, HCBD, HCE, PCBs, dioxins, and furans; the wastes from this process are over 60 percent chlorine by weight, and about 30 percent of the contents are unknown compounds.[47]

The most stable, persistent, and toxic organochlorines—dioxins, furans, PCBs, HCB, hexachlorobutadiene, and OCS—have been detected as by-products in the wastes or products from the manufacture of the full spectrum of organochlorines. It has long been known that these compounds are formed in the synthesis of chlorinated aromatic substances. Chlorobenzenes, for example, are produced by combining chlorine gas with benzene, a process that progressively chlorinates the benzene ring with more and more chlorine atoms until hexachlorobenzene, the form saturated with chlorine atoms, is formed. Chlorobenzenes can never be produced individually but are always manufactured in mixtures, so synthesis of benzenes with only one or two chlorine atoms results in the formation of significant quantities of hexachlorobenzene.[48] Dioxins and furans have also been identified as by-products in almost all congeners of chlorobenzenes and chlorophenols.[49] Dioxin is even expected to form in the synthesis of chlorotoluenes—chemicals that contain chlorine atoms not on the benzene ring but on a carbon side chain—because, according to one analysis, "a small degree of nuclear chlorination can also occur."[50]

More hazardous by-products are formed when chlorobenzenes and chlorophenols are used in the manufacture of other, more complex aromatic organochlorines. For example, extremely high concentrations of dioxins have been found in dioxazine dyes and pigments derived from chlorobenzenes.[51] In the 1980s, EPA identified over 150 pesticides and industrial chemicals that are known or suspected to be contaminated with dioxins and furans, virtually all of which were aromatic organochlorines or were manufactured using aromatic organochlorines as intermediates.[52] The best-known examples are the chlorophenoxy herbicides, including 2,4,5-T and 2,4-D; synthesized from chlorophenol feedstocks, both pesticides are contaminated with considerable levels of dioxin, and even higher levels are found in wastes from their production.[53] Dioxins have even been identified at low levels in wastes from the synthesis of pharmaceuticals, presumably because chlorinated benzenes and phenols are common feedstocks for drug synthesis.[54] As a report for NATO concluded, "All processes involving the manufacture of chlorophenol derivatives are suspected of producing PCDD and PCDF. Most likely all technical chlorophenols and chlorophenol derivatives contain small amounts of PCDD and PCDF. These values show that all aromatic chlorinations are suspect of causing PCDD/F formation."[55]

During the years that dioxins were being discovered in more and more chlorinated aromatic substances, few chemists even looked for them in other types of organochlorines. As a result, most scientists and government officials assumed that the problem was limited to the "aromatic" branch of chlorine chemistry. More recently, however, dioxins, furans, and PCBs have been found as by-products where they were least expected: in the simplest aliphatic organochlorines. An analytical program by German chemists, for example, found detectable dioxins and furans in commercial samples of ethylene dichloride, trichloroethylene, the chemical intermediate epichlorohydrin, and the specialty chemical hexachlorobutadiene. Concentrations of each individual dioxin and furan congener ranged from 13 to 425 parts per trillion. The authors concluded, "These results suggest that the synthesis of short-chain chlorinated hydrocarbons can also lead to PCDD/PCDF formation."[56] In addition, very large quantities of PCBs and dioxins have been identified in the wastes from the manufacture of chloroethylene solvents and ethylene dichloride, the feedstock for PVC plastic, when they are produced by a process called oxychlorination, as I show in detail in the next chapter. These results show that dioxin formation is not limited to one class of organochlorines but is associated with the full range of chlorinated chemicals.

Hexachlorobenzene is also formed in the full range of processes involving organochlorines. HCB occurs as a by-product not only in the manufacture of chlorobenzenes, chlorophenols, and their derivatives but also in the synthesis of vinyl chloride, the solvents trichloroethylene and perchloroethylene, the intermediates dichloropropene, 1,2,3-trichloropropane, and carbon tetrachloride, and the insecticides hexachlorocyclopentadiene and tecnazine.[57] The quantities can be huge. In the manufacture of perchloroethylene, for instance, from 1 to 7 percent of the total reaction output is composed of the highly persistent and toxic by-products HCB, HCBD, and HCE.[58] Technical-grade pentachloronitrobenzene contains up to 1.8 percent HCB.[59] Concentrations like this suggest that organochlorine products probably carry many tons per year of HCB into the environment, and wastes may contain hundreds or even thousands of tons more.

Based on these findings, we must conclude that all quarters of chlorine chemistry produce organochlorine by-products that are extraordinarily hazardous. Not every organochlorine and its production wastes have been analyzed, so it has not been strictly proven that the synthesis of every single organochlorine results in the formation of persistent, bioaccumulative, toxic organochlorines. Nevertheless, dioxins, PCBs, HCB, and other extremely hazardous organochlorines have been positively identified as by-products in the manufacture of a very broad sample of chemical structures, representing the full range of organochlorines. Further, they have been discovered in many different kinds of reactions used in the chemical industry, including direct chlorination (in which chlorine gas is combined with petroleum feedstocks), oxychlorination (in which hydrochloric acid is the chlorinating agent), chlorinolysis (in which highly chlorinated substances are heated to yield substances with lower degrees of chlorination), and other kinds of downstream synthesis reactions using chlorophenols, chlorobenzenes, and at least some chlorinated aliphatics.[60] Most important, no organochlorine has yet been found *not* to generate these by-products. The data thus suggest strongly that the manufacture of any organochlorine results in the formation of extremely hazardous by-products.

The prevalence of dioxins, HCB, and other organochlorine by-products from the manufacture of all different types of organochlorine products underscores why chlorine chemistry is an unusually hazardous branch of industrial chemistry. By-products are formed in any chemical synthesis process, whether or not chlorine is present. But because chlorine is so likely to initiate chemical recombinations with whatever organic matter it makes contact with, the formation of by-products is magnified. Further, because organochlorines tend to be persistent, bioaccumulative, and highly toxic, a soup of unintended organochlorines is even worse than a mixture of unchlorinated hydrocarbon by-products in a chlorine-free process. Create a diverse mixture of random organochlorines, as the chemical industry does every time it makes an organochlorine product, and you are guaranteed to have some intolerably dangerous compounds in the mix.

Uses of Organochlorines

Organochlorines are used in nonreactive contexts that do not result in by-product formation. CFCs circulating in an air-conditioner, for example, do their job of absorbing heat without undergoing any chemical reactions. In nonreactive uses—including most applications of chlorinated solvents, refrigerants, and pesticides—the major hazard is the direct release of these chemicals into the workplace or the general environment. Loss during use is, however, an immense problem, since some 80 to 99 percent of these chemicals are ultimately dispersed into the air, water, or soil, as I show in the next chapter.

Some applications of organochlorines do take place under conditions that initiate chemical reactions, causing by-products to form. For instance, some manufacturing industries remove grease, oil, dirt, and sludges from equipment by mixing chlorinated solvents with alkaline cleaning solutions; this chemical combination creates reactive conditions that convert a portion of the solvent into such by-products as hexachlorobenzene, octachlorostyrene, octachloronaphthalene, dioxins, and furans.[61] When lumber mills treat wood with the preservative pentachlorophenol (PCP) at elevated temepratures and pressures, very large quantities of chlorinated dioxins and furans form.[62] And the use of chlorinated catalysts in the oil refining industry produces dioxins,[63] as does the combustion of leaded gasoline, to which 1,2-dichloroethane is added as a lead scavenger.[64]

More important as a source of by-products is the use of organochlorines in the chemical industry. We have already seen that the synthesis of organochlorines from other chlorinated chemical feedstocks produces very large quantities of by-products. Even when organochlorines are used as intermediates in the synthesis of chlorine-free chemicals—in which most of the organochlorine is converted into hydrogen chloride—some quantity is transformed into organochlorine by-products. Production of propylene oxide from propylene chlorohydrin, for example, generates numerous organochlorine and nonchlorinated by-products,[65] as does the use of phosgene to produce organic isocyanates.[66] In the production of epoxy resins from epichlorohydrin, side reactions produce impurities in quantities high enough to affect product quality in some cases.[67] These findings underscore the fact that organochlorine intermediates introduced into a synthesis process are never entirely converted to harmless products.

There are few data on the formation of dioxins and furans in processes in which organochlorines are used as intermediates in the synthesis of chlorine-free compounds. EPA's list of known or suspected dioxin-contaminated compounds includes a number of chlorine-free substances that are produced by chlorinated phenols.[68] And chemical considerations make clear that the problem is not likely to be limited to aromatic organochlorine intermediates. A report for NATO summarized, "All industrial chemical processes by which chlorine or hydrogen chloride is cleaved from intermediate products must be critically regarded with respect to PCDD/PCDF formation. Such chemicals which contain no chlorine in the end product (in particular those resulting from eliminations, dehydrochlorination and alkaline hydrolyses) should be especially considered."[69]

Intermediates are designed to be chemically reactive, but even the most apparently inert uses of organochlorine products can form by-products if they are subject to accidental fire. The poor combustion conditions associated with burning buildings are particularly conducive to the formation of chlorinated products of incomplete combustion, like dioxins.[70] More than 30 major fires and explosions involving PCB transformers and capacitors in the United States and Scandinavia have been shown to have produced very large quantities of dioxin.[71] Fires in forests, agricultural lands, and wooden structures treated with organochlorine pesticides also generate organochlorines, including dioxins.[72] And as I show in the next chapter, accidental fires in homes, offices, warehouses, and vehicles that contain PVC plastic may release considerable quantities of dioxin into the air.[73]

Finally, some organochlorines—particularly chlorophenols and chlorobenzenes—can be transformed into dioxins and other extremely hazardous organochlorines after they are released to the environment. These transformations can be mediated by purely chemical processes, as when chlorophenols are converted into dioxins by exposure to light.[74] They can also be biologically catalyzed by enzymes: during composting, the dioxin concentration of chlorophenol-contaminated sewage sludge increases by a factor of two to three, even though there is no apparent input of dioxins.[75] If chlorophenols can be transformed into dioxin, it is plausible that other organochlorines with related structures—particularly chlorobenzenes, PCBs, and chlorophenoxy herbicides like 2,4-D—may also be changed into dioxins by environmental processes, but little information is currently available on this subject.

Disposal of Organochlorines

Landfills

The life cycle of chlorine chemistry comes to a close when chlorine-containing products and wastes are disposed of. Organochlorine wastes used to be dumped in pits and landfills. Because they are so long-lived, these substances tend to leach into groundwater; organochlorines thus dominate EPA's list of the nation's most common groundwater contaminants[76] and account for almost half of the substances reported at all Superfund sites, where tricholoroethylene is the most frequently reported contaminant.[77]

The first major controversy over land disposal of hazardous wastes erupted in Niagara Falls, New York. From 1942 to 1952, the Hooker Chemical Company produced large quantities of chlorinated pesticides, solvents, and other chemicals and dumped over 20,000 tons of hazardous waste into Love Canal. In 1953, Hooker covered the dump with a layer of dirt one to four feet thick and sold it to the Niagara School Board for a dollar, with a deed that released Hooker from liability for any injuries. An elementary school was built on top of the dump, and a neighborhood of small homes was built along its edges. In 1976, residents of the neighborhood reported that wastes were seeping into their basements and causing health problems, and they organized the first grass-roots resistance against chemical waste disposal in the United States. Under citizen pressure, the state began to investigate hazards at Love Canal and found that samples of top soil from the site contained 28 volatile and semivolatile organochlorines, which accounted for a stunning 8 percent of the total mass of the soil. Water leaching under the site contained a variety of chlorinated solvents, benzenes, toluenes, benzaldehydes, and pesticides, along with dioxin. By 1978, the New York Department of Health had judged the site "an extremely serious threat and danger to the health, safety, and welfare of those using it, living near it or exposed to it." Residents were moved out and clean-up of the site began.[78]

The widely-publicized controversy at Love Canal led to citizen outcry in other communities where hazardous wastes had been dumped. The resulting political pressure led EPA to promulgate regulations under the Resource Conservation and Recovery Act (RCRA) that banned many organochlorine wastes from land disposal. Because these wastes—particularly chemical industry wastes from the synthesis of organochlorines—are chemically complex mixtures, they generally cannot be recycled. Chemical and biological treatment can degrade some kinds of organochlorines, but because most hazardous wastes are complex and persistent mixtures, these methods do not always work and are very expensive when they do. Regulations under RCRA have thus designated incineration as the "best available treatment" option for virtually all organochlorine wastes—not necessarily because it is a good option but because it is a last resort.

RCRA encouraged incineration in another way. It made waste generators liable for the health and environmental effects of their wastes "from cradle to grave," and it established a tracking system to ensure that wastes in any landfill could be traced back to their generators. Facing billions of dollars of potential liability for land-disposed wastes, generators were eager to turn to incineration, where liability ends at the furnace door. Once wastes are fed to an incinerator, transformed into combustion byproducts, emitted through the smokestack, and dispersed into the atmosphere, there is no way to trace liability back to the generator. Incineration makes a waste producer's liability go up in smoke, which explains why waste generators have been so enthusiastic about a method that costs many times more than the land disposal practices it replaced. In the last two decades, incineration has also come to dominate organochlorine waste disposal in Japan and most of the European nations, but land disposal remains common in many less industrialized nations and in countries with weaker environmental policies, like Great Britain.

The emergence of incineration as the dominant method for destroying organochlorines refutes Dr. Fischer's claim at the beginning of this chapter that organochlorine wastes are a valuable resource that can be recovered for benefical use. In reality, the organochlorine mixtures formed in the production and use of chlorine are useless and intractable chemical cocktails that industry spends immense sums trying to destroy. While incinerators are very good at destroying liability, however, they are not so effective at rendering organochlorine wastes harmless. As we will see, the final scene in the life cycle of chlorine chemistry actually produces a vast number of new organochlorines, introducing a horde of new characters onto the stage just when the curtain was about to fall.

Incineration

Incinerators for trash, medical waste, and hazardous waste—along with other facilities that burn waste chemicals, such as cement kilns and industrial boilers—burn huge quantities of organochlorines. Most of the organochlorines in the municipal and medical waste streams come from PVC plastic, but those in industrial wastes are diverse, including chemical manufacturing wastes, spent solvents, cutting fluids, pesticides, and wood preservatives, along with the sludges and wastes they generate during use. About 35 percent of all wastes burned in U.S. hazardous waste incinerators are organochlorines, and around 13 million metric tons of halogenated incinerable hazardous wastes were generated annually in the United States, according to EPA figures from the late 1980s. Of this quantity, about 10 percent each were spent chlorinated solvents and liquid chemical wastes, and most of the remainder consisted of halogenated solids, which come primarily from the chemical manufacturing industry.[79]

In theory, a properly designed and operated incinerator is intended to convert organochlorines by oxidation into carbon dioxide, water, and hydrogen chloride. Real-world combustion systems, however, never take this reaction to completion for all the compounds fed to them. The majority are completely oxidized, but some small fraction escapes unburned, and a larger portion is converted into new organic compounds, called products of incomplete combustion (PICs). According to EPA's technical review document on hazardous waste incineration, "The complete combustion of all hydrocarbons to produce only water and carbon dioxide is theoretical and could occur only under ideal conditions. . . . Real world combustion systems . . . virtually always produce PICs, some of which have been determined to be highly toxic."[80]

By-products form in incinerators for the same reasons they do in chemical manufacturing: multiple pathways, local optima that lead to stable by-products, and deviations from optimal conditions. In incineration, the problems are particularly acute, because wastes are complex mixtures of diverse materials that can never be uniformly blended. Further, combustion is by nature a random process of bond breakage and formation; at high temperaures, most of the molecules will be completely oxidized, but some will follow alternative reaction pathways and emerge as PICs. Transient variations and upsets are a particular problem with incinerators. Good management can reduce but never eliminate the production of PICs, as EPA's analysis made clear:

[Deviations from optimum] usually are a consequence of a rapid perturbation in the incinerator resulting from a rapid transient in feed rate or composition, failure to adequately atomize a liquid fuel, excursions in operating temperature, instances where the combustible mixture fraction is outside the range of good operating practice, or inadequate mixing between the combustibles and the oxidant. . . . The amount and composition of PICs will depend in a complex and unpredictable way on the nature of the perturbation.[81]

The type of incinerator and how well it is operated will affect the magnitude of the PICs released, but the production of chlorinated PICs, including the most hazardous ones like the dioxins and furans, is a universal and inevitable outcome whenever chlorinated wastes are burned. As the British Department of the Environment noted, "Comprehensive tests have established that all waste incinerators, independent of type of incinerator or waste composition, are likely to produce all of the possible 75 PCDD and 135 PCDF isomers and congeners, as well as about 400 other organic compounds."[82] By-products form as products of diverse and unpredictable reactions not only in the furance but also in the cooler zones, where control over combustion conditions is virtually irrelevant.[83] Dioxins can even form in the pollution control devices or the smokestack itself, where chlorine gas, organochlorine precursors, or hydrochloric acid come in contact with organic compounds in fly ash. This process, called de novo dioxin formation, is greatly accelerated if iron or copper catalysts are present, as they are in most hazardous and municipal wastes.[84]

This means that incinerators not only destroy organochlorines, as they are supposed to, but also manufacture them. EPA estimates that PICs formed in the incineration process number in the thousands.[85] Of these, a small fraction have been characterized, and the rest remain unidentified. Laboratory tests show that burning methane—the simplest possible hydrocarbon—in the presence of a chlorine source produces more than 100 organochlorine PICs; these by-products, ranging from chlorinated methanes to dioxins, are produced by a set of reactions thought to be common to all incineration processes when chlorine is present.[86] It is much more challenging to analyze PICs in the stack gas of real-world incinerators, but over 50 organochlorines or groups of organochlorines have been identified in the emissions of hazardous waste incinerators, ranging from the structurally simple carcinogen carbon tetrachloride to highly persistent and bioaccumulative compounds like chlorinated hexanes, ethers, phenols, naphthalenes, thiophenes, dioxins, furans, and PCBs (See appendix D). Medical waste incinerator emissions also contain chlorinated PICs, including carbon tetrachloride, chloroform, vinyl chloride, chlorobenzenes, chlorophenols, PCBs, and dioxins.[87]

As in many other processes, the identified compounds are just the beginning. In municipal incinerators, the most thorough analysis to date identified several hundred PICs, including 38 organochlorines—chlorinated benzenes, PCBs, methanes, ethylenes, and others—but 58 percent of the total mass of PICs remained unidentified.[88] At hazardous waste incinerators, the most comprehensive research burns have identified about 60 percent of the total mass of unburned hydrocarbons in incinerator stack gases, and most field tests have had far less success in identifying the PICs emitted.[89] There is good reason to be concerned about these mystery compounds, because at least some appear to be in the same toxicological family as the ultratoxic dioxins. German researchers measured the dioxin-like toxicity of trash incinerator fly ash using a biological test; they found that the toxicity was up to five times greater than could be accounted for by the amount of dioxins, furans, and PCBs in the ash.[90] The remaining dioxin-like effect was presumably caused by the scores of other compounds—such as chlorinated naphtalenes, diphenyl ethers, thiophenes, and many more—that can cause similar health effects but were not specifically measured.

The total quantity of PICs and unburned wastes emitted from incinerators is not known precisely, but it appears to be very large. In the United States, hazardous waste incinerators must pass a trial burn that requires them to demonstrate a destruction and removal efficiency (DRE) of 99.99 percent of the organic compounds fed to them, which means that no more than 0.01 percent of several test chemicals fed into the furnace may be measured in stack emissions. But high DREs do not mean that the environment is protected, for several reasons. For one thing, organochlorine wastes are burned in such huge amounts that even if all incinerators achieved 99.99 percent DRE, they would still emit more than 2 million pounds of unburned hazardous wastes into the air each year in the United States alone.[91] In addition to this amount, far larger quantities of unburned wastes are transferred to the land or water where the ash, sludge, and effluent from incinerators are disposed of, because a high DRE reflects not only destruction of waste chemicals but also their removal by pollution control devices; an incinerator with a filter that captures 95 percent of the dioxin in the stack gas deposits 20 times more dioxin in its ash than it emits to the air. Even greater amounts of organochlorines are released as PICs, because "destruction" means only that the chemical tested was transformed into some substance other than the original compound, and PICs are not counted against the 99.99 percent DRE figure. EPA's Science Advisory Board has estimated that the total quantity of PICs that hazardous waste incinerators emit to the air may be up to 1 percent of the organic matter fed to them, an estimate that suggests that U.S. incinerators may emit more than a 100,000 tons of organochlorines into the air each year.[92]

Even these estimates of incinerator emissions, which are based on DREs measured in trial burns, are likely to be unrealistically low. Incinerators in routine operation seldom achieve the kind of performance they reach in a trial, where they burn simplified mixtures of pure chemicals under carefully controlled, closely scrutinized conditions. In daily use, incinerators generally perform less efficiently due to the complexity of real-world wastes and the frequency of upsets, operator error, and equipment malfunction.[93] Further, the standard trial burn protocol allows the measurement of emissions to stop when the feed of waste chemicals to the incinerator stops, but emissions can continue for days, resulting in total emissions of unburned wastes that are orders of magnitude greater—and DREs far lower—than those measured during the trial burn.[94] Finally, DREs are calculated based on an incinerator's performance burning test chemicals that are fed in very high concentrations, but two EPA studies have found that substances in low concentrations burn much less efficiently. Chemicals that are present in wastes in the parts per billion or parts per million range—such as the dioxins, PCBs, and many other byproducts in chemical manufacturing wastes—are subject to destruction efficiencies as low as 99 percent, so significant amounts of these very hazardous substances will escape from incinerators undestroyed.[95] For all these reasons, an incinerator may be certified as achieving 99.99 percent DRE when in fact it is emitting huge quantities of unburned and partially burned wastes into the environment.

Dioxin and Incineration: The Vinyl Institute Strikes Back

We have seen that burning organochlorines produces substantial quantities of dioxin. It is also true that incinerators that burn organochlorines appear to be the world's largest dioxin sources. All inventories prepared by scientists and government agencies agree that incinerators and other combustion facilities are the world's largest sources of dioxin to the air (table 7.1). In the United States, municipal waste incinerators comprise the largest source sector, hospital waste incinerators rank second or third (depending on the inventory), and hazardous waste combustion facilities are important dioxin sources as well. Together these three sectors account for 60 to 90 percent of all dioxin emissions from identified sources. Smelters and incinerators that burn smaller quantities of chlorine-based wastes also make important contributions to dioxin loadings.

In the 1990s, a controversy sprang up over whether organochlorines are actually the reason that incinerators are major dioxin sources. Like some of the other debates in the field, this one is less the product of contradictory or inadequate information than the result of a concerted public relations effort by the chlorine industry. The most important of the dioxin inventories was prepared as part of EPA's "Dioxin Reassessment," first released in 1994.[96] With its alarming new information on dioxin toxicity and the unambiguous conclusion that incinerators burning chlorinated wastes are dominant sources of dioxin, the reassessment posed a real challenge to the chlorine industry. Anticipating the report's release, the Vinyl Institute commissioned the public relations firm Nichols-Desenhall Communications to prepare a "Crisis Management Plan for the Dioxin Reassessment." According to the plan, "EPA will likely conclude that the incineration of chlorinated compounds is the single largest known contributor of dioxin. . . . We believe that PVC will be specifically mentioned and potentially slated for further regulation." In order to "prevent punitive regulation of PVC by EPA, Congress, or the state legislatures," the plan advised the Vinyl Institute how to present itself in the media and "strongly urge[d] VI to aggressively defend the industry's credibility through the use of third party sources to debunk . . . EPA's misleading claims."[97]

The industry took its PR firm's advice. In 1994, the Vinyl Institute's Incineration Task Force hired the consulting firm Rigo and Rigo, Inc., to prepare an "independent" analysis, which found that the amount of dioxin that incinerators released has no relation to the amount of chlorinated organic materials fed to them.[98] The institute also managed to have the report published not as an industry-sponsored study but as a product of the prestigious American Society of Mechanical Engineers (ASME), an independent professional organization. One of the leaders of the Vinyl Institute's Incineration Task Force, Dick Magee, was also an active ASME member; Magee brokered an arrangement in which the Vinyl Institute would hire and fund Rigo to write a report that would be released under the ASME banner. A Vinyl Institute memo makes very clear that ASME's role was to create the illusion of "third-party" authority and that the Rigo report was conceived, carried out, and handsomely rewarded in a spirit of public relations, not unbiased analysis. The memo reads:

> The Vinyl Institute has created an Incineration Task Force in anticipation of adverse EPA actions regarding dioxins and furans. . . . Dick Magee brought forward a proposal from the American Society of Mechanical Engineers to hire Rigo & Rigo Associates, Inc., of Cleveland, OH. The purpose of ASME as the contractor is to provide unassailable objectivity to the study. . . .
>
> The Incineration Task Group interviewed Dr. H. Gregory (Greg) Rigo, principal of Rigo & Rigo Associates, Inc. by phone and found him to be extremely knowledgeable about incineration and to have several proprietary databases VI had not discovered. He is also user friendly, i.e., willing to set his priorities to our needs, and appears to be sympathetic to Plastics, Vinyl, PVC and Cl2. . . .
>
> The ASME proposal calls for $130,000 for the Rigo & Rigo/ASME study. Since there are many unanswered questions regarding dioxins and since VI may want to use Greg Rigo as an expert witness or advocate to talk about the report, I am proposing an additional $20,000 as a contingency fund, for a total of $150,000 to be fully funded by VI.[99]

After releasing the Rigo study in 1995, the chlorine industry aggressively used the report in worldwide lobbying and public relations efforts, at scientific conferences, and in the press. The Rigo report has always been presented as an ASME study, not an industry-funded project. A position paper from the Vinyl Institute and the Chlorine Chemistry Council, for instance, says that "The study conducted by the American Society of Mechanical Engineers (ASME) remains the premier weight-of-evidence review of the role of chlorine in dioxin generation from combustors of all kinds. The study shows that there is no statistical link between the amount of chlorine in combustor feedstocks and the dioxin emitted by that combustor."[100] Many independent bodies have not seen through the disguise. A report for U.S. EPA, for instance, cited "the results of a study commissoned by the American Society of Mechanical Engineeers" without any mention of industry sponsorship, and documents prepared by the Swedish EPA, the United Nations Environmental Program, and the European Community have all referred to the study as if it provided impartial evidence that dioxin emissions from incinerators are not linked to the burning of organochlorine wastes.[102]

The questionable motivations behind the Rigo study—and the way it has been deployed in the public arena—do not tell us whether the study's conclusions are correct, although they should certainly raise our suspicions. A close reading, however, reveals that the Rigo report's methods were no less flawed than its origins, undermining the reliability of its claim that burning organochlorines is not related to dioxin formation.[103] The study was not experimental, so it could not directly refute the existence of a chlorine-dioxin link. Instead of generating new data, the authors compiled existing trial burn measurements from a large number of incinerators, statistically analyzed the relationship between indicators of chlorine feed and dioxin releases, and concluded that there is no statistically significant relationship between the two. A statistical analysis of this type—an epidemiological study on machines, in a sense—is particularly sensitive to design problems: if the putative cause and effect are not measured accurately or confounding factors are not taken into account, then a meaningful relationship is likely to go undetected. Rigo's methods were deeply flawed on several counts, suggesting that the failure to find a link between burning organochlorines and dioxin generation is an artifact of bad study design, not a meaningful finding that no relationship exists in the real world.

The first problem with Rigo's approach was his failure to take account of differences among facilities. We know that chlorine input is not the only factor that determines the magnitude of dioxin emissions from incinerators: combustion conditions, the types and quantities of substances in the waste, and other variables also affect the amount of dioxin that will be released. Because of constant fluctuations in many of these factors, statistical relationships between stack emissions and indicators of waste input or combustion conditions can seldom be established, even at individual incinerators.[104] Massively compounding this problem, Rigo used data from a large number of incinerators operating under widely variable conditions, but he did not control or adjust for differences in facility type, waste type, operating parameters, or any other important variables. There is no reason to expect that a statistical summary of data from many different facilities, with no attempt to control or adjust for confounding factors, would have detected a "signal" within all this "noise." Even a very strong relationship between organochlorine input and dioxin output is likely to go undetected in a study designed this way.

Rigo's study was also undermined by its use of data from unreliable sources. The emissions data in Rigo's analysis came almost exclusively from trial burns designed for permitting purposes, without the proper kinds of controls and measurements necessary to evaluate the relationship between chlorine input and output. Moreover, as we have seen, trial burn data are notoriously problematic because they do not accurately represent the way incinerators routinely operate and do not measure the much larger quantity of chemicals that are released after the feed of waste to the incinerator has stopped. In fact, many trial burns have conducted their evaluations of low- or no-chlorine wastes *after* chlorinated wastes have been burned, so the later stack samples become contaminated by continuing emissions from earlier runs. The use of results from trial burns of this sort will thoroughly scramble any relationship that might otherwise have been recognizable between chlorine input and dioxin input.

The final flaw of Rigo's study is its reliance on the wrong kinds of measurements. To investigate a link between the amount of organochlorines burned and the amount of dioxin produced by incinerators, Rigo should have examined the statistical relationship between the mass of organochlorines fed to an incinerator and the mass of dioxins released. Instead, the study tracked surrogate measures for both of these paramaters, tracking the concentration of hydrogen chloride (HCl, the primary byproduct of organochlorine combustion) as a rough indicator for the amount of organochlorines in the feed; as a surrogate for the amount of dioxin released, it examined the concentrations of dioxin in the stack gas.[105] The problem is that concentrations do not accurately represent quantities. For one thing, if the total flow of stack gas increases (as it generally will if more waste, and thus more chlorine, is fed to the incinerator), the concentrations of dioxin in the gas may decrease even if a larger amount of dioxin is being emitted. Further, stack gas measurements omit pollutants directed into fly ash, bottom ash, and scrubber water, so changes in the efficiency of pollution control devices can reduce the concentration of dioxin in stack emissions while total dioxin formation increases, or they can reduce the concentration of HCl while total organochlorine inupt rises. (Because pollution control devices have different capture efficiencies for dioxin and hydrochloric acid, the concentrations of these materials in the stack gas after it passes through this equipment will not reflect the ratios of the amounts that were actually produced by the incinerator.) Another problem lies in the fact that hydrogen chloride can be formed not only by the combustion of organochlorines but also by the burning of chloride salts, further undermining the reliability of HCl as an indicator of organochlorine feed. The variables that Rigo analyzed are thus grossly inappropriate substitutes for the quantities that are truly of interest; Rigo's failure to find a relationship between the surrogates he used says nothing about whether a link really exists between organochlorine input and dioxin generation.

All the flaws we have discussed cripple the "ASME" study's ability to establish a link between chlorine and dioxin. A finding of "no relationship" is only as good as a study's power to detect a relationship, and in this case that power can only be described as pathetically weak. On the basis of Rigo's analysis, no reliable inferences can be drawn about whether a relationship exists between the amount of organochlorines burned and the amount of dioxin formed in an incinerator. The "conclusions" that Rigo and the Vinyl Institute trumpeted so authoritatively could hardly be less conclusive than they truly are.

A Dioxin-Chlorine Link

Rigo's study is by no means the only information that bears on the relationship between burning organochlorines and creating dioxin. A large body of unbiased evidence, derived by superior methods, contradicts Rigo's findings, indicating that chlorine input is related to dioxin output. This is not to say that the organochlorine content of the waste is the *only* factor involved in dioxin formation; facility design, operating conditions, and the presence of catalysts also play major roles. But chlorine is a requirement for dioxin synthesis, and preventing the introduction of organochlorines into incinerators is the best means to prevent dioxin formation.

Dioxin cannot be formed without a chlorine source, so emissions from incinerators must be due to burning organochlorines, burning salt, or some combination of the two. To suggest that organochlorines are not important dioxin precursors requires the combustion of inorganic chloride salts to be the major source of dioxin—another version of Dow's "Trace Chemistries of Fire" theory. Just as several lines of evidence undermined Dow's argument that the natural combustion of salt is a major dioxin source, several other kinds of data refute the view that the burning of salt in incinerators is primarily responsible for their dioxin emissions. First is the historical data on dioxin levels in sediments, which we reviewed in chapter 5. If organochlorines have nothing to do with dioxin emissions, then why were dioxin levels in the environment nonexistent or minuscule before the chemical industry began to produce them? In particular, why were dioxin levels so low during the nineteenth century, when combustion of chloride-containing materials like coal and wood was at its peak? These data make abundantly clear that burning salt in nature or industry is a very minor source of dioxin, virtually all of which is due to the production, use, and disposal of chlorine gas and organochlorines.

Second, a number of well-conducted studies in the laboratory show that burning salt produces little or no dioxin, but adding organochlorines causes dioxin formation to increase by orders of magnitude. Results from the laboratory are particularly convincing, because—unlike trial burns at full-scale incinerators—they allow combustion conditions, input materials, and emissions to be carefully controlled and accurately monitored. The German EPA, for example, has found that burning organochlorine-containing plastics and other chemicals produces dioxin—with concentrations in ash residues ranging from 3.2 to 662 parts per trillion (TEQ)—but combustion of several types of organochlorine-free but chloride-containing paper, wood, cotton, or wool does not produce dioxin above the detection limit of 0.1 part per trillion[106] (table 7.2). Finnish researchers have found that burning perchloroethylene in a laboratory combustion reactor produces orders of magnitude more dioxin, chlorophenols,

Table 7.2
Dioxins in ash from burning organochlorines and chloride-containing materials

Material	Total PCDD/F (ppt)	TEQ
Materials not known to contain organochlorines		
Writing paper	ND	ND
Wood	ND	ND
Cotton	ND	ND
Wool	ND	ND
Polyethylene	2.9	<0.1
Acrylonitrile-butadiene rubber	2	<0.1
Fir wood	21.4	0.65
Materials known to contain organochlorines		
Bleached coffee filters	6.3–7.7	0.15–0.23
PVC plastic	244–2,067	3.2–42.2
PVC flooring material	352–1,847	8.2–14.5
PVC window frame material	7.5–969	8.8–18.1
PVC cables (copper)	669–2,670	11.4–52.6
PVC cables (no copper)	416–843	7.4–16.6
PVC gloves, hose, pipes, tape	158–954	2.5–16.5
PVDC plastic	3,304	14.1
Chloropolyethylene plastic	840	10
Polychlorobutadiene plastic	323–1,096	0.7–4.7
Chloroparaffins	1,049	5.3
Dichloromethane	26,302	478
1,1,1-Trichloroethane	21,746	340
Tetrachloroethane	9,072	132
Trichloroethylene	120,915	149.5
Perchloroethylene	212	0.4
Epichlorohydrin	1,532	36
Chlorobenzene	16,135	0.5
p-Chloronitrobenzene	190,096	21.5
o-Chloronitrobenzene	32,293	216
p-Chlorotoluene	1,033	ND
2,4-D	178,016	361
Linuron pesticide	3,110	32

Source: Theisen 1991.

and chlorobenzenes than burning sodium chloride does.[107] And studies by three separate teams of American researchers have found that adding salt to a combustion reaction has no detectable effect on dioxin formation;[108] another found that formation of dioxin precursors rises as the proportion of organochlorines in the waste increases.[109]

Other laboratory studies have had similar results. When PVC is added to a mixture of chloride-containing coal and bark, dioxin concentrations increase by a factor of 10 to 100; the more PVC added, the higher the dioxin concentration.[110] Adding PVC during combustion of natural chloride-containing wood products increases dioxin levels in the ash by 15 to 2,400 times; when large quantities of inorganic chloride chemical hardeners are added, dioxin levels rise somewhat but are still 3 to 350 times lower than when PVC is included in the mix.[111] Combustion of a mixture of coal and salt produces trace quantities of dioxins and furans in the off-gas, but when elemental chlorine is added to the mix, total dioxin formation increases 130-fold.[112] Finally, burning chloride-containing vegetable matter does not produce detectable PCDD/Fs, but including chlorine gas or PVC plastic along with the plant material does.[113]

In full-scale or pilot-scale incinerators (units smaller than commercial burners but similar in design), the evidence also supports a relationship between burning organochlorines and creating dioxin, but there are some contradictory studies, presumably due to the complexity of analyzing input and output streams and adjusting for fluctuating conditions. The Danish EPA, for example, has found that doubling the PVC content of an incinerator's waste input increases dioxin emissions by 32 percent, while doubling the chloride content increases dioxin emissions by a much smaller margin.[114] Two groups of Finnish researchers have found that dioxin levels in stack gas or fly ash are low when a mixture of coal and chlorine-free plastics is burned, and they rise substantially when PVC is added to the mix.[115] German scientists have found that removing PVC sheathings from copper cables before they are recycled in copper smelters causes dioxin emissions to drop precipitously.[116] And four studies have found that the addition of PVC-containing "refuse-derived fuel" to incinerators burning organic matter like wood chips or peat results in significant increases in dioxin formation.[117] In many of these studies, a relationship was seen in the air emissions but not in the fly ash, or vice versa, reinforcing the difficulty of establishing statistically significant relationships in the complex context of burning real wastes in large incinerators.

Because dioxins are difficult and expensive to analyze accurately in stack gases, chemists often measure emissions of chlorophenols as an indicator of dioxin emissions, because these compounds are considered the primary precursors for dioxins and furans. Two well-conducted research programs indicate that the feed of organochlorines is directly related to emissions of chlorophenols. A 1996 study for the Dutch Environment Ministry reported that when both PVC and chloride-containing compostable matter are removed from municipal waste, emissions of chlorophenols were extremely low. When 20 percent of the original amount of compostables was added back into the mix, chlorophenol emissions did not increase, but when 30 percent of the original amount of PVC was added along with the compostables, chlorophenol emissions approximately doubled.[118] Echoing these results, a series of studies at a pilot-scale incinerator at the University of Florida has documented a clear relationship between the feed of PVC and the emission of chlorophenols. The authors summed up their findings: "These experimental, phenomenological and theoretical studies of toxic emissions from incineration all support the physically intuitive hypothesis that reduction of chlorinated plastics in the input waste stream results in reduction of aromatic chlorinated organic emissions. . . . We are convinced that, when all other factors are held constant, there is a direct correlation between input PVC and output PCDD/PCDF and that it is purposeful to reduce chlorinated plastics inputs to incinerators."[119]

There are a few studies with more ambiguous results, but they do not refute the link between burning organochlorines and making dioxin. For example, a recent Swedish investigation found that dioxin formation is directly related to chlorine content, but only when chlorine levels in the fuel exceed 0.5 percent, as they do in most modern waste streams. Levels of chlorine below this level had no statistically significant effect on dioxin emissions.[120] These results could indicate that there is a threshold below which chlorine has no impact on dioxin levels, but it is just as possible that the failure to find a correlation at low chlorine levels is an artifact of the limits of chemical and statistical analysis: as levels of both chlorine and dioxin decrease, measurement error and statistical fluctuations become more and more important, swamping a fading signal under a growing chorus of noise.

Unlike the studies already discussed, this report found that it did not matter whether the chlorine came in organic or inorganic form; both gave rise to dioxin in approximately equal amounts. This finding, which contradicts all the other studies discussed, should be investigated further, but from a policy perspective it does not really matter if it is right or wrong. If the many other studies are correct that chloride salts result in minimal dioxin emissions, then dioxin output depends on the organochlorine content of the waste: lowering the input of organochlorines is necessary to reduce the formation of dioxin. If, on the other hand, the Swedish study is correct, then dioxin generation depends on the waste's total level of chlorine (organic plus inorganic); lowering the quantity of organochlorines in the waste will reduce the total chlorine level and will again cause dioxin formation to fall. Either way, if we want to prevent dioxin formation in incinerators, we need to stop burning organochlorines.

Two studies, one by the New York Department of Environmental Conservation[121] and the other by the European plastics industry,[121] have found no relationship between dioxin emissions at trash incinerators and PVC content of the waste feed. Neither of these investigations, however, controlled or adjusted for variations in other factors that are also known to affect dioxin emissions, such as operating conditions and waste input composition. A potential relationship between PVC and dioxin may thus have been masked by fluctuations in other factors. Indeed, an EPA reanalysis of the data from the New York study found that when combustion conditions were adjusted for, emissions of dioxins and furans increased as PVC content of the waste rose.[123]

In summary, it is possible that some dioxin can be formed by the combustion of chloride-containing salts, but the available evidence indicates clearly that industrially-produced materials containing organochlorines, PVC in particular, are the predominant causes of dioxin generation in incinerators. Even more important, they are the most readily preventable cause of dioxin formation: salts are naturally ubiquitous, but we can choose to stop making and burning organochlorines. As the Danish Technical Institute has written, "It is most likely that the reduction of the chlorine content of the waste can contribute to the reduction of the dioxin formation, even though the actual mechanism is not fully understood."[124]

Control or Precaution for Chemical Mixtures

We have seen in this chapter that all uses of chlorine result, at one or more points during their life cycle, in the formation of a cocktail of organochlorines that includes substances known to be persistent, bioaccumulative, and toxic. We also understand why this is so: chlorine is so reactive that a vast spectrum of organochlorines is produced whenever chlorine is used, and the most stable compounds—which are also the most hazardous—are always among the by-products. Thus, even the least toxic organochlorine results in the formation of the most toxic organochlorines during its life cycle.

To prevent organochlorine generation, the Risk Paradigm relies on engineering solutions that seek more control over chemical reactions and the movement of pollutants before they are released. But many chlorine-based processes are fundamentally uncontrolled, as in the bleaching of wood pulp for the manufacture of paper, the dispersal of pesticides directly into the environment, or the burning of PVC in an accidental fire. In others—chemical synthesis and incineration, for example—engineers can reduce by-product formation, but they can never prevent it. Because the processes of chlorine chemistry are to a large degree random and autonomous, there can be no technical solution that ensures dangerous organochlorines are not produced. Industry has harnessed the power of chlorine but has not mastered it. The only way to prevent the synthesis of extremely hazardous by-products is not to introduce chlorine into the processes in which they are formed.

To regulate chemicals on a substance-by-substance basis, the Risk Paradigm must make an even more basic assumption: that we know what chemicals we are producing. In fact, we do not even know the chemical names or structures of the majority of the by-products formed in chlorine-based processes, not to mention their toxicity and environmental behavior. Given what we know about other organochlorines, the vast majority of the unidentified compounds will probably turn out to be harmful, and there is a reasonable possibility that some will be as bad as the dioxins.

That we are largely blind to the composition of the chemical cocktails that industry produces fundamentally undermines the chemical-by-chemical approach to environmental regulation. How can we possibly claim to assess and control chemicals that we have not even identified?

Organochlorines are always produced in mixtures, never in isolation, so it is unrealistic to think we can control or even assess them individually. A risk assessment based on the toxicity and environmental behavior of perchloroethylene, for example, is no basis for a decision about whether perc should be produced and used, because the life cycle of this one chemical involves the generation of hundreds or thousands of additional compounds. Since regulations are ineffective unless they consider impacts that occur throughout the entire life cycle of a substance, risk-based limits on individual compounds are an intrinsically unsuitable instrument for environmental policy.

The Ecological Paradigm focuses not on individual compounds but on the chlorine-based processes that produce chemical mixtures. For instance, we cannot regulate each of the thousands of organochlorines discharged by the paper industry to water, air, sludge, and products, but we can eliminate all of them by not introducing chlorine into the bleaching process in the first place. The Ecological Paradigm addresses the input to rather than the output from industrial processes, preventing the generation of organochlorine mixtures altogether and obviating the need to measure and control each one from a variety of discharge points.

Sometimes the best way to stop pollution from one industrial facility is to take action at a distant location. Industrial processes are linked by the materials that flow through them, so upstream decisions about materials determine what pollution will occur later in the life cycle of a chemical. To prevent emissions of dioxins and other chlorinated PICs from a hazardous waste incinerator, for instance, we must address the industries that generated the organochlorine wastes in the first place—such as the dry cleaners that used chlorinated solvents and then shipped their residues to the incinerator. We move one level up the family tree of chlorine use to address the use of materials rather than the output of pollutants, but in the case of incinerators, we come to focus on a material that is fed into an upstream process. In this way, the Ecological Paradigm calls for a materials policy—

a set of environmental criteria for deciding what kinds of materials the economy can safely produce.

As we continue to identify the input materials that cause chlorine-based products and processes to generate organochlorine mixtures, we see that responsibility ultimately lies with the production of chlorine gas. Organochlorine pollution begins the instant that chlorine gas is produced in the chlor-alkali process, and it accelerates inevitably as chlorine and organochlorines are distributed through a plethora of industrial processes. To address the general problem of organochlorine pollution, we must deal with the input of chlorine itself into the industrial economy. The hierarchical view of chlorine chemistry developed in the previous chapter inevitably turns our attention from the tips of the chlorine tree, where individual organochlorine compounds are released, back to the limbs, branches, and ultimately the trunk, where chlorine inputs occur. If there were chlorine uses that did not produce hazardous organochlorines, this shift of focus might not be necessary, but the evidence is clear that all parts of the tree result in the synthesis and release of persistent, bioaccumulative, and toxic organochlorines. The Ecological Paradigm thus seeks ways to reduce the production of chlorine by finding safer substitutes for the various uses of chlorine and organochlorines.

The Ecological Paradigm's upstream focus extends the idea of pollution prevention, now a widely accepted principle for reducing the environmental impacts of individual industrial facilities. Governments, engineers, and environmentalists generally acknowledge that it is more effective to address toxic inputs to an industrial process than to try to control wastes after they have been produced. Waste prevention programs by a number of state and national governments have sought to reduce the generation of hazardous waste by helping firms to use less of the toxic substances that produce these wastes. But pollution prevention has not been extended backward to address the input of materials into the economy as a whole, as the Ecological Paradigm would have it. We lose control of chlorine the second it is produced, so it is both logical and necessary to extend the principle of prevention beyond individual chemicals and processes to the more fundamental level of a chlorine-free materials policy for the economy as a whole.

8

Major Sources of Organochlorine Pollution

What is food to one, is to others bitter poison.
—Lucretius, *De Rerum Naturum*[1]

To this point, I have offered a host of analytic arguments to support a phaseout of organochlorines, but this emphasis is, in a historical sense, misleading. It was concrete experiences in the real world, not abstract reasoning, that first led people to think that there might be something environmentally problematic about chlorine chemistry as a whole. Before the intellectual analysis, there were countless incidents of severe pollution in places where various aspects of chlorine chemistry are practiced. It has long been obvious that a large fraction of local and regional environmental calamities—contaminated ecosystems, polluted groundwater, communities with unusually high chemical exposures and reports of health damage, and so on—are the outcome of one or another aspect of the production, use, and disposal of chlorine and organochlorines. These experiences provided the unifying thread that led to the idea that there might be something intrinsic about chlorine chemistry that causes the generation of large quantities of persistent and very toxic substances, a suspicion that closer analysis has borne out.

In this chapter, I briefly describe some of these experiences, reviewing in more detail than the last chapter allowed the environmental impacts of four major uses of chlorine: pesticides, solvents, PVC plastic, and the bleaching of pulp and paper. A detailed examination of the life cycle of each chlorine application is well beyond the scope of this book, so I treat these four sectors as case studies, which I have chosen for two reasons. First, they account for about 60 percent of all chlorine use in the United

States and much more in some other nations, so they cannot be dismissed as anomalies. Second, they represent familiar chlorinated products and process that touch the daily lives of virtually everyone in the industrialized world. They show in a concrete way the costs of society's very recent dependence on chlorine chemistry to fulfill our most ancient and elementary needs, including food, clothing, shelter, and communication.

Organochlorine Pesticides

Our most basic biological need—one common to absolutely every living thing on earth, from the smallest bacterium to the largest whale—is food. From the matter we eat, we derive energy to fuel the basic processes of life and nutrients from which we build our tissues. For thousands of years, agricultural civilizations dealt with pests by mixing together plants that repel each other's pests, planting naturally pest-resistant crops, encouraging the presence of natural predators, rotating crops, and removing pests by hand. Then, about halfway through the twentieth century, people began to substitute synthetic pesticides for these techniques, dispersing toxic chemicals onto agricultural lands to kill insects, fungi, and weeds.

Pesticides are poisons we release directly and intentionally into the environment and food supply. According to one eminent ecologist, "Pesticides are purposeful environmental contaminants. . . . Only a tiny fraction reaches the target pest, while the remainder, greater than 99.9 percent, is essentially wasted and enters the environment through air drift and soil runoff, by leaching into groundwater, by entrapment into dust and air currents, and by deposition in rain."[2] As modern agriculture has become more and more dependent on pesticides, now applying billions of pounds of these chemicals every year,[3] pesticides have become ubiquitous contaminants of the world's air, lakes, rivers, seas, groundwater, soil, and food chain.

Chlorine consumption for pesticide production is relatively small—about 1 percent of all chlorine use—but there are at least a hundred chlorine-based active compounds used in pesticides today (appendix E).[4] The production of pesticides, like other organochlorines, results in the formation of large quantities of complex, persistent, toxic organochlorine by-products. As I noted in the last chapter, EPA has identified well over 100 pesticides in which dioxins are known or suspected by-products. In many cases, the levels are very high. Wastes from the production of lindane, for instance, contain dioxins at up to 50 parts per million (TEQ).[5] Agent Orange itself contained 2,3,7,8-TCDD, the most toxic dioxin congener, at concentrations in the parts per million range,[6] with even higher concentrations in production wastes; local dioxin contamination is a major problem at dozens of sites where 2,4,5-T and other chlorophenoxy herbicides were produced or formulated.[7] The wood preservative pentachlorophenol (PCP), a highly toxic, bioaccumulative organochlorine in its own right, also contains dioxins in the low parts per million range. Very large quantities of PCP continue to be used in many countries, and EPA estimates that pentachlorophenol-treated lumber carries an astounding 25,000 grams of dioxins and furans (TEQ) into the U.S. environment each year, making PCP by far the largest known source of dioxin in the nation.[8] More recently, dioxins and furans have been detected at lower but still significant concentrations—in the very high parts per trillion range—in the herbicide 2,4-D,[9] one of the most widely used agricultural chemicals in the United States, with production of over 26 million tons per year in that country alone.[10] 2,4-D is a common component of the pesticide formulations that commercial treatment services use on residential lawns.

Because agricultural pesticides are toxic and dispersed directly into the environment, their most severe impacts come during their use. Consider the story of Agent Orange. During the Vietnam War, the U.S. military sprayed about 20 million gallons of chlorophenoxy herbicides—containing a stunning 230,000 grams of 2,3,7,8-TCDD—on about 10 percent of the land area of the former South Vietnam.[11] Around the time that they were exposed, U.S. servicemen who had contact with Agent Orange are believed to have had body burdens of TCDD 13 to 200 times higher than the general U.S. population. TCDD levels in the mother's milk of Vietnamese women living in areas sprayed with Agent Orange were measured at up to 500 times the U.S. average. Twenty years later, TCDD levels remained 8 to 13 times higher in the blood of people in sprayed areas compared to those in unsprayed areas, and dioxin levels in food from South Vietnam are 10 times or more than those in food from industrialized nations.[12] As we saw in chapters 3 and 4, a number of studies suggest that exposed men and animals in Vietnam and in the U.S. military have suffered severe health damage.

Table 8.1
The dirty dozen pesticides

Aldicarb
Aldrin,* dieldrin,* and endrin*
Chlordane* and heptachlor*
Chlordimeform*
Dibromochloropropoane (DBCP)*
DDT*
Ethylene dibromide (EDB)
Lindane,* hexachlorocyclohexanes,* and benzene hexachloride (BHC)*
Paraquat
Parathion and methyl parathion
Pentachlorophenol*
2,4,5-T*
Toxaphene*

Note: Asterisks indicate organochlorines.
Source: Pesticide Action Network North America 1995.

2,4,5-T is just one of 13 pesticides (or groups of related pesticides) that comprise the "dirty dozen"—those that are so dangerous and have become such severe environmental contaminants that most industrialized nations have banned or severely restricted their use (table 8.1).[13] Nine of the dirty dozen are organochlorines, including most of the first-generation pesticides that Rachel Carson linked to massive local and regional contamination and severe damage to bird populations in her classic book, *Silent Spring*.[14] The dirty dozen continue to be used, primarily in developing countries, and chemical companies in Europe and the United States continue to produce them for export to these nations. These and other pesticides are responsible for some 25 million cases of acute occupational pesticide poisonings each year.[15] The vast majority of these cases occur in developing countries, where 3 percent of agricultural workers suffer one or more episodes of pesticide poisonings each year, according to the World Health Organization.[16]

In the 1970s, when many industrialized nations restricted the dirty dozen, the chemical industry introduced new second-generation chlorinated pesticides to replace them. Today, a significant fraction of pesticides in widespread use are organochlorines (appendix E). The industry promoted these second-generation pesticides as nonpersistent, nonbioaccumulative, and far less toxic than the substances that had been taken off the market. They were certainly less hazardous, but they too have created massive environmental hazards. Of 38 "priority pesticides" that EPA has listed as major groundwater contaminants, 16 are organochlorines, and these substances account for 78 percent of the total quantity of priority pesticides applied to agricultural land.[17]

Two groups of second-generation chlorinated pesticides are worth examining in some detail: the triazines (atrazine, cyanazine, and simazine) and the acetanilides (alachlor, metolachlor, and acetochlor). Used as weedkillers on agricultural land, these chemicals are by far the two largest-selling groups of pesticides, accounting for more than 80 percent of all U.S. herbicide use, and their use continues to grow, rising 6 percent between 1992 and 1994 alone.[18] Despite their reputation, the chlorinated pesticides in the second generation are also persistent and toxic, though less so than the first generation. As we saw in chapter 1, atrazine can persist in air and water for very long periods of time, particularly in cold climates. One test found that after 70 days, less than 1 percent of atrazine in water degraded fully, 80 percent remained intact, and the remaining 19 percent was transformed into other chlorinated breakdown products.[19] Alachlor lasts only about 6 to 10 weeks in soil, but that is long enough for it to enter groundwater, where its half-life is up to 5 years.[20] Given their moderate persistence and huge volume, both compounds have become ubiquitous in surface water, groundwater, rainwater, and fog, as we saw in chapter 1.

Because they are used so widely, atrazine and alachlor contaminate drinking water in alarming concentrations. A 1994 study by Physicians for Social Responsibility and the Environmental Working Group found that chlorinated triazine or acetanilide herbicides were present in drinking water samples from all but one of 29 U.S. cities studied. Cyanazine exceeded the federal maximum contaminant level (MCL) in 18 cities, and atrazine exceeded the MCL in 13 cities. Both alachlor and atrazine are known animal carcinogens and probable human carcinogens, and they cause endocrine disruption and reproductive toxicity in laboratory animals,[21] so the daily exposure of millions of adults and children represents a real public health problem.

Solvents and Dry Cleaning

In the 1930s, Europeans and Americans began to use synthetic chemicals to clean fabrics. The first chemical dry cleaners—so called because they did not use water—used hydrocarbon solvents, immersing clothing in tubs of petrochemicals that removed oils and dirt. These chemicals were highly flammable, so they were replaced by the less flammable but more toxic organochlorine solvents, including carbon tetrachloride, trichloroethylene, and perchloroethylene. The first two of these are seldom used for dry cleaning today; perc has been the dominant solvent in the dry cleaning industry since the 1960s. Many fabrics that people have cleaned with water for centuries—linens, silk, and cotton, for example—now bear tags instructing the wearer to "dry clean only."

Oil-soluble, highly volatile, and generally toxic, chlorinated solvents are used not only for textile cleaning but also for cleaning and degreasing industrial equipment and as a base for paints and coatings. Some organochlorine solvents—the CFCs and HCFCs—are also used as refrigerants in air-conditioners and refrigerators, as aerosol propellants, and as foam-blowing agents. Far greater quantities of these ozone-depleting compounds are used as solvents.[22]

The release of organochlorines during the life cycle of solvents begins with their production. The chemical industry synthesizes chlorinated solvents by three different methods, and all three generate huge quantities of hazardous wastes. When solvents are produced by *direct chlorination*—combining chlorine gas with ethylene or other hydrocarbons—hexachlorobenzene, hexachloroethane, and other organochlorines are formed as by-products in large amounts.[23] The second, more common approach is to add more chlorine to organochlorine compounds that have been formed as by-products in the manufacture of other chemicals. For example, trichloroethylene is often produced by *oxychlorinating* ethylene dichloride (EDC), which means reacting EDC with hydrochloric acid in the presence of oxygen.[24] This technique results in the formation of huge quantities of tarry wastes, a complex mixture of by-products that contain 10 to 15 percent of the carbon and 2 to 10 percent of the chlorine fed to the process.[25] The third and most popular method for making chlorinated solvents is the most hazardous. In this process, called *chlorinolysis*, existing organochlorines with higher degrees of chlorination—usually wastes from other organochlorine production processes—are heated in the absence of oxygen until they begin to degrade, producing the desired compound along with huge quantities of accidental by-products. When perchloroethylene is produced by chlorinolysis, for instance, up to 10 percent of the total yield consists of hexachloroethane, hexachlorobenzene, and hexachlorobutadiene.[26]

Chlorinated solvent production appears to be an important source of dioxins. Low concentrations of dioxins and furans have been identified in commercial preparations of trichloroethylene and dichloroethane.[27] Much larger quantities are found in the wastes from solvent production, as the most detailed available analysis of chlorinated solvents production, conducted in 1996 at an ICI plant in Runcorn, England, shows. At Runcorn, tri- and perchloroethylene are produced by oxychlorination of wastes called "heavy ends" that are formed in the production of ethylene dichloride. There are already dioxins in the heavy ends, but dioxin levels in the wastes from the solvent synthesis process, which are disposed in underground caverns, are up to ten times higher than in the heavies themselves. ICI and the UK government estimate that the manufacture of 100,000 tons of solvents at its plant generates about 500 grams (TEQ) of dioxins each year.[28] This quantity of dioxin is greater than the air emissions of all but three source categories in EPA's inventory of dioxin sources in the United States, trailing only all trash incinerators combined, all secondary copper smelters combined, and all uses of pentachlorophenol.[29] If ICI's experience is typical of other facilities that produce solvents by oxychlorination, then this process is likely to be one of the world's most important—but hitherto unrecognized—generators of dioxin.

When organochlorine solvents are used, extraordinarily large quantities are released into the air, resulting in global contamination of the atmosphere and stratosphere. Chlorinated solvents are so volatile that some 80 to 90 percent of their total quantity evaporate during their use or, if they are washed into wastewater, during effluent treatment.[30] The bulk of refrigerants also escape into the air at some point during their lifetime through leaky seals, during recharging, or when the air-conditioners, refrigerators, and automobiles that contain them are finally junked.[31] Because organochlorine solvents begin to volatilize immediately, they pose a particular health hazard to people who work with them: 401,000 U.S. workers experience direct exposures to trichloroethylene and 680,000 more are directly exposed to perchloroethylene, both of which are carcinogenic and neurotoxic.[32]

At the end of their life cycle, chlorinated solvents are major sources of dioxins and other organochlorine by-products. Spent solvents that have not escaped into the air eventually become hazardous wastes. A small percentage of these wastes are recycled for further use, but the majority are directed to incinerators, where they are an important source of chlorine input.[33] At this point, a large quantity of the solvent is destroyed, some is released to the environment through air emissions or residuals, and some is converted to toxic products of incomplete combustion.

Perchloroethylene is a particular problem, because its use as a solvent in thousands of small dry cleaning shops means that it pollutes the homes and workplaces of millions of people at short range. In 1996, about 50 million pounds of perchloroethylene were used at U.S. dry cleaners.[34] Depending on the type of machine, 29 to 80 percent of the perc consumed escapes directly into the air, with most of the remainder becoming hazardous waste destined for disposal.[35] The first people to breathe the fumes are the workers. In 1998, U.S. EPA estimated that 119,000 to 278,000 workers are exposed to perc in dry cleaning shops. Typical exposures are high enough to pose calculated cancer risks of around 1 in 100 and exceed EPA's reference dose for noncancer effects by a factor of 4 to 37. Risks were in this range even in shops using modern "dry-to-dry" machines, which obviate the need to transfer perc-soaked clothes between machines, reducing opportunities for the chemical to evaporate.[36]

Because many dry cleaners are located in residential neighborhoods—and sometimes in residential buildings—air releases of perc subject nearby residents to very high exposures. Perc enters homes and business through windows, doors, seams, and joints or by passing directly through floors and walls. One investigation in New York found that perc levels in apartments in buildings with ground-floor dry cleaners were an average of 50 times higher than in control residences with no direct dry cleaning exposure; concentrations exceeded the state guideline for perc in indoor air by up to 500 times. Even in apartments above shops using dry-to-dry machines, perc levels were 5 to 50 times higher than in control residences.[37]

In the Netherlands, perc exposures in the vicinity of dry cleaners have been found to exceed the government's standard for a "maximum tolerable risk,"[38] and U.S. EPA has concluded that "cancer concerns may also extend to residents living in co-location with dry-cleaning establishments, particularly if they live in such dwellings for more than several years. Noncancer effects may also be a concern."[39]

Because perc is strongly attracted to fats, it migrates from the air into fatty foods, like dairy and meat products. A study by the U.S. Food and Drug Administration in Washington, D.C., compared the levels of perc in butter purchased from groceries near dry cleaners to that in butter from "control" groceries not in the vicinity of dry cleaners. Every butter sample from stores located next to a dry cleaner contained elevated levels of perc, and several contained the chemical at concentrations 20 times or more higher than levels in butter from stores in the control group.[40]

Dry-cleaned clothes always contain residues of perc, which off-gas from the fabric once clothes are brought home. EPA has found that putting freshly dry-cleaned clothing in a closet results in high levels of perc in the closet, the bedroom, and even the adjacent room; the concentrations exceed New York guidelines for indoor perc exposure by a factor of 5 to 190.[41] Even clothes cleaned using the most modern machines absorb and then off-gas appreciable quantities of perc into homes and vehicles.[42] Exposure can be reduced somewhat by airing out dry-cleaned clothes for a day or more before bringing them inside, but substantial off-gassing continues even at this time, giving new meaning to the phrase "dressed to kill."

PVC Plastic

In the 1950s, polyvinyl chloride plastic, also known as vinyl, began to play a major role in the modern house. Today, with annual production of almost 20 million tons, vinyl is the largest use of chlorine in the world.[43] Few materials have infiltrated modern life as thoroughly as PVC. Since its introduction in the last 1930s, vinyl—the only major plastic that contains chlorine—has taken the place of wood, metal, ceramics, and paper in a range of products—including pipes, window frames, exterior siding, floor tiles, wall coverings, furniture, upholstery, appliance casings, toys, shower curtains, and other household items. Vinyl is also widely used in automobile and other vehicle components, office supplies, packaging, and medical devices. Since PVC is by far the largest use of chlorine, accounting for about 40 percent of world chlorine consumption, the environmental damage it causes deserves an extended treatment.

PVC's life cycle consists of several stages (figure 8.1). The first step is the synthesis of the feedstock ethylene dichloride (EDC, also known as 1,2-dichloroethane) from ethylene and either chlorine gas or hydrogen chloride. EDC is then converted into vinyl chloride monomer (VCM, the chemical name of which is chloroethylene), by a reaction called pyrolysis. In the polymerization stage, VCM molecules are linked together to yield polyvinyl chloride, a white powder. This pure PVC is then mixed with other chemicals to yield a usable plastic with desired properties. The formulated plastic is molded to produce the final product—a bottle, window frame, or pipe, for instance. The product is then sold for use, the duration of which may be very short, as in PVC packaging, or rather long, as in PVC window frames. Finally, the product is disposed of, typically in incinerators or landfills. In each stage, organochlorines are formed and/or released to the environment. When its entire life cycle is considered, PVC appears to cause the formation of more extremely hazardous by-products than any other product, a fact that should come as no surprise, since more chlorine goes into vinyl than any other application.

PVC Production

PVC manufacture begins with the chlor-alkali process, which produces dioxins and other persistent organochlorines, as we saw in the last chapter. Organochlorine pollution becomes much more severe during the production of EDC and VCM. The chemical plants that produce these feedstocks typically release huge quantities of them directly into the environment. Globally the PVC industry releases at least 100,000 tons each of EDC and VCM into the air each year, plus over 200 tons of EDC and 20 tons of VCM into surface water.[44] The actual total may be quite a bit higher, because this estimate is extrapolated from emissions at a single facility in Norway, a relatively modern and well-regulated facility with more advanced pollution control equipment and careful plant operation than is typical of manufacturers in many other nations.

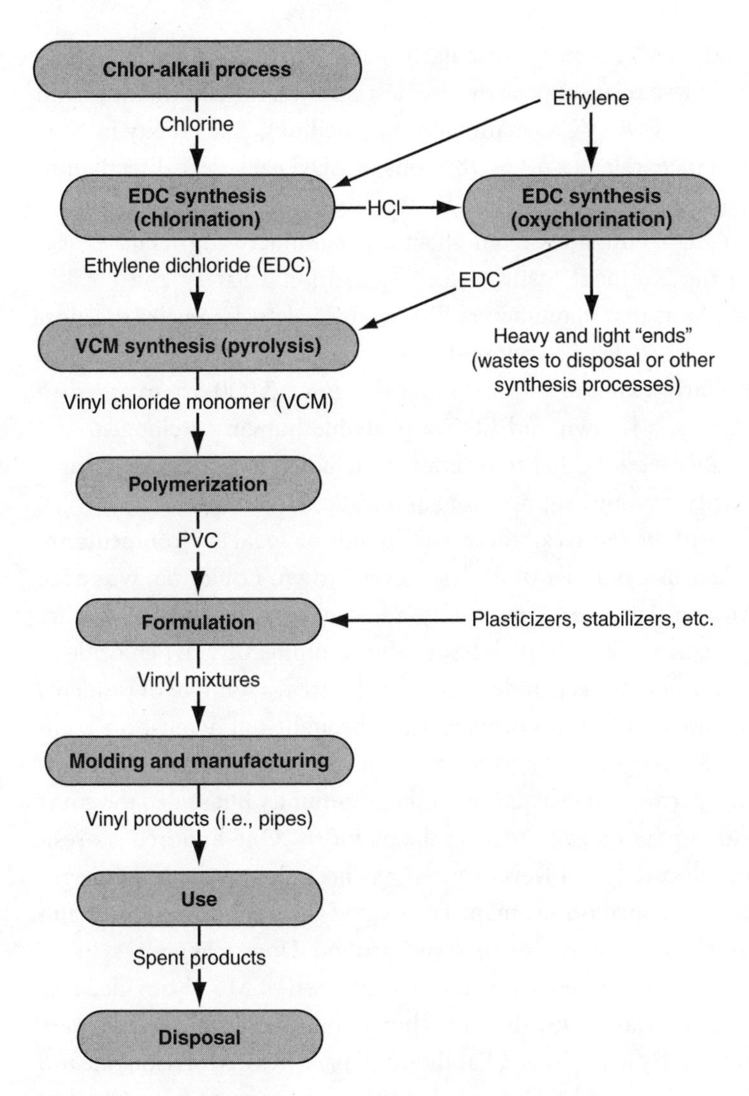

Figure 8.1
Stages in the "life-cycle" of PVC. At virtually every stage, highly persistent and toxic by-products are generated. (see text).

VCM and EDC are not particularly persistent, but both are highly toxic. Thus, these releases pose the greatest hazards for communities and ecosystems near EDC/VCM manufacturing facilities. The facility in Norway, for instance, releases 40 to 100 tons of EDC each year directly into the local atmosphere.[45] In the United States, some 12.5 million persons are exposed to EDC emissions from chemical manufacturing facilities, according to the National Institute for Occupational Safety and Health. Workers in plants that manufacture PVC or its feedstocks receive the most severe exposures to these compounds in workplace air; 81,000 U.S. workers are regularly exposed to vinyl chloride, and 77,000 are exposed to EDC.[46] VCM is a known and EDC a probable human carcinogen, and both cause a variety of other toxic effects, including liver damage, neurological toxicity, immune suppression, and testicular damage.[47]

The effects of the release of these compounds on local environments are painfully clear in a number of places. Reveilletown, Louisiana, was once a small African American community adjacent to a VCM/EDC facility owned by Georgia-Gulf. In the 1980s, after a plume of vinyl chloride in groundwater began to seep under homes in the area, a number of residents began to complain of health problems and brought suit against the company. In 1988 Georgia-Gulf agreed to an out-of-court settlement that provided for the permanent evacuation of the community but sealed the court records and imposed a gag order on the plaintiffs. One-hundred six residents were relocated, and Reveilletown has been demolished, the neighborhood all but wiped off the map. The next year, as concern over air and groundwater pollution began to grow around Dow Chemical's EDC/VCM facility just five miles away in the small town of Morrisonville, near Plaquemine, Louisiana, Dow began to buy out and relocate citizens there in a program to avoid exposure, liability, and bad press. Morrisonville too is now all but abandoned.[48] On the other side of the state, in Lake Charles, Louisiana, PPG and Vista Chemical manufacture EDC and VCM, which now contaminate water and sediments in the Calcasieu estuary, along with several by-products of their synthesis.[49] Here, residents continue to occupy their homes, drink local water, and eat fish from the area's polluted bayous.

EDC/VCM synthesis generates huge quantities of persistent, bioaccumulative by-products. There are two ways that EDC is made: ethylene can be chlorinated with chlorine gas, or it can be oxychlorinated with hydrogen chloride formed as a waste in other synthesis processes. (Almost all EDC producers use both methods in a linked cycle, because chlorination of ethylene generates hydrogen chloride as a by-product, which can then be used in oxychlorination.) Both processes yield a complex mixture of reaction products, which are then distilled to yield three batches of materials: the distilled EDC product, the light ends (those substances more volatile than EDC), and the heavy ends, which are less volatile than EDC. The quantity of wastes is quite large—about 2 kilograms each of heavy and light ends for each ton of EDC produced. Based on these figures, world EDC synthesis by the oxychlorination process produces at least 30,000 tons per year each of light and heavy ends.[50] In general, the "heavies" are discarded and the light ends reprocessed in other chemical reactions. The EDC goes on to be pyrolyzed—heated in the absence of oxygen—to yield vinyl chloride monomer; by-products of this process include chlorobutadiene, chlorobenzene, chlorinated ethanes, ethylenes, methanes, and large amounts of complex but uncharacterized waste tars.[51] According to one industry source, the total production of chemical wastes produced in the various processes involved in EDC/VCM synthesis is about 3 percent of the VCM yield, or some 570,000 tons—more than 1 billion pounds—of by-products each year.[52]

The heavy ends contain most of the persistent and toxic by-products. No academic or government studies have sought to identify the compounds present in these wastes, but there are some data from industry and environmental groups. In 1990 Dow Chemical analyzed its EDC heavy ends and found they were about 65 percent chlorine, including 302 parts per million PCBs, 0.3 percent hexachloroethane, 1.2 percent hexachlorobutadiene, and 30.6 percent unidentified compounds.[53] If this analysis is representative of heavy ends in general, then EDC oxychlorination results in the worldwide production of a stunning 20,000 pounds of PCBs each year, even though these compounds were banned from intentional production in the late 1970s.[54] In 1993 chemists from Greenpeace's laboratory at the University of Exeter analyzed material from a number of European EDC/VCM manufacturers. Soil and gravel samples taken near a Swedish oxychlorination reactor contained a wide variety of persistent organochlorines in the high parts per million range, and HCB and

HCBD were present at the remarkable levels of 1.9 and 0.6 percent by weight.[55] These results are particularly important because they argue against the claim that contaminants are contained within the production equipment and are never released to the environment.[56] The next year, Greenpeace obtained samples of heavy ends from several U.S. EDC/VCM manufacturers and had them analyzed by the Exeter laboratory. In one sample from Borden Chemical, 174 organochlorines were identified, including a wide variety of highly chlorinated complex aliphatic and aromatic substances.[57]

With PCBs and hexachlorobenzene in the wastes from PVC production, it is no surprise that we find dioxins and furans as well. Dioxins have been detected in the wastes from VCM synthesis and in the incineration of wastes from this process,[58] but they are formed in particularly large quantities in the production of EDC by oxychlorination. As the British chemical company ICI made clear in a submission to the government, the formation of dioxin in this process is inevitable and unpreventable: "It has been known since the publication of a paper in 1989 that these oxychlorination reactions generate polychlorinated dibenzodioxins (PCDDs) and polychlorinated dibenzofurans (PCDFs). The reactions include all of the ingredients and conditions necessary to form PCDD/PCDFs, i.e., air or oxygen, a hydrocarbon (ethylene, etc.), chlorine or hydrogen chloride, a copper catalyst, an ideal temperature, and an adequate residence time. It is difficult to see how any of these conditions could be modified so as to prevent PCDD/PCDF formation without seriously impairing the reaction for which the process is designed."[59]

The 1989 paper to which ICI was referring was the work of a group of chemists at the University of Amsterdam, who simulated the oxychlorination process in the laboratory and found dioxin formation at a rate that would make this method of producing EDC one of the world's largest sources of dioxin, if not the largest.[60] This research generated considerable concern, so the vinyl industry began its own sampling program. In 1993, the Norwegian PVC manufacturer Norsk-Hydro confirmed that its EDC/VCM synthesis plant produced dioxins but claimed the quantities were hundreds of times lower than the Dutch study had predicted.[61] How much dioxin is actually formed remains uncertain, because both studies have advantages and disadvantages. On one hand, the Dutch analysis may be a more accurate indicator of total dioxin generation, because the researchers were able to capture and analyze *all* the material outputs from the oxychlorination process; the Norwegian report, like any other study of a full-scale facility, inevitably missed some of the by-products, which are directed into too many different wastes, products, catalysts, recirculating materials, and equipment surfaces to be completely assessed. On the other hand, the Dutch study was a laboratory simulation and the industry analysis took place at a real production facility, and there may be something about the simulation that caused more dioxin to form than during the industrial-scale synthesis of EDC.

Whatever the exact quantities, there can be no doubt that dioxin generation occurs in amounts that are far from negligible. In 1994 government scientists found dioxins at high concentrations (up to 414 parts ber billion TEQ) in sludges from a fully modernized EDC/VCM plant in Germany, refuting the claim that only outdated EDC/VCM technologies produce dioxin.[62] The same year, ICI Chemicals and Polymers found that its vinyl chloride plant in Runcorn, United Kingdom, was producing large quantities of dioxin—not as much as the Dutch studies predicted but more than Norsk-Hydro had estimated. Most of the dioxins at ICI were deposited in heavy-end wastes, and smaller quantities were released directly into the air and water.[63] In the United States, wastes from Vulcan Materials' EDC plant in Louisiana have been found to contain dioxins and furans at the extraordinary concentration of 6.4 parts per million (TEQ), which makes them among the most dioxin-contaminated wastes ever discovered, on a par with wastes from the manufacture of Agent Orange.[64]

These extremely dangerous wastes go to one of two places. In some facilities, they are used in the manufacture of chlorinated solvents, in which case the contaminants end up in the wastes or products from those processes. In others, the wastes are disposed of, usually by incineration. The vinyl industry has argued that we should not be concerned about dioxins and other dangerous chemicals in heavy ends because they are destroyed by burning. But as we saw in the last chapter, the ability of incinerators to destroy hazardous wastes completely has been drastically overestimated; in fact, incinerators are quite inefficient at destroying chemicals that are present in low concentrations, like the dioxins in EDC wastes. Additional amounts of toxic and persistent organochlorines are formed in the incineration process as products of incomplete combustion. With these problems in mind, it is hardly reassuring that the chemical industry burns the huge quantities of hazardous wastes it produces in the manufacture of PVC feedstocks.

Not all of the by-products of EDC/VCM synthesis end up in the hazardous wastes; some escape directly into the environment. Dioxins have been detected in wastewater discharges and air emissions from a number of EDC/VCM plants,[65] and local and regional contamination of water, sediments, and shellfish has been linked to EDC/VCM manufacturers in both Europe and the United States.[66] For example, severe dioxin contamination of sediments in Italy's Venice Lagoon has been linked to an EDC/VCM manufacturing facility.[67] In the Netherlands, levels of dioxins in sediment samples in the River Rhine jump dramatically just downstream from an EDC/VCM manufacturing plant;[68] the levels are so high, in fact, that the majority of dioxins in Rhine sediments downstream from the plant, all the way to the river's mouth, and in the entire North and Wadden Seas, appear to be attributable to the facility.[69]

Other by-products are also present near chemical plants that make PVC feedstocks. In Lake Charles, Louisiana, the National Oceanic and Atmospheric Administration (NOAA) has found very high levels of persistent organochlorines in the water, sediment, and fish of bayous near EDC/VCM facilities owned by PPG and Vista Chemical. According to NOAA, the geographical pattern of contamination indicates that PPG is the primary cause of very high levels of organochlorines in the water and sediment. In one portion of the estuary near PPG's facility, concentrations of HCB, HCBD, and hexachloroethane exceeded 1,000 parts per million in sediment; in some samples, these three by-products represented from 0.1 percent to a mind-boggling 4.8 percent of the sediment's total mass.[70]

Use of PVC Products

PVC is not bioavailable, so the polymer itself is not toxic during its use. But vinyl products are not pure PVC: they contain both accidental contaminants and chemical modifiers that are added to the plastic on purpose, and some of these are quite hazardous. Moreover, PVC products often encounter reactive conditions—accidental fires in particular—that transform the plastic into very hazardous by-products.

Consider first by-products within the plastic itself. A small portion of the dioxins formed in EDC/VCM synthesis end up in the PVC product itself. In May 1994, the Swedish EPA found that pure PVC plastic from two Swedish producers contained dioxins, furans and PCBs in the low parts per trillion.[71] In 1995, the UK government found dioxins and furans in the same range in PVC food packaging items, including cling film and bottles for oils and beverages.[72] Subsequently the U.S. Vinyl Institute and the European plastics industries conducted their own studies; both identified trace quantities of some dioxin congeners in some samples of PVC plastic.[73] The levels were very low, but *any* quantity of dioxin in consumer products is a matter of concern.

Chemical additives are even more problematic, because they are present in very large amounts. In its pure form, PVC is not particularly useful; it is rigid and brittle, and it gradually catalyzes its own decomposition when exposed to ultraviolet light. For PVC to be made into useful products, additives must be mixed with the polymer to make it flexible, moldable, and long lasting.[74] PVC additives include a range of toxic compounds, but the most important of these are the phthalate plasticizers and lead stabilizers. Phthalates are organic chemicals used to make vinyl plastic flexible, and they can make up a large portion—up to 60 percent by weight—of the final product.[75] Lead is used to extend the life of PVC products exposed to light, and it is typically present in lower but still significant concentrations. About 4.3 million tons of phthalates and 124,000 tons of lead are used in the worldwide production of PVC every year.[76]

The additives are not chemically bonded to the PVC polymer but are merely mixed into the plastic during its formulation. Over time, they leach out of vinyl products, entering the air, water, or other liquids with which the product comes in contact. Phthalates, which are suspected carcinogens and endocrine disrupters, are especially likely to evaporate, creating the familiar smell we associate with a new vinyl shower curtain. Based on Swedish figures, some 42,000 tons of phthalates are emitted into the air from PVC products in the world each year.[77]

Phthalates are somewhat persistent and quite bioaccumulative, so they have become ubiquitous environmental contaminants, present in water, air, fish, and human tissues on a global basis. Because vinyl accounts for more than 90 percent of the total consumption of phthalates, these problems are overwhelmingly caused by PVC.[78] Phthalate accumulation also takes place closer to home. When PVC containers and films are used to hold food products, plasticizers migrate out of the plastic and accumulate in foods, especially fatty ones like cheese and meats.[79] The common practice of storing blood and drug formulations in PVC bags causes phthalates to leach into the contents of the bag, which can result in substantial short-term phthalate exposures for the recipient.[80] Newborn infants that receive a single blood transfusion have been found to have extremely high levels of phthalates in their systems.[81] Lead too is released from PVC products. Significant lead releases have been documented from PVC window blinds,[82] and lead is also known to leach into water carried in PVC pipes that contain lead stabilizers.[83]

Given the presence of toxic additives in vinyl products, it is remarkable that so many toys are made of vinyl, including many teething toys. In the late 1990s, government ministries in Denmark and the Netherlands found that substantial quantities of phthalates are released from PVC into saliva. These countries, along with ministries in Belgium, Austria, and Spain, sought bans on the use of soft PVC in toys. In 1997 a number of European toy retailers and manufacturers suspended sales of PVC teething rings or announced plans to eliminate all vinyl from their toy lines. In late 1998, when a wave of publicity on the issue hit the U.S. press, Toys-R-Us, Mattell, and several other U.S. retailers and toy makers also announced they would stop selling or making some kinds of vinyl toys.[84] Subsequently the Consumer Product Safety Commission called on the toy industry to voluntarily stop making vinyl toys that were made to be chewed and contained phthalates.[85] But babies chew and such on virtually anything, not just toys intended for that purpose, so these gestures do not go far enough.

Another major hazard associated with the use of PVC products is the possibility of fire. PVC is ubiquitous in modern buildings and vehicles. When vinyl burns, substantial amounts of dioxin and other organochlorines form as products of incomplete combustion,[86] along with huge amounts of hydrochloric acid, which is highly toxic to firefighters and can cause massive damage to computers and other equipment.[87] The combustion conditions in an accidental fire—where gases do not mix thoroughly and materials cool rapidly as they escape from the flame—are perfectly suited to the rapid production of dioxins.[88] As a result, all accidental fires in modern buildings are likely to generate dioxins and other persistent, bioaccumulative organochlorines. After a fire in a German kindergarten that contained substantial quantities of PVC, scientists measured dioxin levels in indoor soot at concentrations of 45,000 parts per trillion (TEQ)—almost 300 times greater than the German government's health standard. This situation required the building's interior to be completely stripped—of all floors, ceilings, wall coverings, furnishings, and so on—sandblasted, and remediated by hazardous waste experts before children were allowed to enter again.[89] Dioxins have also been identified in the residues from burning automobiles, subway cars, and railway coaches.[90]

Even a very small amount of dioxin from each of the 621,000 structural fires and 421,000 vehicle fires in the United States every year could make a substantial contribution to dioxin contamination of the environment.[91] The German EPA and the German environment ministers have called for the use of substitutes for PVC in all areas susceptible to fire, but PVC use in cars and construction continues to grow on a global basis.[92] As a result, a stockpile of PVC, waiting to burn, is accumulating in staggering quantities. Worldwide, over 400 million tons of PVC are in stock—that is, in use in various applications, mostly construction related, and susceptible to fire at some point.[93]

While many small fires taken together may constitute an important source of organochlorines, a single fire at a PVC factory, warehouse, or disposal site can create very large quantities of pollutants. A home contains at most a few hundred kilograms of PVC,[94] but a warehouse or landfill may have hundreds of tons on site. After a fire at a plastics warehouse in Binghamton, New York, for example, dioxin levels in soil on the site were found to be over 100 times greater than other samples from the same community.[95] In Hamilton, Ontario, after some 200 tons of PVC burned at a plastics recycling facility, samples of soot, ash, and tree leaves from the fire contained greatly elevated quantities of dioxins.[96] Fires are also regular events at landfills and waste storage sites where large quantities of PVC are present; although data on dioxin releases from these fires are limited, EPA estimates that they may be one of the largest sources of dioxin emissions in the United States.[97]

PVC fires not only create dioxins and other organochlorines; they also release additives held in the plastic. The world stock of PVC contains a staggering 3.2 million tons of lead and 83 million tons of phthalates.[98] Lead cannot be destroyed by combustion, so accidental fires represent a major potential source of lead exposure, a hazard that looms larger as more and more PVC accumulates worldwide in building applications.

Disposal of PVC Products

The final stage of PVC's life cycle creates the most severe organochlorine pollution. About 30 to 50 percent of the PVC produced annually—some 6 to 9 million tons per year worldwide—ends up in the trash stream.[99] The remainder goes into longer-life uses like pipes, siding, and flooring—products that ultimately end up in the trash, too. A significant portion of the thrown-away PVC goes to landfills, where it will last for centuries, and almost all the rest is burned; the exact proportions going to land disposal and incineration vary from one country to another.

One thing is true everywhere: very little PVC is recycled. Even in Denmark, a nation with an ambitious and effective recycling program, only 0.6 percent of postconsumer PVC is recycled.[100] In the United States, the American Association of Postconsumer Plastics Recyclers announced in 1998 that its attempts to recycle PVC had failed and that it would henceforth view vinyl products as unrecyclable contaminants in the municipal waste stream.[101] Recycling PVC is difficult because vinyl products are mixtures of PVC and additives, and each specific formulation is uniquely suited to its application. In recycling, many formulations are mixed together, which destroys the special properties of each one. This means that recycled PVC is always of lower quality than the original material, so it can be used only in products without strict material requirements, such as fence posts and speed bumps. Since recycled PVC cannot be used to make a new version of the original product, it is not recycling at all; it would be more accurate to call it "downcycling."[102] An example of true recycling is the reprocessing of paper: the old fibers are used to make new paper products, and a new tree does not need to be cut down. In contrast, a new vinyl bottle or floor tile must be made of new plastic. In this way, downcycling does not reduce the amount of PVC produced each year or the total quantity of PVC building up on the planet. The illusion of recycling actually increases the global PVC burden by finding new uses for old PVC while creating a positive image for a product that can be neither safely disposed nor truly recycled.

Huge amounts of PVC go to incinerators. In municipal incinerators, PVC contributes at least 80 percent of the organically-bound chlorine and 50 to 67 percent of the total chlorine (organochlorines plus inorganic chloride) in the waste stream, although it makes up only about 0.5 percent of the trash stream by weight.[103] In the United States, an estimated 200,000 to 300,000 tons of PVC is incinerated in trash burners every year.[104] Vinyl is even more important in the medical waste stream. It is the most commonly used polymer in medical packaging and devices, totaling 700 million pounds per year in the United States, with an annual growth rate of over 6 percent,[105] and it accounts for 5 to 18 percent of all medical waste.[106] According to a report for New York City, PVC products account for over 90 percent of the organic chlorine and over 80 percent of the total chlorine content of medical waste.[107] In hazardous waste incinerators, the vinyl life cycle is also an important chlorine source, because wastes and tars from the synthesis of EDC and VCM are among the most abundant chlorinated wastes produced and burned by the chemical industry.

There is no doubt that burning vinyl is a source of dioxin, because laboratory combustion tests involving pure PVC (or pure PVC in the presence of metal catalysts) produce considerable amounts of dioxin.[108] No one has attempted to identify the full range of organochlorines formed when PVC burns, but 45 organochlorines—including persistent and toxic benzenes, styrenes, PCBs, and PCDFs—have been found in the combustion products when the closely related plastic polyvinylidene chloride (PVDC, commonly known as Saran Wrap) is incinerated.[109]

Incinerators also release additives contained in PVC products into the environment. Over 45,000 tons of lead stabilizers in PVC enter the world's municipal trash each year.[110] Because lead cannot be destroyed by incineration, all the lead fed to an incinerator ultimately enters the environment, via stack emissions, ash, scrubber effluent, or wastewater sludges. In nations where leaded gasoline is banned, incinerators are now the largest source of lead emissions to the environment, and PVC is responsible for about 20 percent of the lead in the waste stream, according to Swedish figures.[111] Vinyl thus appears to be a major cause not only of dioxin but of lead pollution, as well.

Not all burning of vinyl takes place in high-tech incinerators, of course. In developing countries and rural areas of industrial nations, open burning of waste is a common way to get rid of trash. A recent study by U.S. EPA and the New York Department of Environmental Conservation indicates that backyard burning of trash in barrels can result in massive emissions of toxic chemicals, including chlorinated methanes, benzenes, phenols, and dioxins and furans. Emissions of dioxins and furans per pound of waste burned, in fact, were 12,000 to 75,000 times higher than emissions from a modern trash incinerator; EPA estimated that open burning by just three households would produce as much dioxin as an incinerator large enough to serve about 40,000 people. Further, when more PVC was burned, average releases of all chlorinated PICs, including the dioxins, rose substantially.[112] These findings suggest that the rapidly expanding use of vinyl in developing countries, where expensive means of waste management are not available, has the potential to cause a huge increase in worldwide emissions of dioxins.

Some spent metal products that contain PVC are reprocessed in smelters, and these facilities are also major dioxin sources. Secondary copper smelters, for example, recover copper from PVC-coated wire and cable and PVC-containing telephone cases; very high dioxin emissions have been measured at these facilities, which are considered major dioxin sources in most inventories.[113] Most important, removing some of the vinyl sheathing before cables are fed to the smelters reduces dioxin emissions considerably.[114] Secondary steel smelters have also been found to emit very large quantities of dioxin, primarily because they recover metal from scrap automobiles that contain PVC.[115] Secondary lead smelters release dioxin and other organochlorines, too, due to the feed of lead automobile batteries with internal PVC separators. In the United States, however, PVC has been recently phased out of this application, so EPA no longer considers lead smelters an important dioxin source.[116]

In summary, the PVC life cycle presents one opportunity after another for the formation and environmental discharge of organochlorines. The plastic that is all around us, and expanding its role in our lives all the time, turns out to be one of the most hazardous materials on earth, when its production, use, and disposal are considered together. In fact, PVC is the major chlorine source in the majority of the combustion facilities that dominate inventories of dioxin sources; the production and use of PVC also appear to be important but largely unrecognized causes of dioxin pollution. Apparently harmless, essentially inert and nontoxic, vinyl appears to create more persistent, toxic organochlorines than any other single product. Like the Romans who sipped from lead cups, ran drinking water through lead pipes, and bathed in lead basins, we have built our house of poison, unaware of the consequences.

Pulp and Paper Bleaching

Unbleached paper has a beige or brown tint, the traces of the natural lignin found in the pulp of the wood or other plant material from which the paper was made. In the nineteenth century, pulp mills began to use bleaching powder to make paper white, and early in the twentieth century, they adopted chlorine gas for this purpose.[117] Today about 100 million tons of wood pulp are bleached for paper production each year, consuming several million tons of chlorine and chlorine-based bleaches in the process.[118]

Chlorine gas combines with the lignin and other organic substances in wood pulp, producing at least 1,000 organochlorine by-products, about 300 of which have been identified.[119] A single average-sized pulp mill using elemental chlorine generates about 50 tons of organochlorines each day.[120] On a global basis, the paper industry has been the largest source of organochlorine releases directly to waterways, discharging around 4 million tons of organochlorines per year during the 1980s and early 1990s when most bleached pulp was produced using chlorine gas.[121] Studies for the Swedish and Canadian government indicate that pulp mills in those nations are responsible for as much as 90 percent of all organochlorines that have been discharged directly into the Baltic Sea and the Great Lakes.[122]

Chlorine bleaching pulp mills also release organochlorines into the air. The bleaching process produces substantial quantities of the volatile organochlorines EDC, dichloromethane, chloroform, carbon tetrachloride, 1,1,2-trichloroethane, trichloroethylene, and perchloroethylene. Some portion of these compounds remains in the effluent—chloroform, for instance, has been detected in waterways up to 100 kilometers downstream from bleached pulp mills[123]—but the majority will evaporate into the air. Worldwide emissions of carcinogenic chloroform from the paper industry were around 30,000 tons per year when chlorine gas was the industry's dominant bleaching agent.[124]

Large quantities of organochlorine by-products are also concentrated in the sludge from pulp mill water treatment processes, which contains organically-bound chlorine at levels up to 4 percent by weight.[125] In the United States, about 80 percent of the sludge is buried in landfills or spread directly on the land. The remainder is incinerated, destroying some organochlorines but producing new ones as products of incomplete combustion. Pulp mill sludge incinerators have been found to release PCBs, dioxins, furans, chlorinated phenols, catechols, guaiacols, and benzenes into the environment.[126]

Finally, a considerable portion of the organochlorines formed during bleaching ends up in the bleached pulp and paper products themselves, which can contain as much as 1 percent organochlorines by weight.[127] Diapers, cigarette paper, tampons, coffee filters, cosmetic tissues, and bleached milk cartons contain low concentrations of dioxins and furans, and other products prepared from bleached pulp, including detergents and liquid "soft soaps," are contaminated at much higher levels.[128] Dioxins have been shown to leach out of bleached paper cartons into milk and out of bleached filters into coffee, although the resulting exposures are quite low compared to the quantity ingested in the diet.[129]

With such abundant and diverse toxic releases to the environment, it is not surprising that pulp mills have caused severe environmental damage. Organochlorines, including dioxins, have built up in local and regional environments, injuring fish, invertebrates, and aquatic plants. An extensive research program for the Swedish EPA detected chlorinated bleaching by-products as far as 1,400 kilometers downstream from the mills in which they originated. Pulp mill discharges—organochlorines in particular—were linked to a variety of severe health problems in fish, including damage to the skin and gills, deformed jaws and fins, smaller gonads, hormonal changes, impaired reproduction, changes in mating behavior, liver disorders, and changes in population structure. The effects, were far more severe in ecosystems near bleached pulp mills than near unbleached pulp mills, and were observed as far as 40 kilometers away from the discharge points of bleached pulp mills. Marine invertebrates also suffered developmental toxicity, altered behavior, and changes in the genetic structure of populations, and aquatic plant colonies were damaged, as well. The authors were unable to determine any safe exposure level and concluded that "regional and possibly large-scale" damage to fish and aquatic ecosystems may have occurred throughout the Baltic ecosystem due to organochlorine discharges from the paper industry.[130]

The damage is not limited to Scandinavia but occurs wherever the pulp and paper industry is located. Fish in Canadian rivers into which pulp mill effluent is discharged display a similar suite of toxic effects to those found in the Swedish studies.[131] On the coast of British Columbia, dioxin contamination in the early 1990s forced the closure of fishing grounds around 11 of the 14 pulp plants in the region.[132] The Maine Department of Health recommended in 1994 that women who are pregnant, nursing, or of childbearing age not eat fish from several rivers or lobster livers from coastal waters due to dioxin contamination, which was caused primarily by the state's seven chlorine-bleaching pulp mills.[133] In Florida, a pulp mill owned until recently by Procter & Gamble has caused severe dioxin contamination of water and fish in the Fenholloway River, into which it discharges some 50 million gallons of effluent every day. The river flows out to the Gulf of Mexico and also recharges an aquifer that provides drinking water for thousands of people. The company has been forced to provide bottled water to the community, because water from many wells in the area is now brown and foul tasting.[134]

In the early and mid-1990s, many mills in North America and Europe switched from chlorine gas to chlorine dioxide bleaching, a process the industry calls elemental chlorine free (ECF). Chlorine dioxide combines the oxidizing potential of chlorine and oxygen in a single molecule, so that mills can achieve equal brightness with a lower total chlorine input. By 1997, half of the world's total production of bleached pulp was bleached with chlorine dioxide, 44 percent with chlorine gas, and the remaining 6 percent with totally chlorine free methods.[135]

The shift from elemental chlorine to chlorine dioxide has reduced organochlorine discharges significantly, but they remain massive. A complete switch from chlorine to chlorine dioxide can reduce organochlorine formation by about 80 percent.[136] The oxygen in chlorine dioxide also helps to break up some, but not all,[137] of the benzene-based molecules in the lignin, making them more soluble in water and reducing the formation of dioxins and other compounds still further. But chlorine dioxide reacts with water in the bleach plant to produce elemental chlorine (Cl_2) and hypochlorous acid (HOCl), each of which then reacts with organic matter to produce organochlorines.[138] The name *elemental chlorine free* thus turns out to be misleading; elemental chlorine gas is not added on purpose but is produced in the process of chlorine dioxide bleaching. The result is that so-called ECF bleaching produces the same types of organochlorines formed in chlorine bleaching, though in smaller quantities.

The by-products formed in chlorine dioxide–bleaching mills are even less well characterized than those from plants using chlorine gas. The identified compounds include chlorophenols, chloroform, chlorinated acids, and highly bioaccumulative compounds, such as chlorinated cymenes, cymenenes, fluorenes, phenanthrenes, naphthalenes, and sulfones, which are known to be taken up into the tissues of fish.[139] Dioxins and furans have been identified not only in the wastewater but also in the air from mills that have converted to ECF bleaching.[140] Of the massive unidentified fraction, a large portion—about 50 percent of the organically-bound halogens in the effluent—is high-molecular-weight material, which degrades in the environment into more persistent and toxic organochlorines.[141] Chlorine dioxide bleaching also produces large amounts of chlorate, a powerful herbicide that kills both plants and fish if not fully removed from the effluent.[142]

Despite the shift to chlorine dioxide, the pulp industry retains its position as one of the largest source of organochlorines into the environment. Some of the most useful quantitative data on by-product formation from chlorine dioxide bleaching come from a report for the trade group of pulp manufacturers that have switched to chlorine dioxide.[143] Based on that study's data, we can estimate that if every pulp mill in the world were converted to chlorine dioxide bleaching and equipped with state-of-the-art pollution control equipment, the paper industry would discharge about 140,000 tons per year of organochlorines into waterways, plus additional releases to air, land, and products.[144] Of this, about 2,000 tons per year would consist of persistent and bioaccumulative compounds.[145] In the United States, EPA has estimated that if the entire industry converted to ECF bleaching, mills that produce kraft pulp (the strongest kind of paper) would discharge over 22,000 tons per year of organochlorines, including substantial quantities of chloroform (8.1 tons per year), chlorinated phenolics (9.4 tons per year), and dioxins and furans (11.1 grams per year of the two congeners 2,3,7,8-TCDD and 2,3,7,8-TCDF alone).[146]

Organochlorines are also present in the pulp product bleached with chlorine dioxide, in concentrations from 250 to 780 parts per million. These figures suggest that if the world pulp industry entirely converted to chlorine dioxide bleaching, its paper products would contain about 40,000 tons of organochlorines each year.[147]

Organochlorines in effluents from pulp mills that bleach with chlorine dioxide remain a significant environmental problem. The Swedish EPA research program in the Baltic Sea had particularly important results relevant to ECF bleaching. Laboratory tests and field investigations showed that chlorine dioxide effluents were less toxic to fish than chlorine-bleached effluents but more toxic than effluent from mills that used no chlorine-based bleaches.[148] Two subsequent studies have confirmed these findings, showing that totally chlorine-free bleaching effluents are less toxic to fish and microorganisms than those from mills that bleach with chlorine dioxide.[149] These reports leave no doubt that it is beneficial to remove *all* chlorine-based compounds from the bleaching process, not to stop halfway.

This is not to say that organochlorines are the only harmful compounds in pulp mill effluent, because they are not. Discharges from chlorine-free mills contain a number of compounds known to be toxic to fish, as one would expect, given the fact that tree tissues contain large quantities of natural pesticides, acids, and feeding deterrents. These compounds are liberated during the pulping process and can cause a variety of biochemical and physiological effects.[150] Advocates of chlorine-based bleaching have sought to portray this finding as an acquittal for organochlorines in the effluent, as if the conclusion that organochlorines cause environmental damage were a case of mistaken identity.[151] But the presence of additional toxic compounds in bleaching effluents in no way exonerates the organochlorines, which are clearly an environmental hazard. After the organochlorines are removed, other toxic substances are still present, but this does not imply that the organochlorines were harmless in the first place. Removing chlorine compounds from the bleaching process is necessary but not sufficient to attain the ultimate goal of environmentally sound paper production.

The toxicity of chlorine-free effluent to fish reinforces why it is so important to eliminate wastewater discharges from pulp mills altogether by creating closed-loop bleaching facilities that recycle their effluents. As I discuss in chapter 10, the effluent loop can be closed safely and effectively only if chlorine and chlorine compounds are eliminated from the bleaching process, because the resulting by-products are too corrosive to be recycled through the plant. The hazards of chlorine-free effluents thus become another reason to convert pulp production to totally chlorine-free bleaching practices. The paper industry supplies a product on which our civilization truly depends—one that could be made in a sustainable fashion from a renewable resource. With chlorine and chlorine-based bleaches a continuing centerpiece of pulping, however, the industry remains far from the model of environmentally sound production it can and should be.

Public Access to Information

Any framework for making environmental policy needs some way of discovering health and ecological damage and linking it to its sources. Unfortunately, the kind of information I have discussed in this chapter on the types and quantities of toxic substances released by industrial facilities is often very difficult to come by, for two major reasons. First, as we saw in chapter 1, analytical methods are limited in their ability to identify and quantify all the components in complex waste mixtures. Second, for most industries, no studies have even tried to identify all the substances produced and discharged to the environment. To the extent that data are available, most of the information has been generated by the industries themselves, so it cannot be confirmed.

Why are there so few independent studies of the pollutants released by major industries? The answer is that industries guard their privacy and autonomy carefully, and they do not welcome the prying eyes or instruments of government and academic scientists. A reliable audit of by-product formation and release would require extensive information about and access to all aspects of a company's production processes, and few industries welcome this kind of inspection, especially when the findings may result in more stringent regulations. Conceivably, access could be compelled by law, but the authority of the U.S. federal government to gather information from within the boundaries of an industrial facility is limited, and EPA often avoids exercising the powers it possesses.

The available knowledge about the sources of organochlorine pollution thus reflects a political rather than a purely scientific reality. The lack of information necessary to make fully enlightened policy is the result of a political framework that limits the ability of government and independent scientists to gather data about pollution sources. In a democracy, scientific and political judgments cannot be based solely on information provided by one interested party. Democratic decision making requires that the public have full access to information and materials associated with industrial processes that affect health and the environment. This principle may sound obvious, but it is not practice in the Risk Framework. Public right-to-know laws represent an improvement but do not go far enough, because the information is submitted by the industry to the government with little or no independent verification. More alarming, there are a number of egregious cases in which government has obtained important information but for political reasons has kept it from the public eye, or filtered it in a way that deliberately obscures the public's understanding of the industrial sources of pollution.

Consider first the actions of U.S. EPA and Dow Chemical in the early 1980s concerning the sources of dioxin in the Great Lakes. In 1980 Canadian researchers discovered dioxin in gull eggs in the Great Lakes, with particularly high levels in birds from Saginaw Bay, near Dow's facility in Midland, Michigan. After the Canadian government requested the U.S. government to investigate the source of this pollution, EPA's regional office in Chicago took the initiative and began to study the problem. By 1981 the office had prepared a draft report concluding that dioxin levels in Great Lakes fish, particularly from the Saginaw area, posed a hazard to consumers and recommended that fish consumption from the area be prohibited. The report found very high concentrations of dioxin in the bodies of water receiving Dow's effluent; it concluded, "Dow Chemical of Midland, Michigan, has extensively contaminated their facility with PCDDs and PCDFs and has been the primary contributor to contamination of the Tittabawassee and Saginaw Rivers and Lake Huron."[152]

EPA submitted a draft of the report to EPA headquarters for approval. According to the sworn testimony of EPA Region V officials, the agency's administrators rejected the report, informing the region that it would "inflame the public, which is already sensitive to the dioxin problem. . . . , implicate industry, and suggest that a health problem is evident. . . . Moreover, there is some concern over the new political atmosphere in Washington." Scientists at headquarters told the region, "No one here disagrees with your conclusions," but said that "putting in print your conclusions could inflame the public." Headquarters insisted that the report be limited to a presentation of the quantities of dioxin in the Great Lakes, with no assessment of the health hazard of fish consumption and no inference about the sources of the pollution. Meanwhile, in a highly unusual action, EPA assistant administrator John Hernandez gave the draft report to Dow and instructed the region to take comments from the company's representatives in preparing a final draft. Dow reviewed the report in detail, making clear in a conference call "what lines and what sections are acceptable to the company and what should be preferably deleted."

By the time the report was finally released, it had been, in the words of its authors, "fairly well stripped of information." The recommendation to prohibit public consumption of fish from the area was removed, as were all assessments of the human health hazard posed by dioxin in the Great Lakes. Headquarters also replaced the region's summary of dioxin's toxicity, intended to be comprehensible to the public, with a detailed and much less accessible scientific review. All mention of Dow as a source of dioxin—indeed, all reference to Dow, period—was deleted. The report, in fact, was published without any conclusions at all.

The gutting of the Great Lakes dioxin report was revealed two years later, when Congress began to investigate corruption in the Reagan administration's EPA under administrator Anne Gorsuch. In the midst of this controversy, a draft of the region's original, unedited report was leaked to a Toronto newspaper, and Congress opened a new round of hearings on the cover-up. For his part in the scandal, Hernandez was forced to resign, following in the footsteps of Gorsuch and her assistant Rita Lavelle, who had been removed for their misconduct on a number of chemical pollution and hazardous waste issues. Were it not for the leak and the hearings, the information linking dioxin contamination to Dow Chemical might never have come to light.

The next example of suppressed information involves the paper industry.[153] In the mid-1980s, under order from Congress, U.S. EPA began to investigate dioxin pollution around the nation. The agency discovered substantial contamination in fish downstream from pulp and paper mills in a number of states, suggesting for the first time that pulp mill bleaching was a major dioxin source. Particularly high levels were found in fish near a pulp mill owned by Boise Cascade's in International Falls, Minnesota, near the Canadian border. Several state agencies also began to investigate the problem, and in 1986, EPA's Great Lakes regional office got into the act again. Under the authority of the Clean Water Act, the region requested access to Boise's mill in order to take samples for dioxin analysis. Alarmed by the region's aggressiveness, the American Paper Institute (API), the trade and lobbying organization of the pulp and paper industry, asked EPA headquarters to take over all dioxin investigations at pulp mills and suggested a joint EPA-industry study, to be conducted in secret.

EPA consented, entering into an unprecedented written agreement with the API that provided for an unpublicized study in which the industry would design the sampling plan for each site and contract with outside laboratories for chemical analysis.[154] Under the agreement, the industry would have "input into the development of the final report," which would be limited to a "technical document responsive to the study objectives"—not, that is, a recommendation for any regulatory action on the paper industry. Most important, EPA agreed that the results of the sampling could be treated as confidential trade secrets and that there would be no "public release of these results without first discussing the situation with industry officials."[155] For over a year, EPA and industry pursued its joint study in secret, finding that chlorine-based pulp mill bleaching was a major source of dioxin but never announcing the results to the public. In 1987, the secret documents were leaked to Greenpeace, and media coverage forced the information on dioxin contamination into the open.

This story seemed to be replaying itself in the 1990s, with the PVC industry in the lead role. The Vinyl Institute's 1994 "Dioxin Reassessment Crisis Management Plan," to which I alluded in the last chapter, suggested how the industry should portray technical information to "avoid deselection of PVC by major customers and to prevent punitive regulation of

PVC by EPA, Congress, or the state legislatures."[156] The strategy advised the industry to enter into joint scientific activities with EPA to avoid future regulation. According to the document:

> The short-term objective of the plan is to mitigate the effects of potential negative press coverage by positioning the vinyl industry as a proactive and cooperative entity, working in tandem with EPA to characterize and minimize sources of dioxin. . . . The vinyl industry must actively and aggressively communicate with the media its commitment to working with EPA to characterize and minimize any dioxin in the PVC lifecycle while at the same time asserting that based on reliable data available to date, the industry believes its contribution is minimal. Cooperative positioning is the key element.

In its draft reassessment, EPA concluded that it did not have enough data to estimate how much dioxin is generated or released by the processes of PVC manufacture. The agency noted that the synthesis of ethylene dichloride appears to produce dioxin, but it said only that there was controversy over the amount formed: environmentalists claimed the amounts were significant, and the industry said they were trivially small. EPA chose not to take account of analytical results from laboratory studies and full-scale PVC plants in Europe, arguing that facilities on that continent might not be comparable to those in the United States, though it did not specify any significant design differences between American and European technologies. In any case, one would think it would be simple enough to address the lack of domestic data: gather samples from U.S. EDC/VCM plants and analyze them.

EPA did no such thing. Soon after the reassessment was released, EPA and the Vinyl Institute announced a plan by which the industry would "self-characterize" dioxin emissions from PVC production. The industry successfully positioned itself as "cooperating" with EPA and pre-empted the agency from gathering any independent information. Ultimately, the product of this cooperative venture, a report released by the Vinyl Institute in 1998, did not provide the necessary information to estimate dioxin generation or releases associated with the life cycle of PVC.

In the study,[157] the PVC industry decided where and when samples were taken from its own plants, collected them, analyzed their dioxin content, chose which data to present, interpreted them, and then submitted the results to EPA. The submission was reviewed by an independent panel, but the industry chose which data the panel saw. Information about the samples, including the facility they came from, was completely confidential, so neither reviewers nor the public had the opportunity to determine whether sampling times and locations accurately represented typical dioxin releases. Most important, no one was able to independently evaluate, confirm, or act on the information.

The self-characterization was designed to leave out critical information by ignoring the most important parts of the industry that needed characterizing. The study analyzed several potential pathways for dioxin release, finding low to moderate quantities of dioxins and furans in one or more samples of EDC, PVC products, air emissions, wastewater, and the sludge from effluent treatment. But the pathways that contain the largest amounts of dioxin, along with many PVC-related processes that are major dioxin sources, were completely ignored. No data, for example, were gathered on dioxin contamination of chemical streams that recirculate in the manufacturing process, of light ends and other wastes used in other synthesis processes, and—most important, because these are known to be so severely contaminated—heavy ends, tars, and other hazardous wastes that are sent to disposal facilities. Nor did the program address what is apparently the largest PVC-related dioxin source: the burning of vinyl in incinerators, smelters, and accidental fires.[158]

There is some irony to the image of openness and cooperation the industry created by proposing this "self-characterization" in 1994, because it knew of the problem many years earlier. It was in 1989, after all, that Dutch scientists reported at an international scientific conference that the production of EDC by oxychlorination could produce large quantities of dioxin. Company correspondence discovered in lawsuits against the vinyl industry show that the international vinyl industry quickly became aware of these results and arranged a series of internal meetings to formulate a response. One official of the Dutch PVC manufacturer Akzo wrote in a memo to other industry leaders, "By itself the formation of the above mentioned compounds [PCDD/Fs] during the oxychlorination is well known, at least since 1983 when the quench water effluent of the Akzo Botlek VCM-plant was analyzed."[159] Exactly when the U.S. industry first learned of the problem is unclear, but documents show that the Vinyl Institute and its members met to discuss the Dutch research in January 1990, more than four and a half years before it proposed the dioxin self-characterization.[160]

Society needs independent oversight of any industry with potential impacts on public health and the environment. To leave all knowledge-generating activity to the industry that is causing the pollution is breath-takingly naive. A popular misconception about science is that its knowledge is objective—that properly conducted studies will find the same answer regardless of who conducts them. But who does scientific work has an undeniable impact on the outcome, because there are many opportunities at which the interests and perspectives of scientists and their sponsors inevitably affect the design of research and the reporting of results. Even scientists who do not have a stake in the outcome of their research typically design their experiments to maximize the likelihood of finding a meaningful or interesting result; the ability to do so is the mark of a successful scientist. A scientist with a personal or institutional interest in getting one answer or another can easily design research to increase the likelihood of getting a favorable outcome. Further opportunities to influence results emerge in the interpretation stage. Scientists never report all their data unfiltered; even disinterested scientists sift through a mass of noisy data, selecting those they believe are reliable, portray a meaningful pattern, and allow an interpretation that is interesting and important. A scientist with an interest in a specific outcome can easily select those data that support a desired conclusion and shade his or her interpretation in favor of that result.

Science is not a blind activity performed by automatons; it is a form of intellectual and practical work done by thinking human beings. The researchers who write scientific papers and reports are real people, with full-fledged perspectives and motivations. Most scientists endeavor to make their work as honest and fair as possible, but scientific research can never be entirely stripped of the interests and viewpoints of the person who creates it. This is no real problem as long as those who read and make use of the products of science are aware of it. If we allow all our knowledge to be created by those with a specific interest in the outcome, we can be sure that our information will be neither accurate nor fair. Only a willing fool would pretend that the fox's study of the henhouse will give the same results as the farmer's.

III

The Solution: A Chlorine Sunset

9

An Ecological Policy on Chlorine Chemistry

As a society, we cannot continue protracted debate while the actual or even suspected injury to living species continues to occur. Yet this is precisely what occurs and will continue to occur until governments address classes of chemicals rather than a few specific chemicals at a time.

—International Joint Commission, *Eighth Biennial Report on Great Lakes Water Quality*[1]

In 1983, President Ronald Reagan unveiled the Strategic Defense Initiative (SDI), a trillion-dollar plan for a vast antimissile defense system designed to protect the United States from nuclear attack. The proposal, dubbed Star Wars by the media, called for a network of lasers and other paraphernalia that could shoot down every single incoming missile, potentially numbering in the thousands, in an attack on U.S. cities and military installations. Not only was the SDI extraordinarily expensive, it was risky. If it failed to destroy even one incoming missile, the impacts would be intolerable. The Star Wars plan bet millions of human lives on the infallibility of its technology and treated efforts to prevent nuclear war as unnecessary and outmoded. In the late 1980s, in the face of public opposition and near-unanimous agreement that the system would never accomplish its stated goal to intercept and destroy all incoming warheads, Congress finally restricted funding for the SDI.[2]

The problem of global organochlorine contamination raises an analogous issue: is the appropriate focus of policy at the narrow, reticulated level of managing individual compounds or the more general one of addressing the technologies of chlorine chemistry as a whole? Potential remedies to the organochlorine problem exist somewhere on a spectrum of options. At one end stands continued use of the Risk Paradigm to assess, regulate, capture, and dispose of organochlorines on a substance-by-substance basis. At the other is an extreme precautionary policy that would ban chlorine chemistry immediately and entirely. Between the two poles stand a number of other options, including bans on individual chemicals and a gradual, planned phaseout of chlorine chemistry.

In this chapter, I analyze the policy implications of the information covered in the first two parts of this book, summarizing why organochlorines cannot be effectively managed under the Risk Paradigm. Scientific, engineering, and administrative considerations all indicate that organochlorines should be treated as a class and prevented at the source. The proper policy is not an outright ban on chlorine chemistry but a gradual phaseout of the production and use of organochlorines in a planned process called a chemical sunset. Chemical sunsetting is a practical way to implement the Ecological Paradigm, providing maximal protection of the environment while minimizing technological, economic, and social disruption.

Two Frameworks Revisited

Industry and some scientists have argued that the current approach of chemical-by-chemical assessment and control can address the environmental hazards of organochlorines. The Chlorine Chemistry Council (CCC), for instance, suggests using risk assessments to determine acceptable discharge limits that will "reduce risk from exposure to specific chlorinated compounds."[3] An article the CCC distributed argues, "Regulations should target specific substances whose environmental harm has been clearly demonstrated through rigorous scientific studies."[4] Dow Chemical takes a similar tack, recommending testing to identify individual organochlorines that are "persistent, toxic, and bioaccumulative," followed by installation of control technologies and efficiency improvements to "reduce or eliminate emissions" of these substances.[5]

On the other hand, the American Public Health Association,[6] the International Joint Commission,[7] and a plethora of environmental organizations have called for a preventive policy that would address organochlorines as a class, gradually eliminating the industrial and agricultural practices that lead to the production and use of chlorinated organic chemicals. Several international agreements, including the Paris Convention on the Northeast Atlantic[8] and the Barcelona Convention on the Mediterranean Sea,[9] have pointed to organochlorines as a priority in programs to eliminate sources of persistent, bioaccmulative toxic substances.

In evaluating these kinds of proposals, the first order of business is to learn from experience. Since the 1970s, industrialized countries have implemented scores of environmental laws and regulations. The vast majority have been firmly based in the Risk Paradigm of chemical-by-chemical assessments, discharge limits, and pollution control technology. A few exceptions have taken an approach based in the concept of pollution prevention; that is, they have banned or phased out the production or use of chemicals so that they do not need to be licensed, controlled, captured, and disposed. Our analysis should begin by examining which policies have worked, which ones have failed, and why.

The Risk Paradigm has been especially well tested in the United States, where a complex regulatory bureaucracy enforces a byzantine set of standards at a total cost of billions of dollars per year. Despite this vast commitment of bureaucratic and economic resources, the Risk Paradigm has failed to accomplish its goals. Barry Commoner and his colleagues at the Center for the Biology of Natural Systems have reviewed trends in pollutant levels in the United States since the nation's major environmental laws came into force in the 1970s. Commoner's analysis found that pollution control regulations—countless discharge limits and requirements to use specific control technologies—have reduced discharges from individual facilities to individual environmental media, but environmental levels of the major substances regulated by this approach have not declined substantially; some have actually increased.[10] Commoner concluded, "Congress has mandated massive environment improvement; the EPA has devised elaborate, detailed means of achieving this goal; most of the prescribed measures have been carried out, at least in part; and in nearly every case, the effort has failed to even approximate the goals. In both the columns of statistics and everyday experience, there is inescapable evidence that the massive national effort to restore the quality of the environment has failed."[11]

Commoner also found a handful of regulatory success stories, but all were of a fundamentally different sort. In particular, DDT, PCBs, CFCs, and major uses of lead were addressed not by control but by prevention. Instead of trying to limit the release of these chemicals into the environment, governments restricted their production or use. Subsequently, releases of these compounds to the environment plummeted, and environmental levels and human exposures have gradually followed suit. (That emissions have declined does not mean that contamination is no longer a problem. Levels of these substances in the environment remain unacceptably high, because they are so persistent and because the actions taken have generally addressed only some uses in some nations.[12] Nevertheless, the improvement compared to what would have ensued had production continued without restriction is undeniable.) As Commoner points out, the unique success of chemical phaseouts illustrates the commonsense point that if you do not make a substance, it will not end up in the environment; if you do make it, at least some will eventually make it out into the ecosystem.

If the Risk Framework has resoundingly failed to reduce pollution by the few dozen substances it has addressed, it certainly cannot solve the massive problem of global organochlorine contamination. Organochlorines have become global pollutants, in large part, during the period in which pollution control regulations have been in force. Maintaining the status quo will not solve the problems created by the current system. It is time to learn from our mistakes and our achievements and build a new set of policies around prevention, precaution, and clean production.

The Failure of Control

Why has the Risk Paradigm failed so miserably? A central problem is its focus late in the pollution process, at the end of the pipe. The Risk Paradigm puts no restrictions on the production or use of chemicals, relying instead on tacked-on pollution control devices to capture, treat, and dispose of them after they are generated, just before they are discharged into the environment.

This approach inevitably fails for four reasons. First, many chemicals, organochlorines in particular, are deliberately dissipated into the environment or the economy. Once a pesticide is sprayed onto agricultural land, a paint stripper sold to a handyman, or PVC pipes installed in a home that may one day catch fire, even the most efficient devices attached to factories become irrelevant. In these cases, the hazard comes from the product, not a waste, and the only way to prevent the release of these substances or their ultimate by-products into the environment is not to use them at all. When the product is the poison, pollution control regulations are powerless: end-of-pipe controls do not work when there is no pipe to control.

Second, control devices merely shift chemicals from one environmental medium to another. Installing a filter on an incinerator smokestack may reduce dioxin emissions to the air, for example, but the captured pollutants are then concentrated in the filter residue, which must be disposed of in a landfill, from which they will ultimately escape and contaminate groundwater. The most effective precipitators, scrubbers, filters, evaporation tanks, and landfill liners merely change the time or place in which organochlorines enter the environment. Eventually captured pollutants make their way into the ecosystem in one place or another, in one form or another.

Third, control and disposal technologies seldom perform as well in reality as they are supposed to. Human error, aging equipment, and fluctuations in operating and environmental conditions can all result in unexpectedly large releases of chemicals to the environment during capture or disposal. Landfill liners decay and leak, incinerators undergo upset conditions and explosions, chemicals are spilled, and so on. As the use of chlorine-based products and processes expands globally, especially in developing countries where regulatory and technological infrastructures are less developed, the scenario of optimally designed, operated, supervised, and maintained control and disposal technologies becomes little more than a fantasy.

Finally, the goal of pollution control is to reduce the quantity of emissions per unit product. In a capitalist economy, however, production—measured as either the amount of product made at each facility or the total number of factories in operation—is always expanding. So long as economic growth of this sort continues, emissions will grow too, eventually overwhelming improvements in the rate of emissions per unit product. In theory, perpetual increases in pollution control efficiency could avoid this problem; in reality, after an initial reduction in discharges, the cost of improved control increases exponentially, so it quickly becomes prohibitively expensive to achieve further reductions. Like the Red Queen in *Alice in Wonderland,* who had to run faster and faster just to stay in the same place, companies must spend more and more on pollution control just to maintain a constant rate of environmental pollution. If industrial societies are to continue developing and not create an ever-increasing pollution burden, they will have to do more than capture and control pollutants after they are produced.[13]

These considerations make obvious why phaseouts have been successful. If a hazardous substance is not produced at all, it cannot enter the environment, either by direct dispersal or after a shellgame among discharges to air, water, and land. Eliminating the input of toxic materials into industrial processes also reduces reliance on control and disposal devices that must always work at or near perfection, so problems of less-than-optimal technological design, control, and maintenance become moot. Finally, phasing out toxic chemicals at the source brings the ratio of pollutant to unit product all the way to zero; economic output can grow with no increase in pollution, because zero times any rate of production will always be zero.

Bridges and Ecosystems

The second major problem with the Risk Paradigm is that its tools for controlling and assessing pollutants were not designed to deal with synthetic chemicals, so they are fundamentally ill suited to preventing the kinds of damage that these substances can cause. The current system of environmental regulation emerged as a matter of historical contingency from the tools and ideas that happened to be around in the 1970s when governments began to regulate synthetic chemicals. The Risk Paradigm was cobbled together, uncritically for the most part, from elements that were designed to address other types of environmental and engineering problems.

Consider the centerpieces of the Risk Paradigm: assimilative capacity and acceptable discharges. These two constructs were originally applied to pollutants of biological origin, like oil and grease, human and animal waste, and other naturally biodegradable pollutants. Permitting discharges works fine for substances like this, which ecosystems can break down and assimilate in limited amounts. It works for acids and bases too, which can be absorbed in finite quantities with little or no change in the pH of an ecosystem.[14] The first phase of pollution control regulations in the 1970s, which targeted these kinds of traditional, visible pollutants, successfully made many rivers and lakes cleaner in both appearance and reality.

When regulators extended the assimilative capacity concept to synthetic chemicals, however, they never considered the fundamental difference between degradable pollutants and man-made substances that accumulate in the environment, the food web, and our bodies. These chemicals, including a very large number of organochlorines, cannot be integrated into natural cycles and do not break down to any appreciable extent. Discharged in even very small amounts, substances like this build up over time to higher and higher concentrations in the environment and in living things. Given enough time, acceptable discharges ultimately reach unacceptable levels.

The Risk Paradigm borrowed its other major element, risk assessment, from engineers and economists, who have long used the technique to calculate the odds that a structure or an investment will fail. A civil engineer cannot say with certainty whether a bridge actually will or will not collapse, but he can calculate with confidence the probability that it will do so. These predictions are reasonably reliable, because the system has been built by humans and is thus well-characterized. Further, only a limited number of factors affect the integrity of the bridge. The system is classically mechanical and linear: the effect of weakness in one part of the bridge on another part is well understood, and the probability of individual events can be added or multiplied to yield the probability of an overall outcome. Although there are always uncertainties, these too can be defined and even quantified. Finally, the impacts of failure are local and immediate, and each project can be considered in isolation, since the collapse of a bridge in Delaware has no effect on the integrity of a bridge in Illinois.

This model is fine for engineering applications, but it is totally inappropriate for ecosystems and organisms, which are not built objects but are alive, unpredictable, densely interconnected, hierarchical, and largely uncharacterized complex systems.[15] In systems like this, quantitative predictions are unreliable. As we saw in chapter 2, the first barrier to reliable prediction is missing data. The vast majority of the thousands of organochlorines now in production have not been subject to even rudimentary toxicology testing, and the full range of toxicological impacts on all potentially affected organisms is understood for absolutely none of these substances. When the initial values that should be put into a predictive model are not known, we can have no confidence whatsoever in the output.

Second, there is limited sensitivity. Organochlorine doses once thought to cause no effect frequently turn out to be the level at which available methods of measurement and analysis cannot detect impacts. As the sensitivity of tools improves, lower doses are found to cause effects, revealing the danger of environmental exposures once thought to be safe. The inability to distinguish toxicological fact from artifact means that we can never be confident that an exposure level predicted to be harmless will actually be safe.

Third, there is ignorance. No one can predict the impacts of perturbing a system unless the system itself is well characterized. Our understanding of how organisms and ecosystems develop and function, however, remains rudimentary at best. Every revelation in molecular biology, for instance, raises new questions about how individual genes and proteins interact to regulate processes at the level of the whole organism. The complex dynamics that characterize relationships among species in an ecosystem remain even more elusive. Until we can fathom how organisms and ecosystems function, there is no way we can reliably predict the ultimate effects of the multiple, simultaneous changes we inflict on these systems.

Fourth, there is complexity. Prediction works well in simple physical systems with determinate, linear paths of causality; each cause produces one effect, and the total result of numerous small causes is simply the sum of all their individual impacts. In general, machines and bridges work this way. Organisms and ecosystems, however, are characterized by multifactorial and circular causality—negative and positive feedback loops, redundancy, multiple functions, critical periods of extraordinary sensitivity, and so on. Circular webs of causality mean that the system is buffered against some changes, but it is extremely sensitive to others. Multiple tiny changes can cause runaway or synergistic effects, resulting in a major reorganization or breakdown of the system. Disrupt a hormone or reduce the population of a keystone species at a critical time, and a cascade of effects may occur, magnifying and spreading the impact of the original intervention. Other impacts may be more subtle, degrading the performance or adaptability of the system without the obvious signs of failure. Further, because causality is not additive, the impact of simultaneous exposure to thousands of individual chemicals cannot be predicted based on the known effects of each substance in isolation.

Finally, there are surprises—forms of damage of which we are not even aware but may occur in the future. When toxicologists establish that a substance does not cause any of several effects that they have investigated, the Risk Paradigm declares it harmless. But the chemical may turn out to cause an effect that they did not investigate, or one they never even dreamed of. CFCs were once considered miraculously safe because they were largely nonflammable and nontoxic; it came as a total surprise when, almost fifty years later, they were found to cause severe damage to the stratospheric ozone layer. Similarly, chemicals were not tested for hormonal activity until recently, when evidence began to build that organochlorines, legally discharged into the environment for decades, had caused large-scale endocrine disruption in wildlife and people. When risk assessment is used for its original engineering and economic purposes, the outputs to be estimated are always well defined: what are the chances that the bridge will fall down, the reactor vessel corrode, the pool of capital dwindle to nothing before the investment pays off? When we do not know all the impacts to be concerned about, predictions of safety based on a few effects become quite meaningless.

A New Kind of Hazard

Not only are the systems to which risk assessment is now applied entirely different from the method's original subject matter, but the dangers themselves are of an entirely novel sort. The Risk Paradigm was designed to deal with well-defined, local, short-term hazards. But humans now possess technologies that can cause fundamental and irreversible changes in living systems. Created on an immense scale by large corporations and governments, today's hazards, unlike the local risks of the Risk Paradigm, are essentially unbounded in time and space, posing global and multigenerational threats to health and ecosystems. Synthetic chemicals represent a fundamentally new kind of problem, with which current regulatory institutions and concepts cannot cope effectively.[16]

The very concept of risk, in fact, is inappropriate for the kind of health and environmental damage that organochlorines can cause. A risk is a chance, usually a quantifiable one, that something bad will occur; the negative outcome, the probability of which is expressed as a risk, either happens or it does not. If the gambler does not lose his money, he makes some; if the bridge does not collapse, people make it across safely. In contrast, the long-term hazards of chemical pollution are not discrete either-or events, the risk of which can be expressed as so many per million. Instead, pollutant exposure modifies the distribution of functional ability throughout the entire population; virtually everyone exposed will be affected to a greater or lesser degree. Further, the injury emerges slowly in the subtle and diverse symptoms of global environmental decay, so these hazards cannot be reduced to probabilities and are difficult to quantify in any fashion. Global contamination causes universal side effects and systemic damage, not individual risks.

The tunnel-vision approach that attempts to keep individual activities within locally acceptable levels is utterly inappropriate for hazards of this type. Preventing major local effects does not avert the slow growth of global damage, which occurs as the cumulative result of all the technological activity in society. Consider an incinerator permitted to emit pollutants into the environment in quantities that will not cause extreme effects in the immediate surrounding community; now picture thousands of such facilities, each granted permission to pollute on the same basis, with no attention ever paid to the cumulative pollution burden that the universe of these facilities creates. The local focus of the risk-based system is intrinsically at odds with the problem of global accumulation.

Once global injury occurs, the current system's methods for dealing with damage also break down. The scope of this kind of damage—large scale impairment of the health of humans and wildlife population, contamination of the entire food web—is so vast that it can never be cleaned up or repaired. The inability to trace causality to individual actors means that victims cannot be compensated or individual perpetrators held legally responsible. Most important, this system, which requires a demonstration of a causal link before action can be taken to eliminate the cause of a problem, cannot even stop the damage it is doing when it finally becomes obvious; the limits of epidemiology and the lack of local, determinate causality mean that this requirement will never be satisfied. Current institutions become paralyzed by their own unrealistic standards of proof.

To cope with this new type of danger, we need a new scientific and institutional model designed specifically for synthetic chemical pollution. The view of nature on which the Ecological Paradigm is built derives not from engineering but from ecology, which recognizes that ecosystems and organisms are complex systems characterized by intricate interconnections, homeostasis, feedback loops, cascades, and critical periods. In keeping with this model, the Ecological Paradigm focuses on the possibility of large-scale, multigenerational forms of health damage, particularly injuries to the functional and adaptive capacity of organisms and changes in ecosystem structure and dynamics. In making inferences, it integrates information from a variety of scientific disciplines and biological levels—from molecular biomarkers to global changes—and it interprets evidence in a precautionary framework that seeks to minimize false judgments that no hazard exists when in fact one does.

In the realm of action, the implications of the ecological view depart even further from current regulatory practice. With prediction and proof of chemical-specific causal linkages out of the question, the Ecological Paradigm seeks not to calculate the limits of acceptable hazards but to minimize systemic damage of all kinds. It avoids global injury by minimizing the use of synthetic chemicals whenever possible. To implement this program, it places priority on substances known to have hazardous properties, including persistence, bioaccumulation, and the ability to disrupt basic mechanisms of development, regulation, and information transfer in organisms and ecosystems. In these aspects, the Ecological Paradigm is guided by the ethical principles of medicine and public health: hazards that can be prevented are never ignored or accepted. This approach strives for continued reductions in damage to health and the environment, and precaution defines the default state: "First do no harm."

Precautionary Policies

A central tenet of the Ecological Paradigm is the precautionary principle, which entered the world stage in the 1980s. At that time, evidence was building that pollution was inducing large-scale damage in the

North Sea, but action was repeatedly delayed because scientists were unable to establish definitive proof of causality. Germany, where the principle was formulated in the late 1970s, began to advocate a new approach in which action could be taken without cause-effect proof. By 1987 Germany and its supporters, buttressed by growing public concern about the environment, had convinced national representatives at the Second International Conference on the Protection of the North Sea to adopt the precautionary principle as part of international law for the first time. The conference agreed that discharges of substances that are "persistent, toxic, and liable to bioaccumulate" should be prevented at the source, "even where there is no scientific evidence to prove a causal link between emissions and effect."[17]

Since then the precautionary principle has been adopted in a number of environmental agreements and conventions and has emerged as a general and legitimate part of international law.[18] As formulated by the United Nations Conference on Environment and Development, the principle states, "Environmental measures must anticipate, prevent and attack the causes of environmental degradation. Where there are threats of serious or irreversible environmental damage, lack of scientific certainty should not be used as a reason for postponing measures to prevent environmental degradation."[19] The precautionary principle states that chemicals that *may* cause harm should not be released into the environment. Precaution is designed to prevent injury before it happens; it does not require even a suggestive demonstration that pollution has already damaged the environment or human health.

Some representatives of industry, government, and academia have criticized the precautionary principle as unrealistic or unworkable on a number of counts.[20] Some of these critics raise honest intellectual objections, while others are motivated by financial interest. In 1994, for instance, the Chlorine Chemistry Council hired the public relations research firm Mongoven, Biscoe and Duchin (MBD) to gather intelligence on environmental and public health organizations that were involved in the chlorine issue and recommend strategies to counter their activism. MBD and its principals have helped corporate clients oppose efforts by citizens' groups on such issues as the aggressive marketing of infant formula in developing countries, apartheid in South Africa, nuclear energy, endangered species, oil spills, and consumer safety, according to the Center for Media and Democracy, a nonprofit organization that monitors the public relations industry.[21] In its report to the CCC, MBD's "main recommendation [was] to mobilize science against the precautionary principle. . . . Engage a broad effort on risk assessment within the scientific community . . . and take steps to discredit the precautionary principle within the more moderate environmental groups as well as within the scientific and medical communities."[22]

Whatever the motives, it is important to address the substance of the arguments. Behind diverse forms of rhetoric, there are two major criticisms of the precautionary principle. The first is that the principle is so vague that it does not lead to any specific policies and is thus an ineffectual piece of international law.[23] The second argues that the principle is absolutist: because no activity can ever be proved strictly safe, this position states, the precautionary principle would ban virtually everything, leaving no room for setting priorities or balancing the benefits of regulation with its economic and social costs.[24]

These are reasonable criticisms of the precautionary principle *in isolation.* As for the first objection, we should remember that the precautionary principle is only a principle. If viewed as a rule or a standard, it is hopelessly vague, doing nothing to define the policies that should flow from it. This vagueness also leaves the principle open to redefinition by those who would like to weaken or co-opt it as its influence in national and international policy grows. For instance, the Chemical Manufacturers' Association has argued that the precautionary principle requires all regulations to be based on risk assessment, because this approach focuses not on existing harm but on the probability of future injury.[25] Others suggest that the principle merely requires a "margin of safety" between permitted pollution levels and the predicted assimilative capacity[26] of the environment and public health, or that it requires that all companies use the best available technology for pollution control.[27] Following definitions like these, one law professor has gone so far as to argue that virtually all U.S. regulations already embody the precautionary principle.[28]

These definitions clearly violate the principle's spirit. Precaution has its historical and logical roots in a recognition of the complexity of ecosystems and organisms, the unreliability of scientific predictions, the failure of the assimilative capacity approach, and the need to prevent pollution at the source. In its original German formulation, the precautionary principle is based on a responsibility to future generations to avoid irreversible damage entirely and a profound awareness of the limits of science to diagnose environmental damage and its causes. It also calls explicitly for a government-directed program of industrial development to introduce cleaner technologies in all sectors of the economy.[29]

A truly precautionary policy thus requires much more than a mild strengthening of the existing policy framework. It represents a fundamental departure from the assumptions and methods of the current regulatory regime. But the words of the principle as it has been adopted in international agreements do not make this deep change entirely clear. The precautionary principle, after all, says what we should *not* do—wait for proof of cause-effect linkages—but it does not specify what we *should* do or even when we should do it. Three additional concepts provide more specific rules and guidance, forming the pillars of a truly preventive paradigm for environmental policy.

The first rule is Reverse Onus, which specifies how scientific information should be evaluated for determining appropriate action.[30] The default state of the Risk Paradigm is to allow chemical discharges; the burden is on the public to provide evidence justifying restrictions on any source of pollution. This laissez-faire system allows a lack of scientific evidence to serve as evidence of safety; uncertainty favors those who would pollute the environment, delaying action until definitive evidence is established. Reverse Onus shifts the burden of proof onto those whose actions may damage the environment. Before a polluting technology is approved, proponents are required to demonstrate to a reasonable degree of certainty that their activities are safe and necessary.

Reverse Onus simply means erring on the side of caution. Where the potential consequences of a mistake are serious or irreversible, it is common sense to take precautions to avoid the possibility of harm. In fact, most nations already do this in regulating pharmaceuticals, requiring drug companies to conduct extensive testing to provide reasonable assurance that a new chemical is safe and effective before granting a license to market it. Why are synthetic chemicals, which are introduced into people's bodies involuntarily through the environment, treated with less stringency? Given the limits of science and the magnitude of the injury that global chemical pollution can cause, it is clearly appropriate to place the burden of proof on those who would use and produce synthetic chemicals.

The second principle is Zero Discharge of substances that are persistent or bioaccumulative. First articulated by the U.S. and Canadian governments in the Great Lakes Water Quality Agreement of 1978, this idea was later adopted in several other international agreements.[31] Zero Discharge is based on the fact that persistent substances, by definition, are incompatible with the ecosystem's natural processes of material breakdown and recycling. Substances that resist natural degradation or accumulate in fat gradually build to higher and higher levels in the environment and in our bodies; if we merely reduce discharges of these substances, they accumulate more slowly, but eventually they will reach unacceptable levels. For these chemicals, the ecosystem's assimilative capacity can in fact be quantified: it is zero.

Discharges of synthetic chemicals that are persistent *or* bioaccumulative synthetic chemicals must thus be eliminated altogether.[32] This position is broader than many other discussions of global or regional pollution, which express concern only about substances that are both persistent *and* bioaccumulative.[33] But accumulation in either the ambient environment or the bodies of organisms is a problem. A substance that degrades in the environment but is drawn quickly into the food chain will ultimately result in unacceptable exposures, and so will persistent chemicals that are not bioaccumulative. Seen in this light, bioaccumulation is really a particularly troublesome form of persistence. In the Ecological Paradigm, then, a chemical need only be persistent or bioaccumulative—not both—to trigger the Zero Discharge rule.

Zero Discharge has two practical implications. The first is that no releases of these compounds may be allowed, obviating the need for risk assessments and permits for acceptable discharges. The second is that because pollution control devices can never be 100 percent effective—a consequence of the second law of thermodynamics—persistent or bioaccumulative substances and the processes that produce them must be phased out, at the source, entirely. As the International Joint Commission put it,

> Zero discharge means just that: halting all inputs from all human sources and pathways to prevent any opportunity for persistent toxic substances to enter the

environment as a result of human activity. To prevent such releases completely, their manufacture, use, transport and disposal must stop; they simply must not be available. Thus, zero discharge does not mean less than detectable. It also does not mean the use of controls based on best available technology, best management practices, or similar means of treatment that continue to allow the release of some residual chemicals.[34]

The last pillar of the preventive framework specifies how Zero Discharge can be achieved, and it offers an alternative to pollution control technology. The concept of Clean Production, first described in 1989 by the United Nations Environmental Program, is "a conceptual and procedural approach to production that demands that all phases of the life-cycle of a product or of a process should be addressed with the objective of prevention or minimization of short- and long-term risks to human health and to the environment."[35] The most important aspect of Clean Production is an emphasis on upstream solutions: whereas pollution control would capture and manage pollutants after they have been produced, Clean Production prevents the generation and use of hazardous chemicals in the first place. As its name suggests, Clean Production focuses on the technology and materials of production itself; it requires redesigning products and manufacturing methods to eliminate the *input* of toxic substances (or of feedstocks that result in toxic by-products).[36] For example, current policies ask how we can control perchloroethylene wastes and vapors produced in the dry cleaning process; Clean Production seeks a method of cleaning clothing that does not use toxic solvents at all, such as those based on water, steam, or carbon dioxide.[37]

Clean Production is the concept behind toxics use reduction laws, passed by a number of states, which encourage industry to reduce hazardous waste generation by minimizing chemical inputs into manufacturing processes. The Massachusetts Toxics Use Reduction Act, for instance, seeks a 50 percent reduction in toxic waste generation by requiring industry to submit plans to reduce chemical use, providing technical assistance, and sponsoring research and development in less hazardous alternatives.[38] National environmental law in the United States now contains language that makes prevention a priority, but most regulations continue to rely on control methods, avoiding intervention at the level of products and processes. The Swedish Chemicals Action Program implements Clean Production with more bite by incorporating into national law the substitution principle: organizations and individuals must always use the least toxic available product or process. The program also mandates the phase out of chlorinated solvents, chlorinated paraffins, and dozens of pesticides (including atrazine and 2,4-D, which remain in use in most other industrialized nations).[39]

Ultimately Clean Production is a total environmental approach that addresses more than the use of toxic chemicals. It also seeks to reduce the consumption of energy, water, and raw materials and to use renewable resources as sustainably as possible. The principle of Clean Production is thus valuable on two different scales: it offers practical guidance for eliminating specific sources of chemical pollution, and it provides a vision of an economy that at once fulfills human needs and is compatible with ecological processes.

Reverse Onus, Zero Discharge, and Clean Production together make quite clear what the Ecological Paradigm would actually do. It would entirely eliminate the production of synthetic chemicals that are known to be persistent or bioaccumulative. It would then place the burden of proof on industry to justify the use of other synthetic chemicals, removing the nearly impenetrable shield of scientific uncertainty that now protects decisions about hazardous technologies from public intervention. And it would achieve its goals by requiring the most ecologically sustainable production methods to be used, moving from after-the-fact solutions to prevention targeted at the root causes of pollution. An ecological policy is thus far more than a minor tweaking of the current system of managing emissions. It represents an ambitious program of industrial development and conversion, with a stick to eliminate the most egregiously damaging practices and a carrot to guide the implementation of clean alternatives.

From Chemicals to Classes

The central question in the debate over chlorine chemistry is whether organochlorines should be regulated one by one or as a class of related substances. Although the chemical industry has called the latter approach radical, thoughtless, and unprecedented, it is none of these. When society wants to solve a problem, it always has to decide the appropriate focus of action. Intervention is often ineffective at the level of individual entities, particularly in a complex system with large numbers of independent agents. Attempts to mitigate such problems as traffic congestion and the spread of pests and infectious disease focus instead on the systemic causes of these problems. It is also standard fare to regulate classes of things rather than specific types; for instance, policies do not individually target the production and sale of whisky, or gin, or tequila, or vodka, but instead apply uniform rules to alcoholic beverages as a class.

Class-based regulation is not so foreign in environmental policy either. Although the majority of chemicals that are controlled at all are regulated individually, others have been addressed in groups. For instance, there are over 200 kinds of PCBs, all of which have been banned together in a single policy. Lead-containing substances have been addressed as a class: rules restricting leaded gas successfully prevented emissions of a huge number of lead-containing compounds from automobile tailpipes. The Montreal Protocol phased out all chlorofluorocarbons and numerous other compounds as a single class of ozone-depleting substances. In these cases, related substances were regulated as groups because they contributed to a common problem, caused harm by the same toxicological mechanisms, or were always produced together in complex mixtures.

Treating organochlorines as a class would extend this precedent to a much larger group of substances. Organochlorines have seven characteristics that make a chemical-by-chemical approach utterly impractical. First, there are 11,000 of these substances in commerce, plus thousands more produced as unintended by-products. Only a small fraction have been subject to basic toxicity testing, and new chemicals are brought on the market much faster than toxicologists can catch up. As of the early 1990s, U.S. EPA had prepared health assessments for fewer than 100 chemicals, issued air emission standards for fewer than 10 and established effluent guidelines for 128.[40] EPA's thorough risk assessment for just one organochlorine, dioxin, is still not complete after more than a decade of study, hearings, review, and redrafts. Even with a massive commitment of resources, it would take centuries to assess and regulate each existing organochlorine on a chemical-specific basis, to say nothing of the vast number of new ones that come on the market each year.

Even if it were possible to develop data and regulations for each individual organochlorine, chemical-by-chemical regulations would still fail to address their hazards. The second problem is that organochlorines are always formed as complex mixtures of thousands of compounds, and the great majority of the by-products formed in chlorine-based processes have not been chemically characterized. There is no way to assess and control on an individual basis substances that have not even been identified. Moreover, one cannot regulate complex mixtures as if they were composed of neatly packaged capsules of pure, individual chemicals. Organochlorine pollution is caused by industrial processes, not individual products. The point at which we may intervene effectively is when chlorine and organochlorines are introduced into these processes. Why focus on the myriad of individual organochlorines when we could much more easily direct our attention upstream, to the much smaller number of processes in which organochlorines are produced? Since chlorine chemistry is organized as a tree of applications, it is far more efficient to concentrate on the tree's major limbs and branches than on each of the thousands of organochlorine products and by-products at the tips.

Suppose we decided to eliminate only the most persistent, toxic, and bioaccumulative organochlorines: even then, we would need to treat organochlorines as a class. The third reason that chemical-specific regulations are doomed to fail is that chlorine chemistry cannot be practiced without generating persistent, bioaccumulative, highly toxic by-products. The continuing discharge of tons of PCBs into the environment each year as by-products of other chlorine-based industrial processes, decades after the intentional production of PCBs was phased out, provides ample testimony to the failure of chemical-by-chemical policies. Even dioxin and related compounds appear to be formed at some point during the life cycle of absolutely all chlorine-based products and process. To prevent the generation of the single most hazardous organochlorine, we must phase out the entire field of chlorine chemistry.

Fourth, the limits of toxicology and epidemiology make the chemical-by-chemical approach unsuited to protecting health and ecosystems. We have seen that the thousands of organochlorines in the environment—products, by-products, and the results of their environmental transformation—cause toxic effects in mixtures rather than as isolated compounds. Toxicologists and epidemiologists will never be able to pinpoint which organochlorine is causing health problems, because the fact is that groups of these compounds are always responsible. The framework of chemical-specific assessment, control, and monitoring is intrinsically unable to deal with the actual effects of organochlorine mixtures. In contrast, a focus on the technologies that cause organochlorine pollution eliminates the formation of these mixtures, immediately dispensing with the problems of unidentified compounds and synergistic effects.

The fifth problem with chemical-by-chemical regulations is economic. We have seen that the chlor-alkali industry must, given its financial imperative, ultimately circumvent chemical-by-chemical regulations. To maintain growth in markets for chlorine and its more profitable coproduct caustic soda, chlorine producers must always compensate for declines in one chlorine use with growth in another. The industry has accomplished this goal by replacing one banned organochlorine with a new one, directing chlorine into applications previously filled by traditional materials, or aggressively expanding organochlorine markets in developing countries. Thus, despite regulations on a number of specific organochlorines, total worldwide production of chlorine and organochlorines has continued to grow. Any policy to address global organochlorine pollution must foil this strategy by reducing the generation and use of all organochlorines, not just a few.

Sixth, it makes no sense to presume that the thousands of organochlorines not yet tested will somehow turn out to be safe, when virtually every organochlorine investigated to date causes one or more toxic effects. Reversing the onus for members of the class of organochlorines is consistent with what we have learned about their characteristics: chlorination almost always increases the toxicity and bioaccumulation of organic chemicals, and it frequently makes them more persistent, too. It is logical and consistent with existing knowledge to address the class of organochlorines by presuming its members hazardous unless specific information suggests otherwise.

The final advantage of treating organochlorines as a class is an ethical one. Chemical-by-chemical regulations presume that each organochlorine is safe until it is proven hazardous. This a patently ridiculous assumption, given what we know about all the organochlorines that have been assessed, but it hides an unjustifiable moral stance, too. Placing the burden of proof on the public to demonstrate a hazard allows a vast—though poorly organized—program of toxicological experimentation in which the ecosystem and our bodies are contaminated by novel chemicals, the effects of which are not well known but are likely to be harmful. People, not chemicals, have the right to be presumed innocent until proven guilty. People also have the right not to be experimented on without informed consent; no one has ever had the opportunity to grant or deny their consent before being exposed to the organochlorine burden that now contaminates us all.

What about the rights of businesses? Chemicals may not hold legitimate rights, but shouldn't companies be free to make and use chemicals as they see fit? The right of businesses to pursue their enterprises freely must surely stop at the point that their activities harm workers, consumers, the public, or the environment. At a minimum, a company's liberty should be subject to the same constraints that limit the freedom of an individual to take actions that may injure others, like striking another person with a weapon or placing a dangerous animal in a public square. This analogy between corporations and people suggests that it is quite reasonable for society to limit the liberty of businesses to use technologies that may affect public health and the environment, such as the production and use of synthetic chemicals. We should take our analysis even further, however, because corporations are in fact institutions, not people; they have no natural rights, although current law has granted them many of the rights (but few of the responsibilities) of a human being. A corporation is an institution created and licensed by society to perform a certain function. There is absolutely no moral, logical, or political reason that this license should be unlimited. Corporations should have the right to pursue their approved line of business, but never in a way that endangers the well-being of the society that sanctioned its existence in the first place.[41]

Hesitant to "interfere" in industrial decisions over materials and technology, environmental bureaucracies in most nations, the United States in particular, have granted to chemicals rights that properly belong to people. Society has decided, wittingly or not, that it prefers to err on the side of pollution and disease rather than on the side of health and a clean environment. The decision was not made consciously; it derives from neither any scientific principle nor a thoughtful consideration of ethics. It came into being piecemeal, at a time when the potential for toxic pollution to harm public health and the environment was poorly understood. That it continues to this day illustrates how difficult it is to reform entrenched institutions as knowledge evolves, particularly when industries that wield considerable political power vigorously oppose any change.

These seven reasons show why organochlorines can never be adequately managed on a chemical-by-chemical basis. They show why even highly restrictive policies that target individual substances will never be adequate to prevent health and ecological damage. At the heart of the Ecological Paradigm is a shift of focus from the micromanagement of individual substances to the macromanagement of major classes of pollutants and production practices. With this gesture, the new model seeks to gain control of a practice—industrial chemistry—whose complexity and diversity has far outstripped the ability of current regulatory tools to manage it.

It is essential that the classes on which the Ecological Paradigm focuses be understood as groups of substances related by their chemistry and the technologies that produce them, not as groups of substances that have been shown to cause a certain kind of health effect, like endocrine disruption or cancer. A preventive policy on endocrine disrupters would require first that we identify which chemicals are endocrine disrupters. That is, it would immediately return us to the chemical-by-chemical approach, which is doomed to fail, for all the reasons just discussed. To address the chemicals, known and unknown, that disrupt the endocrine system or cause other important health effects, we must focus on the classes of substances and processes that include and create these chemicals.

An Ecological Policy on Chlorine Chemistry

The first step in applying the Ecological Paradigm to organochlorines is to reverse the burden of proof, applying Reverse Onus to this class of substances. Members of the class of chlorine-based products and processes should be presumed to be unsafe and therefore candidates for phase out unless demonstrated otherwise. *Otherwise* means that industry would have the opportunity to show to a reasonable degree of certainty that any specific aspect of chlorine chemistry is not associated with the generation of persistent or bioaccumulative or toxic substances during its life cycle, or that it serves a compelling social need for which there are no chlorine-free alternatives, the criteria for which I discuss in more detail below. This proposal echoes that of the American Public Health Association, which concluded in a 1993 resolution that there should be a "rebuttable presumption" that organochlorines are hazardous and should be phased out unless industry can provide a reasonable demonstration that they are not hazardous or that no safer substitutes are available.[42]

Addressing organochlorines as a class does not mean an outright ban. Because chlorine chemistry has invaded every corner of the economy, it would be disastrous to end the production and use of chlorine and organochlorines overnight. The only practical approach is a planned process to replace chlorine-based technologies gradually with cleaner alternatives. This transition should include democratic mechanisms to set priorities, evaluate and select the best possible alternatives, and minimize and compensate for any economic dislocation that it causes.

The International Joint Commission has called this kind of process a chemical sunset, defined as "a comprehensive process to restrict, phase-out and eventually ban the manufacture, generation, use, transport, storage, discharge, and disposal of a persistent toxic substance. Sunsetting may require consideration of the manufacturing processes and products associated with a chemical's production and use, as well as of the chemical itself, and realistic yet finite time frames to achieve the virtual elimination of the persistent toxic substance." Since 1991, the IJC has repeatedly recommended that the U.S. and Canadian governments begin consultative processes that will lead to sunsetting organochlorines as a class by phasing out the use of chlorine as a feedstock in industrial processes.[43]

My proposal is essentially that the IJC's recommendation be fleshed out and adopted on a global basis. Given the movement of organochlorines across the planet and the rapid expansion of the chlorine industry into developing nations, action must be thoroughly global to be successful. Legally binding international agreements, like those on ozone depletion and global warming, provide a political model for a common approach by all nations to a global environmental problem. The international agreement on persistent organic pollutants (POPs) currently being negotiated under the auspices of the United Nations Environmental Program is an appropriate instrument under which organochlorines could be sunset on a global basis, although its current language does not even approach such an ambitious scope.

A chlorine sunset would first need to set time lines and priorities for the phase out of major applications of chlorine and organochlorines. These decisions should be based on two criteria. The first is the degree of environmental hazard posed by each use: chlorine applications that result in the most massive production of persistent or bioaccumulative organochlorines should receive high priority for rapid phase out. Prime candidates by this criterion are PVC plastic, paper bleaching, pesticides, chlorinated solvents and paraffins, and the use of chlorine and chlorinated compounds in combustion-based processes like iron and steel smelting. Medium-priority applications would include other chlorinated plastics, chlorinated intermediates used in the chemical industry, and water treatment. Low-priority uses would probably include organochlorine pharmaceuticals and inorganic chlorides that are produced from chlorine gas, like chlorides of hydrogen, aluminum, and phosphorous.

The second criterion for priority setting is the availability of alternatives. As I show in the next chapter, cleaner, effective substitutes are available now for the great majority of chlorine uses. There are a few minor applications, however, that serve compelling social needs for which no alternatives are currently available. But what is a "compelling social need"? Exemptions should be granted for applications whose loss would cause direct and demonstrable effects on public health, and the burden of proof should rest with those seeking an exemption. The most important chlorine uses in this respect are certain kinds of pharmaceuticals and some aspects of drinking water disinfection that cannot currently be carried out without chlorine; banning these applications before alternatives are available would cause real damage to public health. Processes like this should be exempted from phase-out while research and development continue, until viable alternatives are available. Exemptions should not be granted for chlorine uses that merely produce luxury or convenience items or are profitable for their producers or users. Unnecessary benefits like these should not trump the risk of irreversible injury to the health of millions of people and the environment that sustains us all.[44]

Decisions about the implementation of a chlorine sunset should be made democratically. One way to enable democratic control would be for the conversion to a chlorine-free economy to be governed by a transition planning board that includes representatives of all affected stakeholders, such as workers, communities, environmentalists, the general public, and the businesses that use and produce chlorine-based products. The board's responsibility would be to determine how society should go about meeting the goal of sunsetting chlorine chemistry within a fixed period of time. This transition planning process should address priorities, time lines, exemptions, and mechanisms for the conversion to a chlorine-free economy. It should also be responsible for determining the best substitute available, so that one hazardous chemical or process is not replaced with another when a better alternative is waiting in the wings. Transition planning thus becomes a way for society to exercise democratic control over the technologies it deploys to meet its needs. In this way, the Ecological Paradigm provides the tools to implement what the political philosopher Langdon Winner has called "a process of technological change disciplined by the wisdom of democracy."[45]

The most powerful tool in a chemical sunset is and always will be the mandatory phaseout, but it is not the only mechanism available for encouraging the transition to Clean Production. Society can speed the sunset of organochlorines by establishing disincentives against the continued use of organochlorines, such as stronger liability laws, taxes on chlorine or organochlorines, and negotiating voluntary phaseouts negotiated with industry. Tax and credit incentives should also favor the adoption of chlorine-free alternatives. Governments are among the largest consumers of many products—paper, construction materials, vehicles, and so on—so procurement policies that favor the purchase of chlorine-free products and materials can also encourage the use of substitute production processes. At the same time, public education programs can increase private demand for these products. Governments should also fund research in the development of cleaner production processes and provide training and technical assistance to ease their implementation. When policies like these complement the strongest action—a required phaseout by a certain date—the prohibition often does not have to be enforced at all. A well-articulated sunset policy provides a clear horizon for investors and industrial decision makers, who know that chlorine chemistry will no longer be a profitable avenue and that alternatives offer better opportunities. By the time the date comes, technology has already shifted away from the banned product.

Protecting Workers and Communities

The transition planning process must address one of the most difficult issues raised by chemical sunsetting: the displacement of workers and communities now highly dependent on chlorine chemistry. As I show in the next chapter, the long-term economic effects of phasing out chlorine are likely to be positive, but there are some businesses—specifically the ones heavily invested in the production of chlorinated chemicals—that may not do well in the transition. The transition planning process must minimize the economic dislocation experienced by workers and communities that depend on these businesses and, when dislocation does occur, compensate and assist affected people.

A planned conversion process might accomplish this goal in a number of ways. The most compelling proposal—put forward separately by both environmentalists and labor unions—is for a tax on chlorine and organochlorines, which would provide revenue to establish a transition fund for workers and communities.[46] The tax would serve as a financial disincentive against chlorine consumption, and the transition fund would be used to minimize economic dislocation and compensate affected individuals and communities when it does occur. To keep people employed, funds would be spent to reward companies that invest in chlorine-free alternatives in the same locations in which chlorine-based processes were previously used, and to encourage additional economic development in these communities, as well. When jobs in a community are lost, funds would be used to ensure that displaced workers receive income, health care, meaningful opportunities for higher education and training, and help finding a comparable or better-paying job. Workers and communities should not bear the economic burden of the transition to a nontoxic economy. It is only appropriate that those who have profited from the use of hazardous chemicals provide the funds for conversion.

The idea that the polluter pays for the economic costs of sunsetting should apply across national boundaries. Although the chlorine industry is still concentrated primarily in wealthy nations, some organochlorines, pesticides and PVC in particular, are used widely in developing countries, which may lack the financial, administrative, and information resources to eliminate organochlorines without external assistance. Many alternatives to chlorine chemistry are financially beneficial but require an initial investment; others may be viable only with special training and capacity building. A global agreement should include commitments for shared responsibility among nations, especially financial and technical assistance from wealthy countries to those with fewer resources. A global chlorine tax could serve this purpose as well.

Precaution Made Reasonable

I turn now to the second objection against the precautionary principle: that it is inflexible and unworkable, banning virtually every activity with no opportunity for weighing costs against benefits. Is a sunset policy for chlorine chemistry, which I have said demonstrates how precaution can be put into action, absolutist after all? It is in fact quite flexible. This approach begins with the rebuttable presumption that chlorine-based products and processes are hazardous, but it provides the opportunity to revise this judgment for any member of the class if information to support such a rebuttal is available. This approach is no less flexible than the current system, which presumes all chemicals safe and then allows them to be demonstrated hazardous.

Some critics have charged that it is impossible to prove any chemical entirely safe, and they are strictly correct. The basis for reversing the presumption of hazard should not be proof of safety but *a reasonable demonstration of no hazard.* To meet this standard, a proponent of chemical production and use would have to show that a substance and its associated by-products and breakdown products are neither persistent nor bioaccumulative and that they are not carcinogenic, mutagenic, disruptive of intracellular signaling (by hormones, neurotransmitters, growth factors, cytokines, and so on), or toxic at low doses to development, reproduction, immunity, or neurological function. To make such a case, testing should be carried out on a variety of sensitive vertebrate and invertebrate species, should include transgenerational effects, and should involve examination of both whole animals and cultured cells. Testing like this would be rather demanding, but no more so than the extended laboratory and clinical trials required before new drugs are given a clean bill of health.

The chlorine sunset I have described also refutes the criticism that the precautionary principle does not allow problems to be prioritized or economic impacts to be considered. The Ecological Paradigm I have described *is* absolute about some things—specifically that persistent or bioaccumulative substances must be phased out altogether and that the cleanest available method of satisfying human needs must always be used. These are quite reasonable things to be absolute about, however; the former is an ecological imperative, the latter a social and ethical one. As an economy-wide program of technological conversion, however, the chlorine sunset would prioritize the most hazardous and easily replaced technologies, allow exceptions when necessary to protect public health and welfare, and take specific steps to address the socio-economic impacts of conversion. In these ways, sunsetting allows great flexibility in the specific steps taken to achieve its imperatives.

A chlorine sunset serves as an ambitious case study that shows how an ecological policy would actually work. In so doing, it answers the major objections raised against the precautionary principle. Because phasing out chlorine is just an example, however, a chlorine sunset does not define the full range of activities to which precautionary action should be applied. Would the Ecological Paradigm subject every synthetic chemical to a preventive approach, or just some?

There are two reasonable answers to this question. The first is a practical version of precaution, which requires a specific prima facie case that an activity (or class of activities) might cause serious or irreversible damage to health and the environment before preventive action is taken.[47] That is, only when evidence is available to suggest the possibility of injury is the onus reversed and a process or product subject to substitution. This approach avoids the objection that the precautionary principle would ban absolutely everything, but it still allows a lack of evidence to serve as evidence of safety. If a class of substances or technologies has not been tested (at all or for some specific types of damage), then there would be no data to support a prima facie case, although the activity may in fact be highly hazardous. The weakness of this approach is that it fails to deal with the problem of ignorance, one of the original motivating forces behind the precautionary principle.

The other way of looking at precaution is more complete. It seeks to avoid not only hazards we know about but also those of which we are still ignorant. This strong version of precaution does not require a specific prima facie case but applies a precautionary approach to all technologies in which synthetic chemicals are used or produced. The Ecological Paradigm thus requires progressive reductions in the use of all chemicals by gradually converting the entire economy to clean production processes. Within this very large mandate, classes of substances and processes would be prioritized according to current understanding of the hazard they pose (based on data showing persistence, bioaccumulation, toxicity, degree of uncertainty and knowledge gaps, and so on).

By these criteria, the latter approach would prioritize the substances and processes for which a prima facie case could be made in the first version, but it would continue on after these items had been replaced with cleaner substitutes to address other synthetics, as well. In practice, it will take decades to address the high-priority, prima facie environmental problems, so there is virtually no difference in the foreseeable future between the two approaches. In theory, we can view the strong version of precaution as proceeding based on a prima facie case against all synthetic chemicals. The Science Advisory Board to the International Joint Commission has developed this prima facie case based on based on two facts. First, the majority of synthetic substances that have been tested are toxic (and many are persistent or bioaccumulative too). Second, the introduction of novel substances into integrated complex ecosystems and organisms is more likely than not to be harmful. The board concluded that there should be a rebuttable presumption that all synthetic chemicals are hazardous—precisely the requirement for phaseout under the limited definition of precaution.[48]

In my view, the strong version is superior for two reasons. First, it takes a consistent position toward both known and unknown hazards, an important advantage given the limits of science. Second, by establishing a long-term, economy-wide agenda for clean production, this approach would more effectively accomplish its goals than the weak version. A statement by society that the majority of synthetic organic chemicals should be gradually phased out would stimulate investment in clean technologies in all industrial sectors, including those that have not yet come up on the priority list or had a prima facie case established. The articulation of a more comprehensive horizon for investors would hasten progress on a larger set of environmental hazards, ensuring continuing improvement with less compulsory management as technology and society evolve.

No matter which definition of precaution's scope one prefers, however, chlorine chemistry must be sunset. For those who favor the more limited version, the first eight chapters of this book establish an unambiguous prima facie case against the class of chlorine-based chemicals and processes. For those who favor the strong approach, this same information puts organochlorines at the top of the priority list. Whatever chemicals will be on the agenda 50 or 100 years in the future, we have more than enough information to address the hazards of chlorine chemistry now.

10

Beyond Chlorine: The Alternatives

Technology moves on but not on our lines; it proceeds but not to our goals. It is urgent that science and technology be given goals of significance and value to us, lest the sorcerer's apprentice be converted from a literary symbol into a terrifying reality.

—René Dubos, *Science and Man's Nature*[1]

If organisms and ecosystems can't live with chlorine chemistry, can society live without it? The existence of viable alternatives is a critical issue in the debate over chlorine. If there are ways of making essential products that do not involve chlorine chemistry, society can and must begin the process of conversion at once. If, on the other hand, safer substitutes do not exist, then phasing out chlorine could result in economic havoc and a failure to fulfill basic human needs.

Chlorine and organochlorines are now widespread in the economy. They are involved in the manufacture of a great variety of products, from pipes to paper, cars to circuit boards. The chemical industry has made society's current dependence on chlorine chemistry the centerpiece of its defense of chlorine. For example, when U.S. EPA proposed in 1994 to study the viability of a national strategy to "prohibit, substitute, or reduce" the use of chlorine in just four major use sectors (PVC, solvents, pulp bleaching, and water treatment), the Chlorine Chemistry Council launched an aggressive counter-campaign. In a summary of "talking points" for its members to use in lobbying Congress, the press, and other officials, the CCC's first point was this:

Chlorine and chlorinated compounds are essential to modern society. They are used to meet the most vital needs of modern life, including protecting the water supply, in 85 percent of all medicines, 96 percent of all crop-protection chemicals, in hospital and food-handling cleanliness, keeping swimming pools safe and food fresh and free of contamination. . . . 1.3 million U.S. jobs depend on the chlorine industry—equal to the number of jobs in Oregon. Wages and salaries total more than $30 billion a year. Further, almost 40 percent of all U.S. wages and income depend in some way on chlorine and the products of the chlorine industry. Two hundred twelve industries use chlorine and related compounds, generating 45 million jobs and $1.6 trillion in economic activity.[2]

The Vinyl Institute went even further in its criticism of EPA's plan. "Essentially, this proposal declares war on modern society," said Bob Burnett, the institute's executive director.[3] How a mere study could declare war on modern society is not clear, unless knowledge itself poses a threat, but the industry's campaign quickly forced the White House to withdraw the proposal. Since then, the rhetoric about our dependence on chlorine has not died down. In a 1996 report, the industry-funded Competitive Enterprise Institute (CEI) came to this conclusion: "The end of chlorine would spell the end of modern civilization itself."[4]

Are these predictions reasonable? There is no doubt that industrial economies, guided by the quest of individual firms to maximize profits in the short term, have adopted chlorine-based technologies for a great variety of applications. The question, however, is not how much we use chlorine but how much we *must* use it. If viable alternatives exist, then our dependence on chlorine is not inescapable.

Chlorine chemistry has been a widespread part of modern life since about 1950, not 1850. Before chlorine, we had food, cars, electricity, telephones, airplanes, refrigeration, radios, films, antibiotics, vaccines, beautiful clothing, and sturdy, heated homes. Chlorine chemistry has penetrated today's production technologies, but it is hardly the foundation of modern life as we know it.

In this chapter, I show that safe and effective alternatives exist now for the vast majority of chlorine uses. There is no single substitute; each application requires a specific substitute—sometimes a drop-in alternative material, sometimes a change of process, sometimes a reformulation of the product itself. A complete discussion of all the alternatives to chlorine chemistry would require a multi-volume encyclopedia of explanation. I thus restrict this chapter to a general discussion of the issues in implementing alternatives and a brief treatment of the viability of substitutes in six major applications of chlorine: pesticides, pulp and paper bleaching, solvents, PVC plastic, chemical intermediates, and water treatment. Together these sectors account for almost 90 percent of total chlorine use; they also represent a fair sample of the universe of chlorine applications, including those with readily available substitutes and those for which alternatives are more technically or economically challenging to implement. (Table 10.1 provides a more complete list of chlorine uses and their alternatives.)

For each sector, I address two questions: are safe alternatives to chlorine technically satisfactory, and would adopting them impose an unacceptable economic burden on society? In fact, safer substitutes for most chlorine uses are already in commercial use and are quite technically effective.[5] For a handful of smaller uses, less hazardous alternatives are not yet available, but the rapid pace at which industry has been developing substitutes suggests that the technical barriers to a chlorine phaseout are surmountable. All of the alternatives—from chlorine-free plastics to wooden window frames, for instance—have environmental impacts of their own, but they are less hazardous than the chlorine-based processes that they replace. If they are not less hazardous, I do not propose them as viable substitutes; it would do little good to phase out an organochlorine and replace it with another synthetic chemical with equally hazardous qualities, like an organobromine. As for the economics of the matter, many chlorine alternatives are quite affordable, because—although they may require an initial investment to install and more care to operate and maintain—they pay off in reduced expenditures on chemicals, pollution control, waste disposal, liability, and regulatory compliance. In the long run, a well-planned investment in chlorine-free technologies is likely to yield economic benefits. Phasing out chlorine thus represents a major step on the road toward a sustainable economy based on clean materials and production techniques.

This is not to say that chlorine can be sunset overnight, without careful planning, implementation, and the devotion of significant resources. The many chlorine-based technologies now in use throughout the modern industrial base cannot be eliminated with a wave of the hand. Conversion to a nontoxic economy is a major technical and economic undertaking, of magnitude equal to or greater than the post-cold war demilitarization of the civilian economy in the United States. It will take commitment, money,

Table 10.1
Summary of major chlorine uses and alternatives

Chlorine use	Application	Alternative	Share of U.S. chlorine use, 1997 (%)[a]
PVC	Construction, packaging, automobile components, others	Wood, metal, glass, textiles, paper products, chlorine-free plastics	41.1
Propylene chlorohydrin	Manufacture of propylene oxide	Tert-butyl peroxide process	7.8
Phosgene	Synthesis of polycarbonates, isocyanates	Carbonylation by carbon monoxide, carbon dioxide, diphenylcarbonate, dimethylcarbonate	7.8
Chlorinated solvents and refrigerants	Cleaning, coating, and extraction	Aqueous and semiaqueous cleaning and coating processes, hydrocarbon solvents, dry coatings, supercritical carbon dioxide	7.3
	Refrigeration and airconditioning	Hydrofluorocarbons, hydrocarbons, ammonia, thermoacoustic cooling, zeolite/water-based cooling	
Pulp bleaching	Lignin removal and whitening	Use of unbleached paper whenever possible; oxygen, ozone and peroxide-based bleaching	6.2[b]
Water treatment	Disinfection of drinking water, wastewater	Slow sand filtration, ozone, ultraviolet light, biologically active carbon filters	4.4
Epichlorohydrin	Manufacture of epoxy resins	Non-epichlorohydrin-derived epoxies, other plastics (phenolic and urethane based), alternative epoxidation processes, alternative synthesis process to reduce chlorine use	3.3

Titanium dioxide production	White pigment	Sulfuric acid process;[c] other white pigments, less white pigment	3.2
Hydrochloric acid	Steel pickling, pH adjustment, oil well acidulation	Nitric and sulfuric acids[d]	3.0
Other chlorinated plastics (PVDC, polychloroprene, chlorinated polyolefins)	Packaging, miscellaneous	Chlorine-free plastics and rubbers	2.1
Hypochlorites	Bleaches and disinfectants	Sodium perborates, hydrogen peroxide	2.1
Pesticides	Agriculture	Alternative agriculture, including introduction of natural predators, improved choice and rotation of crops, mechanical/manual removal of weeds and pests, use of biological pesticides and resistant strains of crops	1.4
Chlorinated paraffins and other flame retardants	Cutting oils, lubricants, fire retardants	Over-based sulfonate cutting oils and lubricants, metal anhydride and phosphorous-based flame retardants	0.5

[a]From table 6.3.
[b]Does not include chlorine dioxide consumed in pulp bleaching.
[c]The sulfuric acid process produces very large quantities of acid and metal-bearing wastes. These wastes can be neutralized or recycled, but they represent a significant environmental hazard if not treated properly.
[d]Some minor uses require other compounds. For instance, choline hydrochloride, an animal feed additive, can be replaced by choline bitartrate, a significantly more costly compound.
Sources: Environment Canada 1997, Charles River Associates 1993, Plimke et al. 1993.

and further research and development. My aim is not to show that organochlorines can be phased out *easily;* rather, it is to show that, with a well-planned transition, chlorine can be sunset without causing major economic harm to society.

Mindless Bans and Mindful Sunsets

After EPA announced its proposed study of chlorine, the Chlorine Chemistry Council convened a meeting of industry executives to outline their public relations strategy. A source who attended described the meeting to the newspaper *In These Times:* "We were told that we shouldn't talk about reducing or substituting, but that we should portray the EPA plan as an inflexible ban. The Chlorine Champions communications program would use the public distrust of government and regulations to get this message across. We were told to make sure to draw the connection between the environmentalists, the EPA, and the government—to talk about them as one group—and to emphasize that once again the government is trying to restrict and kill jobs."[6]

This strategy served as the foundation for CCC's estimates of the social and economic costs of phasing out chlorine that I quoted above. The industry's figures come from a report the Chlorine Institute commissioned from Charles River Associates (CRA), a private consulting firm.[7] Released in 1993 as part of the CCC's response to the International Joint Commission's (IJC) chlorine sunset recommendation, the CRA report catalogued the many uses not only of chlorine and organochlorines but of caustic soda as well—along with every downstream industry that uses a product in which these substances have ever been involved. Through a sector-by-sector analysis of chlorine use, CRA asserted that the IJC's recommendation would cost the U.S. and Canadian economies $102 billion per year, result in the loss of 1.4 million jobs, and prevent society from meeting basic human needs, including nutrition, health care, and clean water.

The central flaw in CRA's analysis is that it estimates the costs of an immediate ban, not a sunset. CRA's calculations assume that the transition to a chlorine-free economy will occur instantaneously by bureaucratic fiat, without any intelligent thought or planning. But a gradual, planned phase-out—in which priorities are set, the best alternatives used, and exceptions granted where no substitutes are available—would have much lower costs, for a number of reasons.

First, making exceptions for essential uses has the potential to reduce phaseout costs massively. Of the $102 billion annual price tag that CRA predicts, over half is projected to come from a single sector: the pharmaceutical industry, which uses less than 1 percent of all chlorine produced. According to CRA, phasing out chlorine in pharmaceuticals would cost $335,000 per ton of chlorine—about a hundred times more than the price of all the other sectors combined. Why so much? Because CRA assumes that absolutely all pharmaceuticals now made with any kind of chlorinated solvent, extractant, neutralizer, feedstock, or intermediate would be immediately banned and removed from the market; diseases would become more debilitating, hospital stays would be longer, doctor bills higher; and so on. This scenario would never occur under an intelligent sunsetting procedure, which would make exceptions for medicines that could not be manufactured without chlorine.

Setting priorities will shrink the price tag further. CRA's own analysis shows that a few minor chlorine applications account for the bulk of the cost of phasing out chlorine, but the vast majority of chlorine use could be sunset for a much lower price. According to CRA's own cost estimates, over half of all chlorine use in the United States and Canada could be phased out at the very modest cost of about $4 billion per year. Eighty-five percent of all chlorine use could be sunset for just $11.6 billion per year, and 95 percent could be eliminated for just $17 billion. A remarkable 97.6 percent of all chlorine used in the United States and Canada could be eliminated for less than $22 billion—about a fifth the total cost that the industry publicized so widely as the overwhelming price of a "chlorine ban" (figure 10.1).[8]

Phasing out chlorine on a time line rather than overnight will also make an immense difference in the total cost. If chlorine were simply banned, new equipment would have to be installed in many industries overnight. But industrial equipment typically lasts one or two decades and then needs to be replaced anyway. If phaseout dates are harmonized with industry's investment cycles, then the net capital costs of conversion become not the cost of chlorine-free equipment but the difference between that equipment and a new generation of chlorine-dependent machinery. Thus, when the

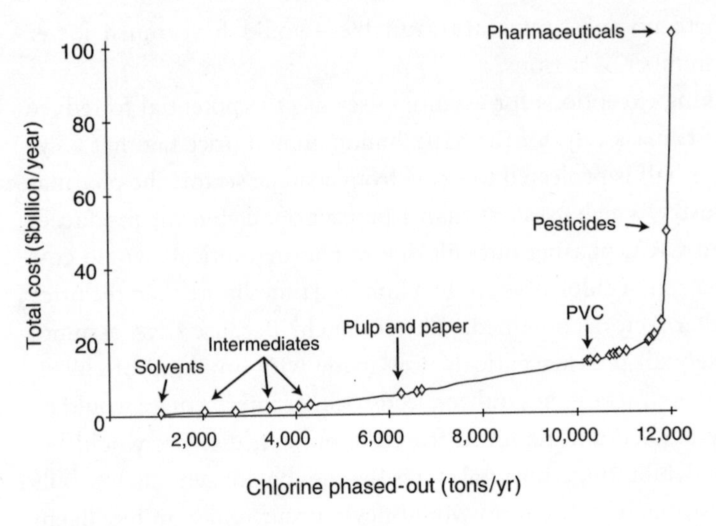

Figure 10.1
Cost of phasing out chlorine in the United States and Canada according to the Chlorine Institute, ordered by increasing cost. Over 97 percent of chlorine could be sunset for just one-fourth the cost of a total chlorine ban. Sectors, from least to most expensive, are: industrial solvents, propylene oxide, HCl (excluding steel pickling), isocyanates, epichlorohydrin, HCl (steel pickling), pulp and paper, titanium dioxide, flame retardants, PVC plastic, chlorinated polyolefins, chloroprene, hypochlorites, polycarbonates, fluoropolymers, PVDC plastic, refrigerants, wastewater treatment, extractants, dry cleaning, silicones, drinking water, pesticides, pharamaceuticals. (Source: Charles River Associates 1993.)

German state of Hessen asked the independent consulting firm Prognos AG to estimate the costs of phasing out chlorine in four major use sectors (PVC plastic, phosgene, and intermediates for the production of propylene oxide and epoxy resins), Prognos assumed a phase-in period of 5 to 20 years, depending on the expected lifetime of chlorine-based equipment in each industry. Considering only chlorine-free technologies that were already available on a commercial scale and did not cost significantly more, Prognos found that more than 50 percent of the chlorine use in Germany (about 63 percent of the amount studied) could be phased out with no net increase in capital costs and a meager rise of just 1 percent in operating costs. If society were willing to spend some money on the transition, Prognos found that it was technically possible to substitute for 80 percent of the total chlorine use assessed.[9]

Planning to ensure that the best substitutes are adopted as chlorine is phased out will also lower the costs of the transition. CRA assumes that expensive alternatives with poor performance—and often major environmental effects—will take the place of chlorine-based processes in a phaseout. If we use the best and cleanest production processes, however, the alternatives are in most cases quite reasonably priced. For instance, CRA assumes that dry cleaners will replace perchloroethylene with Stoddard solvent, a flammable and toxic chemical that will require replacement of all existing equipment at a cost of $4.6 billion, making this one of the most expensive sectors on a per ton basis for the phaseout of organochlorines.

But there is a much safer and cheaper alternative. A recently developed method called wet cleaning uses spot cleaning, steam, water, and gentle soaps and detergents to clean clothes of all types, including delicate, synthetic, and shrinkable fabrics, as well as mixed-fabric garments like suits and lined jackets. There are two forms of wet cleaning: one, called multiprocess wet cleaning, is done primarily by hand; in the other, machine wet cleaning, advanced machines use minimal agitation to clean clothes in water and then dry them with computerized control over moisture content to prevent shrinkage.[10] The methods are now in use at over a hundred successful "wet cleaners" or "green cleaners" in cities across North America, many of which have completely eliminated the use of perchloroethylene.[11] When U.S. EPA studied multiprocess wet cleaning in 1993, it found that its performance was equal to or better than that of perc-based dry cleaning. In blind tests, customers were *more* satisfied with wet-cleaned clothes than those that had been sent to traditional perc-based cleaners. Moreover, EPA found that wet cleaning required 42 percent less capital to install, cost about the same to operate, resulted in a 5 percent increase in profits, and offered a 78 percent better return on investment than dry cleaning with perc.[12] Tests of machine wet cleaning are just as impressive: performance and customer satisfaction are as high with wet cleaning as dry cleaning, and EPA estimates that machine wet cleaning costs only a third as much in both capital and operating expenses as perc-based dry cleaning.[13] Additional alternatives—including the use of liquid carbon dioxide, a gentle and extremely effective cleaner—have been developed, will soon appear on the market, and are expected to be technically and economically competitive.[14]

Another example of predicting that less-than-optimal alternatives will be adopted emerges in CRA's answer to an important question in the chlorine phaseout: how to fulfill industry's demand for caustic soda, the vast majority of which is currently produced by electrolysis in the chlor-alkali process. Reducing chlorine production will reduce caustic soda production as well, affecting numerous industries that use it. In fact, the majority of the industries that CRA lists as chlorine dependent are actually alkali dependent: they use sodium hydroxide but not chlorine or organochlorines. There are alternative ways of satisfying industry's demand for caustic soda, one of which is the production of so-called chemical caustic from soda ash. Soda ash (sodium carbonate) is easily derived from an ore called trona (sodium sesquicarbonate), which is present in deposits in Wyoming, Turkey, and Africa that are large enough to meet current world caustic demand for centuries.[15] CRA assumes that a chlorine ban will require 100 percent of demand for sodium hydroxide to be replaced with chemical caustic, which would entail the construction of new mining, processing, and transport operations at a cost of over $2.5 billion per year. Environment Canada took a similar approach and found that a total replacement of electrolytic caustic with chemical caustic in Canada would cost about a third as much on a per-ton basis—about $90 million per year in that country.[16] Environment Canada put these numbers in economic context: the switch to trona-derived caustic would raise the market price of caustic by about 14 percent but increase the total price of the final products in which caustic plays a role by just 0.4 percent.

The problem with both these analyses is that chemical caustic is not the best substitute. Trona-derived sodium hydroxide should be the last in a hierarchy of environmentally and economically preferable alternatives. The best option is to reduce the demand for caustic soda through recycling and improved efficiency; a number of studies suggest that major users such as the paper, aluminum, and petroleum industries could use 50 to 90 percent less caustic than they currently do. [17] The next best alternative is to use other chemicals in place of sodium hydroxide. Where caustic soda serves as an alkali, as it does in the neutralization of acidic wastewater and in many chemical reactions, other basic materials—including soda ash, lime, or hydroxides of calcium, magnesium, or potassium—can often fulfill the same function.[18] For example, Dow Chemical, one of the world's largest single users of caustic soda, has substituted lime for sodium hydroxide in the production of propylene oxide, reducing caustic demand by hundreds of thousands of tons per year.[19] Where caustic soda serves as a source of sodium ions, sodium sulfate and sodium carbonate are often viable options. Only after reduction and substitution have substantially reduced demand for caustic soda should sodium hydroxide be manufactured by other methods, like the causticization of trona or—even better—the electrodialysis of sodium sulfate, a substance abundant in the wastewater of many industries.[20] With these superior methods in use, society would have to call on the relatively expensive and energy-intensive option of trona-derived caustic to fulfill only a fraction of current alkali demand, reducing the price tag of conversion by a considerable amount.

Pesticides

It is easy enough to paint a frightening picture of technological and social collapse if one assumes that the absolute worst will happen in each aspect of the chlorine sunset. Consider the impacts of phasing out chlorine-based pesticides on the agricultural economy. CRA predicts a whopping $24 billion per year cost from banning the 1 percent of chlorine used to produce pesticides, making this minor chlorine use the second most expensive application to sunset. CRA derives this number by assuming that a pesticide phaseout will result in a 20 to 70 percent decline in agriculture yields, the forced cultivation of tens of millions of additional acres to offset massive crop losses, and significant increases in the market price of food.

But chemical pesticides are only one way of managing pests, and they do not necessarily represent the best method. Despite massive pesticide use, some 37 percent of all U.S. crop production is lost to pests—significantly more than was lost before the advent of synthetic pesticides.[21] If farmers use other pest management methods, both food production and farm economics can remain relatively stable or even improve as pesticide use decreases. The U.S. National Academy of Science's Board on Agriculture surveyed experience with alternative agriculture, a set of techniques that reduce or eliminate pesticide use in favor of improved choice and rotation of crops, mechanical methods of weed and pest control, introduction and maintenance of natural predators and parasites, and use of biological pesticides like bacteria, viruses, and natural chemicals. Alternative agriculture combines modern technology and information with an approach that was dominant before chemicals began to replace human labor on farms after World War II, and it turns out to be remarkably effective. In every case the academy studied, yields on farms that had reduced or eliminated pesticide use increased or stayed the same, while production costs declined and profits increased. The academy concluded, "Reduced use of these [chemical] inputs lowers production costs and lessens agriculture's potential for adverse environmental and health effects without necessarily decreasing—and in some cases increasing—per acre crop yields and the productivity of livestock management systems."[22]

When farmers stop using pesticides, they have to add something else if their farms are to succeed. That something else is information and work. Alternative agriculture requires farmers to have detailed knowledge of crops, pests, and ecosystems, and it requires more labor than chemical-intensive agriculture. In the first years after a transition to alternative agriculture, farmers often experience decreases in yield of 5 to 30 percent during a period of innovation and learning. Yields then tend to recover to a large extent as the new system stabilizes and the agricultural ecosystem—soil organisms, natural predators on pests, and so on—recovers from previous chemical-intensive practices.[23] In Germany, for example, a long-term study of 44 farms has found that yields of wheat, oats, and rye have steadily increased over a 17-year period following the farmers' transition to strictly organic agriculture.[24] Numerous studies in the United States have found that pesticide reduction programs have caused no substantial decrease—and sometimes increases—in agricultural yield.[25] A comprehensive review of experience with alternative agriculture in the United States and Europe concluded that yields remain stable or decline slightly (by 5 to 10 percent)—not a significant problem, since most governments now pay farmers billions of dollars annually to reduce their output or buy up their surplus production.[26]

While yields stay more or less the same, the impact on farmers' profits can be very positive. Pesticides account for as much as 20 percent of the variable costs of crop production, and chlorine-dependent pesticides alone carry an annual price tag of about $8 billion dollars in the United States and Canada alone.[27] Pesticide-free alternatives eliminate these costs entirely, offering massive savings to farmers. They also increase profits by reducing the health costs and lost productivity associated with pesticide poisoning. One study of rice farming in the Philippines, for instance, found that the health costs of chemical-intensive agriculture ate up over 60 percent of farmers' profits; of four pest management strategies, the one that did not involve chemical pesticides produced the highest net profits.[28] In addition, farms practicing alternative agriculture are typically more diversified than pesticide-intensive farms, so their owners reduce their financial risk from variable year-to-year returns.[29] The National Academy did not attempt to calculate the potential cost savings of a nationwide program to adopt pesticide-free agriculture. It did note that a program to reduce pesticide use on just nine major crops in a 15-state area yielded a net increase of $578 million in returns to farmers, suggesting that the potential savings from a larger program may be in the billions of dollars.

These studies have focused on farm productivity and economics in the United States and Europe. Since a chlorine phaseout must be a global endeavor, the international viability of alternative agriculture becomes a critical question, particularly given a growing world population that requires an ever-increasing food supply. According to a comprehensive review of worldwide experience with alternative agriculture by Jules Pretty of the International Institute for Environment and Development,[30] sustainable farming methods are in fact the *only* way to ensure a stable and adequate food supply and a healthy agricultural economy in the coming century.

According to Pretty, industrial-style agriculture—with its combination of high-yield crops, synthetic pesticides, and chemical fertilizers—increased global food production by about 7 percent from the 1960s to the 1980s. The benefits of this productivity did not touch the world's very poor nations, where farmers and governments cannot afford the rather expensive technological packages of modernized farming; in these countries, hundreds of millions of people remain undernourished. Today, even in the countries where the green revolution—the introduction of chemical pesticides, fertilizers, and high-yield crops during the 1960s and 1970s—increased the food supply considerably, there are now serious questions about the sustainability of that growth. Since the 1980s, the average annual increase in agriculture yield has slowed in every nation where the green revolution has had a significant impact, and in some countries cereal yields have begun to fall from their once-maximal levels.[31]

These and other signs suggest that chemical-intensive agriculture is beginning to hit its limits. In the United States, the share of crop yields lost to insects has nearly doubled from 1950 to 1990, despite a more than tenfold increase in the use of pesticides.[32] The data for specific crops are even more convincing. In the early 1940s, insecticides were generally not used on corn, and less than 4 percent of production was lost to insects; since then, pesticide use has increased over 1,000-fold, but losses to insects have more than tripled.[33] In California, cotton yields fell 1.6 percent annually from 1960 to 1980, despite annual increases of 12 percent in the application of pesticides.[34] Among the causes are declining soil quality, altered agricultural ecosystems, reduced use of other pest management practices, and the evolution of pests that are resistant to chemical pesticides. There are now at least 480 species of pesticide-resistant insects and chelicerates, 113 species of weeds, and 150 species of fungi and bacteria.[35] Further, pesticides often kill off pests' natural predators, including birds and beneficial insects, which, because of their smaller populations and slower generation times, cannot evolve resistance nearly as quickly as the pests themselves. As a result, pesticides sometimes backfire in the long run: they kill pests like magic in the first year of their use but change the agricultural ecosystem in a way that sets the stage for uncontrollable pest outbreaks some years later.

While the returns of chemical-intensive methods are diminishing, alternative agriculture offers tremendous international opportunities for sustainable growth in agricultural productivity and economic health. Pretty reviewed dozens of reports of actual experience with sustainable farming methods and concluded that alternative agriculture has the potential to increase yields in poor countries by a factor of two to three. In countries where green revolution farming has been dominant, alternative agriculture can be implemented with no reduction or even a moderate increase in yields. One famous example comes from Indonesia. In the 1980s, after researchers at the United Nations Food and Agriculture Organization discovered that pesticides were radically exacerbating outbreaks of the brown planthopper, a major rice pest, the national government banned 57 of the 63 pesticides that had been used on rice and began an ambitious program of alternative agriculture. In the next six years, rice yields increased by 10 percent.[36]

Many other nations have made significant strides toward reducing their use of synthetic pesticides, and Denmark, the Netherlands, Norway, Sweden, and several Canadian provinces have programs to reduce agricultural pesticide use by 50 percent by the year 2000.[37] In most countries, however, the policies of national governments and international lending institutions encourage farmers to rely on chemical-intensive measures.[38] An ambitious program to assist rather than hinder farmers in the transition to alternative agriculture could drastically reduce the use of synthetic pesticides and make the sunsetting of one major and particularly hazardous group—the organochlorines—a quite feasible proposition.

One small nonagricultural use of pesticides deserves special mention: the use of DDT to control the insect vectors that carry malaria and other infectious diseases, which are major causes of death and sickness in tropical and equatorial regions of the world. Although DDT is banned in most industrialized nations, about 30,000 tons per year are still used for vector control in developing countries. Some nations, however, have developed systems of integrated vector management (IVM) to control insect populations and minimize the use of synthetic pesticides. The techniques are similar to those of alternative agriculture: improving sanitation to eliminate insect breeding sites; using biological insecticides, predators, and parasites; introducing sterile insects to reduce population growth; and installing traps and bed nets within homes. When outbreaks do occur, chemicals may be necessary, so IVM relies on less persistent insecticides, including insect growth regulators, pyrethrins, and organophosphates. In a thorough study of IVM programs in Africa, Latin America, India, and the Philippines, the World Wildlife Fund (WWF) found that these methods have been just as effective as the spraying of DDT while preventing environmental contamination and the evolution of pesticide-resistant insects. WWF concluded that it is technically feasible for the world to maintain or improve malaria prevention by following the example of Mexico, which has pledged to eliminate the use of DDT for vector control by the year 2007.[39]

Pulp and Paper

One of the largest consumers of chlorine and chlorinated chemicals is the pulp and paper industry, which uses chlorine gas and chlorine dioxide to bleach pulp bright white. As governments and environmentalists in the United States and Europe began to scrutinize chlorine bleaching as a major water pollution source in the late 1980s and early 1990s, pulp companies took two different paths. In Europe, environmental concern and government policies increased demand for totally chlorine-free (TCF) paper, and a number of Scandinavian and other European pulp producers invested in equipment to fulfill this demand. In TCF mills, pulp is bleached with oxygen-based chemicals, such as oxygen gas, ozone, and hydrogen peroxide; other processes are usually modified as well, such as extended cooking of wood chips and improved delignification of pulp before bleaching begins.[40]

There can be no doubt that TCF bleaching is technically and commercially viable. By 1996 there were 56 mills in the world—47 in Europe, 7 in Canada, and 2 in the United States—producing totally chlorine-free bleached pulp.[41] With rapid improvements in bleaching techniques during the 1990s, TCF pulp now meets the market's most stringent standards for brightness and strength.[42] Today major book publishers—including the publisher of this book—retail catalogues, government offices, and major magazines in Canada and Europe—and smaller magazines in the United States—are using TCF paper for technically and aesthetically demanding uses. In 1993 TCF paper accounted for 30 percent of the European market for printing and writing paper; by 1996, it was expected to account for 60 to 70 percent of the market in the Germanic and Nordic countries. On the supply side, 60 percent of all chemical pulp production in Scandinavia is expected to be TCF by the year 2000.[43]

Most paper companies in North America, along with some in Europe, invested in chlorine dioxide bleaching. These firms chose chlorine dioxide because, although expensive to install and operate, it can serve as a "drop-in" substitute for chlorine gas; that is, unlike TCF methods, it requires no major redesign of the bleaching process. Under pressure from investors demanding immediate returns, these pulp manufacturers chose chlorine dioxide, which one trade journal referred to as a comparatively "low-risk, low-capital" way to comply with environmental regulations.[44] Once committed to chlorine dioxide, these companies have had to protect the value of their investments, which take 15 to 20 years to be fully amortized. They have thus resisted market and regulatory developments that would promote TCF bleaching and make chlorine dioxide bleaching obsolete. American paper companies and trade groups, for example, lobbied successfully to remove any requirement for TCF bleaching in EPA's revised regulations for the pulp and industry by arguing that TCF bleaching would bankrupt them. The industry even forced the White House to scuttle a rather harmless executive order that would have helped TCF markets develop by favoring chlorine-free paper in government procurement policies.[45]

Is TCF in fact prohibitively expensive? There have been two independent analyses of the costs of pulp mill conversion, one by the Radian Corporation and one by a joint task force of the Environmental Defense Fund (EDF), academic researchers, and a number of major commercial paper consumers. In a detailed application of the methods and findings of these two reports, the Center for the Biology of Natural Systems (CBNS) at Queens College estimated the capital and operating costs of converting the nine bleached pulp mills in the Great Lakes region from existing processes to one of four options: totally chlorine free, drop-in chlorine dioxide (with no process improvements over current practice except the substitution of chlorine dioxide for chlorine gas), "modern" chlorine dioxide (extended pulp cooking, oxygen delignification, and two chlorine dioxide bleaching stages), or "advanced" chlorine dioxide (all the features of the modern process, but with the first chlorine dioxide stage replaced by ozone). CBNS found that neither the Radian nor the EDF method indicated a major difference between TCF and chlorine dioxide in costs for operation and maintenance: the Radian estimate predicted that TCF bleaching would cost $7 per ton more than drop-in chloride dioxide, while the EDF estimate predicted a savings of $6 per ton (table 10.2).[46]

The real issue is the capital cost of equipment for TCF bleaching. Overall, it costs slightly less or about the same—depending on whose analysis one uses—to build a brand new TCF pulp mill than to build one that uses chlorine dioxide.[47] But new pulp mills are so expensive that only one has been built in the 1990s in Europe or North America—a TCF facility in western Finland that cost $17 million less than a comparable chlorine

Table 10.2
Difference in cost to convert Great Lakes pulp mills to alternate bleaching methods

	Capital cost($million)		Average change in cost per ton of pulp ($)		
	Total	Annualized	Capital	Operating and Maintenance	Total
Drop-in chlorine dioxide	ND/160	ND/21	ND/+11	ND/+7	ND/+19
Modern chlorine dioxide (oxygen delignification)	150/285	20/21	+6/+16	–3/+0	+4/+16
Advanced chlorine dioxide (oxygen delignification, ozone bleaching, effluent reduction)	225/440	30/58	+9/+18	–3/+2	+6/+20
Totally chlorine free (oxygen delignification, ozone and hydrogen peroxide bleaching, effluent reduction)	225/450	30/60	+9/+19	+7/+1	+17/+20

Note: The first number in each column is based on the conversion methodology of the Radian Corporation, the second on that of the Environmental Defense Fund's Paper Task Force. ND = no data available.
Source: Commoner et al. 1996.

dioxide mill.[48] The debate focuses instead on the costs of converting existing plants to TCF bleaching. The cost of upgrading a bleach line to TCF can be large, ranging from $10 to $50 million, depending on the size of the plant and its current technology, according to CBNS. But this is not much more than the $11 to $26 million it would cost to install chlorine dioxide generators and other equipment for drop-in substitution of chlorine dioxide into old-fashioned chlorine bleach lines. And it costs about the same to install TCF as it does to upgrade an existing bleach line to the "modern" or "advanced" options, which represent the current standard for chlorine dioxide bleaching.[49]

Based on these projected investments, CBNS calculated the total increase in production costs—operating and maintenance costs plus capital outlay annualized over a 15-year period—associated with each bleaching method. The results were strikingly unimpressive. The production costs of TCF bleaching were higher than those associated with drop-in chlorine dioxide bleaching by just $1 to $17 per ton of pulp, depending on whether the EDF or Radian method was used. And compared to "modern" or "advanced" chlorine dioxide technology, TCF would cost the same or slightly more, with an excess cost ranging from nothing to $13 per ton of pulp (table 10.2). In the long run, as soon as the initial investment is paid off, TCF's lower operating costs would begin to offer substantial net savings, even in comparison to drop-in chlorine-dioxide bleaching.[50]

What do these numbers mean? Would costs of this magnitude really bankrupt the industry? CBNS put them in perspective relative to the current revenues, costs, and investments of the nine Great Lakes mills. The total increase in costs associated with converting to TCF would amount to just zero to 4 percent of the $460 average cost of producing a ton of pulp. It would be well within the normal range of variation in production costs among mills, typically somewhere between $35 and $170 per ton. The increase would be even smaller in relation to the difference between the cost of producing pulp and its market price, which has been as high as $350 per ton. The required investments, annualized at $9 to $19 per ton of pulp, would be minor compared to the mills' average capital expenditure of $88 per ton. And in light of total revenues of $2.6 billion dollars per year ($887 per ton of pulp), the costs of conversion would be truly trivial. CBNS thus concluded that it is well within the industry's resources for all existing mills in the Great Lakes to convert to TCF bleaching. These results are consistent with the experience of Scandinavian companies, which report that they have already made money by investing in the production of TCF paper.[51]

Once a pulp mill eliminates chlorine bleach, it can then pursue even greater savings by creating a closed-loop or totally effluent-free system. Traditional chlorine-bleaching mills cannot recycle their effluents, which are rich in organochlorines and chlorides that would build up in the system and corrode plant equipment. Advanced mills that use chlorine dioxide only in the last bleaching stage can recycle the effluent produced prior to this stage, but they must discharge the wastewater from the final bleaching step, for the same reason. Some companies that have committed themselves to chlorine dioxide bleaching are attempting to develop a closed-loop pulp mill without eliminating chlorine bleaches, but there is a fundamental flaw in their technology. The bleaching effluent must first be filtered to remove organochlorines and chlorides, which are then incinerated.[52] As we saw in chapters 7 and 8, incineration releases organochlorines to the air and creates new ones, including dioxins, in the combustion process. Shifting pollutants from one environmental medium to another cannot legitimately be considered a closed loop. Further, effluent recycling does nothing to stop the deposition of the organochlorine by-products of chlorine dioxide bleaching in the mill's sludge and in the pulp and products themselves.

Eliminating all chlorinated bleaching agents is thus a prerequisite for the establishment of a truly effluent-free mill. Several TCF mills in Scandinavia have already recycled 85 percent or more of their effluents, and a number are in trial stages of total closure.[53] The Louisiana-Pacific corporation has converted its pulp mill in Samoa, California, to TCF bleaching and is seeking to become the first closed-cycle mill in North America.[54] The Union Camp Corporation, a pioneer in reducing effluent volumes with an advanced chlorine dioxide process, has now developed a TCF bleaching method in order to achieve complete effluent recycling.[55]

The totally effluent-free mill offers massive environmental and economic advantages. As we saw in chapter 8, all pulping processes liberate chemicals derived from the natural pesticides in wood pulp, which can be toxic to fish when discharged. A closed-cycle mill eliminates this problem entirely. Closed-loop mills also save significant amounts of energy, consume far less water, and reduce their demand for many kinds of chemicals, which can be recycled along with the effluent; they can also eliminate major expenditures on wastewater treatment and regulatory compliance. Industry analysts have calculated that an effluent-free mill can reduce its operating costs by $35 per ton of pulp, considerably more than the cost of converting to TCF.[56]

The paper industry and environmentalists alike agree that closed-loop mills are the technology of the future for both environmental and economic reasons. Converting to TCF is a prerequisite to the truly effluent-free mill, an initial and necessary step on the path toward the most modern and environmentally sound production methods.

Solvents and Refrigerants

Chlorinated solvents are one of the largest and most easily phased out of all chlorine applications. Just as perchloroethylene can be replaced in dry cleaning by water, steam, or carbon dioxide, chlorinated chemicals now used to clean equipment and as a medium for coatings and paints can be phased out in favor of a number of safer, affordable alternatives.[57] In fact, as the Montreal Protocol on ozone depleting substances has taken effect—along with bans on organochlorine solvents in some countries and changes in U.S. environmental and liability law—major manufacturers in a number of industrial sectors have adopted these methods, demonstrating their technical and commercial viability.

The most common use of chlorinated solvents is for cleaning: oil-soluble material dissolves in chlorinated methanes, ethanes, and ethylenes, so these chemicals are used to degrease industrial machinery or to remove excess flux, the material applied to printed circuit boards before soldering. The first step in reducing solvent use is to prevent the mess in the first place, so that less or no cleaning is required. For instance, engineers have developed flux formulations that are so low in solids that they require no cleaning at all. When cleaning is necessary, aqueous solutions (water mixed with soaps and detergents) and semiaqueous preparations (water mixed with alcohols or natural chemicals, such as terpenes extracted from citrus fruit or tree tissues) have proven to be effective cleaning agents and more economical, as well.[58] One analysis found that "it is technologically feasible to employ aqueous cleaning in 80 to 90 percent of all metal cleaning and degreasing applications and 90 percent of all non-surface mount electronics and assembly cleaning operations, at costs comparable to solvent-based cleaning costs."[59] Other cleaning methods include chlorine-free chemical solvents (such as ketones, pyrrolidones, and other chemicals), ultrasonic cleaning (in which sound waves create abrasive bubbles in an aqueous solution), and supercritical carbon dioxide cleaning, in which carbon dioxide is heated and pressurized to a state in which it has the properties of both a liquid and a gas, making it a very effective and nontoxic solvent, although it must be handled with care to minimize occupational hazards.[60] For particularly demanding applications, emerging technologies include plasma cleaning with charged gaseous particles, combined use of ultraviolet radiation and ozone gas, and cleaning with tiny ice particles.[61]

Chlorinated solvents evaporate quickly, so they are used as a base for paints, adhesives, and coatings. Clean alternatives—primarily dry coatings and water-based formulations—are well developed and are already in use for many applications. Nonhalogenated chemical solvents can be used when necessary, and supercritical carbon dioxide can also serve as a high-quality medium for the application of paints and coatings.[62] In paint stripping, chlorinated solvents can be replaced by abrasion, high- and low-temperature techniques, and nonchlorinated organic chemicals.[63] Chlorine-free chemicals, water, and supercritical carbon dioxide have replaced chlorinated solvents as media for a number of reactions in the chemical industry.[64]

Major companies in a range of industries have reduced or eliminated the use of chlorinated solvents, including 3M, Northern Telecomm, Raytheon, IBM, Rockwell International, Motorola, Hitachi, IBM, GE, and MEMC Electronics.[65] It would be difficult to argue that chlorinated solvents are irreplaceable, when so many companies have successfully stopped using them. In Sweden chlorinated solvents are slated for complete phaseout by the year 2000.[66]

Chlorine-free alternatives are not only technically adequate but also economically beneficial. In a famous case study, a medium-sized Swedish manufacturer of lighting fixtures successfully replaced solvents for degreasing, painting, and coating with aqueous cleaning methods and powder-based paints. By reducing costs for chemical procurement and waste disposal, the company saved more than $300,000 per year.[67] Even CRA's report found that industrial solvents are the least expensive of all chlorine uses to eliminate.[68]

Some portion of chlorinated solvents—specifically the CFCs and HCFCs—is used as a coolant in air-conditioners and refrigeration systems. Because HCFCs also contain chlorine and deplete the ozone layer, technologies are rapidly being developed and deployed to replace them with chlorine-free cooling systems. The most common replacement is the hydrofluorocarbons (HFCs), which contain no chlorine. HFCs are an improvement over the ozone-depleting HCFCs, but they are far from unproblematic, because they are potent global warming gases and are manufactured using organochlorine intermediates.

A number of safer refrigerants have proven effective and are already in widespread or limited use. One technology that has grown especially rapidly is the use of a mixture of the hydrocarbons propane and isobutane in domestic refrigerators. After their introduction in Germany in 1991, energy-efficient hydrocarbon-based refrigerators garnered 80 percent of the domestic refrigeration market in Europe by 1997, and they were also in use in China, Latin America, and Australia. One concern about hydrocarbons is that they are flammable, but the refrigerator contains only about as much of these materials as a typical table-top cigarette lighter, so it is certified safe for household use. Chlorine-free hydrocarbons are also used in 3 to 4 percent of all industrial refrigeration units and are suitable for use in home air-conditioners, as well.[69]

In large systems, nonflammable refrigerants may be more appropriate. About 34 percent of all commercial refrigeration systems—and 81 percent of all cold storage and food processing systems in the United States—currently use ammonia, a powerful heat-absorbing compound, so this is clearly a viable option. Ammonia is also used in some 200,000 domestic air-conditioners in the United States and in home refrigerators in Europe and Japan. It can even serve as a drop-in substitute in some cooling equipment that now uses CFCs or HCFCs, and most equipment that does not accept ammonia can be retrofit to do so.[70]

A number of other promising refrigeration methods are rapidly becoming available. A particularly attractive option is the water-zeolite absorption system. Zeolites are natural crystals of aluminosilicate; the crystals contain millions of tiny pores that take up water, absorb its heat, and then expel the cooled water. They are very efficient refrigerants that can operate at high or low temperatures and contain no toxic, flammable, or volatile materials whatsoever. Another technology is the evaporative air-conditioner, which uses water as the working fluid: hot air is passed across a porous wet surface that is cooled by the evaporation of water; the hot air passes its energy to the porous surface and is then returned as cool air. Also in development is thermoacoustic cooling, in which high-frequency sound waves are passed through a tube filled with a stack of metal plates and helium gas; the waves compress and expand the stack, which results in the extraction of heat from one end of the tube—which becomes cool—and its transfer to the other, which becomes the hot or "exhaust" end.[71]

In summary, technologies thus exist now to phaseout most uses of organochlorine refrigerants, and even better alternatives are rapidly becoming available. The economics of their implementation is more complex. There are millions of refrigerators and air-conditioners in the world, and few of the substitute systems can be simply dropped into existing equipment. Most alternatives would require considerable investment in new equipment, so substitutes should be phased-in on a time line that minimizes premature loss of capital. It is important, however, that we not mistake the difficulty of a quick phaseout with the inability to phaseout organochlorine refrigerants at all.

PVC Plastic

By far the largest single chlorine use, vinyl appears in a wide variety of applications, from packaging to water pipes, floor tiles to medical devices. There is no single best substitute for all uses of PVC; the appropriate alternative depends on the qualities required for each application.

Not so long ago, virtually all products now made from PVC were made from some other traditional material that functioned perfectly well. Until the 1960s, home siding was predominantly wood, stucco, or aluminum; window frames and profiles were wood or metal; floors were wood, ceramic tile, linoleum, or carpet; clothing and upholstery were cloth or leather; pipes were cement or metal; containers and packaging, when used at all, were made of glass, metal, or cardboard. In most of these uses, PVC made inroads not because it offered some uniquely desirable technical quality but because it was cheaper or required less expertise to handle than the material it replaced. In many cases, the most sensible alternative to PVC is the original traditional material whose place vinyl took. For instance, cement remains a technically excellent material for sewer pipes, and ductile iron works well for water mains. Wood and metal window frames and ceramic and hardwood flooring offer performance superior to that of PVC.[72]

Some PVC uses cannot or are not likely to be replaced with traditional materials. For better or for worse, plastics have become a major part of modern life. In these applications, chlorine-free plastics can substitute for PVC. Although no synthetic plastic is environmentally benign, virtually all chlorine-free plastics are clearly superior to PVC because they result in much lower hazardous by-products during their manufacture and disposal. In a study of the environmental impacts of manufacture and disposal of all major packaging materials, the independent Tellus Institute found that PVC is by far the most environmentally damaging of all plastics.[73] A life cycle analysis by the Danish EPA found that the common plastics polyethylene, polypropylene, polystyrene, polyethylene terephthalate (PET), and ethylene-propylene diene synthetic rubber are all clearly preferable to PVC in terms of resource and energy consumption, accident risk, and occupational and environmental hazards.[74]

The most appropriate alternative plastic depends on the application. For instance, polyethylene pipe is rapidly gaining market share at the expense of PVC, while polypropylene and acrylonitrile-butadiene-styrene (ABS) polymers are the most common replacements for PVC in automobile components.[75] PVC coatings for electrical cables can be substituted with cross-linked polyethylene, ethylene-vinyl acetate copolymer, polyamide, or silicone, depending on the type of cable and where it will be installed.[76] A wide range of chlorine-free polymers is now available for medical packaging and devices that were once exclusively PVC, including examination gloves, overshoes, aprons, mattress covers, diapers, wound dressings, syringes, drainage and solution bottles and bags, tubing, connecting parts, scalpel handles, catheters, and respiratory masks.[77]

The best measure of the technical viability of PVC alternatives is the extent to which companies and governments have succeeded without them.

In construction, the Berlin Museum of Jewish Culture, the Channel Tunnel between France and England, and the Oberpallen School in Luxembourg have all been built without PVC. Tarkett, a major flooring manufacturer, has announced that it will eliminate all PVC from its products, and IKEA has a long-standing policy of not selling PVC furniture and other products. Evian, Vittel, and the Body Shop no longer use PVC packaging, and Nike and the electronics manufacturers AEG and Sony-Europe all have PVC phaseout policies. In transportation, the new Vienna subway system was built without PVC, and BMW, Opel, and Volkswagen have committed to eliminating PVC in their automobiles. In medicine, several European hospitals, including the Vienna Ost-SMZ hospital, are reported to have virtually eliminated their use of PVC-containing products.[78]

There are some minor PVC uses that cannot be eliminated in the short term. The American Public Health Association (APHA), for instance, noted that "alternatives are currently available for many, but not all health care uses of chlorinated plastics." When the APHA resolved in 1996 that hospitals should "reduce or eliminate their use of PVC plastics" because of the production of dioxins and other organochlorines during the PVC life cycle, it urged suppliers to develop and bring to market substitutes for the handful of products for which none were yet available.[79] Progress has already been made, with a polyolefin IV bag coming into commerce, along with substitutes for PVC in formerly difficult-to-replace types of tubing and connection parts.[80] As the Enquete Commission, an expert advisory group of the German parliament, concluded, "Except for special products with particularly characteristic requirements, such as blood packs or electrically conducting flooring for clinically clean rooms, it is safe to assume that there are materials available to replace PVC in all its uses, these in many cases being marketed by the same manufacturers as those producing PVC."[81]

What would be the economic impact of phasing out PVC? There is no denying that vinyl products are cheap, subsidized as they are by the higher price of caustic soda. But the cost of replacing PVC is generally relatively modest. According to Environment Canada, for instance, aluminum window frames are 5 percent more expensive to produce and install than PVC, but wood frames are 10 percent cheaper. Replacing PVC pipe with iron, cement, polyethylene, or ABS pipe would reduce the cost of pipe by

1 percent or increase it by up to 6 percent, depending on the application. Chlorine-free plastic substitutes for PVC wire and cable coatings are 11 percent more expensive than PVC, but polyethylene food wrap is 11 percent cheaper.[82]

Some of the most detailed information comes from a consultant's report to the International Joint Commission, which examined the cost of PVC and its alternatives in the Great Lakes region.[83] The report found that technically effective substitutes are available at equal or lower market price for 26 percent of all PVC demand. For about 65 percent of PVC applications, the alternatives are slightly or moderately more expensive, and only the last 9 percent or so of PVC use would be quite expensive to phase out, with substitutes costing more than $4 per kilogram more than vinyl. Overall, the report found that 89 percent of the PVC could be replaced for a net *savings* of 2 cents per kilogram (a total of $13 million), and over 95 percent could be sunset for a net cost of just 25 cents per kilogram (143 million dollars) (figure 10.2); only the last 5 percent would be very expensive, raising the cost of a total vinyl ban to about 50 cents per kilogram. Although the study considered only the market price of substitutes and not their fully installed costs, it made its point quite clearly: careful planning, with priorities, time lines, and new product development for uses that are currently expensive to replace, would make a PVC phaseout economically feasible.

With such small price differences for most applications, the overall cost of phasing out PVC is rather moderate. Environment Canada's analysis included the full installed cost of PVC substitutes, and it found that alternative materials would cost just 6 percent more than vinyl. More important, using substitutes would increase the price of all end-use products that are now made with PVC by an average of just 0.2 percent. In construction applications, replacing all PVC with chlorine-free substitutes would increase the cost of home building by just 0.4 percent, raising the price of a typical new house from $150,000 to $150,600.[84]

Chemical Intermediates

One of the most diverse of all chlorine applications is the use of chlorinated intermediates to produce chlorine-free compounds. Each reaction is

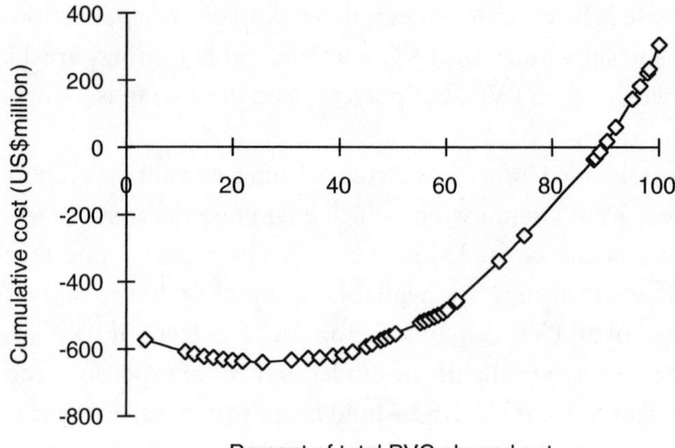

Figure 10.2
Net cost to purchase alternatives to PVC in the Great Lakes region. The graph shows the net cumulative costs of a PVC phaseout in the Great Lakes region, implemented in order of increasing price. The majority of PVC could be replaced with substitutes for a net price reduction. Estimates are for market price only and do not include installation costs. Sectors, from least to most expensive, are: window frames, storm sewer pipe, water main pipe (six sizes, each charted separately), sanitary sewer pipe, wire and cable, film packaging, sheet packaging, bottles packaging, water main pipe (two sizes), irrigation pipes, upholstery, health care products, appliances, wall coverings, appliances, compounders and resellers, plugs and other electrical products, footwear, medical tubing, garden hose, solution bags, other pressure pipe, water main pipes (one size), electrical conduit, drain waste vent pipes, flooring, siding and accessories, electrical conduit (one size), vinyl adhesives and sealants, and electrical conduits (seven sizes). (Source: Hickling 1993.)

unique and requires its own specific alternative. For many of the hundreds of chlorine-dependent chemical synthesis processes, no research has been published on safer alternatives. But despite the diversity, three major intermediates—propropylene chlorohydrin, phosgene, and epichlorohydrin—dominate this sector, accounting for the vast majority of chlorine consumed in chemical intermediates. For these and some other smaller chlorine uses, it turns out that viable alternatives are either available now or in development by the industry.

The largest single chlorine application in this sector is the synthesis of propylene oxide via propylene chlorohydrin. Propylene oxide is used to make a wide variety of chemical and plastic products, including brake fluids and polyurethane plastics. Not only is it possible to make propylene oxide without chlorine, it has been done for years. Dow Chemical makes less than half the propylene oxide in the United States through the chlorohydrin route, but Arco, Shell, and Texaco produce the rest through two processes that directly combine oxygen or air with chlorine-free hydrocarbons.[85] According to Prognos AG's report on chlorine alternatives, the chlorine-free route costs 40 to 50 percent less to operate and has significant ecological advantages over the chlorohydrin method. Prognos concluded that a complete change-over to this process could be accomplished in five to ten years, with substantial economic benefits to society and industry.[86]

Phosgene is the next largest use of chlorine as an intermediate. An extraordinarily toxic chemical, phosgene serves as an intermediate in two major processes: it combines with bisphenol-A to create polycarbonate plastics (the plastic from which compact discs and baby bottles are made), and it reacts with nitrogenated chemicals to produce isocyanates, which then serve as a feedstock for polyurethane, pesticides, and some specialty chemicals. Because phosgene is so dangerous, chemical engineers have made the development of substitutes a priority. By the early 1990s, DuPont and Monsanto had developed catalytic processes to produce isocyanates from carbon monoxide or carbon dioxide rather than phosgene.[87] Both GE Plastics and Asahi have commercialized their methods for making polycarbonate from diphenylcarbonate rather than phosgene, a change that actually improves the quality of the product.[88] And the Italian firm Enichem now replaces phosgene with dimethylcarbonate, a much less toxic chemical synthesized from methanol, carbon monoxide, and oxygen.[89]

Epoxy resins are chlorine-free polymers used as adhesives and coatings. About 75 percent of the epoxy resins on the market are made using epichlorohydrin as an intermediate.[90] There are, however, numerous ways to reduce the use of chlorine in this application. The first and most obvious point is to substitute alternative epoxy resins not manufactured with organochlorines for those derived from epichlorohydrin. According to Prognos, at least 15 percent of epichlorohydrin could be eliminated in favor of available alternative resins that have clear ecological advantages. In addition, an alternative process for making epichlorohydrin can radically reduce the amount of chlorine consumed in the synthesis of this chemical. Prognos concluded that these two steps alone could reduce chlorine use in this sector by 60 percent, with a net rise in costs of just 5 percent.[91] Additional substitution could be achieved in some applications by substituting other kinds of resins and plastics instead of epoxies.[92] Finally, just as it has been possible to produce other compounds through chlorine-free routes, it should be possible to develop chlorine-free epoxidation processes for the production of these substances.

There are numerous smaller uses of organochlorines in the chemical industry, but information on alternative processes is available for only a handful. For many, it is likely that no substitutes are currently available, but the pace at which chlorine-free substitutes have been developed suggests that these uses could be eliminated as well, given time for research and development. Monsanto, for instance, has developed a method to produce nitrogenated aromatic chemicals without chlorinated benzenes and phenols.[93] And Enichem engineers have found that dimethylcarbonate can replace methyl chloride as a methyl donor in a wide range of chemical reactions; potential uses include the synthesis of phenolic ethers, quaternary ammonium compounds, silicones, and methyl cellulose.[94]

Even in the pharmaceutical industry, chlorine use could be significantly reduced. CRA asserted that 85 percent of drugs are "chlorine-dependent," but a close look at its analysis reveals that only 20 percent of active ingredients are actually organochlorines, while another 6 percent contain inorganic chloride groups in their formulations. Of all the supposedly "chlorine-dependent drugs," about 69 percent contain no chlorine at all; chlorinated chemicals are merely used in their production as solvents, extractants, intermediates, reaction media, or neutralizers.[95] Many of these process could be carried out with chlorine-free substitutes. For instance, fully 20 percent of "chlorine-dependent" drugs use hydrochloric acid or caustic soda as neutralizing agents for pH control, a role that other acids or bases can fill. Organochlorine solvents are used for applying coatings to tablets, but a report by the Jacobs Engineering Group for the Metropolitan Water District of Southern California found that water-based substitutes are viable in 50 to 100 percent of these cases, and nonhalogenated solvents can take care of almost all the remainder.[96] Organochlorines are also used as extractants to separate reaction products or remove desired substances from fungi, plant, and animal tissues, but supercritical carbon dioxide has emerged as a very effective and environmentally benign substitute, according to the Jacobs report. Overall, Jacobs offered the conservative estimate that up to 64 percent of chlorinated solvent use in the pharmaceutical industry could be replaced by aqueous or other nonchemical alternatives, and chlorine-free solvents could eliminate much of the rest.[97] As for intermediates used in pharmaceutical production, it should be possible to develop chlorine-free routes to produce chlorine-free medicines like those to produce chlorine-free industrial chemicals. The firm Sipsy has a chlorine-free epoxidation process for production of some pharmaceuticals,[98] and Italian researchers have developed a dimethylcarbonate-based substitute for chlorinated methanes in the synthesis of nonsteroidal anti-inflammatory drugs.[99]

In the last analysis, we should remember that some of the chlorine-free chemicals now made with organochlorine intermediates are environmentally hazardous themselves. For instance, polycarbonate plastics and many epoxy resins are polymers of bisphenol-A, an estrogenic compound that has been found to leach out of containers made of these plastics into the solutions they hold. Isocyanates are used to make polyurethane, which itself can cause a range of health effects, including cancer. The best solution in cases like these is not to find a better way to make bad materials but to make less hazardous products. Thus, traditional materials and safer plastics, like polyolefins, should be used whenever possible. In the long run, work by the chemical industry to develop biopolymers—plastics composed of long chains of molecules derived from plants—becomes especially valuable. The recent growth of research in the field of "green chemistry," the purposeful design of safer chemical products and processes, shows that chemical engineering can be used for the very positive purpose of developing new materials that are safe for health and the environment.[100]

Water Treatment

For many people, "water purification" is the first chlorine application that comes to mind. Only a small fraction of chlorine (1.4 percent in

Canada and 5 percent in the United States), however, is used to disinfect water, and the majority of this amount is applied to wastewater, not drinking water.[101]

People must have safe drinking water. Waterborne infectious diseases have the potential to kill millions, and in areas of the world with inadequate sanitation systems they are a major cause of death, especially among children. If the question is whether to use chlorine or no disinfection at all, we must choose chlorine. The real choice, however, is between chlorine and alternative disinfection systems. For all the major types of disinfection, effective alternatives are already in use around the world, providing safe water without dangerous by-products.

In wastewater disinfection, which accounts for about 80 percent of chlorine used in water treatment, eliminating chlorine is relatively easy. In the 1970s and 1980s, chlorine and its disinfection by-products began to cause massive fish kills downstream from sewage plants, so EPA and other national regulatory agencies issued regulations that drove some systems to seek alternative disinfectants.[102] The best and fastest growing chlorine substitute is ultraviolet light (UV), which uses radiation in a narrow band of wavelengths to damage the DNA of microbes, killing them or rendering them unable to reproduce. In a UV treatment plant, water runs through shallow troughs where tube-shaped lamps emit a strong dose of ultraviolet light into the water. UV is an extremely effective disinfectant against bacteria and viruses, so it is a practical option for the treatment of sewage and industrial wastewater.[103]

UV has already replaced chlorine at hundreds of wastewater treatment plants in the United States, Canada, New Zealand, and Australia, ranging from fairly small facilities to very large urban plants. Currently only 5 percent of wastewater in the United States is treated with UV, but that number is expected to grow to 25 percent by 2007.[104] The disinfection and operating experience has been quite positive. UV treatment produces no chemical by-products, and it eliminates the accidental chlorine leaks and spills that occur during chlorine transport, storage, and handling. In most cases, UV is even cheaper than chlorine to operate, because there are no dangerous chemicals to buy or store. The Water Environment Research Foundation, an independent engineering research group, has concluded that UV disinfection is "both efficient and cost-effective."[105]

In drinking water disinfection, alternative methods are already providing safe drinking water to millions of people without organochlorine by-products. The three major alternatives to chlorine are slow sand filtration, ozone, and UV treatment, all in use in water treatment systems throughout the world. In the first method, water is filtered very slowly through a bed of very fine sand, imitating the very effective filtration that makes most natural groundwater safe to drink without disinfection (except where it has been contaminated by leaking septic tanks or sewage lines). The sand bed is, in fact, an engineered and carefully maintained ecosystem. At its interface with the water, it contains millions of tiny organisms that consume pathogens and organic matter. As the water passes through the sand, physical removal processes take over, filtering out organic matter that remains in the water and, as the temperature of the water falls significantly, killing any remaining pathogens.[106]

Low tech as it sounds, slow sand filtration is an extraordinarily effective disinfectant against bacteria, viruses, and protozoans, rendering the water almost completely pure. Officials at the American Water Works Association call slow sand filtration a "timeless technology for modern applications . . . , one of the surest ways to provide safe, reliable water."[107] Slow sand filters have long been the treatment method of choice in thousands of communities around the world, including major cities like Amsterdam, Antwerp, and Madras.[108] In the United States, slow sand filtration is used in several hundred communities, mostly small ones seeking effective treatment without the expense of chlorination.[109] Because the sand beds are inexpensive to build and easy to maintain, they are particularly appropriate for many settings in developing nations.[110] Operated properly, the systems are quite effective at removing Giardia, Cryptosporidium, and other protozoa that are difficult to treat by many other disinfection methods.[111] The only major drawback of slow sand filtration is that it requires a fairly large amount of space, which can present problems in large, crowded cities without well-developed systems for transporting drinking water.

Ozone, which made its debut in the disinfection of drinking water in 1893 in the Netherlands, is now used at thousands of facilities in Europe; in North America, fewer plants use ozone, but the number is rising rapidly as concern about chlorination by-products grows.[112] Ozone is an extremely powerful oxidant, providing more effective disinfection than chlorine or UV against bacteria, viruses, and protozoan pathogens of all types. It also breaks down organic compounds in water, degrading harmful contaminants and reducing unpleasant tastes and odors.[113]

But ozone also has two potentially negative qualities: it is more expensive than any of the other major methods, and it results in the formation of oxygenated by-products, including ketones, aldehydes, and carboxylic acids. Ozonation appears to be a clear improvement over chlorine disinfection, because the by-products it generates are significantly less persistent and less mutagenic than chlorine by-products.[114] Experiments have shown that, as long as an adequate dose of ozone is supplied, ozonation causes no measurable increase in the mutagenic activity of water; chlorine treatment, in contrast, increases mutagenicity by a factor of 12 to 14.[115] Ozone by-products have not been thoroughly evaluated, however, so we cannot rule out the possibility that they may pose some other kind of health hazard. When suitable alternatives that do not form chemical by-products are available, we should use these instead of ozone.

One example is UV light, as effective a disinfectant for many kinds of drinking water as it is for wastewater. UV treatment was first used to sanitize drinking water in 1955 in Switzerland and Austria. Today about 1,500 European facilities rely on UV, and there are about 500 such systems in North America.[116] Extensive testing on groundwater and surface water in the Netherlands has shown that UV is a very effective disinfectant against bacteria, viruses, and most protozoa, that it produces few or no by-products, tastes, or odors, and that its price is comparable to that of a chlorine system, requiring slightly more capital to install but offering lower operating costs. According to the engineers who headed the Dutch evaluation, "UV is a very promising technique for disinfection of all types of drinking water sources . . . at relatively low costs."[117] More recently a U.S. EPA study of groundwater disinfection found that UV treatment is a "safe, effective, and cost-competitive" option for disinfection of drinking water from underground sources, especially in small systems.[118]

This is not to say that UV can be installed everywhere, without thought or planning. UV is not as effective when water is murky as when it is clear, so water sources of poor quality require either higher doses of UV or filtration prior to disinfection. Systems where water is contaminated with encysted protozoans like Giardia and Cryptosporodium present a special challenge, because neither UV nor chlorine can kill these organisms reliably. In these cases, UV treatment must be preceded by fine membranes or carbon filters that remove the microbes from the water.[119] Because of this limitation, ozone or slow sand filtration, depending on the setting, may be the most cost-effective disinfectants for water systems with these problems.

Chlorine has one advantage that neither UV nor ozone can provide. While disinfected water is in transit from the treatment plant to the tap, residual chlorine remains in the water, providing continued disinfection that prevents regrowth of bacteria and other organisms. In groundwater systems, the water is usually so low in organic matter that there is no food on which bacteria can grow, so residual chlorine is not necessary and is usually not used. Surface waters are typically rich in organic matter, however, so regrowth presents a real problem. For this reason, many systems that treat surface water with ozone or UV apply a small, final dose of chlorine as a residual. If this is the best we can do—replacing chlorine as a primary disinfectant but continuing its use as a residual—we will have reduced chlorine use in water treatment substantially. But the chlorine residual produces toxic by-products, so we should seek methods to eliminate it without compromising water quality.

There are better ways of preventing regrowth. Recent research, in fact, indicates that a chlorine residual, despite its disinfecting power, often fails to protect water from recontamination during distribution. In systems with high levels of organic matter, bacteria can aggregate into "biofilms" on the inner surfaces of delivery pipes, where they are protected from disinfectants in the water and multiply rapidly. When the biofilm sloughs off the pipe, pathogens enter the water in such abundance that the typical chlorine residual cannot kill them all. In fact, biofilms are thought to have caused contamination of water supplies in a number of systems, despite the presence of chlorine.[120] A much more effective way to prevent regrowth is to filter surface water to remove most of the organic matter with activated carbon filters or very fine membranes; the filtered water is so low in food that pathogen populations cannot grow during distribution. This is precisely why many systems that use groundwater or treat surface water with slow sand filtration do not need a chlorine residual.[121] Some drinking water systems in Europe now use filtration methods, in combination with UV or ozone, to eliminate the need for a chlorine residual entirely.[122] Combining treatment methods increases costs, of course, but it provides the safest drinking water, from both a biological and chemical perspective. (Even with low organic matter, water distribution lines must be maintained in good condition to prevent sewage pipes, septic tanks, or other sources of pathogens from contaminating drinking water during transport from the treatment plant to the tap. Systems plagued by leaky distribution lines or other causes of "cross-contamination" will need to maintain a chlorinated residual until they can address these issues.)

Thus, it is technically possible in most cases to provide safe treatment of drinking water and wastewater without chlorine disinfection. Implementing alternative technologies is a more complex matter, because the infrastructure for water treatment is so diffuse, with more than 60,000 drinking water treatment plants and 10,000 wastewater treatment facilities in the United States alone.[123] For wastewater, UV disinfection costs about the same as or less than chlorine systems; one study of a facility in Indiana treating 8 million gallons of wastewater per day found that the capital and operating costs of UV treatment would be 36 percent and 32 percent less, respectively, than those for chlorine disinfection.[124] In drinking water, slow sand filtration is substantially cheaper, UV is about the same, and ozone is more expensive than chlorine treatment. Whatever the operating costs, converting so many systems to disinfection alternatives will take considerable time and money, so a phase-in should be harmonized with existing investment cycles. According to Environment Canada, reequipping all Canadian drinking water plants with ozone and all wastewater treatment plants with UV disinfection would require capital investments of $583 million and $154 million respectively. With the investment annualized over 20 years, the total difference in capital and operating costs would represent just 3.3 and 1.7 percent of the total price Canadian customers pay for water treatment.[125] If chlorine-free alternatives are installed as new facilities are built or old ones upgraded or maintained, it should be possible to move away from our current dependence on chlorine treatment for a manageable price.

In developing countries, where contaminated water claims millions of lives every year, water disinfection presents a different kind of challenge. Here the major issue is not replacing existing disinfection facilities with new technology but installing safe water systems in the first place. Wealthy nations finance military buildups, immense dams, and costly power plants in developing countries, but they direct little aid to fulfilling the immediate need for safe drinking water. As water systems are installed in the developing world, they should use the most effective, environmentally benign methods available, including dry sanitation, UV treatment of sewage, and slow sand filtration, UV, or ozone disinfection of drinking water. While UV and ozone are normally considered technologically demanding methods, engineers at the Lawrence Berkeley National Laboratory have developed an inexpensive, portable, low-maintenance UV unit for use in rural settings in developing countries. The unit, which can serve several thousand people, attaches to a community water pump or tank and can be powered by a car battery, solar panel, or small wind turbine; it costs about 7 cents per person per year.[126] Where chlorine systems already exist, they should be used until they can be replaced with safer substitutes. In cases of emergency contamination—as in floods and refugee camps, for instance—hypochlorite is a convenient and effective portable disinfectant, so this compound should remain available for these episodes.

For swimming pool disinfection, a very small but familiar use of chlorine, several chlorine-free alternatives are on the market. The oldest, ozonation, is extremely effective and is well suited to pools that serve large numbers of swimmers. Ozone was, for example, the pool disinfectant at the 1996 Summer Olympics in Atlanta.[127] The relative expense of ozonation, however, is a drawback for individual pool owners. Another option, a chlorine-free chemical disinfectant called polyhexamethylene biguanide (biguanide for short), is a more powerful disinfectant than chlorine. Originally developed as a water-soluble antimicrobial for medical uses, biguanide kills bacteria but has no effect on mammalian cells, so it is apparently nontoxic to people. Combined with a chlorine-free algacide, biguanide provides very effective disinfection for most kinds of pools.[128]

Perhaps the most intriguing choice for pools is the catalytic disinfector, a technology adapted from the system that NASA developed to disinfect drinking water on manned space flights.[129] The disinfector is a small unit that attaches to the pool's circulation system. As water enters the disinfector, it passes over a bed of silver crystals, which catalyze dissolved oxygen molecules in the water to split into free oxygen atoms. Oxygen atoms are extremely potent, killing virtually all the bacteria and viruses in the water within five seconds. The units then release small quantities of copper and silver atoms into the water for further defense against bacteria and algae. Catalytic disinfectors do not destroy organic matter in the water the way chlorine and ozone do, so other chemicals—like much smaller quantities of chlorine, sodium persulfate, or potassium peroxymonosulfate—are necessary to "shock" the pool. According to the leading manufacture of these units, thousands of pools around the world are now using catalytic disinfection, reducing or entirely eliminating their use of chlorine and chlorinated chemicals.[130]

Jobs and Chlorine

These examples show that safe and effective alternatives exist now for the majority of chlorine uses; for those applications for which substitutes are not yet viable, there is every indication that alternatives could be developed, so long as regulatory and economic policies provide incentives to innovate. Although converting to a chlorine-free economy will cost money, it appears possible to substitute for most uses of chlorine for a reasonable price.

The economic effects of technological change go beyond mere costs, however. There are also matters of jobs, the productive use of resources, and their equitable distribution. Modernization, with its emphasis on "efficiency," has tended to replace labor with chemicals and machinery. Although this strategy may increase profits and competitiveness against other firms that take the same approach, it is debatable whether it offers compelling benefits to the economy as a whole or to the bulk of the people the economy is supposed to support. A technology that costs more but creates jobs and spurs further economic activity may be economically preferable to one the market has adopted because it is currently the cheapest.

To evaluate the employment effects of phasing out chlorine, we must address first and foremost chlorine's largest user, the PVC industry. Of the many jobs that CRA listed as "chlorine dependent" in its report for the chlorine industry, the vast majority are involved in the manufacture of plastic products that include PVC resin, like car seats, window frames, and toys. In Environment Canada's analysis, three-fourths of all chlorine-related jobs are of this type. A chlorine sunset will require companies in this sector to use alternative materials in place of PVC, but it will not pose fundamental challenges to the work they do. For instance, as Volkswagen and Tarkett phase PVC out of their products, workers will continue to produce cars and floor tiles, but they will make them out of different materials. To the extent that PVC will be replaced with chlorine-free plastics, many plastic formulators—firms that mix PVC with additives for specific uses—will also have the opportunity to prepare alternative products and mixtures, requiring some modification of equipment but no fundamental change in the nature of business.

We thus might expect that new employment in substitutes for vinyl will cancel out the loss of employment in PVC itself. In fact, the net effect on jobs is likely to be positive, because alternatives to PVC generally require more labor in production or installation than vinyl does. Environment Canada estimates that chlorine-free substitutes for vinyl create 87 percent more jobs per dollar of sales than PVC does; a PVC phaseout would thus result in the net creation of more than 5,500 new jobs in Canada alone. Extrapolating based on Canada's share of worldwide PVC production, a global PVC phaseout would create over 200,000 new jobs.[131] The calculations by which these numbers are derived are highly inexact, so the numbers should be taken with a grain of salt; they do demonstrate, however, that investing in chlorine-free technologies would have considerable positive effects on employment.

Alternatives to chlorine in sectors other than PVC sometimes follow this pattern, but not always. For instance, pesticides also substitute chemicals for labor, reducing farm employment significantly; alternative agriculture spends a greater portion of its costs on people. Similarly, EPA's study of multiprocess wet cleaning found that dry cleaners that avoid perchloroethylene require more labor but keep their total costs constant, increasing the number of jobs by 21 percent and wages by 38 percent, because the jobs in wet cleaning require more skill. Based on these figures, a program to convert all dry cleaners in the United States to the multiprocess wet cleaning method could result in the creation of more than 33,000 new jobs and a net increase in wages of over $600 million per year.[132] In some other sectors, however, replacement is neutral. For instance, substituting a hydrocarbon solvent for an organochlorine solvent or a chlorine-free intermediate for a chlorinated chemical is not likely to cause much net change in employment. Overall, then, the employment effects of a total chlorine phaseout are less dramatic than in the PVC industry alone, but they remain substantial: Environment Canada found that in a near-total chlorine sunset, substitutes would create 62.5 percent more jobs per dollar of revenue than the chlorine-dependent products they would replace.[133]

In addition to permanent jobs producing chlorine-free products, the technological conversion to a chlorine-free infrastructure will provide a considerable economic stimulus, employing people in the manufacture and installation of alternative equipment. Exactly how many jobs would be created is difficult to predict, but there are statistics on the average number of jobs created per dollar invested in manufacturing industries. Based on these figures and the $67 billion investment that CRA estimates as the capital cost of a chlorine ban, the conversion to a chlorine-free economy would create about 925,000 job-years of new employment—or an average of more than 45,000 constant jobs in the United States and Canada alone during a 20-year transition period.[134] If less capital is required, the number of jobs created will be proportionally lower.

Although the overall effects of a chlorine sunset on employment will probably be positive, there will be some economic dislocation. At the earliest stages of production of chlorine-based chemicals, conversion to alternative processes is much more difficult. Phasing out chlorine really does threaten the essential business of chemical companies that manufacture chlorine and organochlorines. This does not mean that all firms will be unable to adapt: for example, Dow Chemical, North America's largest producer of organochlorines, has invested heavily in chlorine-free substitutes for chlorinated solvents and in metallocene-catalyzed polyolefin plastics, which it has positioned as replacements for PVC.[135] Other firms, however, have not demonstrated the will or resources to diversify, so a phaseout of organochlorines could cause significant economic impacts on these companies, their workers, and the communities where they are located.

It is difficult to estimate the numbers of workers around the world whose jobs are truly dependent on chlorine chemistry, but Environment Canada's report on the economics of chlorine chemistry in that country provides some basis for global extrapolations. If the ratio of jobs to production volume in each sector is the same in other countries as in Canada, then there are about 30,000 workers in the chlor-alkali process worldwide, 10,000 or fewer in the production of EDC and VCM feedstocks, and 10,000 to 15,000 more in the polymerization of PVC resin. The number of chemical industry workers involved in the manufacture of other chlorinated chemicals is unknown, but if the ratio of jobs to the amount of chlorine consumed is similar to that in the synthesis of PVC feedstocks, then it is roughly 10,000 to 20,000. Based on these estimates, the total number of workers potentially subject to displacement by a global chlorine phaseout is probably somewhere between 50,000 and 100,000.[136]

The number who actually will be displaced is likely to be considerably lower, given the opportunities for some companies to expand their involvement in alternatives to PVC and other organochlorines. Transition policies to encourage companies to invest in the production of chlorine-free alternatives in the same communities where chlorine and organochlorines were once produced could shrink the number of affected workers further. If the phaseout is spread out over a period of years, the resulting job dislocation would not be particularly significant from the perspective of the economy as a whole.

From an individual standpoint, however, these numbers are not trivial. Thousands of human beings could lose their livelihoods, with real implications for their physical and mental well-being and that of their families and communities. These people are the compelling reason that a chlorine sunset must include provisions to compensate and assist workers who lose their jobs in the transition to a nontoxic economy.

Markets for Clean Production

The Clean Production approach invests in technologies that make efficient and sustainable use of resources and energy, protect the health of workers and the public, and avoid contamination of the environment. In many cases, these technologies require initial investments but ultimately cost less to operate. The United Nations Environmental Program summed up the reasons nicely:

The systematic avoidance of wastes and pollutants increases the efficiency with which goods and services are produced and improves product quality. It also reduces risks to humans and their environment. The prevention of pollution at source reduces the cost of inputs (materials, energy and water), and the cost and liability of waste treatment and disposal. . . . The cleaner production strategy is therefore one of win-win outcomes which provide attractive economies as well as environmental benefits.[137]

The total potential for savings is quite large. EPA estimates that pollution control and disposal will cost U.S. industries $170 billion annually by the year 2000, an economic drag that employs few people and produces no useful products.[138] Preventing pollutant generation at the source obviates the need for this type of expenditure entirely. In the six years since the Massachusetts Toxics Use Reduction Act took effect, manufacturers in that one state decreased their use of listed chemicals by 24 percent, reduced their waste generation by 34 percent, and lowered their toxic releases to the environment by 73 percent. In the process, they saved $91 million.[139]

If Clean Production really is an economic win-win situation, then companies in a free market system should be spontaneously adopting all the cleanest technologies, and mandatory phaseouts of toxic chemicals should be unnecessary. Why can't we simply rely on the market to bring safe substitutes to chlorine into widespread use? There are several reasons. First, in a highly competitive capitalist system, companies tend to focus on the next quarter, not the next decade. Under pressure from buyers seeking low prices and investors demanding immediate returns, many companies reject putting substantial amounts of capital into technologies that take time to pay off.

Second, not all Clean Production technologies result in savings for the companies that adopt them; some cost more to install and operate and will not succeed on their own in the free market. Clean production is not always less expensive than dirty production techniques, because market prices do not accurately reflect a product's total costs. Chlorine for example, usually sells for a fraction of the price of caustic soda, even though the two cost exactly the same amount to produce. Vinyl, chlorinated solvents, and other bulk chemicals with a high chlorine content are thus effectively subsidized by the high price of caustic soda.

Prices are also artificially low because they do not include externalities, the diffuse economic, social and environmental costs imposed by a product throughout its life cycle. The price of PVC packaging, for instance, does not include the environmental cost of dioxins that may be formed in an incinerator where the plastic is burned, nor do PVC-coated electrical cables reflect the cost of corrosion damage to electronic equipment in the event of an accidental fire. The price of pesticides does not include even the direct environmental costs of pesticide use—accidental pesticide poisonings, evolved pest resistance, local fishery and wildlife losses, and so on—which have been estimated at $2 to $4 billion per year in the United States, not including more diffuse health and environmental damage.[140]

No one knows the total externalized costs of toxic chemical use, but their magnitude is certainly astonishing. The bill for remediating contaminated chemical sites in the United States alone is expected to total about $750 billion.[141] Chronic and occupational diseases in the United States—at least a small fraction of which is caused by exposure to toxic chemicals—cost about $1 trillion every year in treatment and lost productivity.[142] In light of costs of this magnitude, even CRA's inflated estimate of the cost of banning chlorine overnight begins to seem like a bargain.

If externalities could be fully represented in the cost of products, could the market solve our environmental problems after all? The answer is still no. First, all of a product's environmental costs can never be included in its price. In chapters 3 and 4, I showed that ecology and epidemiology are unable to untangle the roles that each individual chemical plays in large-scale health and ecological damage. There is thus no way to apportion the costs of environmental injury to specific products and rationally include externalities in market prices. Further, there is a fundamental contradiction between the market's basic mechanism and what ecosystems actually require. The market's function is to use prices to establish an optimum level of a given activity. High prices will reduce the production and use of a toxic technology, but because the optimal level reaches zero only at an infinite price, it will never eliminate it absolutely. On the other hand, we have seen that the ecosystem's capacity to assimilate persistent toxic substances is exactly zero. For this reason, improved economic decision making may encourage the use of cleaner technologies, but it will never satisfy the ecological imperative of Zero Discharge.

There is one final problem with the idea of using the market or other forms of economic decision making, like cost-benefit analysis, as the sole basis for environmental policy: the effects of persistent toxic pollution, like other forms of ecological damage, can never be adequately expressed in hard currency. What price tag should we put on a mother lost to breast cancer? What is the cost of infertility to a couple? When half the dolphins along the Atlantic coast die in mass mortalities, how much money have we lost? What is the cost of worldwide declines in the cognitive ability of children? The attempt to reduce human and environmental injury to purely financial terms is truly obscene. When adequate technologies are available to prevent severe and widespread damage to human health, they must be used, whether or not they cost more. As abstract as Clean Production may sound in principle, as technical as it may be in its applications, it is a moral imperative. In criminal law, we never accept profit or convenience as justification for murder. Money does not become a legitimate excuse simply because the damage and suffering inflicted on human beings and other species is mediated through the workplace, the water, or the food supply.

Rethinking Progress

The Competitive Enterprise Institute's (CEI) report on chlorine concludes, "The war over chlorine . . . is between those who believe increased wealth and technological progress are the best means of improving health and safety and those who do not."[143] Exaggerated as it is, CEI's statement does point directly to a core issue in the debate over chlorine: what technological progress is, how it can be achieved, and who should make decisions about doing so.

CEI adopts a definition of progress with deep roots in the culture of industrial society. In this view, progress is the process by which one technology replaces another; when one technology makes another obsolete, we have moved forward. As René Dubos wrote, progress thus becomes an end in itself, an interminable march of technological "advances" without specific direction. The degree of progress is measured not by motion toward some destination but by the sheer rate of change.[144]

In my view, progress is a meaningless concept unless it implies steps toward some specific end. A lost man wandering randomly through the desert does not make progress, though he may cover many miles. Technology is never a goal in itself but is a means to an end, a way of fulfilling human needs and desires. We achieve true technological progress when we invent techniques that fulfill human needs—health, nutrition, housing, love, beauty, self-realization, and so on—without jeopardizing the ability of future generations to do the same. Enriching ourselves at the expense of others, our grandchildren, or the natural world on which our lives and happiness depend can hardly be considered progress.

By this definition, a chlorine phaseout clearly represents progress. Sunsetting chlorine is a program of technological modernization, not unthinking opposition to technology as a whole. Replacing polluting technologies with efficient, ecologically sound production techniques—whether they are high-tech processes like TCF pulp bleaching and supercritical carbon dioxide extraction or, like wet cleaning and organic agriculture, they substitute expertise and labor for chemicals—is no Luddite position. The Ecological Paradigm's program of Clean Production embodies not opposition to all technological change but a commitment to technical advancement for a worthy goal.

Each definition of progress implies a political program. If progress is measured by the rate of technological change, then the most progress occurs when those who develop and implement technologies are allowed to do so unfettered. Progress undefined by direction requires a laissez-faire system for the agents of progress, which means private industry. The Risk Paradigm, with its refusal to interfere in decisions about the technologies of production, is precisely such a system. Thus, in CEI's view, the precautionary principle, which would force industry to justify on environmental grounds the technologies it chooses to develop and utilize, represents an unprecedented threat to progress.

Defining progress as a project with specific social goals has nearly opposite political implications. In an era of immense corporations and industrial technologies that can alter ecosystems on global scales, decisions about what processes to use affect the lives of millions of people. If the goal of technology is to improve the social conditions for human life, then it is the job of society to ensure that it does so. Society must not abdicate technological decisions to the market or to a small group of individuals or institutions who have a very different goal: maximizing profits. We must begin to steer the ship of technology, not let it blow across the economic seas unguided. The Ecological Paradigm gives society the tools to direct technological decisions, ensuring progress toward the socially defined goal of production systems that are truly sustainable in the long run.

The CEI concluded its discussion of precaution as the enemy of progress this way: "Environmentalists see rapid technological advancement as a threat to human dignity." In fact, a chlorine sunset is designed to be exactly what CEI says environmentalists oppose: a program of rapid technological advancement, albeit one guided by democratic mechanisms and environmental criteria. Technological change is a threat only when unbound from its proper goal of maximizing human dignity and well-being, as it has been in the current economic and policy paradigm. With appropriate structures in place to ensure that engineering and economic decisions serve this end, technological advancement becomes a means for achieving progress of a truly human kind.

11

Good Science, Good Politics

[We must establish] a sort of scientific priesthood throughout the kingdom, whose high duties would have reference to the health and well-being of the nation in its broadest sense, and whose emoluments would be made commensurate with the importance and variety of their functions.

—Sir Francis Galton, nineteenth-century eugenicist and anthropologist[1]

It has become commonplace—a shibboleth, even—for the chemical industry to dismiss calls for restrictions on chlorine chemistry with the remonstrance that environmental policy must be based on "good science" or "sound science." This sounds obvious and inoffensive, if it means we should use the best information and rigorous logic as we make decisions about the environment. Who wants policy to be based on bad science?

It is not clear, however, that well-informed and carefully reasoned decision making is what the advocates of "sound science" really mean. Consider a few examples that show how the phrase has been used in the chlorine debate. Dow Chemical's brochure on chlorine chemistry, for instance, asserts that "a sound science approach to policy development" must be based on "cost/benefit analysis" and "individual consideration of substances."[2] CanTox's report on organochlorine contamination for the Chlorine Chemistry Council identifies chemical-by-chemical assessment of toxicological thresholds as a basic "scientific principle."[3] David Meeker, executive vice president of Edward Howard and Company, a public relations consulting firm hired by the Vinyl Institute, told a plastics industry conference, "Safety, health, and environmental effects must be analyzed using sound science and then be weighted for their significance in a risk/benefit context."[4] And Jack Mongoven of MBD, the Chlorine

Chemistry Council's public affairs consultant, recommended that the industry "accelerate the program to bring about an agreed-upon risk assessment policy and the deployment of vehicles of sound science."[5]

Clearly "sound science" does not refer to a method of asking questions but is a stand-in for a specific answer: the use of chemical-by-chemical discharge limits derived by quantitative risk assessment. In just two words, the phrase implies that any other approach—particularly that embodied in the Ecological Paradigm—is based on bad science, emotion, ideology, fear, or some other suspect motive. In a later article, Mongoven took his argument even further:

> A massive attack is underway in the United States against products made through chlorine chemistry; and one of its most troubling aspects is its rejection of accepted scientific method. Activists . . . plan to replace established scientific method with the Precautionary Principle. . . . American industry should take the precautionary principle seriously, and develop a strategy to deal with it. Whatever that strategy is, it must include development and public acceptance of a rational risk-assessment policy.[6]

Is a precaution-based policy really bad science? Is risk assessment the only way to make environmental decisions on a sound scientific basis? What are the implications of deploying the prestigious name of science to advocate a specific method of making public policy? With this chapter, I conclude my argument by discussing the scientific quality of the two paradigms for environmental policy, the political use and misuse of scientific authority, and science's proper role in decisions about technology and the environment in a democratic society.

Good Science?

Sound science's advocates seldom explain clearly why chemical-by-chemical risk assessment is its only possible manifestation, while a precautionary or ecological approach is somehow unscientific or even antiscientific. Defining science is a difficult and surprisingly slippery exercise. On close examination, the various criteria that scientists and philosophers have put forth to demarcate science's boundaries over the years turn out to be problematic and inconsistently applied, excluding some work most people would regard as science or excluding others that are considered non-science or pseudoscience.[7] Nevertheless, there are certain qualities, like a deference to empirical data and a skeptical attitude toward received wisdom and supernatural explanations, that most people would agree are characteristic of good scientific thinking. If we grant that science is a unique way of knowing—despite the difficulty defining it—then there are two ways to justify the charge that the Risk Paradigm is scientific and the Ecological Paradigm unscientific. The first is that the risk-based approach might be more consistent with science's most up-to-date information and theories about nature. The second is that there might be something about the methods or attitude of risk assessment that is more scientific than an ecological policy. Neither turns out to be true.

Consider the first possibility: that the risk-based approach to regulation is superior because it is more fully informed by science's current understanding of the world. As we have seen throughout this book, however, the Risk Paradigm is based on a set of assumptions that are refuted by what science has established about ecosystems, organisms, and chemicals. This framework assumes, contrary to science's actual findings, that ecosystems have assimilative capacities for substances that persist or bioaccumulate, that toxicity thresholds exist and are discoverable for all toxic effects, that the hazard posed by the total burden of chemical exposure is the sum of the effects of each individual compound in isolation, that risk assessments focused on human effects predict the potential impacts on all the diverse species and ecosystems in nature, and that pollution control technologies are an effective way of preventing chemicals from getting into the environment. The Risk Paradigm steadfastly ignores what we have learned about these assumptions from experience, observation, and experiment: that they are false. In contrast, Ecological Paradigm is designed specifically to take account of these findings. By this criterion, risk assessment is clearly not sound science, but the Ecological Paradigm is.

The second possibility is that there is something about the epistemological structure or attitude of the Risk Paradigm that is more scientific than the approach I have described. Dow's brochure ventures this position, arguing that "some parties have also suggested that every chemical containing chlorine poses the same risk to the environment. . . . This assumption is not justified by science." The Society of Toxicologists makes a similar argument: "All chlorine-containing compounds are not equally hazardous.

Therefore, SOT takes the position that a broad-based ban of the class of chemicals containing chlorine or any other element for that matter, would be both irresponsible and unscientific."[8] I have explicitly *not* argued that every organochlorine poses the same environmental hazard, because we know that they do not; the basis of my argument is that chlorination, for well-understood reasons, virtually always amplifies one or more of the potentially dangerous qualities of organic chemicals, including toxicity, persistence, and bioaccumulation. As a result, virtually all organochlorines cause one or more severe toxic effects, and all chlorine-based products appear to result in the formation of extremely hazardous by-products at some point in their life cycle. Every organochlorine need not be the same to justify the conclusion that the practice of chlorine chemistry represents an environmentally unacceptable set of activities. I have argued that *for the purposes of environmental policy,* organochlorines and the processes that produce them should be regarded as hazardous unless demonstrated not to possess a specific set of undesirable properties.

Although this position is different from the straw-man position that Dow and the SOT shoot down, it does make a generalization about organochlorines, raising the question whether a presumption about a chemical class might be intrinsically unscientific. Scientists usually seek to be precise in their language and logic, so it is tempting to think that specificity is some kind of scientific virtue, requiring each substance to be assessed individually and forbidding generalizations about chemical classes. In fact, the value science places on specificity is no higher than the value it gives to generality. The primary goal of science is to explain diverse observations in terms of unifying causes, patterns, or mechanisms. As Karl Popper has argued, the scientific value of a theory is directly related to its boldness and inclusivity; a general or universal theory does much more to advance scientific knowledge than does a hypothesis that explains only local phenomena.[9] A theory of unrelated singular effects, in fact, is not a scientific theory at all. Science seeks the maximum degree of generality that can be supported by the evidence. It is no more unscientific to presume for policy purposes that organochlorines are undesirable and should be replaced by safer substitutes whenever possible than it is to immunize all children against infectious diseases—whether or not they would have contracted the disease without the vaccine—in order to prevent epidemics.

Skepticism is a long-held virtue in science, so perhaps it is unscientific to make a generalization about the class of organochlorines until all have been tested and we know the generalization is true. A sunsetting approach to the class of organochlorines, however, does not conclude that every organochlorine is dangerous but instead uses the *working hypothesis* that any chlorine-based product or process is dangerous unless it has been demonstrated not to meet a certain set of criteria. The Risk and Ecological Paradigms make generalizations of equal breadth: the risk-based framework assumes that individual organochlorines are safe until they are demonstrated hazardous, an exact mirror image of the working hypothesis on which the ecological approach relies. Neither paradigm is any more skeptical than the other; the two are equally scientific *in structure.* Once we begin to consider the data, however, the scale tips against the Risk Paradigm, because virtually all organochlorines that have been tested have turned out to be hazardous. To justify the presumption that the other members of the class of organochlorines are safe, we must ignore all the evidence we have, a gesture that is unscientific in almost anyone's book.

Scientific skepticism relates to the chlorine debate in a more general way, as well. If skepticism is a stubborn reluctance to accept unsupported beliefs and assertions—the refusal to pretend that we know more than we do—then the Risk Paradigm's pretense to establish safe exposure levels and calculate the number of people who will develop cancer due to a single pollutant source is anything but skeptical. We have seen throughout this book that the prediction of chemical impacts is undermined by uncertainty and ignorance: data gaps vastly overshadow the available information, organisms and ecosystems remain poorly understood, and prediction is essentially impossible in complex, nonlinear systems. Nevertheless, the Risk Paradigm asserts hard and certain judgments in its risk estimates and discharge limits: "facility x will pose a cancer risk of 1.7 per million to the most exposed individual," or "exposure to less than 1 microgram per kilogram of body weight per day of pesticide y will cause no adverse reproductive or developmental effects." I can think of no less skeptical a position than the confident assertion of safe exposure levels based on sparse data and limited understanding. In contrast, the Ecological Paradigm is designed explicitly to take account of the limits of our knowledge; it avoids risky bets made on scientific predictions, erring on the side of caution by minimizing as much as possible the environmental impacts of human technologies.

The final and perhaps the most important criterion of the "scientificness" of a theory—the one that scientists most frequently use to define science and the scientific method—is its testability. This definition is derived from the philosophy of Karl Popper, who showed that theories cannot be proved, only refuted, by empirical testing. In Popper's view, science is a trial-and-error process not of proving or verifying "truths" but of proposing, refuting, and refining hypotheses. A theory is scientific to the extent that it can be refuted by testing. Thus, the statement, "The lights I saw in the sky that disappeared without a trace were ships from outer space" is not scientific, because no amount of empirical observation could ever refute it.

Can the predictions made by the Risk Paradigm be tested? Suppose a risk assessment for an incinerator predicts that the local community will experience exposures that pose cancer risks of one per million and are well below the threshold for other chronic effects. Suppose further that the prediction turns out to be grossly wrong, with cancer risks 10,000 times higher than expected, a moderate increase after a decade or more in the incidence of low birthweight and developmental impairment, and reduced fertility in several local wildlife populations. Because epidemiology and ecology are poorly equipped to detect subtle, long-term effects, these impacts will probably never be detected using standard criteria of statistical significance. In fact, they will probably never be studied at all, since most places are never the subject of epidemiological or ecological studies. If the effects are somehow discovered, they will never be definitively linked to the emissions from the incinerator, because the contribution of that facility can never be clearly distinguished from that of other contributing factors, including chemicals from other kinds of sources, other incinerators in the region, and other incinerators halfway across the world. Testing a risk assessment's prediction in the real world would require a mode of environmental monitoring and causal analysis that is far beyond the reach of any foreseeable science. Even a weather forecast turns out to have been observably right or wrong, allowing meteorologists to refine their methods of prediction with each passing day; whether a risk assessment was perfectly accurate or grossly off-base we will never know.

Science is a process of trial and error, but the Risk Paradigm includes no realistic provision for discovering error and making necessary adjustments. In contrast, the Ecological Paradigm provides a way to revise judgments and policies based on feedback from nature. Recall that the precautionary principle was formulated in negotiations among European governments during the 1980s precisely to address growing evidence that pollution-induced ecological damage was taking place in the North Sea—evidence that was inadmissible under the Risk Paradigm's requirement for strict proof. By removing the requirement to prove causal links before taking preventive action, the ecological approach allows society to respond to evidence of environmental degradation and revise existing environmental policies to prevent further injury.

Political Science

If chemical-by-chemical risk assessment is inconsistent with science's current understanding of nature and with basic criteria of scientific method, how can it be the only manifestation of "sound science"? The answer is that "sound science" is a political term rather than an empirical or epistemological one. It does not express the scientificness of a method or a theory about nature but instead reflects the role of scientific institutions and scientists in the process of making decisions about public policy. "Sound science" means that scientists appear to be given the lead in determining what chemicals, exposures, and technologies are permissible in society.[10] In two words, the phrase also manages to deploy all the political and cultural authority of science on one side of a contested debate about social policy.[11]

When technologies are approved solely based on whether they will pose risks that exceed a standard of acceptability, then the person who calculates the magnitude of the risk becomes the only one with meaningful input into the decision. The risk assessor's product determines whether discharges from a technology will be permitted as proposed or whether additional pollution control devices are required to bring releases to the acceptable level. Indeed, because control devices can always bring local releases below acceptable discharge standards, risk assessment need never say no to a technology or facility. To my knowledge, no chemical has ever been banned, no facility ever denied a permit based on a risk assessment. Risk assessment is a method designed to determine *how* to permit pollution sources, not whether they should be permitted. The action that will be taken—outright approval or requiring one or another type of device—depends primarily on the result of the risk assessor's calculations. The Risk Paradigm is "scientific" because scientists carry out the only aspect of the policy process in which there is any degree of freedom.

Kip Howlett, the Chlorine Chemistry Council's managing director, is explicit that sound science means a dominant role for scientists. In 1995 he asked members of the American Chemical Society to support the CCC in its efforts to avoid environmental restrictions:

> We need scientists, such as yourselves, to speak out as advocates for chlorine chemistry. . . . I recently saw a bumper sticker with the saying, Environment is one-tenth science and nine-tenths politics. And I thought—that is exactly the problem; that is exactly the opposite of how we should approach our common goal of environmental protection. With your help, I hope we will soon see the day when public policy decisions about our environment are nine-tenths science and only one-tenth politics.[12]

It is perfectly understandable that many scientists would be attracted to an appeal of this type and respond by advocating a regulatory framework that grants scientists a privileged political position. Scientists need not be power hungry to seek an influential role in public life; they may just as well be motivated by the desire to use their expertise to serve society. But it remains in the institutional interest of scientists to favor a method of making policy that maximizes their influence. In its resolution for continued chemical-by-chemical risk assessment of organochlorines, the Society of Toxicologists (SOT) was remarkably clear that its advocacy was based in part on its desire to maintain the authority and central role of its members in environmental decisions:

> The toxicologist is specially trained to examine the nature of the adverse effects of chemical and physical agents on living organisms and the environment. . . . A responsibility of the toxicologist is to define the potential toxic effects that chemicals can induce and to determine the conditions of use that minimize or prevent these effects so that the beneficial attributes of chemicals can be realized safely. . . . The determination of unacceptability should be based on scientific data that document the adverse effects of exposure and a weighing of the risks vs. benefits of using the chemical in question. Indeed, based upon sound principles of toxicology, rational and effective assessments of the potential toxicity of chemicals, including chlori-

nated chemicals, are currently taking place and rigid standards exist for registration of new products to which people will be exposed.[13]

The SOT's suggestion that science should be the sole basis for decisions about health and the environment is misguided and misleading. Political decisions always incorporate not only scientific inputs but also ethics, values, opinions, and the interplay of conflicting interests and perspectives. Only in authoritarian societies do decision makers claim that public policy can or should be derived entirely from science; in those societies, these claims serve primarily to mystify and conceal the political motives and mechanisms by which decisions are actually made. A democratic society must subject these issues to explicit, popular consideration and resolution.[14]

The risk-based policy framework is extraordinarily effective at hiding the political judgments that lie behind its use. By reducing a decision to a single question—does the calculated risk exceed the acceptable risk standard?—risk assessment takes as settled or unworthy of attention a horde of political and moral questions. How much risk really is acceptable? Who gets the benefits and who suffers the risk of the practice in question? Do we really need this product at all? Are there less hazardous ways of satisfying human needs? Who should bear the burden of proof in decisions about technologies? Why does society assume that firms have a right to produce, use, and discharge toxic chemicals at all?

"Scientism" means the deployment of scientific authority to justify political or cultural positions that are not solely determined by science.[15] By scientizing environmental regulation, risk assessment protects those whose interests might be threatened by a broader political discussion of the basis for environmental decisions. And this is why the Ecological Paradigm must be portrayed as unscientific. The principles of precaution and clean production open the decision process to ethical and political questions, placing at center stage issues of the appropriate standard and burden of proof and the availability of alternatives. Critics call this approach bad science because it threatens a mode of policymaking that deploys the authority of science for specific political and economic ends.

These considerations suggest deep and irreparable political flaws—to go with the scientific problems we have already discussed—in the Risk Paradigm. By shielding ethical and political issues of extraordinary consequence from public debate, the risk-based approach undermines democratic rights and processes. In the mid-twentieth century, the German philosopher and social critic Jurgen Habermas warned that society was moving toward a state of technocracy, in which scientists would play the central policymaking role and ordinary citizens would be increasingly excluded from the technical debates that determine political decisions.[16] At this point in history, Habermas appears to be wrong on one count: although scientists and scientific institutions provide the bulk of the information on which decisions are made, they have not replaced business executives, government legislators, and bureaucrats as the primary decision makers in society. Corporate managers still decide what production technologies they will use and where they will use them, and government officials write the laws and policies that determine the methods by which specific regulatory judgments will be made. Scientists advise and inform politicians and CEOs, and their evaluations are central to determining the outcome of individual decisions, but they have not supplanted the political and economic elite at the top of society.[17]

The dominance of risk assessment, however, suggests that environmental politics tend in other important ways toward technocracy. Quantitative risk assessment has been used since its debut to blunt democratic input. In 1983, EPA administrator William Ruckleshaus first called for risk assessment to become the centerpiece of regulatory policy in the United States. His appeal came in what he called a "troubled period for pollution control": Ruckleshaus had just taken over the agency from administrator Anne Gorsuch, who, along with several of her deputies, was forced to resign when a congressional investigation revealed that EPA had deliberately not enforced the Superfund hazardous waste clean-up law and had allowed the agency's scientific assessments of the hazards and sources of dioxin to be determined by the chemical industry's economic interests and the agency's own public relations concerns.[18] Meanwhile, the public, alarmed by Love Canal, was calling insistently for increased restrictions on chemical use and disposal. In a speech to the National Academy of Sciences, Ruckleshaus cautioned that the political environment of the time threatened to set up "a grim and unnecessary choice between the fruits of advanced technology and the blessings of democracy."[19] Against this background, he sought to revive the credibility of the Risk Paradigm by appealing to the authority of science.

Ruckleshaus called for a "government-wide process for assessing and managing environmental risks":

> Many communities are gripped by something approaching panic, and the public discussion is dominated by personalities rather than substance. . . . I believe that part of the solution to our distress lies with the idea that disciplined minds can grapple with ignorance and sometimes win: the idea of science. . . . The polls show that scientists have more credibility than lawyers, businessmen or politicians, and I am all three of those. I need the help of scientists.

Risk assessment promised to scientize environmental policy in a way that would quiet panicked communities and restore the authority and credibility of EPA. Ruckleshaus did not get the congressional legislation that he sought, but quantitative risk assessment gradually became the centerpiece of the regulatory decision process at EPA and numerous other federal agencies.[20]

Eight years later, EPA was going through another troubled period. In 1990 popular environmental pressure in the United States reached a new high with the twentieth anniversary of Earth Day and growing activism in communities faced with hazardous and municipal waste incinerators and landfills, pulp and paper mills, and other pollution sources. In a speech at the National Press Club,[21] EPA administrator William Reilly again called for a "revolution" in agency policy that would make risk assessment not just a key tool in pollution permitting but the basis on which the agency would set all its priorities. In Reilly's proposal, quantitative risk assessment would provide firm numerical answers that would determine which problems were worthy of the agency's attention. In justifying this expansion of risk-based decision making, Reilly recounted how one senator had cautioned him,

> "Above all, don't allow your agency to become transported by middle class enthusiasms." What he meant was—respect sound science. Don't be swayed by the passions of the moment. . . . As we gear up to deal with the environmental problems of the 1990s and beyond, I think the time has come to start taking aim before we open fire. In short, we have to find a better way of setting environmental priorities. Risk is a common metric that lets us distinguish the environmental heart attacks and broken bones from indigestion or bruises. . . . Sound science is our most reliable compass in a turbulent sea of siren songs.

This is the classic language of technocracy: contempt for democratic input and an appeal to science as the sole and proper basis for making social decisions (even manifestly political ones about issues like government priorities). By marginalizing discussion of the central political issues that inform environmental policy, the Risk Paradigm reduces the regulatory process to a debate about toxicological extrapolations, cancer potencies, quantitative exposure predictions, and the financial costs of pollution control devices. The dominant language of environmental governance is now the obscure jargon of experts. The highly technical and apparently objective process of risk assessment denies most citizens the means, will, and confidence to participate in environmental decisions. We thus have all the symptoms of technocracy without its expected distribution of power.

This is clearly an intolerable situation for anyone who believes in democracy. Choices about technology and the environment have a profound effect on every person's health, work, and quality of life. Deciding what hazards are acceptable and what production techniques should be used to meet society's needs are matters that should not be left solely in the hands of scientists, bureaucrats, or corporate managers. These issues are eminently appropriate for citizens in a democratic society to discuss and settle for themselves. With this in mind, the Ecological Paradigm moves decisions about industrial technology from the hands-off zone to their proper sphere of democratic politics, making explicit the social, ethical, political, and technological issues that are taken for granted in the "sound science" framework.

Knowledge and Democracy

If an ecological policy would reduce the undemocratic deployment of scientific authority, does it banish science altogether from environmental governance?[22] The answer is a resolute no. In the Ecological Paradigm, decisions will have to be made, and the best scientific information should continue to play a critical role in informing them. Society will still have to evaluate prima facie cases about chemical classes proposed for sunsetting, grant exceptions to individual products and processes, evaluate alternative technologies, and monitor the quality of the environment to detect new problems and evaluate the effectiveness of the policies it implements. The Ecological Paradigm elevates the importance of political and ethical issues, but that does not mean it reduces the importance of scientific knowledge, except in a relative sense.

How is the best science to be acquired? The Risk Paradigm is based on a traditional model of the relationship of science to policy: scientists first determine what is true and then submit this judgment to policymakers, who can then act on it. This idea is enshrined in a distinction that Ruckleshaus made, now widely accepted, between risk assessment and risk management. "It is important to distinguish these two essential functions," Ruckleshaus explained. "Scientists assess a risk to find out what the problems are. The process of deciding what to do about the problem is risk management. . . . Nothing will erode the public's confidence faster than the suspicion that policy considerations have been allowed to influence the assessment of risk."[23] Reilly's insistence that "sound science is our most reliable compass" in a sea of "middle-class enthusiasms"[24] is a more poetic way of asserting that science is independent, objective, and apolitical.

As Ruckleshaus pointed out, this distinction is essential to maintaining the authority of science over environmental decisions. Some opponents of a chlorine sunset policy have argued that the precautionary principle violates the apolitical nature of science, because it filters scientific information through a politically- and ethically-based screen. Albert Fischli, president of the International Union of Pure and Applied Chemistry, complained in 1996 that the debate over chlorine had reversed the proper order of things, which begins with the authority of scientists to establish the truth, from which policy choices then inevitably flow:

> This [chlorine] issue has gone too early beyond the borders of the scientific community, before facts have been proven and evaluated. As a result, it has been treated in inappropriate and emotional ways. . . . It is the responsibility of the scientific community to develop this science base, of the media to help them to inform the public in an understandable and unbiased way, of the public authority to take the relevant decisions on the basis of sound science and not on emotional reactions.[25]

At first glance, the distinction between matters of science and those of policy seems quite reasonable. Risk assessment does appear to be technical and objective. How can a process that gives a result like "a lifetime cancer risk of 1.7 per million" be political and subjective? In fact, it is the very hardness of the result that disguises the political qualities of the science itself. The first source of subjectivity in risk assessment lies in the fact that it is "underdetermined" by data. A quantitative risk assessment involves at least 20 steps in which the magnitude of discharges and their persistence, environmental transformation, transport in the environment, bioaccumulation, intake by the organisms, distribution within the body, and toxicity are all predicted.[26] None of these variables is precisely known; all must be estimated, sometimes by choosing a value from a range of empirical data, sometimes by pure assumption, as in the tenfold "uncertainty factor" used to extrapolate from rodent toxicity to humans. At each stage of the process, the scientist must make a choice that is not precisely dictated by data. The entire process of risk assessment is thus highly subjective, reflecting the personal views, preferences, or interests of the person who makes these choices. As Roy Albert, the former head of EPA's Carcinogen Assessment Group, once wrote, "The larger the range of uncertainty the more room for institutional or purposeful bias. Individuals who have very different institutional loyalties and pressures can produce very different risk assessments for the same materials, where large uncertainties exist."[27] Whether the scientist makes these choices honestly or with the intent to derive a higher or lower risk estimate, the process is neither objective nor determined by knowledge about nature.

Out of this persistently subjective process comes a precise number, which, if other reasonable choices were made, could have been so different as to be unrecognizable. The value chosen at each step is multiplied by those in the next steps, so error and uncertainty propagate, reaching extraordinary magnitude by the end of the process. When a group of scientists at EPA, for example, used four accepted models to calculate the cancer risk posed by trichloroethylene in drinking water, they found that the resulting risk estimates varied by a factor of 100 million. The authors concluded, "These estimates provide a range of uncertainty equivalent to not knowing whether one has enough money to buy a cup of coffee or pay off the national debt."[28]

The second source of subjectivity in the scientific assessment process is that facts never speak for themselves. Scientific evidence can never be synthesized and presented for political use except through a political lens. Data in themselves do not provide a basis for action; a purely "scientific" conclusion about the magnitude of a hazard always involves filtering information through a set of criteria based on political and ethical judgments.[29]

The most obvious example is the weight of evidence required before some type of decision may be recommended.[30] In American criminal law, a person is considered innocent until proven guilty beyond a reasonable doubt and to a moral certainty as determined by a jury of peers. Society has decided to place an extremely high value on preventing the incarceration or execution by the state of innocent persons, at the explicit cost of letting some guilty individuals go free. In contrast, most types of civil litigation between private parties require only a preponderance of evidence; society has decided it has no stake in which party wins, so neither side should bear a greater burden than the other. As long as the evidence admits of any uncertainty, the same facts may yield different decisions in the two different legal contexts.

In medical practice, ethical and goal-oriented judgments also inform the evaluation of data. Consider a baby boy born to a mother who has a mild fever, but he has no symptoms of infection. Most hospital pediatricians will have the baby undergo intravenous antibiotic treatment as a precautionary measure. This course of action is appropriate not because the physician *knows* that the child has an infection or even believes that he is *probably* infected; the doctor gives antibiotics because an infection during the first days after birth can be fatal. If the pediatrician guesses that the baby is not infected but is wrong, the consequences are much greater than those of a false positive guess, in which the cost is the monetary expense of unnecessary treatment and some mild discomfort. The doctor's sworn responsibility is not to give the correct diagnosis but to choose the course of action that will minimize the likelihood of harm to the baby.

When making a decision about any kind of action, the goal of weighing evidence is not to determine which description of the world is correct or even probably correct. The object is to draw conclusions that inform a course of action that will minimize the likelihood of significant harm. What constitutes a significant harm is always an ethical and political decision. Which do we value more highly, protection of citizens from false imprisonment or the incarceration of all criminals? Disease prevention or saving money and antibiotic supplies? Protecting health and the environment or maximizing the liberty of industry to choose its technologies?

The content of scientific knowledge is thus shaped by a political process that has already settled the proper ends and methods for both science and policy. This explains why the effort to establish scientific truth breaks down without some kind of natural or enforced political consensus.[31] Thus, one group of scientists can look at the information on breast cancer and the environment and conclude that there is no evidence that pollution contributes to the disease, while another group—using a different standard of proof—agrees that there is a sound basis for concern and preventive action.[32] The political context for a scientific evaluation determines what information is admissible and how conclusions will be drawn from it.

The risk assessment framework makes hidden assumptions not only about the appropriate standard of proof but also about which kinds of health damage are worth assessing and avoiding. When considering health damage, are we worried only about human cancer and liver damage, or also about subtle damage to fertility and intelligence? Do we care about effects on people, or also on other mammals, on frogs, on fish, on mollusks, on insects, on trees? Until these issues are resolved, a scientific evaluation of harm or safety is impossible. The Risk Paradigm offers very limited answers to these questions. In the context of a risk assessment, the only relevant information pertains to the chemical-specific quantification of cancer risk and the evaluation of exposures from a single pollutant source in relation to the threshold of systemic toxicity. Effects that cannot be described probabilistically or by reference to a threshold cannot be treated in a risk framework and have thus been ignored by all risk assessments to date. If we reject the political judgment that acceptable cancer risk and threshold doses are the only appropriate benchmarks for environmental decisions, however, then risk assessment immediately becomes mired in issues that question the method's competence to provide adequate scientific information. These include the possibility of no-threshold effects other than cancer, the synergistic action of multiple chemical exposures, and the health impacts of exposure to the cumulative pollution burden caused by hundreds of chemicals released by thousands of facilities, each assessed in isolation from the others. Once these issues enter the discussion, risk assessment's precise numerical answer no longer appears to be a firmly established fact. Risk assessment seems objective and clear-cut only because the scientific discussion, like the political discussion, has been narrowed by a prior political framework.

In his speech to the National Academy of Sciences, Ruckleshaus was explicit that one of the purposes of the risk system he advocated was to restrict the input of certain types of scientific information into policy. At the time of his speech, researchers had recently established that there was no threshold for some health effects, including cancer, suggesting the possibility that widespread exposure to low levels of toxic chemicals might pose a public health threat. Ruckleshaus described the problem:

> We must now deal with a class of pollutants for which it is difficult, if not impossible, to establish a safe level. These pollutants interfere with genetic processes and are associated with the diseases we fear most: cancer and reproductive disorders, including birth defects. The scientific consensus is that any exposure, however small . . . , embodies some risk of an effect. Since these substances are widespread in the environment, and since we can detect them down to low levels, we must assume that life now takes place in a minefield of risks from hundreds, perhaps thousands of substances. We can no longer tell the public that they have an adequate margin of safety.

At the time, U.S. environmental statutes required EPA to provide a margin of safety for public health, so this information put powerful pressure on Ruckleshaus and his agency to restrict a large number of chemicals:

> One thing we clearly need to do is ensure that our laws reflect these scientific realities. The administrator of EPA should not be forced to represent that a margin of safety exists for a specific substance at a specific level of exposure where none can be scientifically established. This is particularly true where the inability to so represent forces the cessation of all use of a substance without any further evaluation.

In the climate of the new Reagan administration, banning chemicals was not a palatable course of action, whatever science and the law said. The agency's credibility, always derived in part from the authority of the science it relied on, was thus threatened by scientists' own inability to provide definitive safe exposure levels that would dictate decisions that were also politically viable. Ruckleshaus's risk framework was designed to impose a new goal for environmental policy, replacing margins of safety with acceptable risk levels, limiting the assessment process to a narrow set of potential harms. Ever since, evidence has continued to mount that low-dose exposures have the potential to cause effects without a definable threshold, but these impacts remain inadmissible in the risk assessment framework.

If prior political judgments determine the information that may enter the scientific assessment process and the conclusions that emerge from it, then scientific knowledge is not objective or apolitical. I am not suggesting that the individual scientists who participate in the assessment process compromise their objectivity, manipulating or stretching evidence to serve some desired political purpose (although, as we have seen, that happens too). The dynamic I am describing operates primarily at the level of institutions, not individuals. The key issue is the structure of inquiry, which directs a scientist conducting a risk assessment or serving on an expert evaluation committee to answer a specific question in a specific way. Even the most scrupulously disinterested scientist, if he provides the information that the policy process deems relevant, puts forth a conclusion that has been limited, filtered, and informed by a prior political settlement.

Throughout this book, I have presented a number of examples that show how scientific knowledge is never a purely objective picture of nature. Sometimes the subjectivity comes from pure dishonesty, the product of manipulation to satisfy some political end, as appeared to be the case for some of the epidemiological studies of dioxin. In other cases, it is the result of institutional commitments to long-held assumptions that are not supported by the data, as when toxicologists claim that thresholds are not artifacts but are always real properties of organisms. In still other cases, subjectivity arises from the culture of the time—the limited concepts, categories, and definitions available to scientists as they attempt to put together a coherent picture of the world. For example, the possibility that chemicals might cause large-scale, multigenerational effects on reproduction and development never occurred to most epidemiologists and wildlife biologists until the past decade or so, because people thought of ecosystems as local and resilient rather than global and subject to subtle, long-term damage. When scientists in these fields concluded that low levels of chemicals were not causing obvious damage to ecosystems, it was because the limited cultural vision of the time kept them from ever looking for a different kind of impact.

By arguing that science is never purely objective, I am not denying that most scientists are honestly seeking to understand nature on its own terms. Nor, by showing that scientific knowledge is in part institutionally and culturally determined, am I arguing that science is a purely social construct. There is no doubt that well-produced scientific knowledge contains a large quantity of empirical content. One theory may be far better than another at describing the patterns and processes of nature, and science's hypothesis-testing approach remains a powerful tool for making and testing theories about the world. My point is that this knowledge is never purely objective; in addition to its empirical content, it bears the marks of the interests, social positions, cultural context, and political environment of the people who created it.[33] In this way, the science that supports the Risk Paradigm—the same science that assures us that low-level contamination does not threaten public health—is intrinsically linked to that framework's political values, assumptions, and commitments.

Good Politics

I have argued that decisions about technologies and health hazards are properly social matters that should be made in a fully democratic process. The Ecological Paradigm makes this vision real in two ways. First, it opens to public consideration the political and ethical issues that currently serve as the unstated foundation of environmental governance. Second, the sunsetting policy I have described places these broader issues under the direction not of appointed or career bureaucrats but of a transition planning body that includes representatives of all social groups that have a stake in the outcome. Not government officials, not scientists, and not industry alone, but also workers, farmers, health advocates, community members, and so on, would determine how chlorine should be phased out: the time lines, the exceptions, the production technologies to be implemented as alternatives, the recipients of funds to minimize and address economic dislocation. Of course, even a representative body can come to operate in undemocratic ways, so the transition planning body should be governed by rules of democratic process, including requirements to take broad public input, operate in a transparent manner, and be held accountable at all times to the general public and the mandate it has been given.

Ultimately the Ecological Paradigm should be applied not just to chlorine chemistry but should become the general method for making environmental decisions. In this larger context, a democratic approach like the transition planning process could be extended to evaluate other classes of substances and technologies as candidates for phaseout and to implement sunsetting procedures for the ones that are deemed ecologically undesirable. Potential foci for action in the Ecological Paradigm include technologies that produce other persistent pollutants (including other organohalogens, long-lived radioactive materials, and toxic metals), those that radically accelerate the depletion of natural resources (such as factory trawlers in the fishing industry and clear-cut forestry methods), and those that disperse genetically engineered organisms into the environment.

I am not arguing that all technologies must be subject to a protracted environmental assessment process. There is no reason, for instance, that a software engineer's decision to use one or another programming language needs to be evaluated on environmental grounds. Rather, the Ecological Paradigm would allow concerned citizens, including scientists, to bring a claim before a democratically constituted body, based on a prima facie case, that a technology or class of technologies may reasonably be presumed to pose a hazard to health and the environment. Most technologies—such as a new suspension system for cars or a new shape for bicycle helmets—would never even be considered for sunsetting, while some others that citizens did bring up for review would presumably pass the test and remain in unrestricted use. Constituted in this way, the Ecological Paradigm would increase the opportunity for social decision making about technologies, but it would not always exploit that opportunity; it would impose restrictions on technological choice only when justified on environmental grounds.

I should also be clear that democratizing environmental decisions addresses only one of many political issues associated with technology. In his book *Democracy and Technology,* Richard Sclove makes a compelling argument that technologies play a major role in determining not just environmental quality but the character of life in society—the nature of work, personal relationships, economic security, community cohesion, and politics.[34] If technology is such a major factor in determining the social order and everyone's quality of life, Sclove argues, decisions about technologies should be made democratically not just on environmental grounds but on social, political, and economic grounds, as well. *Democracy and Technol-*

ogy shows in detail what kinds of institutional changes are necessary to achieve popular determination of technological decisions. These structures and processes, of course, go well beyond the kind of environmental evaluation and sunsetting procedures I have described. My approach is not meant to take the place of Sclove's general vision but to describe in detail what could be just one aspect of a broad program of technological democratization.

There may seem to be some tension between my call for more democracy on one hand and my advocacy, on the other side, of specific actions and criteria, such as Zero Discharge and Clean Production. What if stakeholders, swayed by economic exigencies or other factors, decide to allow the continued use of persistent substances, even when alternatives are available? We should see principles like Zero Discharge, Reverse Onus, and Clean Production as ecological imperatives that provide broad guidance for specific decisions, akin to political imperatives like free speech, due process, and equal rights. These higher principles are constitutionally established in most democracies, providing the goals for day-to-day decisions, but remaining protected from infringement by majority decisions that might be driven by the crises or prejudices of the day. Ecological imperatives should play a similar role in technology policy: they represent the proper ends that environmental policy should achieve, but how best to get there must be open to democratic debate and determination. If the Ecological Paradigm were adopted in a democratic fashion, then Zero Discharge and Clean Production would represent analogous self-imposed principles to protect basic aspects of health and the environment in the context of a democratic approach to technological governance.

Science and Democracy

Democratizing the range of issues considered and the process by which environmental decisions are made would go a long way toward remedying the technocratic nature of environmental policy in the current paradigm. To address the issues we have explored in the relationship between science and politics, however, we need to go a bit further. If scientific knowledge plays a major role in public decision making—as it does and will continue to do—and if that knowledge is determined in part by social and political factors, then it is no more legitimate in a democracy for scientists to monopolize the input of knowledge into policymaking than it is for them to impose their values directly on the rest of society. Granting scientists full authority over knowledge that is ineluctably subjective gives them undue influence over policy. There is no way that members of a single institution—professional scientists—could ever adequately represent society, with all its social, cultural, economic, and demographic diversity. Further, institutions with the resources and connections to control or direct the activity of large numbers of scientists—corporations and the military come to mind—can deploy science's cognitive authority in order to gain an undemocratic hold over decisions that are made on the basis of science.

It thus becomes essential to democratize the process by which scientific knowledge is produced for the purpose of making social decisions. Because science has such a profound effect on public decisions, it should be subject to much wider scrutiny and participation than current practice allows. The data that science produces never enter the political process unfiltered: government scientists and advisory boards of externally-employed scientists now play critical roles in synthesizing the results of scientific studies for the purpose of political decision making, drawing conclusions and making recommendations for policy. These groups should be opened up to more democratic participation. Including informed and concerned citizens—local residents, workers, fishermen, and so on—along with scientists would add needed demographic, political, cultural, and epistemological diversity to the process of evaluating technologies and the condition of the environment. More participation would also make for better science, because it would make use of the rich first-hand knowledge that these people possess about health impacts, industrial processes, and the structure and status of ecosystems—knowledge that is now unavailable to "experts-only" science.

Are citizens up to this task? Can we really expect nurses, machinists, housewives, and farmers to give meaningful perspectives on toxicology, epidemiology, and other subjects? This question I can answer with an unambiguous yes. I have known scores of citizens, untrained in science but motivated by their experience with local environmental hazards, who have educated themselves so thoroughly that they can fully engage in discussion and debate about the meaning and use of scientists' data. Similarly, health activists infected with HIV rapidly gained such scientific literacy in immunology and epidemiology that they became productive and important players in the U.S. government's design and evaluation of AIDS treatment methods.[35] In Denmark, citizen "science tribunals" play an important role in the government's policymaking process. In a science tribunal, a panel of ordinary citizens hears scientific testimony on a policy issue, discusses what it has heard, and makes a recommendation to the government on the course of action it favors. By all accounts, these panels display enormous sophistication in their ability to understand and evaluate scientific claims and their political implications.[36]

To broaden public participation in scientific evaluation does not undermine the special role of scientists as creators of knowledge. The daily production of basic science need not be democratized to ensure that public decisions policy are truly democratic; it is no more necessary for citizens to be involved in determining the sequence of genes than it is for them to participate in the activity of novelists or historians. Scientists are in a unique position, because of their training, professional roles, and access to resources, equipment, and professional networks, to advance knowledge in their field. But citizens need not *be* scientists to evaluate the use of science in public policy, to critically evaluate the interpretation of scientific data, and to play a role in deciding what science it is in society's interest to pursue. We can accept the specialized roles of those who do scientific work without granting them the authority to do all of society's thinking.

Prospects and Progress

To conclude this book, I should touch on the small matter of political reality. I have called for a multibillion dollar segment of the chemical industry to be eliminated and for scores of industrial techniques and materials to be replaced with safer substitutes. I have advocated a major overhaul in how environmental regulations are made and a fundamental change in the political governance of technological decision making, and I have even called for changes in the way knowledge is produced for political purposes. Is there realistic hope that such a program will be adopted?

Many of the signs are discouraging. In particular, the political barriers to greater social control over technological decisions are immense. Large corporations are enormously influential, holding virtual veto power over government decisions in some nations. In the United States, the overwhelming dependence of political candidates on financial contributions from corporations and the growth of antienvironmental rhetoric has made the current political climate less amenable than ever to proposals that threaten corporate profits and autonomy. The *Detroit Metro-Times,* after a review of Federal Election Commission records, found that Dow and Occidental Chemical, the two largest chlorine producers in the United States, each gave politicians running for national office more than $1 million in PAC and "soft" money between 1983 and 1993, and the top U.S. chlor-alkali producers gave more than $1.4 million to dozens of key congressional campaigns in 1992 alone.[37] Further, as international barriers to trade and investment are negotiated away, national environmental standards are often driven to the level of the nation with the least stringent controls. Now that capital can flow more easily from one nation to another, corporations often move their facilities to countries with the lowest standards. Even the threat of capital flight weakens the power of governments to impose environmental standards on industry within its borders.

On the other hand, public and scientific awareness of global environmental damage due to chemical exposure is growing. Even the rather conservative scientific community as a whole is moving closer to acknowledging the existence of chemically-induced reproductive, neurological, and behavioral damage to wildlife and possibly humans. As I conclude this book, representatives of all the world's nations are meeting to negotiate the first international treaty on persistent organic pollutants (POPs). Although the POPs agreement deals with just 12 chemicals, many of which are already restricted in most nations, the emergence of an international legally binding instrument to deal with toxic pollution is a major step forward. As the power of national governments to regulate industrial enterprises wanes, global agreements provide a way for international society to address global environmental problems. The POPs treaty is a legal framework that could in principle be extended in future negotiations to deal with classes of chemicals rather than a few individual substances.

Will the nations of the world establish a global ecological policy on chlorine chemistry? There is a chance that they will, if the scientific evidence continues to accumulate, if public demands for protection of health and the environment grow, and if societies resist further weakening of democratic institutions. In the meantime, it is likely that steps toward a chlorine phaseout will be taken, and progress has already begun. Many nations—the more environmentally-minded European countries in particular—have adopted policies to restrict or reduce major uses of chlorine, including pulp bleaching, PVC, solvents, and pesticides. As the economies of these nations continue to prosper and public concern grows, other nations are likely to follow suit, and the leading countries will probably extend their policies further. Meanwhile, many industries are quietly eliminating their reliance on organochlorines, investing in the development and deployment of alternative processes in anticipation of future regulation, public concern, and the need to modernize technologies. By the time a global agreement on chlorine chemistry might finally be adopted, I expect that the use of these technologies will already have shrunk radically.

What are the prospects for my larger goal, the democratization of science and technological politics? On one hand, science continues to bear a cloak of objectivity and unassailable authority in the minds of many people. Citizens remain largely alienated and intimidated by technical jargon, and the institutions of science seldom even try to speak to the broad public. On the other hand, there is a long-term trend toward demystification. Science is continually producing findings that contradict its earlier assertions of the safety of various technological practices, a pattern that undermines public confidence in the infallibility of science and the automatic desirability of all new technologies. Media coverage of political and scientific conflict over the hazards of tobacco, chemical pollution, and global warming has made obvious that scientists from industry, government, and public interest groups often reach opposite conclusions based on the same data, revealing that science—in a political context, at least—is not the pure and objective practice it was once made out to be.

All these factors are leading to what the German sociologist Ulrich Beck has called the "reflexivization of science."[38] The myth of the white-coated objective scientist, bearing gifts of pure truth and life-improving technology, is gradually being replaced by a more mature view of science as a social way of knowing about the world. Reflexive science means that people understand that scientific knowledge is a cultural product that reflects not only the reality of nature but also the social conditions of its production and the political purposes to which it will be put. It means that people are coming to see technologies as human creations that sometimes threaten social well-being and may be adopted or refused. We are not entirely there yet, but the signs are all around us, in citizen opposition to polluting facilities, in widespread distrust of scientists-for-hire, and in the social, health, and environmental movements that have claimed cognitive authority equal to that of the science deployed by their opponents.

If Beck is right, and I think he is, the evolution of a reflexive view of science is a wholly positive development, since the authority of science has so often been used to justify practices that threaten both democracy and the environment. By encouraging an enlightened view of the status of scientific knowledge and its political use, reflexivity has set the stage for the advent of a new approach to environmental policy. In turn, the Ecological Paradigm promises to carry the process of reflexivization further by exposing the political issues behind technological decisions and involving the public more broadly in the creation of scientific knowledge.

This book, I hope, contributes to that process. My central project has been to describe a pressing environmental problem, its causes, and a solution to it. That argument is also a case study in the consequences of extending claims of scientific certainty and authority beyond what is clearly known. I have tried to show how recent discoveries in science undermine claims of complete scientific knowledge and control about chlorine chemistry and its impacts on natural systems. The findings I have reviewed in toxicology, epidemiology, ecology, molecular biology, and environmental and industrial chemistry all suggest that nature is far more complex and elusive than current institutions, policies, and ways of knowing make it out to be. These discoveries are a powerful antidote to the illusion that we can predict or control nature's response to changes induced by human technologies. This realization is reflexivity in action—science's products revealing its own limits, undermining its claims to omniscience and omnipotence.

I have also tried to illustrate the political considerations that lurk behind the deployment of environmental science in the sphere of public policy: burdens of proof, the rights of chemicals versus people, the authority of society and private firms in decisions about production technologies, and the role of institutional affiliation and political perspective in scientific work. I have done so in the hope that scientific authority and its role in politics can be further demystified, so that these underlying political issues can be debated openly and democratically. Reflecting this concern, I have tried to be explicit about the ethical and political views that form the foundation of my own analysis of the scientific information. These gestures, too, are reflexivity in action. A detailed view of the strategies and mechanisms by which scientific knowledge is transformed into political power and decisions can only contribute to an awareness that is prerequisite to achieving more democratic forms of knowledge and governance.

I began this book with a personal perspective—an unusual gesture for a book about science and policy published by an academic press. But if all science is in truth a cultural product that reflects the life and world of the person who created it, it is only appropriate to avow that my own knowledge begins from experience that is personal, contingent, subjective.

Our baby, a month from birth when I began writing, is now a year old. He is beautiful, full of joy and demands and love, endlessly occupied with his efforts to walk, speak, and understand the world around him. He is utterly remote from the abstract world of science or the contentiousness of politics. On the other hand, he is the most concrete reason I can imagine why society must take control over its path of technological development. Anyone who has a child, a grandchild, a mother, a father, or a friend—anyone who has somewhere a beautiful lake or river or mountain valley that they love—has a compelling reason, one that transcends political compromise or economics, to know that we must turn off the flow of life-altering chemicals that we are pouring into the world around us. About a century ago, we opened a Pandora's box of poison. Now we see its contents and their terrible consequences. We cannot turn back the hands of time, but we can learn from our experience. Having endangered everything we know and love by opening the box, it is time we choose now to reach out and shut it.

Appendix A

Organochlorines in the Tissues and Fluids of the General North American Population

Substance	Present in	Reference
Aliphatic organochlorines		
bromochlorotrifluoroethane	adipose	11
bromochlorotrifluoroethane	breath	19
carbochlorodithioic acid, S-methyl ester	adipose	11
carbon tetrachloride	milk, blood, breath	8,13,7,17,14
chlordecane	milk, breath	8,19
chloroethane	milk	8
chloroform	adipose, milk, blood, breath	1,8,13,7,17,14
chlorohexane	milk, breath	8,17
chloromethane	milk, breath	8,19
chloropentane	milk	8
chlorotrifluoromethane	milk	8
1,1-dichloro-1-propene	adipose	11
1,4-dichlorobutane	adipose	11
dibromochloromethane	milk	8
dibromodichloromethane	milk	8
dichloroacetylene	breath	17
dichlorodifluoromethane	milk, breath	8,7
dichlorofluoromethane	breath	19
1,2-dichloroethane	breath	7
1,1-dichloroethylene (vinylidene chloride)	adipose	11
dichloroethylene	milk	8
dichloromethane	milk, blood, breath	8,13,7

Appendix A (continued)

Substance	Present in	Reference
1,1-dichloro-1-nitromethane	blood	13
dichloronitromethane	breath	17
dichloropropene	milk	8
trichlorotrifluoroethane	milk, breath	8,7
hexachloro-1,3-butadiene	adipose	18
1,1,1,2-tetrachloroethane	breath	14
1,1,2,2-tetrachloroethane	adipose, breath	1,14
tetrachloroethylene	adipose, milk, blood, breath	1,8,13,7,17,14
1,1,1-trichloroethane	adipose, milk, breath	1,8,7,17,14
1,1,2-trichloroethane	breath	7
trichloroethylene	adipose, milk, breath	15,8,7,14
trichlorofluoromethane	breath	7
tris(2-chloroethyl)phosphate	adipose	1
tris(dichloropropyl)phosphate	semen	10
Chlorinated aromatics		
chlorobenzene	adipose, milk, breath	1,8,7,14
1,2-dichorobenzene	adipose, milk, breath	1,25,7,32
1,3-dichlorobenzene	adipose, milk, breath	11,25,14
1,4-dichlorobenzene	adipose, milk, breath	1,25,17
dichlorobenzene (congener unidentified)	adipose, blood	1,13
1,2,3-trichlorobenzene	adipose, milk	32,25
1,2,4-trichlorobenzene	adipose, milk	1,25,32
1,2,5-trichlorobenzene	adipose	32
1,3,5-trichlorobenzene	milk	25
trichlorobenzene (congener unidentified)	adipose, milk, breath	1,8,19
1,2,3,4-tetrachlorobenzene	adipose	32
1,2,3,5-tetrachlorobenzene	adipose	32
pentachlorobenzene	adipose, milk	32,25,18
hexachlorobenzene	adipose, milk, blood, semen	31,16,2,6,18
chloroethylbenzene	milk	9
pentachlorophenol	semen	6
tetrachlorophenol	semen	6
trichlorophenol	semen, urine	6,37
3,5,6-trichloro-2-pyridinol	urine	37

Appendix A (continued)

Substance	Present in	Reference
octachlorostyrene	adipose	32
4-chloro-2-(phenylmethyl)-phenol	adipose	11
1-chloro-4-(methylsulfonyl)-benzene	adipose	11
2-chloro-6-methylbenzonitrile	adipose	11
(chlorophenyl)phenylmethanone	adipose	11
hexachloronaphthalene	semen	6
tetrachlorodiphenyl ether	semen	6
hexachlorodiphenyl ether	adipose	27
heptachlorodiphenyl ether	adipose	27
octachlorodiphenyl ether	adipose	27
nonachlorodiphenyl ether	adipose	27
decachlorodiphenyl ether	adipose	27
Pesticides		
aldrin	blood, adipose, milk	36,31,4
alpha-chlordane	adipose, milk	32,25
gamma-chlordane	adipose, milk	32,25
dieldrin	adipose, milk, blood	31,25,2,37,32,4,1
o,p′-DDT	adipose, blood, milk	32,37,25,4
p,p′-DDT	adipose, blood, milk	1,2,37,25,32,4
p,p′-DDE	adipose, blood, semen, milk	1,2,12,37,25,32,4
2,4-D	urine	37
endrin	blood	35
alpha-hexachlorocyclohexane	adipose, milk, blood	18,16,35,25
beta-hexachlorocyclohexane	adipose, milk, blood	1,16,2,37,25,18
lindane	adipose, milk	18,16,25
heptachlor	milk	25
heptachlor epoxide	adipose, milk, blood	31,25,2,37,32,1
mirex	adipose, blood	1,25,34,32
photomirex	milk	25
cis-nonachlor	adipose, blood	32,37
trans-nonachlor	adipose, milk, blood	31,16,2,25,1
nonachlor-III	adipose	33
oxychlordane	adipose, milk, blood	32,25,2,37
pentachlorophenol	urine	37
toxaphene	adipose	1

Appendix A (continued)

Substance	Present in	Reference
Polychlorinated dioxins (PCDDs)		
2,3,7,8-TCDD	adipose, milk, blood	1,22,20,30,37
1,2,3,7,8-PeCDD	adipose, milk, blood	1,22,20,30,37
1,2,3,4,7,8-HxCDD	adipose, milk, blood	1,22,20,30,37
1,2,3,6,7,8-HxCDD	adipose, milk, blood	1,22,20,30,37
1,2,3,7,8,9-HxCDD	adipose, milk, blood	1,22,20,30,37
1,2,3,4,6,7,8-HpCDD	adipose, milk, blood	1,22,20,30,37
1,2,3,4,6,7,9-HpCDD	blood	37
OCDD	adipose, milk, blood	1,22,20,30,37
Polychlorinated dibenzofurans (PCDFs)		
2,3,7,8-TCDF	adipose, milk, blood	1,22,20,30,37
1,2,3,7,8-PeCDF	adipose, milk, blood	1,22,20,30,37
2,3,4,7,8-PeCDF	adipose, milk, blood	1,22,20,30,37
1,2,3,4,7,8-HxCDF	adipose, milk, blood	1,22,20,30,37
1,2,3,6,7,8-HxCDF	adipose, milk, blood	1,22,20,30,37
2,3,4,6,7,8-HxCDF	adipose, milk, blood	1,22,20,30,37
1,2,3,7,8,9-HxCDF	adipose, milk, blood	1,22,20,30,37
1,2,3,4,6,7,8-HpCDF	adipose, milk, blood	1,22,20,30,37
1,2,3,4,7,8,9-HpCDF	adipose, milk, blood	1,22,20,30,37
OCDF	adipose, milk, blood	1,22,20,30,37
Polychlorinated biphenyls (PCBs)		
2-monochlorobiphenyl	semen	12
2,2′-dichlorobiphenyl	semen	12
2,3-dichlorobiphenyl	semen	12
2,3′-dichlorobiphenyl	semen	12
2,4-dichlorobiphenyl	semen	12
2,4′-dichlorobiphenyl	semen	12
2,5-dichlorobiphenyl	semen	12
2,2′,3-trichlorobiphenyl	semen	12
2,2′,4-trichlorobiphenyl	semen	12
2,2′,5-trichlorobiphenyl	semen	12
2,3,3′-trichlorobiphenyl	semen	12
2,3,4′-trichlorobiphenyl	semen, milk	12,26
2,3′,4-trichlorobiphenyl	semen	12
2,3′,4′-trichlorobiphenyl	milk, blood, adipose	24,28,26,32
2,4,4′-trichlorobiphenyl	adipose, milk, blood	32,21,28,37, 30,26
2,4′,6-trichlorobiphenyl	semen	12
3,4,4′-trichlorobiphenyl	milk, blood, adipose	32,28

Appendix A (continued)

Substance	Present in	Reference
2,2′,3,4-tetrachlorobiphenyl	milk, adipose	24,32
2,2′,3,5′-tetrachlorobiphenyl	semen, milk, adipose, blood	12,24,28
2,2′,3,6′-tetrachlorobiphenyl	milk	26
2,2′,4,5-tetrachlorobiphenyl	milk	26
2,2′,4,5′-tetrachlorobiphenyl	semen, adipose, milk, blood	12,32,24,28
2,2′,5,5′-tetrachlorobiphenyl	semen, milk, blood, adipose	12,24,28,32,37
2,3,3′,4′-tetrachlorobiphenyl	milk, blood	26,37
2,3,4,4′-tetrachlorobiphenyl	milk, adipose, blood	24,28,37
2,3′,4,4′-tetrachlorobiphenyl	semen, blood, milk, adipose	12,28,24,32
2,3′,4′,5′-tetrachlobiphenyl	milk	26
2,3′,4′,5-tetrachlorobiphenyl	milk	26
2,4,4′,5,-tetrachlorobiphenyl	adipose, blood, milk	32,28,24,37, 30,26
3,3′,4,4′-tetrachlorobiphenyl	milk, adipose, blood	35,37,30
3,4,4′5-tetrachlorobiphenyl	blood	37
2,2′,3,3′,4-pentachlorobiphenyl	semen	12
2,2′,3,4′,5-pentachlorobiphenyl	milk, blood, adipose	32,28
2,2′,3,4,5′-pentachlorobiphenyl	semen, milk, adipose, blood	12,26,28
2,2′,3′,4,5-pentachlorobiphenyl	semen	12
2,2′,3,5,6′-pentachlorobiphenyl	semen	12
2,2′,4,4′,5-pentachlorobiphenyl	milk, blood, adipose	24,28,32,37, 30,26
2,2′,4,5,5′-pentachlorobiphenyl	semen, milk, adipose, blood	12,32,23,37
2,3,3′,4,4′-pentachlorobiphenyl	adipose, milk, blood	32,24,28,37,30
2,3,3′,4′,5-pentachlorobiphenyl	milk	26
2,3,3′,4′,5′-pentachlorobiphenyl	milk	26
2,3,3′,4′,6-pentachlorobiphenyl	milk, blood, adipose	24,28
2,3,4,4′,5-pentachlorobiphenyl	milk	26,24
2,3′,4,4′,5-pentachlorobiphenyl	adipose, blood, milk	32,28,24,37, 30,26
2,3′,4,4′,6-pentachlorobiphenyl	milk	26
3,3′,4,4′,5-pentachlorobiphenyl	milk, adipose, blood	35,37,30
2,2′,3,3′,4,4′-hexachlorobiphenyl	milk, blood	26,30
2,2′,3,3′,4,5′-hexachlorobiphenyl	milk, blood	26,37

Appendix A (continued)

Substance	Present in	Reference
2,2′,3,3′,5,6′-hexachlorobiphenyl	milk	26
2,2′,3,4,4′,5′-hexachlorobiphenyl	adipose, milk, blood, semen	32,21,28,12, 37,30
2,2′,3,4,4′,5-hexachlorobiphenyl	milk	24
2,2′,3,4,5,5′-hexachlorobiphenyl	milk, adipose	24,32
2,2′,3,4′,5,5′-hexachlorobiphenyl	milk, blood	26,37
2,2′,3,5,5′,6-hexachlorobiphenyl	semen, milk, blood, adipose	12,26,28,23
2,2′,4,4′,5,5′-hexachlorobiphenyl	semen, blood, milk, adipose	12,28,21,32, 37,30
2,3′,4,4′,5,5′-hexachlorobiphenyl	milk, blood	26,37
2,3,3′,4,4′,5,-hexachlorobiphenyl	adipose, milk, blood	32,24,28,37,30
2,3,3′,4,4′,5′-hexachlorobiphenyl	adipose, milk, blood	32,28,26,37
2,3,3′,4,4′,6-hexachlorobiphenyl	milk	26
3,3′,4,4′,5,5′-hexachlorobiphenyl	adipose, milk, blood	35,37,30
2,2′,3,3′,4,4′,5-heptachlorobiphenyl	milk, adipose, semen, blood	12,26,29,37,30
2,2′,3,3′,4,4′,6-heptachlorobiphenyl	milk, semen	26,12
2,2′,3,3′,4,5,5′-heptachlorobiphenyl	milk, blood	26,37
2,2′,3,3′,4′,5,6-heptachlorobiphenyl	milk, blood	26,37
2,2′,3,3′,4,5,6′-heptachlorobiphenyl	milk	26
2,2′,3,3′,4,6,6′-heptachlorobiphenyl	semen	12
2,2′,3,3′,5,6,6′-heptachlorobiphenyl	semen	12
2,2′,3,4,4′,5,5′-heptachlorobiphenyl	adipose, milk, blood	21,28,32,37,30, 24,26
2,2′,3,4,4′,5′,6-heptachlorobiphenyl	adipose, milk, blood	32,24,37,30
2,2′,3,4,5,5′,6-heptachlorobiphenyl	adipose, milk, blood	26,32,30
2,2′,3,4′,5,5′,6-heptachlorobiphenyl	adipose, blood, milk	32,28,24,37,30
2,3,3′,4,4′,5,5′-heptachlorobiphenyl	milk, semen	12,24,32
2,3,3′,4,4′,5′,6-heptachlorobiphenyl	milk, blood, adipose	24,32
2,3,3′,4′,5,5′,6-heptachlorobiphenyl	adipose, milk, blood	32,24,37
2,2′,3,3′,4,4′,5,5′-octachlorobiphenyl	milk, blood, adipose	24,28,32,37
2,2′,3,3′,4,4′,5,6-octachlorobiphenyl	milk, blood	26,37
2,2′,3,3′,4,4′,5′,6-octachlorobiphenyl	milk, adipose	26,29
2,2′,3,3′,4,4′,5,6′-octachlorobiphenyl	blood, semen	37,12
2,2′,3,3′,4,5,5′,6′-octachlorobiphenyl	semen	12
2,2′,3,3′,4′,5,5′,6-octachloobiphenyl	milk, blood, adipose	24,28,32
2,2′,3,3′,4,5′,6,6′-octachlorobiphenyl	blood	37
2,2′,3,3′,5,5′,6,6′-octachlorobiphenyl	milk	26

Appendix A (continued)

Substance	Present in	Reference
2,2′,3,4,4′,5,5′,6-octachlorobiphenyl	milk, blood, adipose	24,28,32,37
2,3,3′,4,4′,5,5′,6-octachlorobiphenyl	milk	26
2,2′,3,3′,4,4′,5,5′,6-nonachlorobiphenyl	milk, blood, adipose	24,28,32,37
decachlorobiphenyl	milk, blood, adipose	24,28,1,37

References: (1) Stanley 1986, (2) Murphy and Harvey 1985, (3) Kutz et al. 1979, (4) Jensen 1983, (5) Savage et al. 1981, (6) Dougherty et al. 1981, (7) Wallace et al. 1984, (8) Pellizzari et al. 1982, (9) Jensen 1987, (10) Hudec et al. 1981, (11) Onstot, Ayling, and Stanley 1987, (12) Bush et al. 1986, (13) Laseter and Dowty 1978, (14) Wallace et al. 1988, (15) Onstot and Stanley 1989, (16) Takei et al. 1983, (17) Krotoszynski et al. 1979, (18) Mes et al. 1982, (19) Krotoszynski and O'Neill 1982. (20) Schecter and Gasiewicz 1987, (21) Schecter et al. 1989a, (22) Kahn et al. 1988, (23) Schecter et al. 1989b, (24) Mes and Marchand 1987, (25) Davies and Mes 1987, (26) Safe et al. 1985, (27) Stanley et al. 1990b, (28) Mes and Weber 1989, (29) Williams and LeBel 1990, (30) Schecter 1994, (31) Kutz et al 1991, (32) Mes et al. 1990, (33) Dearth and Hites 1990, (34) Moysich et al. 1998, (35) Djordevic et al. 1997, (36) Mossing et al. 1985, (37) Anderson et al. 1998.

Appendix B

Organochlorine Contaminants in the Great Lakes Ecosystem

Polynuclear aromatic organochlorines
1-chloronaphthalene
2-chloronaphthalene
decachloropyrene
dichloronaphthalene
1,2-dichloronaphthalene
1,7-dichloronaphthalene
2,3-dichloronaphthalene
nonachlorpyrene
tetrachloronaphthalene
1,2,4-trichloroanthracene
trichloronaphthalene
trichlorophenanthrene
pentachloroterphenyl
hexachlorodibenzofuran
2,3,7,8-tetrachlorodibenzo-p-dioxin (dioxin)
PCB-1232
PCB-1242
PCB-1248
PCB-1254
PCB-1260
PCB-1262
2,2′,3,4,5,5′,6-heptachlorobiphenyl
hexachlorobiphenyl (congener unidentified)
2,2′,3,3′,4,4′-hexachlorobiphenyl
2,2′,3,4,4,5,5′-hexachlorobiphenyl
2,2′,3,4,4′,5′-hexachlorobiphenyl
2,2′,4,4′,5,5′-hexachlorobiphenyl
2,2′,3,4,4′,5′-hexachlorobiphenyl
2,2′,3,4,4′,6,6′-hexachlorobiphenyl
2,2′,3,3′,4,4′,5,5′-octachlorobiphenyl
pentachlorobiphenyl (congener unidentified)
2,2′,3′,4,5′-pentachlorobiphenyl
2,2′,3,5′,6-pentachlorobiphenyl
2,2′,4,5,5′-pentachlorobiphenyl
2,2′,3′,5′-tetrachlorobiphenyl
2,2′,4,5′-tetrachlorobiphenyl
2,2′,6,6′-tetrachlorobiphenyl
2,3,4,4′-tetrachlorobiphenyl
2,3,4,5-tetrachlorobiphenyl
2,3′,4′,5-tetrachlorobiphenyl
2′,3,4-trichlorobiphenyl

Other aromatic organochlorines
chlorobenzene
1,2-dichlorobenzene
1,3-dichlorobenzene
1,4-dichlorobenzene
hexachlorobenzene
pentachlorobenzene
1,2,3,4-tetrachlorobenzene
1,2,3,5-tetrachlorobenzene
1,2,4,5-tetrachlorobenzene
1,2,3-trichlorobenzene

Appendix B (continued)

1,2,4-trichlorobenzene
1,3,5-trichlorobenzene
dichlorophenol
pentachloroanisole
pentachlorophenol
2,3,-trichlorophenol
2,4,5-trichlorophenol
2,4,6-trichlorophenol
2-chlorotoluene
4-chlorotoluene
2,4-dichlorotoluene
2,6-dichlorotoluene
hexachlorotoluene
2,3,4,5,6-pentachlorotoluene
2,3,5,6-tetrachlorotoluene
2,3,6-trichlorotoluene
2,4,5-trichlorotoluene
trichlorotrifluorotoluene
m-chlorotoluenetrifluoride
2-chlorotoluenetrifluoride
pentachlorostyrene
heptachlorostyrene
octachlorostyrene
pentachloropyrrolidine
polychlorinated benzoic acids
polychlorinated benzyl alcohols
chlorobenzotrifluoride
3-chloro-5′-hydroxylbenzophenone
3′-chloro-3-methyl-1,3-bis (phenyl methyl)benzene
2,3-dichlorotrifluoromethylbenzene
2,4-dichlorotrifluoromethylbenzene
3,4-dichlorotrifluoromethylbenzene
2-chloro(triflouromethyl)benzene
3,4-dichloro-4′-(triflouromethyl)benzophenone)
4-chloro(trifluoromethyl)benzene
4′,5′,2-tetrachloro-3-methyl-1,3-(bisphenylmethyl)benzene
3,5,4′-trichlorodiphenylmethane
4′,5′,4-trichloro,3-methyl-1,3-bis(phenylmethyl)benzene
trichloro(methylphenyl)phenylmethanes
3′,4′,5′,2,4-pentachloro-3-methyl-1,3-bis(phenylmethyl)benzene
4-chloro-4′-methyldiphenylmethane

Aliphatic organochlorines
bromochloromethane
bromodichloromethane
bromotrichloromethane
carbon tetrachloride
chlorodibromomethane
chloroform
chloromethane
1,1-dibromo-2-chloroethane
1,1-dichloro-2-bromoethane
1,2-dichloro-1-bromoethane
dichlorofluoromethane
1,1-dichloroethane
1,2-dichloroethane
dichloroethane
dichloromethane
hexachloroethane
1,1,2,2-tetrachloroethane
tetrachloroethylene
1,1,1-trichloroethane
1,1,2-trichloroethane
trichloroethylene
trichlorofluoromethane
trichlorotriflouoroethane
1,2-dichloropropane
1,3-dichloropropene
tetrachlorobutadiene
1,1,2,3,4,4-hexachloro-1,3-butadiene
hexachlorocyclopentadiene
3-chloro-1-propynyl-cyclohexane

Chlorinated pesticides
aldrin
chlordane

Appendix B (continued)

cis-chlordane	p,p-methoxychlor
trans-chlordane	cis-nonachlor
oxychlordane	trans-nonachlor
p,p-DDD	octachlor epoxide
o,p′-DDD	toxaphene
o,p′-DDE	2,4-D
p,p-DDE	DCPA
p,p′-DDMU	dichlorprop
p,p-DDT	MCPA
o,p′-DDT	mecoprop
dieldrin	silvex
endosulfan	2,4,5-T
a-endosulfan	mirex
b-endosulfan	kepone
endosulfan sulfate	2,8-diH-mirex
endrin	10-monoH-mirex
heptachlor	photomirex
heptachlor epoxide	atrazine
isodrin	cyanazine
lindane	simazine
a-HCH	alachlor
b-HCH	metolachlor
HCH (mixed)	

Source: GLWQB 1987.

Appendix C

By-products from the Chlorination of Drinking Water and Wastewater

Aliphatic organochlorines
Bromochloroacetonitrile (1,5)
Bromodichloromethane (1,5)
Carbon tetrachloride (5)
Chloral (trichloroethanal) (1)
Chloral hydrate (5)
Chloroacetaldehyde (1)
Chlorodibromomethane (5)
3-Chloro-4-(dichloromethyl)- 5-hydroxy-2(5H)-furanone (1)
Chloroform (1,5)
Chloropicrin (nitrochloroform) (1,5)
Chlorinated methyl ester (5)
Cyanogen chloride (5)
Dichloroacetaldehyde (1,5)
Dichloroacetamide isomer (#513) (5)
Dichloroacetonitrile (1,5)
3,3-Dichloro-2-butanone (5)
1,1-Dichloro-2-butanone (5)
1,1-Dichloropropanone (5)
Hexachloroacetone (1)
3,3,3-Trichloroepoxypropane (5)
1,1,3,3-Tetrchloroacetone (1)
Tetrachloroethylene (5)
Trichloroacetaldehyde (5)
1,1,1-Trichloroacetone (1)
Trichloroacetonitrile (1,5)
Trichloroethylene (5)
1,1,1-Trichloropropanone (5)
1,1,1-Trichloro-2-butanone (5)

Chlorinated dibenzofurans
1,2,7,8-TCDF (2)
2,3,7,8-TCDF (2)
1,2,8,9-TCDF (2)
1,2,3,7,8-PeCDF (2)
2,3,4,7,8-PeCDF (2)
1,2,3,4,7,8-HxCDF (2)
1,2,3,6,7,8-HxCDF (2)
1,2,3,7,8,9-HxCDF (2)
1,2,3,4,6,7,8-HpCDF (2)
1,2,3,4,7,8,9-HpCDF (2)
OCDF (2)

Chlorinated phenols/phenoxyphenols
2-Chlorophenol (1,5)
4-chloro-2-methoxyphenol (3)
4-chloro-3-methoxyphenol (3)
6-chloro-2-methoxyphenol (3)
6-chloro-3-methoxyphenol (3)
Chlorotyrosine (4)
2,4-Dichlorophenol (1,5)
2,6-Dichloro-4-methoxyphenol (3)
4,6-Dichloro-2-methoxyphenol (3)
Dichlorotyrosine (4)
Pentachlorophenol (3)
2,3,4,6-Tetrachlorphenol (3)

Appendix C (continued)

2,4,5-Trichlorophenol (3)
2,4,6-Trichlorophenol (1,5)

Chlorinated acid compounds

Chloroacetic acid, methyl ester (5)
Trichloroacetic acid, methyl ester (5)
Dichloroacetic acid, methyl ester (5)
Bromochloroacetic acid, methyl ester (5)
Butanedioc acid (succinic acid) (1)
Chloroacetic acid (1)
cis-Chlorobutenedioc acid (1)
Chlorobutanedioic acid, dimethyl ester (5)
Chlorobutanedoic acid (1)
Chloro-methoxy-dicarboxylic acid, dimethyl ester (5)
Chlorinated methyl ester (GAC #29) (5)
Dichloroacetic acid (1,5)
2,2-Dichlorobutanoic acid, methyl ester (5)
cis-Dichlorobutenedioc acid (1)
trans-Dichlorobutenedioic acid (1)
Dichloropropanedoic acid (1)
2,2-Dichloropropanoic acid (1)
3,3-Dichloropropenoic acid (1)
3,3-Dichloropropenoic acid, methyl ester (5)
2,2-Dichlorosuccinic acid (1)
Dichloro-methoxy-benzoic acid, methyl ester (5)
Monochloroacetic acid (5)
Propanoic acid, 2,2-dichloro-, methyl ester (5)
Propanoic acid, 2-chloro-, methyl ester (5)
Trichloroacetic acid (1,5)
2,3,3-Trichloropropenoic acid (1)

Unknown compounds

47 unidentified chlorinated organic compounds (5)

References: (1) NRC 1987, (2) Rappe et al. 1989, (3) Onodera et al. 1989, (4) Jolley 1986, (5) Stevens et al. 1990.

Appendix D

Organochlorine By-products from Hazardous Waste Incinerators

Bromochlorothiophene (6)
Carbon tetrachloride (1,2,3,4,5)
Chloroanthracene (6)
Chlorobenzene (1,3,4)
2-chloro-4-bromophenol (6)
1-Chlorobutane (4)
Chlorocyclohexanol (1)
1-Chlorodecane (4)
Chlorodibromomethane (3)
2-Chloroethyl vinyl ether (3)
Chloroform (1,2,3,4,5)
1-Chlorohexane (4)
Chloromethane (3,5)
Chloromethylphenol (6)
Chloronephthalene (6)
1-Chlorononane (4)
1-Chloropentane (4)
Chlorothiophenol (6)
Chlorotoluene (6)
Dichloroacetylene (2)
Dichlorobromomethane (3)
Dichlorobenzene s(6)
1,2-Dichlorobenzene (4,5)
1,4-Dichlorobenzene (4,5)
Dichlorobenzoylchloride (6)
Dichlorobenzylalcohol (6)
Dichloroquinoline
1,1-Dichloroethane (5)
1,2-Dichloroethane (3,4,5)
1,1-Dichloroethylene (3,5)
Dichlorodifluoromethane (5)
Dichlorodiphenylethane (6)
Dichloromethane (1,3,4,5)
Dichloronaphthalene (6)
Dichlorophenols (6)
2,4-Dichlorophenol (5)
Hexachlorobenzene (2,5,6)
Hexachlorobutadiene (2)
Pentachlorobenzene (6)
Pentachlorophenol (6)
Polychlorinated biphenyls (PCBs) (2,6)
Polychlorinated dibenzo-p-dioxins (PCDDs) (2,5,6)
Polychlorinated dibenzofurans (PCDFs) (2,5,6)
Tetrachloroaniline (6)
Tetrachlorothiophene (6)
Tetrachlorobenzenes (6)
Tetrachlorophenols (6)
1,1,2,2-Tetrachloroethane (4,5)
Tetrachloroethylene (1,2,3,4,5)
Trichlorobenzenes (6)
1,2,4-Trichlorobenzene (4,5)
1,1,1-Trichloroethane (1,3,5)
1,1,2-Trichloroethane (5)
Trichloroethylene (1,2,4,5)

Appendix D (continued)

Trichlorofluoromethane (3)	Trichlorophenols (6)
Trichlorotrifluoroethane (4)	2,3,5-Trichlorophenol (5)
Trichloronaphthalene (6)	Vinyl chloride (3,5)

References: (1) Trenholm 1986 (eight full-scale hazardous waste incinerators), (2) Dellinger 1988 (turbulent flame reactor), (3) Trenholm 1987 (full-scale rotary kiln incinerator), (4) Chang 1988 (turbulent flame reactor), (5) EPA 1989 (review of available data at varied units), (6) EPA 1987a, EPA 1987b (two full-scale rotary kiln incinerators), (6) Wienecke et al. 1995 (full-scale hazardous waste incinerator).

Appendix E

Chlorine-Containing Pesticides

Acetochlor
Alachlor
Aldrin
Amitraz
Anilazine
Atrazine
Azinophos methyl
Barban
Bentazon
Captan
Carbofuran
Carbosulfan
Carboxin
Chloramebn
Chlorbufan
Chlordane
Chlorfenson
Chlorbenzilate
Chloroneb
Chloropicrin
Chlorothalonil
Chloroxuron
Chlorpropham
Chlorpyrifos
Chlorsulfuron
Cloethocarb
Crotoxylphos
Cyanazine
2,4-D
Dichloran
DDT
Dicamba
Dichlofluanid
1,3-Dichloropropene
Dichlorprop
Dichlorvos
Diclofop-methyl
Dicrotophos
Dieldrin
Dienochlor
Dimethoate
Disulfoton
Endosfulfan
Endrin
Ethion
Etriziaol
Fenchlorphos
Ferban
Ferriamicide
Folpet
Fonofos
Formetanate hydrochloride
Hetpachlor
Hexachlorobeznene
Hexachlorphene
Imazalil
Imidazole
Ioprodione
Isofenphos
Lindane
Mancozeb
Maneb
MCPA
Metalzyl
Methidathion
Methoxychlor
Metiran
Metolachlor
Metribuzin
Mevinphos
Naled
Oxamyl
Petnachloronitrobenzene
Pentachlorophenol
Phosmet
Phosphamidon
Phthalimide
Piperalin
Piperazine
Prochloraz
Propargite
Propiconazole

Appendix E (continued)

Pyrimidine	Thiodicarb	Trichlorofon
Sulprofos	Thiram	Triclopyr
TCA	Toxaphene	Trifluralin
Tedion	Triallate	Vincolozolin
Tetrachlorvinphos	Triazole	Ziram
Tetradifon		

Source: SRI 1993.

Appendix F

Oil Solubility of Selected Organochlorines and Non-Chlorinated Analogues

Substance	Oil solubility (Kow)	Ratio of K(ow) to that of chlorine-free analogue	Fold increase in oil solubility per chlorine atom
Benzene	135	—	—
Chlorobenzenes	151–692	1.1–5.1	1.1–5.1
Dichlorobenzene	2,754–3,388	20.4–25.1	10.2–12.6
Trichlorobenzenes	10,471–15,488	77.6–115	25.9–38.3
Tetrachlorobenzene	50,119	371	92.9
Pentachlorobenzene	147,911	1,095	219
Hexachlorobenzene	204,174	1,512	252
Phenol	29	—	—
Monochlorophenols	141–295	9.9–10.1	4.9–10.2
Dichlorophenol	437–4,786	15.1–165	7.6–83.0
Trichlorophenols	4,898–18,621	169–642	56.6–215
Pentachlorophenol	131,826	4,546	914
Styrene	891	—	—
Monochlorostyrene	3,802	4.3	4.3
Octachlorostyrene	1,949,845	2,188	273
Dibenzofuran	13,183	—	—
2,3,7,8-TCDF	660,693	50.1	12.5
1,3-cyclopentadiene	78	—	—
Hexachlorocylopentadiene	9,772	1,253	21.0
Butadiene	98	—	—
Monochlorobutadiene	115	1.2	0.9
Hexachlorobtuadiene	79,433	810	98.1

Appendix F (continued)

Methane	12	—	—
Chloromethane	8	0.7	0.7
Trichloromethane	93	7.8	2.5
Tetrachloromethane	676	56.3	13.7
Ethane	65	—	—
Chloroethane	27	0.4	0.4
1,1-dichloroethane	79	1.2	0.6
1,1,1-trichloroethane	309	4.8	1.6
1,1,1,2-tetrachloroethane	457	7.0	1.8
Ethylene	13	—	—
Chloroethylene	4	0.3	0.3
1,1-dichloroethylene	72	5.5	2.7
Trichloroethylene	407	31.3	10.1
Tetrachloroethylene	2,512	193	46.6
Acetic acid	0.7	—	—
Chloroacetic acid	2	2.9	2.5
Dichloroacetic acid	8	11.4	6.2
Trichloroacetic acid	21	30.0	10.5

Note: K(ow) is the octanol-water coefficient, the ratio of a substance's solubility in octanol to its solubility in water, a standard measure of oil solubility.
Source: Converted from logK(ow) values reported in Hazardous Substances Databank, 1997.

Notes

Introduction

1. Havel 1984.

2. Davis and Magee 1979.

3. EPA 1998c.

4. Braungart (1987) provides a list of all 11,000 organochlorines in commerce in Germany. Dow Chemical (1995) has estimated there are 15,000 deliberately produced organochlorines, but the basis for this estimate is unknown, so I have used Braungart's number.

5. I take this comparison from a "risk communication" manual prepared for the Chemical Manufacturers Association (Covello et al. 1988), which presents this and other analogies to show how pollutant concentrations in the environment can be expressed in ways that make them seem negligible. Used to describe the toxicity of the same substances, the comparison seems to have the opposite effect.

6. I use the word *paradigm* in the sense of a total way of looking at the world, "a philosophical and theoretical framework of a scientific school or discipline within which theories, laws, and generalizations and the experiments performed in support of them are formulated," as the *Merriam-Webster Dictionary* puts it. The word was originally used to mean an example that illustrates a pattern. Since the writings of Wittgenstein and Thomas Kuhn, however, *paradigm* has been extended to mean a conceptual structure that orders thought and perception.

7. With this term, I refer specifically to the calculation of acceptable releases and exposures based on probabilistic calculations of cancer risk and reference doses for other toxic effects. I do not use the term in the more general sense of gathering information on the potential dangers of pollution and evaluating it qualitatively, an approach better called *hazard evaluation*.

8. In naming the Ecological Paradigm, I had as many reasonable options as I did in naming the Risk Paradigm. I could have called it the precautionary paradigm, or the prevention paradigm, or the clean production paradigm, and so on. But *ecological* sums up more of the matter than any other word. First, the overwhelming goal of this framework is to protect ecosystems and the organisms that live within it, so the name resonates with the aims and values that underlie it. Second, while the Risk Paradigm is based on a mechanistic view of how natural systems are organized and how causality works within them, the Ecological Paradigm borrows its model of nature as integrated and complex from the science of ecology; it is this view that illuminates the limits of scientific tools for prediction and diagnosis, leading directly to the Precautionary Principle and Reverse Onus as alternatives to the

current forms of environmental management based on scientific proof and prediction. Third, Clean Production reorients production technologies toward ecologically sound design. Fourth, the rule of Zero Discharge also has ecological roots: it is derived from the conclusion that compounds that are incompatible with ecological cycles of material degradation should not be released into the environment at all. Although the word *ecological* does all this quite well, it fails to connote one crucial aspect of the new framework: the political and ethical considerations that are fundamental to its vision, in particular the emphasis on democratic determination of whether "risks" are acceptable and which production technologies society deems most appropriate for fulfilling its needs.

9. Other classes of compounds that might be worthy priorities in an Ecological Paradigm include the organobromines and other organohalogens, which have similar properties to the organochlorines, although they are currently produced in much smaller quantities; metals that are toxic in low doses (like lead, cadmium, and mercury); and phthalates and alkyl phenols (organic compounds that are fat-soluble and, in at least some cases, endocrine disruptors or carcinogens).

10. In 1994, the CCC's operating budget was $12 million, and the council's chair, J. Roger Hirl of Occidental Chemical, estimated that the chlorine industry was contributing "people and resources equal to at least 10 times that amount" (Hilleman et al. 1994).

11. Council of Environmental Advisors 1991.

Chapter 1

1. Lawrence 1966.

2. Tatsukawa and Tanabe 1990. BCFs for striped dolphins were 13 million for PCBs, 37 million for DDT and breakdown products, and 37,000 for hexachlorocyclohexanes.

3. Cortes et al. 1998. Half-lives are based on atmospheric concentrations in the Great Lakes region.

4. ATSDR 1994.

5. Behnke and Zetch 1993.

6. Catabeni et al. 1985.

7. Barrie et al. 1997.

8. Tatsukawa and Tanabe 1990.

9. Brun et al. 1991.

10. Fuhrer and Ballschmitter 1998.

11. Glotfelty et al. 1987.

12. Goolsby et al. 1997.

13. DeLorey et al. 1988.

14. HSDB 1995.

15. Plumacher and Schroder 1993.

16. Schroder 1993.

17. Frank et al. 1993a.

18. Frank et al. 1993b, Plumacher and Schroder 1993.

19. Frank et al. 1993a.

20. Simonich and Hites 1995.

21. Simonich and Hites 1995, Oehme 1991, Norstrom and Muir 1994.
22. Cohen et al. 1995, Colborn et al. 1991.
23. Gregor and Gummer 1989.
24. Patton et al. 1991.
25. Barrie et al. 1997.
26. Makhijani and Gurney 1995.
27. WMO/UNEP 1996.
28. Makhijani and Gurney 1995.
29. Makhijani and Gurney 1995, WMO/UNEP 1996.
30. These half-lives were calculated by Jeffers et al., (1989) based on observed hydrolysis constants. In aquatic ecosystems, other processes, including biological and light-induced degradation, may proceed at faster rates.
31. EPA 1994b.
32. Jeffers et al. 1989, HSDB 1997.
33. HSDB 1997.
34. Bonsor et al. 1988.
35. Chovanec et al. 1993.
36. Tatsukawa and Tanabe 1990.
37. Wester and DeBoer 1993.
38. Fischer et al. 1991.
39. Barrie et al. 1997, AMAP 1997.
40. Barrie et al. 1997.
41. Kocan et al. 1987, Sodergren et al. 1990.
42. EPA 1988b.
43. EPA 1985b, HSDB 1995.
44. Strictly, compounds are considered dioxin-like if they can assume a coplanar structure like that of TCDD, meaning that the placement of the chlorine atoms allows two benzene rings to lie in a flat rather than a twisted position relative to each other. Coplanar chlorinated aromatics can bind to the same intracellular receptor to which TCDD binds, producing similar toxicological effects, as I discuss in chapter 2.
45. Giesy et al. 1994b.
46. HSDB 1995.
47. Miller and Uhler 1988.
48. Beland et al. 1993.
49. Loganathan and Kannan 1994, Nakata et al. 1995. In the United States, materials with PCBs greater than 50 parts per million are classified as hazardous waste.
50. Taruski et al. 1975.
51. Miles et al. 1992, Olsson et al. 1994.
52. Ono et al. 1987. Organochlorine pesticides and PCBs have also been detected in very high concentrations in Dall's porpoises from the North Pacific open ocean (Subramian et al. 1987) and several species of whale, seal, and dolphin from northern regions of the Atlantic and the Pacific (Willett et al. 1998).
53. Reviewed in Norstrom and Muir 1994.

54. Oehme 1991, Norstrom et al. 1990, Oehme et al. 1988.

55. Muir et al. 1997, AMAP 1997.

56. Norstrom et al. 1990.

57. Norstrom et al. 1990. The problem is not restricted to the Canadian arctic. On the European side of the Far North, polar bears and seals on the arctic island of Spitzbergen are contaminated with parts per million levels of hexachlorobenzene, PCBs, and a number of organochlorine pesticides (Norheim et al. 1992, Oehme 1991, Oehme et al. 1988).

58. AMAP 1997. PCB and dioxin concentrations in bear tissues exceed levels observed to cause adverse effects in mink, seals, monkeys, and humans. Concentrations observed in other arctic species, including arctic fox, harbor porpoise, walrus, several species of whale and seal, peregrine falcon, and sea eagle, also exceed levels associated with toxicological effects in mammalian and bird species.

59. EPA 1994b.

60. HSDB 1995.

61. EWG 1995.

62. Weisel and Jo 1996.

63. Anderson et al. (1998) found that Great Lakes anglers who ate an average of slightly less than one meal per week of Great Lakes fish had body burdens of PCDDs, PCDFs, and PCBs, measured as TCDD-equivalents, that were 2.1 times higher than those of a comparison population from the same region.

64. Dewailly et al. 1989, 1993a, 1993c, 1994a.

65. Dewailly et al. 1993a.

66. Schecter et al. 1990.

67. EPA 1994b.

68. EPA 1994b, Patandin et al. 1999.

69. EPA 1994b. In Germany, exposure estimates are in the same range. According to Papke (1998), the average nursing infant is exposed to 386 picograms of dioxins and furans (TEQ) per day—more than three times the average total adult exposure. In terms of the child's body weight, the average dose, 77 picograms (TEQ) per kilogram of body weight per day, is even higher—about forty-five times the average adult dose. A picogram equals one one-trillionth of a gram.

70. Schecter et al. 1994. The dioxin exposure of an average U.S. infant is 60 picograms TEQ per kilogram of body weight per day (EPA 1994b). This compares to government ADIs that range from 0.006 pg/kg per day in the United States (based on a cancer risk of one per million) to as high as 10 pg/kg per day in some European countries.

71. Schecter and Gasciewicz 1987.

72. Dewailly et al. 1993b.

73. Gilman et al. 1997.

74. Tatsukawa and Tanabe 1990.

75. Norstrom and Muir 1994.

76. See, for instance, Schecter et al. 1990, Jacobson and Jacobson 1993.

77. Gilman et al. 1997.

78. Swain 1988.

79. Stanley et al. 1986a, Dougherty et al. 1980.

80. Onstot et al. 1987.

81. Norstrom et al. 1981, Sodergren 1993.

82. Jarman et al. 1992.

83. For example, many of the dioxin-like compounds—chlorinated phenanthrenes, bibenzyls, and thiophenes, for example—have been identified only in recent years (Giesy et al. 1994b). Chlorinated diphenyl ethers, also highly toxic, were not identified in human tissue until the 1990s (Stanley et al. 1990b).

84. Dow 1995.

85. Chlorine Chemistry Council and Vinyl Institute 1997. See also the comments of Bill Carroll of the Vinyl Institute in Bleifus 1995.

86. See discussions in Loganathan and Kannan 1994 and Goolsby et al. 1997.

87. DeLorey et al. 1988.

88. Makhijani and Gurney 1995.

89. Sanders et al. 1993, Kjeller and Rappe 1995. Both studies found low PCB concentrations in pre-1929 sediments, presumably due—as Sanders et al. point out—to disturbance of the sediment, which would mix newer with older layers.

90. International Joint Commission 1991.

91. Alcock and Jones (1996) provide an excellent review of dioxin trends. Specific papers I discuss include examinations of Great Lakes sediment (Czuczwa and Hites 1984, 1985, 1986; Czucwa 1984), Black Forest lake sediments (Juttner et al. 1997), Baltic sediments (Kjeller et al. 1995), British soils (Kjeller et al. 1991), and British foliage (Kjeller et al. 1996). The quantitative figures I provide are given in TCDD equivalents.

92. Kjeller et al. 1996.

93. Reviewed in Alcock and Jones 1996.

94. Sediment cores from two Black Forest lakes, for example, show a contradictory pattern. One shows that dioxin levels in the layer dated 1985–1992 were lower than in that from the period 1964–1985. The other, however, shows that dioxin levels in 1982–1992 were higher than in 1960–1982 (Juttner et al. 1997).

95. Alcock and Jones 1996. Kjeller et al. (1991) also provide a useful discussion.

96. Allan et al. 1991. Alcock and Jones (1996) review studies that suggest a decline in PCDD/Fs in Baltic wildlife during the same period.

97. Huestis et al. 1997.

98. Stanley et al. 1990a.

99. EPA 1994b.

100. Reviewed in Alcock and Jones 1996.

101. Johansson 1993.

102. Gregor et al. 1995, Barrie et al. 1997.

103. Gregor et al. 1996.

104. Barrie et al. 1997.

105. Loganathan and Kannan 1994. Similarly, PCBs and DDT declined from the 1970s to the 1980s in ringed seals at Holman Island and sea birds in Lancaster Sound—both Canadian arctic locations—but the decline then stopped. In arctic peregrine falcons, metabolites of DDT and heptachlor declined during the 1980s, but other organochlorines (including PCBs, HCB, chlordane, dieldrin, and hexachlorocyclohexane) have held steady (Muir et al. 1997).

106. Muir et al. 1997.

107. Norstrom and Muir 1994.

108. Loganathan and Kannan 1994.

109. Guimond 1993.

110. Joy Allison, who was not a smoker, died of lung cancer in 1998.

111. Farland 1993. EPA insisted that the risk estimates were so high because its assessment methods were unrealistically conservative. In some ways, EPA's study was conservative, assuming, for instance, that prevailing winds would always carry the incinerator's emissions in the direction of the farm. In other ways, it was accurate, assuming that a subsistence farmer would consume only home-grown beef (an assumption that turned out to be true in Joy Allison's case). In many other ways, however, it was highly nonconservative, ignoring all other chemicals released by the facility, all health effects other than cancer, and many other routes of exposure, such as consumption of contaminated fish, milk, or other meats.

112. U.S. District Court 1993.

113. As I discuss in chapter 9, Barry Commoner and his colleagues (Commoner 1987, 1990) have shown how phaseouts of major uses of certain pesticides, mercury, and lead resulted in large-scale declines in levels of these pollutants in the environment.

114. Willett et al. 1998.

115. Negotiations are currently underway to include provisions by which other individual chemicals that meet specific criteria can be added to the treaty over time.

Chapter 2

1. Thomas 1979.

2. As I discuss later in this chapter, the National Research Council (NRC 1984) and the Environmental Defense Fund (Roe et al. 1997) examined the availability of toxicological information for commercial synthetic chemicals in general; they found that detailed toxicological data are available for virtually no chemical, and even basic data are lacking for the vast majority of commercial chemicals. Organochlorines may not be a perfectly representative subset of the larger universe of commercial chemicals, so the precise fractions of substances that fall into each category of data availability cannot be determined from these studies. However, given the fact that the great number of organochlorines in commerce outstrips by far the number of compounds estimated to have complete or basic data, these qualitative generalizations must also apply to organochlorines.

3. The groups of organochlorines on the list include the polychlorinated dioxins, biphenyls, terpenes, and chloroparaffins, each of which comprises scores or hundreds of different compounds, so the total number of identified organochlorine carcinogens is over a thousand.

4. The National Research Council's list of neurotoxic chemicals includes chloromethanes (carbon tetrachloride, dichloromethane, trichlorofluoromethane), chloroethanes and ethylenes (vinyl chloride, trichloroethylene, perchloroethylene, and dichlorotetrafluoroethane), chlorobenzenes (o-dichlorobenzene, and trichlorobenzenes), and pesticides (chlordecone and chlordane) (Landrigan et al. 1992).

5. Landrigan et al. 1992.

6. Landrigan et al. 1992, HSDB 1995.

7. Seegal and Schantz 1994, Seegal et al. 1997.

8. Henschler 1994, CanTox 1994.

9. Shane 1989.

10. Welch and Paul 1993, Mattison et al. 1989.

11. A popular review is Colborn et al. (1996); more technical reviews include Toppari et al. (1996) and Colborn et al. (1993).

12. Shane 1989, Welch and Paul 1993, GAO 1991.

13. Talmage et al. 1992, Repetto and Baliga 1996, Barnett and Rodgers 1994, Exon and Koller 1983, Exon et al. 1984, Kerkvliet 1994.

14. Kilburn and Warshaw 1992, 1994; Repetto and Baliga 1996.

15. The evidence linking organochlorines to autoimmune diseases is preliminary but worthy of concern. Workers exposed to vinyl chloride (VC) are at an increased risk of scleroderma, an extremely painful and sometimes fatal disease in which the immune system attacks tissues throughout the body, causing fibrous tissue to grow in the lungs, skin, liver, and spleen. In one population studied in the 1970s, 18 percent of VC-exposed workers developed this syndrome (Kilburn and Warshaw 1994). Occupational exposure to trichloroethylene has also been linked to systemic scleroderma, an extremely painful and ultimately fatal hardening of the skin and other tissues in the body (Lockey et al. 1987). There are also suggestions of autoimmune disease in people exposed to much lower, but still elevated, levels of organochlorines in the environment. In Tucson, Arizona, residents exposed to trichloroethylene and 1,1,1-trichloroethane in groundwater over a long period of time experience a very high incidence of a number of symptoms of lupus, including antinuclear antibodies—immune proteins that attack the body's own cell nuclei as foreign—in their blood (Kilburn and Warshaw 1992). Antinuclear antibodies were also found in the blood of a large umber of persons exposed to chlorinated solvents in Woburn, Massachusetts (Jennings et al. 1988). A British study of eighteen dioxin-exposed chemical workers found that eight of the men had antinuclear antibodies in their blood, compared to zero in a matched control group of unexposed workers, and the incidence of autoimmune antibody complexes in the bloods was three times higher in exposed than in unexposed workers (Jennings et al. 1988). Similarly, workers exposed to the termiticide chlordane have significantly elevated levels of several types of immunoglobulins and anti-self antibodies ten years after exposure has ceased (McConnachie and Zahalsky 1992). People who reside in log houses treated with pentachlorophenol also show signs of autoimmunity, including antibodies against their own smooth muscle cells (McConnachie and Zahalsky 1991). Other endocrine-disrupting pollutants could also have a role in antoimmunity, but there is little specific information on this subject. It is clear, however, that estrogen exposure, in utero or in adulthood, appears to be involved in the development of lupus, an autoimmune disease that strikes primarily young women, and testosterone seems to protect against lupus (Walker et al. 1996). Further, daughters of mothers who took the estrogenic drug DES during pregnancy utero have a markedly increased risk of lupus when they reach adulthood. DES daughters also suffer from high rates of rheumatic fever—a hyperimmune complication that can follow a bacterial infection—as well as higher blood levels of immune proteins that indicate dysfunction in immune regulation (PB Blair 1992, PB Blair et al. 1992). If DES and endogenous steroids can contribute to these diseases, it is possible that estrogenic and antiandrogenic organochlorines may do so too.

16. Huff 1994.

17. Henschler 1994.

18. HSDB 1997.

19. APHA 1994.

20. IJCSAB 1989.

21. Henschler 1994. The order in which I present Henschler's generalizations is different from his presentation.

22. Although the logic behind deductive reasoning is almost unassailable, there are other complications with deduction. The most important is that we can never be absolutely sure that our general theories are true. For the purpose of my argument here, I bracket questions about the epistemological status of scientific laws, taking as a given the well-established theories of chemistry and biology. For more on deduction, however, interested readers should see Popper (1953) for the classic philosophical critique and Barnes et al. (1996) for an excellent sociological exploration.

23. Mackay 1992.

24. Cheek et al. 1998.

25. More detail on the mechanism of action of these receptors can be found in Gronemeyer (1992) and Schmidt and Bradfield (1996).

26. Wurtz et al. 1996.

27. Colborn et al. 1993. Other hormone-disrupting chemicals that are not chlorinated include the phthalates, used as plasticizers in PVC plastic, and the alkyl phenols, found in some detergents and plastics.

28. Katzenellenbogen 1995.

29. Ballschmitter 1996.

30. Mackay 1992.

31. APHA 1994.

32. Roth and Grunfeld 1985.

33. Anderson et al. 1972, 1973, 1975.

34. Carr and Griffin 1985.

35. Carr and Griffin 1985.

36. Reviewed in Rudell 1997, who provides an excellent critique of oversimplified views of endocrine disruption.

37. Safe 1995, Harper et al. 1991. This difference in biochemical effect (measured as induction of aryl hydrocarbon hydroxylase activity) between dioxin and nonchlorinated chemicals that bind the Ah-receptor is far greater than the difference in the affinity of these compounds for the receptor (Harper et al. 1991), corroborating the idea that persistence is an important factor in dioxin toxicity.

38. CanTox 1994. The clients that funded the study were the Chlorine Chemistry Council, the Chlorine Institute, the Halogenated Solvents Industry Alliance, the European Council of Vinyl Manufacturers, and Eurochlor, the European organization of the chemical industry.

39. Willes et al. 1993.

40. In CanTox's discussion, these are actually two separate principles. They are so closely related, however, that I treat them here as a single idea.

41. Karol 1995. The document was prepared by the SOT's "ad hoc Chlorine Working Group," which was composed of four academic toxicologists and two representatives of consulting firms that have been frequently hired by the chlorine industry (Jellinkek, Schwartz, and Connolly, Inc. and the Institute for Evaluating Health Risks).

42. There are a handful of exceptions. For instance, Giesy et al. (1994b) have attempted to calculate acceptable concentrations of dioxin-like compounds in water based on their toxicity to fish-eating birds.

43. Kamrin (1988) provides a useful description of the typical testing regimen for chronic effects. Toppari et al. (1996) discuss protocols for reproductive toxicity testing.

44. Hattis et al. 1996.

45. This definition comes from Merriam-Webster's *New Collegiate Dictionary.*

46. Fitchko 1986, Pimentel and Levitan 1986.

47. Frank et al. 1993c, Roth 1993. In these studies, trichloroethylene, tetrachloroethylene, chloroform, and trichloroacetic acid caused significant changes in the abundance and activity of nematodes, insects, annelid worms, arachnids, and myriapods.

48. Fitchko 1986. Sodergren et al. (1993) describe the impacts of chlorine-bleached pulp mill effluent on aquatic plants and ecology in the Swedish Baltic Sea.

49. Greisemer and Eustis 1994.

50. Huff 1996.

51. Mattison et al. 1989. Toppari et al. 1996.

52. Henschler 1990. The compounds assessed were carbon tetrachloride, chloroform, dichloromethane, vinyl chloride, trichloroethyle, tetrachloroethylene, 1,2-dichloroethane, and 1,1,1-trichloroethane.

53. Hays 1992, Robbins and Johnston 1982.

54. Robbins and Johnston 1982.

55. Robbins and Johnston 1982.

56. Lane et al. 1968.

57. Hays (1992) presents the sequence of events by which the lead threshold was progressively reduced.

58. ATSDR (1997b) notes that neurodevelopmental effects can now be observed at less than 10 micrograms per deciliter, and Rice et al. (1996) confirm that there is likely to be no threshold whatsoever for lead's effects on development.

59. Yang et al. 1994a.

60. Mumtaz et al. (1994) note that 95 percent of studies of mixtures are of just two chemicals. Thus, the portion of all studies that have investigated mixtures of more than two chemicals is 5 percent of the 5 percent of studies that look at more than one chemical, for a total of 0.25 percent.

61. Simmons 1994.

62. Germolec et al. 1991, Yang 1994b.

63. Portier and Sherman 1994.

64. McKinney 1997.

65. Krishnan and Brodeur 1994.

66. Mehendale 1994. The same effect occurred when chlordecone was mixed with a low dose of chloroform.

67. van Birgelen et al. 1996.

68. Hrelia et al. 1994.

69. Kavlock and Perreault 1994.

70. Oesch et al. 1994, Daly et al. 1994.

71. Yang 1994b.

72. Yang 1994b.

73. Yang's (1994b) calculation is based on the formula that the number of experiments required to investigate the effects of all possible combinations of *N* chemicals is always $2^N - 1$. Using this formula, I have calculated the number of experiments required to study all possible mixtures of N = 11,000 commercial organochlorines.

74. Daly 1991, de Swart et al. 1996, Villeneuve et al. 1994.

75. Willes et al. 1993.

76. Pitot and Dragan 1991.

77. Huff et al. 1996a.

78. Bern 1992, vom Saal et al. 1992.

79. For instance, Darcy Kelley and colleagues have described in detail the effects of androgens on muscle and nerve development in the larynx of maturing males in the frog *Xenopus laevis* (Catz et al. 1995, Fischer et al. 1995).

80. Chappell et al. 1997.

81. Zhou and Waxman 1998.

82. Crain et al. 1997.

83. These figures come from Hansen and Jansen (1994). When concentrations are expressed in moles per liter (a mole is $6.02 * 10^{23}$ molecules), the normal concentration of estradiol is 0.2 to 1 nanomolar, while the concentrations of hormonally active DDT metabolites, hexachlorobenzene, hexachlorocyclohexane and PCBs range from 4 to 60 nanomolar (nanomolar is equivalent to 10^{-9} moles per liter) (Zava et al. 1997).

84. The affinity of hormones for their receptors is expressed as the dissociation constant (Kd), which is equal to the concentration of hormone at which half of available receptors are occupied by hormone. The affinity of most steroid hormone receptors for their natural hormones are 10^{-9} to 10^{-11} moles per liter (Baniahmad et al. 1994). The affinity of dioxin for the Ah receptor is in the range 10^{-9} to 10^{-12} moles per liter, depending on the assay used (Safe 1995). The affinity of many pesticides for the estrogen receptor, on the other hand, is several orders of magnitude lower than these values, often in the range of 10^{-6} molar (see, for instance, Zava et al. 1997).

85. Clark et al. (1985) estimate that there are 10,000 estrogen receptors per cell, 20,000 to 100,000 progesterone receptors per cell, and 5,000 androgen receptors per cell. Similarly, the dioxin receptor is present in quantities ranging from about 10,000 to 30,000 per cell, depending on the cell type studied (Roberts et al. 1991, Safe 1995).

86. O'Malley and Tsai (1992) note that the Kd of the ligand-bound steroid receptor homodimer for its response element is about 10^{-9} moles per liter, but in the presence of another dimer, cooperative binding increases the affinity substantially, lowering the Kd to 10^{-11} moles per liter.

87. Anderson et al. 1972.

88. Measurable biological effects occur at hormone concentrations as much as three orders of magnitude lower than the Kd, a level equivalent to 0.1 percent receptor occupancy (Roth and Grunfeld 1985). Given a total number of receptors of 5,000 to 30,000, this level is equivalent to just 5 to 30 occupied receptors per cell.

89. Roth and Grunfeld 1985.

90. Nonlinearities at this stage may arise if multiple receptors have a synergistic effect on gene expression (as, for example, if receptors dimerize before binding DNA—as most do—or if one dimer facilitates the binding of a second dimer on DNA). But nonlinearity does not create a truly threshold-like dose-response relationship until eight or more molecules are required for a biological process. Moreover, although this kind of requirement for cooperative

binding or synergy may dampen the response below a few molecules per cell, it quickly results in a greatly amplified response above this level (see, for example, Darnell et al. 1990).

91. Tritscher et al. 1994, Kohn et al. 1996, Portier et al. 1996.

92. Peterson et al. 1992.

93. Neubert et al. (1991) documented this effect in primate lymphocytes at TCDD concentrations as low as 10^{-14} moles per liter.

94. Kerkvliet (1994) reports that TCDD concentrations in the spleen as low as $2*10^{-15}$ moles per liter caused immunotoxicity in laboratory rats.

95. Schmidt and Bradfield 1996.

96. Gorski and Hou 1995.

97. The state of knowledge on this large "superfamily" of proteins, called the nuclear receptors, is reviewed by Gronemeyer and Laudet (1995). The superfamily includes receptors for retinoic acids, cholesterol metabolites, thyroid hormones, fatty acids, prostaglandins, and other substances; many others are considered orphan receptors because the natural substances that bind to them have yet to be identified. There are also numerous receptors that play central roles in the development and physiology of arthropods (i.e., insects, crustaceans, spiders and other classes) but are not present in mammals (Thummel 1995). Chemicals that interfere with the action of these proteins could have serious ecological impacts, but they would be undetected by any testing protocols conducted on laboratory rodents or mammalian cells.

98. Stancel et al. 1995, Ekenat et al. 1997.

99. Hryb et al. 1989, Nakhla et al. 1994.

100. Landrigan et al. 1992.

101. Hattis et al. 1996, Landrigan et al. 1992.

102. Weiss 1990.

103. Silbergeld 1990, Goldman et al. 1991.

104. Seegal and Schantz 1994.

105. Landrigan et al. 1992. Silbergeld (1990) has made a similar argument.

106. Kang et al. (1996) found that the DDT, dieldrin, toxaphene, and PCBs reduce GJIC in cultured human breast cells; the effect on GJIC in neurons has not been studied specifically.

107. Hattis et al. 1996.

108. Henschel et al. 1997.

109. Porterfield 1994.

110. Hattis et al. 1996, Silbergeld 1990.

111. Alleva et al. 1995.

112. For a contemporary version, see Karol (1995). Toxicologists made the same gesture in the lead controversy, as Robbins and Johnston (1982) show.

113. Hays 1992.

114. Castleman and Ziem 1989

115. NRC 1984.

116. Greisemer and Eustis 1994. Lucier and Schecter (1998) note that the National Toxicology Program has the resources to evaluate ten chemicals for carcinogenicity and ten for reproductive toxicity each year.

117. Yang 1994b.

118. Roe et al. 1997.

119. CanTox 1994, Mackay 1992.

120. SARs are useful for making preliminary predictions of the potency of a compound to produce a single well-defined effect, if that substance is a member of a chemical group that has been thoroughly studied for that specific impact (see, for instance, Waller et al. 1996). Even then, SARs are suited only for use as a screening tool to identify compounds for more detailed investigation; they are not an adequate basis for a judgment of safety or for establishing exposure standards.

121. Huff et al. 1996c.

122. Katzenellenbogen 1995.

123. Landrigan et al. 1992.

Chapter 3

1. Popper 1953.

2. This error-correction mechanism is an informal one, not codified in statutes or regulations. It is largely a political decision whether EPA and similar agencies in other nations will take more ambitious steps than a risk assessment demands to control or prevent pollution. EPA has the authority to restrict chemicals based on data from laboratory animals, but it has seldom done so, and it frequently responds to calls for more preventative policies with the objection that harm in human and wildlife populations has not been conclusively demonstrated (see, for instance, Crisp et al. 1998). Since risk assessment became a central tool in chemical regulation, there have been few, if any, actions to restrict chemical production and use beyond what a risk assessment would call for.

3. Willes et al. 1993.

4. Crisp et al. 1998. The European Commission (1996) came to a nearly identical conclusion in its examination of endocrine disruption.

5. Kolata 1996.

6. Rothman and Greenland (1998), Fox (1991). Although CanTox (1993) has argued that all the criteria must be met before a causal link can be inferred, the Hill criteria are not intended as an absolute set of requirements. Rather, they are to be used as a framework for assessing evidence in the course of making a judgment about causality, and most epidemiologists recognize that all the criteria will seldom be fulfilled.

7. Rose 1987.

8. Kolata 1996.

9. Rose 1987.

10. CanTox 1993.

11. Jack Weinberg and I have described this approach in Weinberg and Thornton (1994).

12. Miller et al. 1991.

13. IJCSAB 1991.

14. In this argument, I have drawn extensively on the work of the International Joint Commission's Science Advisory Report (IJCSAB 1989).

15. For instance, a 1996 report for the European Union (European Commission 1996) defined endocrine disrupters as chemicals that "cause adverse health effects in an intact organ-

ism or its progeny consequent to changes in endocrine function." By this standard, a chemical known to disrupt the endocrine system (by activating receptors, changing gene expression, or reducing the circulating level of hormones) but for which adverse effects at the organismal level have not yet been discovered would not be considered an endocrine disrupter.

16. The International Joint Commission's Science Advisory Board (IJCSAB 1995) discusses these various definitions of health; I draw extensively on the board's work.

17. Fox 1993. My discussion of biomarkers draws on Fox's argument.

18. vom Saal et al. 1992.

19. Kandel 1998.

20. Some scientists have begun to recognize the importance of functional disruption at the biochemical level (see, for instance, Colborn et al. 1993, 1996; McLachlan 1993; Fox 1992, 1993; Zile 1992), but the view persists in official circles that only frankly negative impacts at the level of the organism are worthy of concern.

21. Toppari et al. 1996, Paulozzi 1999. The historical records are somewhat limited, so a role for improved diagnosis in rising rates of these conditions cannot be ruled out; however, as Toppari et al. conclude, at least some of the reported increase is likely to be due to a real rise in incidence.

22. Toppari et al. 1996.

23. Carlsen et al. 1992.

24. Criticism of this study, summarized in Sherins (1995), was sparked by a reanalysis of the Carlsen data by scientists from Dow Chemical and Shell Oil (Olsen et al. 1995), which suggested that there was no decrease in the last twenty years, a conclusion that the analysis of Swan et al. (1997) clearly refutes.

25. Auger et al. 1995, Van Waeghelem et al. 1996, Pajarinen et al. 1997, Irvine et al. 1996.

26. Bujan et al. 1996, Fisch et al. 1996.

27. Swan et al. 1997.

28. Mattison 1991.

29. Toppari et al. 1996.

30. Sager et al. 1991.

31. Kelce 1994. This and other relevant studies are reviewed in Toppari et al. 1996.

32. Newbold 1995, Toppari et al. 1996.

33. Reviewed in Peterson et al. 1992, Theobald and Peterson 1994. Similar effects have since been found in other species of rodents (Theobald and Peterson 1997, Gray et al. 1997a).

34. DeVito et al. 1995. These figures include dioxins, furans, and PCBs, expressed as TEQ. Body burdens in the Yu-cheng incident were elevated by an even greater margin: approximately three orders of magnitude.

35. Gray et al. 1997a.

36. Guillette et al. 1994, Guillette 1995, Semenza et al. 1997, Crain et al. 1997. Turtles from Lake Apopka have also experienced feminization of the gonads and endocrine system. Laboratory studies show that treatment of turtle eggs with low levels of certain PCBs (as low as 1 ppm in the egg) results in sex reversal, such that incubation of eggs at a temperature that would normally produce only males produces a high frequency of females instead (Crews et al. 1995).

37. Double-crested cormorant eggs in Great Lakes colonies contain five to ten times more PCBs and dioxins than eggs from Lake Winnpegosis, a relatively unpolluted ecosystem in Manitoba (Giesy et al. 1994b).

38. Fox 1992.

39. Fry et al. 1987, Fry and Toone 1981.

40. Fox 1992, Fry 1995.

41. Facemire et al. 1995.

42. I have made these quantitative comparisons using the regression equation ($y = -0.25x + 9.99$), the best linear fit to the data, presented by Subramian et al. (1987) for testosterone levels (the dependent variable, in nanograms per liter) versus the sum of PCBs and DDE in the animals' blood (the independent variable, in micrograms per gram). Comparisons are for the mean, low, and high values for the sum of these pollutants.

43. Whorton et al. 1977, 1979.

44. Lerda and Rizzi 1991.

45. Egeland et al. 1994. There is also evidence of impaired male development from the Yucheng incident. Boys born to mothers exposed to rice oil contaminated with PCBs and PCDFs are just now reaching reproductive age, so their fertility has not been evaluated, but at ages 11 to 14, their penises are significantly smaller than those in a group of matched control boys (Guo et al. 1993).

46. Goh et al. 1998. The reduction in blood testosterone showed a dose-response relationship with increasing trichloroethylene exposure.

47. Bush et al. 1986.

48. Pines et al. 1987.

49. Dougherty et al. 1981.

50. Feichtinger et al. 1991.

51. Welch and Paul 1993.

52. Mattison 1991, Mosher and Bachrach 1996. These data are derived from the National Center for Health Statistics' Family Growth Survey, which gathers data on many aspects of reproductive health and behavior in the United States based on interviews with a very large sample of women in the U.S. population.

53. Konickx et al. 1994.

54. Reviewed in Whitten 1992.

55. Herman-Giddens et al. 1997. In contrast to the studies reviewed by Whitten (1992), this study found no evidence that the age at menarche had declined in preceding decades.

56. Whitten 1992.

57. Lundberg 1973.

58. Cummings and Gray 1989.

59. Allen et al. 1979, Barsotti et al. 1979.

60. Mattison et al. 1989, Walters et al.1993, Gellert 1978.

61. Gray et al. 1995, 1997b; Gray and Ostby 1995; Li et al. 1995.

62. Reinjders 1986.

63. Bergmann and Olsson 1985, Olsson et al. 1994.

64. Wren 1991, Jensen et al. 1977, Heaton et al. 1991, Tillitt et al. 1996.

65. Reviewed in Fox 1993.

66. Rachootin and Olsen 1983, Doyle et al. 1997a, van der Gulden and Zielhuis 1989, Kyyronen et al. 1989. As van der Gulden and Zielhuis (1989) point out, there have also been a number of negative studies, but most of these had methodological flaws that undermined their ability to detect a potential association between exposure and effect.

67. Hsu et al. 1994.

68. Nicholson and Landrigan 1994.

69. Phuong et al. 1989a, Phuong et al. 1989b, and Huong et al. 1989 found, in both case-control and cohort studies conducted at the Ob-Gyn Hospital of Ho Chi Minh City in the south of Vietnam during the early 1980s, that mothers exposed to Agent Orange had greatly elevated risks of molar pregnancies or giving birth to babies with malformations. Potential problems with these studies are discussed in Sweeney 1994, so these results should thus be interpreted with some caution. However, the specific problems identified do not completely undermine confidence in their findings: incomplete records and uncertainty in classification would bias the study towards a finding of no association, and populations need not be representative of the entire nation to indicate a difference in risk between specific exposed and unexposed groups.

70. Baukloh et al. 1985. The inverse correlation of fertility with body burden was seen only in Austrian but not German women at the clinic.

71. Gerhard 1993; Gerhard et al. 1991, 1992, 1993.

72. Leoni et al. 1989.

73. Saxena et al. 1980, 1981.

74. Rier et al. 1995, Endometriosis Association 1993.

75. Konickx et al. 1994.

76. Rier 1995, Endometriosis Association 1993.

77. Rier et al. 1993, Rier 1995. The body burden numbers are from DeVito et al. (1995), who calculate a body burden of 69 ppt in the low-dose rhesus monkeys group, compared to 8 to 13 ppt in the general U.S. population.

78. Cummings et al. 1996, Johnson et al. 1997. These studies found that dioxin increases endometrial growth only at low doses. At higher doses, no effect on endometriosis was observed, presumably due to dioxin's toxicity to the ovaries, an effect that may reduce synthesis of steroid hormones that are required for endometrial growth. This mechanism may explain why another study found that PCBs at relatively high doses had no effect on endometriosis in the monkey (Arnold et al. 1996).

79. Rier et al. 1995.

80. Lebel et al. 1998.

81. Mayani et al. 1997.

82. Mattison 1991.

83. Edmonds and James 1990. In Canada, too, cardiac and urinary tract defects increased substantially from 1979 to 1993 (Johnson and Rouleau 1997).

84. It is clear that the reported incidence of some cardiac abnormalities has increased due to improved screening and diagnosis, so whether there has been a real increase remains unknown. (Ferencz and Villasenor 1991).

85. Reviewed in Theobald and Peterson 1994.

86. Walker and Peterson 1994.

87. Reviewed in Theobald and Peterson 1994.

88. ATSDR 1993b, 1993d, 1995d.

89. van der Gulden and Zielhuis 1989.

90. Miller et al. 1991.

91. ATSDR 1995h.

92. Ludwig et al. 1993, 1996; Giesy et al. 1994a, 1994b; Bowerman et al. 1995; Colborn 1991, Fox 1993; Gilbertson et al. 1991.

93. Ludwig et al. 1996.

94. Bowerman et al. 1995.

95. Reviewed in Giesy et al. 1994a, 1994b; Ludwig et al. 1996.

96. For example, the frequency of hatchling failure and birth defects in turtles collected from a number of locations in the Great Lakes in the 1980s is strongly correlated with the levels of PCBs, dieldrin, and other pesticides in the turtle eggs (Bishop et al. 1991).

97. Mac and Edsall 1991; Mac et al. 1985, 1993; Walker and Peterson 1994; Binder and Lech 1984; Mac 1988.

98. See the discussion in Walker and Peterson (1994).

99. Reviewed in Nicholson and Landrigan 1994.

100. Hsu et al. 1994.

101. Dai et al. 1990; Phuong et al. 1989a, 1989b; Huong et al. 1989. Dai et al. (1990) analyzed infant mortality rates in the first year of life in three Vietnamese villages—two sprayed with Agent Orange and one never sprayed. During the years of spraying, the risk of infant mortality in the sprayed villages was 2.4 times higher than in the unsprayed village; the risk of fetal death was elevated through 1986, although the increase in the final five years of the study was not statistically significant.

102. McDonald 1988, Holmberg and Nurminen 1980.

103. Reviewed in Miller et al. 1991.

104. Ferencz and Villasenor 1991.

105. Ferencz and Villasenor 1991.

106. Garry et al. 1996.

107. Karmaus and Wolf 1995.

108. Berry and Bove 1997, Geschwind et al. 1992.

109. Munger et al. 1997.

110. Garry et al. 1996.

111. Bove et al. 1995.

112. Kramer et al. 1992. An increased risk of low birthweight was also found among the group with high chloroform exposures, but the result did not reach statistical significance.

113. Kanitz et al. 1996. In this study, babies born to women who drank chlorine-disinfected water were at a 6.6-fold risk of reduced birth weight, but the increase did not quite reach the level of statistical significance (95 percent confidence interval = 0.9 to 14.6). All associations persisted after adjustment for mother's income, education level, age, and sex of the child.

114. Waller et al. 1998.

115. Fein et al. 1984, Jacobson and Jacobson 1988. Two later studies did not find physical differences associated with the mother's consumption of fish (Dar et al. 1992, Lonky et al. 1996).

116. Rylander et al. 1995, 1996, 1998.

117. Lane and Hathaway 1985; Koppe 1989; Koppe et al. 1989, 1991.

118. Reviewed in Seegal and Schantz (1994); see also Eriksson and Fredriksson (1996), Goldey et al. (1995), Seegal et al. (1997).

119. Goldey and Taylor 1992.

120. Rice 1999, Schantz et al. 1989, 1991; Seegal and Schantz 1994.

121. Reviewed in Seegal and Schantz (1994); see also Bowman et al. (1989), Schantz (1986).

122. DeVito et al. 1995.

123. Brouwer et al. 1989, Zile 1992.

124. Leatherland 1993, 1998; Fox 1993; Moccia et al. 1986. Although goiter can be caused by iodine deficiency, the temporal and spatial patterns of thyroid disruption in Great Lakes species rule out iodine deficiency as the primary cause. Goiter was most severe when and where organochlorine contamination has been highest.

125. Daly et al. 1989, 1998; Daly 1991, 1993.

126. Hsu et al. 1994, Chen et al. 1994, Brouwer et al. 1995.

127. Jacobson et al. 1990; Jacobson and Jacobson 1988, 1993, 1996; Swain 1988.

128. Lonkey et al. 1996. The authors adjusted statistically for a large number of potential confounding factors, including the mother's health, the characteristics of the infant at birth, the mother's use of alcohol, tobacco, caffeine, pharmaceuticals, and other substances, and demographic and socioeconomic factors, so the neurological effect appears to have been caused specifically by something in the fish, presumably chemical contaminants.

129. Gladen et al. 1988; Rogan et al. 1986; Rogan and Gladen 1991, 1992; Gladen and Rogan 1991. Although cognitive deficits were inversely correlated with the concentration of pollutants in breast milk, the effects appeared to be due to exposure in utero rather than through breastfeeding, because avoiding breastfeeding did not improve scores. The lack of a detectable effect at five years and older in the North Carolina study could be attributable to any of several design differences between the studies, including the use of different measures of performance, the use of breast milk instead of cord serum as a marker of prenatal exposure, or the fact that PCBs may be a more reliable marker of total organochlorine exposure in Great Lakes fish—where all contaminants bioaccumulate through a single food chain—than in the more heterogeneous diet of the North Carolina population.

130. Pluim et al. 1993, Koopman-Esseboom et al. 1994, Huisman et al. 1995. Supporting these findings, scientists at the U.S. Environmental Protection Agency have calculated that PCB levels in the blood of the general human population are in the range that can be expected to cause mild hypothyroidism (McKinney and Pedersen 1987); these studies confirm that hypothesis.

131. Landrigan et al. 1992.

132. Alleva et al. 1995.

133. Reviewed in Talmage et al. 1992, DeVito and Birnbaum 1994.

134. Barnett and Rodgers 1994.

135. Zelikoff 1994.

136. Snyder 1994, Talmage et al. 1992.

137. Burleson et al. 1996. The body burden calculation is from DeVito et al. (1995). Corroborating these findings, a single dose of 3 micrograms of TCDD per kilograms of body weight results in depressed natural killer cell activity in response to viral infection in rats (Yang et al. 1994).

138. Neubert et al. 1992.

139. Barnett et al. 1996.

140. Barnett and Rodgers 1994.

141. Exon and Koller 1983.

142. Exon et al. 1984.

143. Reviewed in Kerkvliet 1994. See also DeVito and Birnbaum 1994.

144. Tryphonas 1995. In this study, an immunotoxic dose of 5 micrograms PCB per kilogram of body weight per day produced a blood level of 10 parts per billion, roughly equivalent to the average in the general U.S. population.

145. Grasman et al. 1996. One mechanism by which ambient levels of organochlorines may have suppressed the immune systems of wildlife is by reducing levels of vitamin A and its metabolites—called retinols—in the body. Retinols are essential for immunity, as well as numerous aspects of fetal development and growth, female reproductive function, spermatogenesis, and vision. Dioxin and other organochlorines are known to disrupt the storage and metabolism of retinols. The severity of immunosuppression in the Great Lakes gulls was statistically associated with reduced levels of vitamin A. Similarly, seals fed fish from the Wadden Sea have significantly reduced levels of retinol compared to control seals fed fish from the northeast Atlantic (Brouwer 1989; see also Zile 1992).

146. Cleland et al. 1989.

147. de Swart et al. 1996. For a discussion of organochlorine contamination and immune suppression in Lake Baikal seals, see Nakata et al. 1995.

148. Lahvis et al. 1995. Some critics of the hypothesis that organochlorines have played a causal role have suggested that the die-offs are due to exposure to brevetoxin, a natural toxin released by algal blooms present along the Atlantic coast (CanTox 1993, for instance). As Lahvis et al. point out, brevetoxin may have played a role in some of the deaths in the Atlantic dolphin population, but it cannot completely account for all the die-offs: it was present in some but not all dead animals in the Atlantic, and it cannot explain the contemporaneous mortality in populations in the Gulf of Mexico and the Mediterranean, where brevetoxin is not known to be a problem.

149. Hsu et al. 1994, Rogan et al. 1988.

150. See the discussion in Tryphonas (1995). There are also several studies that have not found a significant association between PCB exposure and immune suppression, but none had the statistical power to rule out a moderate effect.

151. Tonn et al. 1996. Additional immunosuppressive effects have been found in the Hamburg chemical workers by Ernst et al. (1998).

152. Zober et al. 1994.

153. Faustini et al. 1996.

154. Byers et al. 1988.

155. The immunological changes in the Latvian and Swedish populations were not strictly consistent with each other, which somewhat weakens the case that organochlorines caused both types of effects. However, as the authors pointed out, the immune system is modulated by a variety of factors, and the effects of chemicals will vary depending on the background of other environmental and constitutional factors. Thus, differences in outcome between the Swedish and Latvian populations may reflect different responses to the same stimulus in different contexts.

156. Dewailly et al. 1989, 1993a, 1993b; Birnbaum 1995a. Although dioxins have not been measured in breast milk, their level in Inuits' blood is three to seven times greater than in the south, so presumably they are also significantly elevated in Inuits' milk (Ayotte et al. 1997).

157. Barnett et al. 1996.

158. In a 1993 speech at Boston University, David Ozonoff compared John Snow's actions in stopping the cholera epidemic to the proper course of action on chlorine chemistry. I have adapted my discussion from his use of this analogy.

Chapter 4

1. Tomatis et al. 1990.
2. Davis et al. 1994.
3. Epstein 1993.
4. Davis et al. 1994, Huff et al. 1996b.
5. Davis et al. 1994.
6. Davis et al. (1990) make this point about 1980 cancer statistics, and this pattern remains true in 1990 cancer statistics (IARC 1990) as well.
7. Hoel et al. 1992, Davis et al. 1990, Bailar 1990, Tomatis et al. 1990.
8. Huff 1996.
9. IARC 1998.
10. Tomatis et al. 1990.
11. Davis et al. 1990.
12. Davis et al. 1994.
13. Bailar 1990.
14. Davis et al. 1990.
15. Bailar 1990.
16. Lopez 1990.
17. Devesa et al. 1995.
18. Epstein 1993.
19. Huff 1996.
20. Adami et al. 1993.
21. Zahm and Devesa 1995, Devesa et al. 1995.
22. Bailar 1990.
23. Bailar 1990, Lopez 1990.
24. Davis et al. 1994.
25. Davis et al. 1994.
26. Adami et al. 1993.
27. Axelson et al. 1990.
28. Davis et al. 1992, Blair and Zahm 1995.
29. Reviewed in Blair and Zahm 1995.
30. Davis et al. 1994.
31. WMO/UNEP 1996.
32. Makhijani and Gurney (1995) provide a detailed review of ozone depletion, its causes, and its consequences.
33. UNEP 1996, Montzka et al. 1999.
34. UNEP 1994, 1996.

35. UNEP 1994, 1996.

36. Tomatis et al. 1990.

37. Rigel et al. 1996. The cited increase in nonmelanoma skin cancer refers to squamous cell carcinoma only; see also Glass and Hoover (1989).

38. Rigel et al. 1996.

39. UNEP (1994) noted that each 1 percent decrease in ozone is expected to result in a 0.5 percent increase (or a 0.3 to 0.6 percent increase, depending on the study one uses) in cataract incidence. According to UNEP, there were 17 million cases of cataract in 1985, so—using the 0.5 percent figure—a 10 percent peak reduction in stratospheric ozone would result in 850,000 new cases of cataract each year.

40. UNEP 1994, 1996.

41. Makhijani and Gurney 1995, UNEP 1994. Reduced productivity in phytoplankton would be expected to reduce the numbers of the larger species that prey on these organisms as well. Marine animals can also be directly affected by ultraviolet radiation. Even mild ozone depletion, for instance, is expected to reduce the breeding season of shrimp by about 50 percent. Reduced productivity and ecological changes in marine ecosystems could have severe implications for humans, who, on a global basis, obtain about 50 percent of their protein consumption from the sea (Makhijani and Gurney 1995).

42. Kohlmeier et al. 1990.

43. Kohlmeier et al. 1990.

44. Kelsey and Gammon 1991.

45. Harris et al. 1992, Kohlmeier et al. 1990, Hoel et al. 1992.

46. The statistics on breast cancer incidence, mortality, and risk are summarized in Kelsey and Bernstein (1996), Devesa et al. (1995), and Miller et al. (1993).

47. Reviewed in Kelsey and Bernstein (1996). Dorgan et al. (1997) and Schapira et al. (1991) provide specific evidence linking hormonal status to breast cancer risk.

48. Malone 1993, Newbold 1995.

49. Hsieh et al. 1992, Ekbom et al. 1992.

50. Hajek et al. 1997.

51. Rose et al. 1986, Cohen et al. 1993, Wynder et al. 1986. Within countries, however, the relationship between fat and breast cancer is very weak. A study in sixty-five Chinese countries with very low levels of industrialization found that fat intake varied widely—from 5 to 47 percent of total calories— but breast cancer rates showed much less variation and were only weakly associated with fat consumption (Marshall et al., 1992).

52. See, for instance, Graham et al. (1992), Howe et al. (1990). In contrast, Willett et al. (1992) found no link between dietary fat and breast cancer risk. In addition to fat, alcohol consumption is a modest risk factor, and inadequate dietary fiber and vitamins may also be, but the results of studies have been inconsistent (Kelsey and Bernstein 1996).

53. Davis et al. 1993, Epstein 1993.

54. Kelsey and Bernstein 1996.

55. Claus et al. 1991, Colditz et al. 1994.

56. Tomatis et al. 1990.

57. Harris et al. 1992.

58. Madigan et al. 1995, Davis et al. 1997, Harris et al. 1992.

59. Hahn et al. 1989, Kohlmeier et al. 1990, Miller et al. 1993, White 1987.

60. Harris et al. 1992.

61. Harris et al. 1992, Willett et al. 1992.

62. El-Bayoumy 1992, Dean et al. 1988.

63. John and Kelsey 1993, Kelsey and Bernstein 1996.

64. Gammon and John 1993. The proposed mechanism for this effect is that light and EMFs suppress production of melatonin, a natural inhibitor of estrogen synthesis; the resulting increases in estrogen levels would then increase breast cancer risk.

65. Devra Davis and colleagues (Davis et al. 1993, 1997) have put forth the most detailed arguments on the xeno-estrogen/breast cancer hypothesis.

66. Pujol et al. 1994, Glass and Hoover 1990.

67. Glass and Hoover 1990.

68. Huff 1996.

69. See, for instance, Scribner et al. 1981.

70. ATSDR 1995h.

71. Brown et al. 1998. This effect appeared to be due to dioxin's ability to alter the development of the mammary gland in a way that predisposes it to cancer, increasing the number of cells of one kind that is highly susceptible to cancer and decreasing the number of another kind that is less so.

72. Reviewed in ATSDR 1995h.

73. Chiazze et al. 1977.

74. Kettles et al. 1997.

75. Safe 1995.

76. Manz et al. 1991, Flesch-Janys et al. 1993, Kogevinas et al. 1997. In the last study, IARC found an increased breast cancer risk in workers from the German plant, but there was no significant difference among dioxin-exposed workers in other facilities. No studies of any dioxin-exposed cohorts of women have yet found a significantly reduced incidence of breast cancer. It has been said that breast cancer is reduced in the population exposed to dioxin in the chemical manufacturing accident in Seveso, Italy; in fact, there were slightly fewer cases than expected in some exposure zones and more in others, and none of the results were statistically significant (Bertazzi et al. 1993, 1997, 1998, Bertazzi and diDomenico 1994). Two other studies of dioxin-exposed workers have found neither a significant increase nor a decrease in breast cancer (see the discussion in Ahlborg et al. 1995).

77. Wolff et al. 1996.

78. Walrath et al. 1985.

79. The New Jersey study is reported in Hall and Rosenman (1991), and the New York study (MacCubbin PA, Herzfield PM, Theriault GD. *Mortality in New York State, 1980–1982: A Report by Occupation and Industry.* Albany NY: New York State Department of Health Monograph 21, 1986) is discussed in the same paper. Both studies analyzed women of different races separately; for some cancers, statistically significant increases in mortality were limited to white or African-American women only.

80. Griffith et al. 1989.

81. Melius et al. 1994, Lewis-Michl et al. 1996.

82. Najem and Greer 1985.

83. Rylander and Hagmar 1995. Cancers of the cervix and uterus were also increased in these women, but those results were not statistically significant. The wives of Baltic and west

coast fishermen were similar in every way other than the source of the fish they ate and the fact that Baltic fish tend to be fattier than those from the west coast.

84. Falck et al. 1992.

85. Djordevic et al. 1994.

86. Wolff et al. 1993. The results for DDE in this study were statistically significant, while those for PCBs ($p = .058$) were slightly below the customary level of statistical significance.

87. Mussalo-Rauhmaa et al. 1990.

88. van't Veer et al. 1997. In this study, there was no significant difference between the levels of DDE in cases and controls, and there was no significant trend in the risk of breast cancer with increasing DDE levels. When the trend statistics were adjusted using multivariate analysis, however, there was a negative relationship between DDE and breast cancer risk.

89. Hunter et al. 1997. This study observed a negative relationship between DDE level and breast cancer that was not statistically significant. After its publication in the *New England Journal of Medicine,* editorials and accounts of this study in the media suggested that organochlorines had been cleared of a role in breast cancer risk. As several letters to the *New England Journal of Medicine* pointed out, however, current or past concentrations of a few pollutants in the blood may not be good markers of total past exposure to carcinogenic organochlorines. For example, the specific congeners of DDT and PCBs that were measured in this study do not accurately indicate past exposures to the most estrogenic DDT and PCB congeners and metabolites. Further, an adult woman's body burden does not indicate anything about her prenatal and childhood exposures, which may turn out to be even more important than the total lifetime dose of organochlorines. Finally, there are hundreds of other carcinogenic organochlorines about which these studies say absolutely nothing, so an acquittal of all organochlorines would certainly be premature. The letters appear in the *New England Journal of Medicine* 338 (1998), pp. 988-ff.

90. In addition to the Harvard and European studies, two other reports, one each in Mexico (Lopez-Carillo et al. 1997) and Vietnam (Schecter et al. 1997), found no link between breast cancer risk and blood levels of DDT or its metabolite DDE. These findings do not necessarily contradict a relationship between breast cancer risk and the long-term body burden of DDT and other organochlorines; DDT is still in use in both countries, so current blood levels may reflect recent exposures rather than long-term or long-past exposures. Further, where DDT has been long banned, exposure is primarily through the food chain, so the DDT body burden is a rough marker for the body burden of other bioaccumulated organochlorines that may also contribute to breast cancer. Where it is still in use, however, considerable additional DDT exposure occurs through water, air, and direct contact during and after insecticide applications, so DDT and DDE levels are not markers of anything but themselves.

91. Hoyer et al. 1998. In this study, blood levels of beta-hexachlorocyclohexane, the insecticide that was associated with increased breast cancer risk in the earlier Finnish study (Mussalo-Rauhamaa et al. 1990), were also associated with breast cancer risk, but the results did not reach the level of statistical significance ($p = 0.21$). This study adjusted for weight, height, number of full-term pregnancies, alcohol consumption, smoking, physical activity, menopausal status, household income, marital status, and education. It did not adjust for lactation, a potentially important confounding or distorting variable.

92. Dewailly et al. 1994b. The relative risk figure of 9 refers to women with DDE levels greater than the mean plus one standard deviation, as compared to women with levels less than the mean minus one standard deviation.

93. Krieger et al. 1994. No significant relationship to PCBs was found in any group. There was a strongly positive trend of increasing risk with increasing DDE in white women, which

was not significant, and the trend for black women alone was marginally significant (p = .066). As Savitz (1994) has pointed out, the small size of each ethnic group and the strength of the association (odds ratios of 3.9 in black women with the highest DDE levels and 2.4 among white women with the highest levels) may be more important than the lack of formal statistical significance.

94. Adlercreutz 1995, Murkies et al. 1998.

95. Moysich et al. 1998. This study also showed rising breast cancer risk with increasing levels of DDE and HCB in women who had never lactated, but those results were not statistically significant.

96. Huff 1996.

97. Huff et al. 1996a, Tomatis et al. 1990.

98. Toppari et al. 1996.

99. Boyle et al. 1995, Nomura and Kolonel 1991, Tomatis et al. 1990, Epstein 1993, Toppari et al. 1996.

100. Tomatis et al. 1990, Muir and Black 1996.

101. Buetow 1995, Nomura and Kolonel 1991, Toppari et al. 1996.

102. Newbold 1995, Huff et al. 1996b, Boslund 1996.

103. Newbold et al. 1998.

104. Boslund 1996, Boyd 1996.

105. Huff et al. 1996a.

106. Boslund (1996) and Huff (1996) point out that rodent species are poor models for the study of prostate and testicular cancer. In addition, Huff et al. (1996a) and Huff (1996) note that the standard carcinogenicity assay does not consider the delayed effects of exposure during development, a particular problem for hormonally-induced cancer (Newbold 1995).

107. CCRIS 1997; see also Huff (1996).

108. Blair et al. 1992b, Blair and Zahm 1991.

109. Morrison et al. 1993.

110. Donna et al. 1989.

111. Reviewed in Blair and Zahm (1991). Specific studies with positive associations include McDowell and Balarajan (1984); Mills et al. (1984), and Wiklund et al. (1986).

112. See the discussion in Morrison et al. (1993).

113. Fallon et al. 1993.

114. Tarone et al. 1991. See also the discussions in Hayes (1990) and Boslund (1996). The National Academy of Sciences' Institute of Medicine panel on Agent Orange concluded the data are inadequate or insufficient to determine whether there is an association with testicular cancer (Fallon et al. 1993).

115. Hayes 1990. Notably, these dogs also showed signs of testicular dysfunction and impaired spermatogenesis, strengthening confidence in the link with chlorophenoxy herbicides and dioxins, both of which cause the same effects in laboratory rodents.

116. Moss et al. 1986.

117. Hartge et al. 1994, Tomatis et al. 1990.

118. From data presented in Devesa et al. 1995.

119. Hartge et al. 1994, Weisenburger 1994, Devesa et al. 1995. The quotation is from Weisenburger (1994).

120. Schwartz 1990, Tomatis et al. 1990.

121. Hartge et al. 1994, Epstein 1993.

122. Epstein 1993, Devesa et al. 1995.

123. Vineis and D'Amore 1992, Hartge et al. 1994.

124. Hartge et al. 1994, Tomatis et al. 1990.

125. Erikkson and Karlsson 1992.

126. Thomas et al. 1982.

127. Olin 1978, Olin and Ahlbom 1980, Walrath et al. 1995, Arnetz et al. 1991, Li et al. 1989, Searle et al. 1978, Thomas et al. 1979, Hall and Rosenman 1991.

128. Hagmar et al. 1986.

129. Toren et al. 1996, Band et al. 1997, Wingren et al. 1991.

130. Hartge et al. 1994. The estimate of total cases assumes that there are 53,600 new cases of NHL per year in the United States (Ries et al. 1997).

131. ATSDR 1995h.

132. Enterline et al. 1990.

133. Norman et al. 1981.

134. ATSDR 1995, Spirta et al. 1991.

135. Blair et al. 1990, 1992a, 1993.

136. Negative studies are reviewed in a paper sponsored by the chlorinated solvents industry (Weiss 1995).

137. Vaughan et al. 1997, Ruder et al. 1994, Blair et al. 1990. Occupational exposure to trichloroethylene in the cardboard manufacturing industry has also been linked to increased incidence of kidney cancer (Henschler et al. 1995).

138. Reviewed in Zahm and Blair (1992, 1995); Blair and Zahm (1991); Blair et al. (1992b), Davis et al. (1994).

139. Reviewed in Zahm and Blair (1992), Blair and Zahm (1991), Hardell et al. (1994b).

140. Blair and Zahm 1991, Hardell et al. 1994b, Scherr et al. 1992, Zahm and Blair 1993.

141. Becher et al. 1996.

142. Kogevinas et al. 1995. The trend for increasing risk with increasing exposure to dioxin was statistically significant, while those for 2,4,5-T and pentachlorophenol were marginally nonsignificant ($p = .09$ and .08, respectively).

143. Fallon et al. 1993.

144. Brown et al. 1990.

145. Cantor et al. 1992.

146. Reviewed in Zahm and Blair 1992.

147. Eriksson and Karlsson 1992.

148. Reviewed in Daniels et al. 1997. The study of leukemia and no-pest strips is reported in Leiss and Savitz (1995).

149. Bertazzi et al. 1997.

150. Reviewed in Cantor et al. 1996.

151. Zahm and Blair 1992.

152. Hayes et al. 1991.

153. Byers et al. 1988.

154. Fagliano et al. 1990, Cohn et al. 1994. The statistically significant association was found only in women.

155. Aschengrau et al. 1993.

156. Svensson et al. 1995.

157. Hardell et al. 1996, 1998.

158. Rothmann et al. 1997.

159. Quoted in Morris 1995.

160. The history of water chlorination is given in Morris (1995), Cantor (1994), Mughal (1992), and Koivusalo and Vartianen (1997).

161. Koivusalo and Vartiainen 1997.

162. Komulainen et al. 1997.

163. Komulainen et al. 1997.

164. Melnick et al. 1997.

165. Cantor 1994.

166. I have based my discussion of the studies on drinking water and cancer in the 1970s, 1980s, and early 1990s on the presentations of Cantor et al. (1996) and Morris (1995).

167. Morris et al. 1992.

168. Koivusalo et al. 1994, 1997. Non-Hodgkin's lymphoma was also elevated by 40 percent in those exposed to chlorinated drinking water, but the increase was of borderline statistical significance.

169. King and Marrett 1996.

170. Cantor et al. 1998, Hildesheim et al. 1998, Doyle et al. 1997b. Cantor et al. (1998) found that the risk of bladder cancer was related to the duration of exposure to chlorinated drinking water. THM exposure were significantly related to cancer risk only among men who had smoked, and smoking and THM exposure seemed to have a synergistic effect on cancer risk.

171. Tritscher et al. 1996.

172. James Huff (1994) provides a useful summary of the laboratory evidence of dioxin's carcinogenicity.

173. For a review, see the introduction in U.S. House of Representatives (1990).

174. Zack and Suskind 1980, Zack and Gaffey 1983, Suskind 1984.

175. Bond et al. 1983,1989; Ott et al. 1987.

176. See the discussion in U.S. House of Representatives (1990).

177. See, for instance, EPA (1988c).

178. Luoma 1990.

179. Tschirley 1986.

180. Axelson and Hardell 1986, Hardell et al. 1994a.

181. Hardell et al. (1994a).

182. Hay and Silbergeld 1985, 1986.

183. The testimony of Monsanto's medical director occurred in the civil case *Kemner* et al. *v. Monsanto* (1985), in which Monsanto was ordered to pay millions of dollars in punitive damages to the plaintiffs. EPA chemist Cate Jenkins (1990) first brought the issue to public and governmental attention. The testimony is summarized and reproduced in part by van Strum and Merrell (1990) and Hardell et al. (1994a). Later, the newsletter *RACHEL's Environment and Health Weekly* reported on Jenkins' and van Strum's accounts of the testimony, and William Gaffey, one of the authors of the Monsanto studies, sued *RACHEL*'s editor Pe-

ter Montague for $4 million in a libel suit. The litigation, which many viewed as a bald attempt to intimidate Montague and others in the environmental movement, was dismissed by a federal judge when Gaffey died in 1995 (Montague 1996).

184. Rohleder 1989; see also Yanchinski 1989. This case is also discussed by van Strum and Merrell (1990), who provide a concise and useful review of numerous questionable epidemiological studies on dioxin conducted by corporations and governmnt agencies. The BASF study that Rohleder reanalyzed stands in contrast to an earlier study (Thiess et al. 1982), also by BASF scientists, that found a statistically significant increase in stomach cancer and nonsignificant increases in lung cancer and all cancers combined in workers at the same plant.

185. U.S. House of Representatives 1990.

186. Fingerhut et al. 1991.

187. Manz et al. 1991.

188. Bailar 1991.

189. Bertazzi et al. 1993, 1997.

190. Flesch-Janys et al. 1995, 1998.

191. Becher et al. 1996.

192. Kogevinas et al. 1993, 1997. Men exposed to TCDD or higher chlorinated dioxins had a relative risk of 1.29 compared to workers from the same cohort exposed to phenoxy herbicides and chlorophenols but with minimal or no dioxin exposure. The increase relative to the general population was significantly elevated but to a lesser extent (standard mortality ratio = 1.12).

193. Becher and Flesch-Janys 1998.

194. McGregor et al. 1998.

195. McGregor et al. 1998.

196. Tritscher et al. 1996, Huff 1994.

197. Axelson et al. 1990.

198. EPA 1985a. The general population's risk is based on EPA's estimate that a dose of 0.006 pg/kg per day is associated with a one per million cancer risk, and an average intake of 3 to 6 pg/kg per day TEQ, including dioxins, furans, and dioxin-like PCBs (EPA 1994a).

199. Becher et al. 1998.

200. My calculations assume a U.S. population of 250 million and a seventy year lifetime over which the population's total lifetime risk can be divided to give an approximate annual incidence. Cresanta (1992) notes that 1.1 million cases of cancer are diagnosed annually in the United States.

201. Epstein 1993, Westin 1993.

202. Landrigan 1992.

Chapter 5

1. Porter 1996.

2. Fischli 1996.

3. Howlett 1995.

4. Graedel and Keene 1996. While atomic and elemental chlorine gas are very rare in nature, the form of natural chloride does change in some environments. For instance, when salt from

the sea sinks down through the earth's crust to the mantle, it encounters intense heat, causing sodium chloride to be converted into hydrogen chloride, which is emitted in volcanic eruptions into the atmosphere. This HCl ultimately ends up in the oceans, where it dissolves to yield chloride ions again. Small amounts of chlorine gas may also be produced briefly during volcanic eruptions, and relatively small amounts of chlorine ions can be incorporated into organic matter by some organisms, as I discuss later in this chapter. There is also some evidence that substantial quantities of hydrogen chloride are produced from sea salt at the marine boundary layer; it has been proposed that small quantities of short-lived chlorine gas are also formed by a similar process, but empirical support for this mechanism is lacking (Graedel and Keene 1995, 1996).

5. Howlett 1995.

6. Consider, for instance, the exemplary chain reaction in which methane and chlorine combine to yield chloromethane and hydrogen chloride: the net reaction releases 25 kilocalories per mole of energy, primarily because the Cl-Cl bond has much lower dissociation energy than any of the other bonds involved (McMurray 1992).

7. The German chemist Scheele reported the formation of chlorine gas in 1774, but he did not recognize it as an element. In 1808, the British chemist Davy named the gaseous element chlorine, and Faraday produced it by accident in liquid form in a small laboratory explosion (Porter 1996).

8. McMurray 1992.

9. The bond dissociation energy of oxygen (O2) is 498 kilojoules per mole (kJ/mol) and that of chlorine gas is 243 kJ/mol. At normal atmospheric temperatures (298 degrees Kelvin), the exponential term $e^{-Ea/RT}$ in the rate constant expression for the dissociation of chlorine is a factor of 10^{44} greater than that for the dissociation of oxygen, so chlorine gas will dissociate astronomically faster than oxygen molecules will (Levine 1995).

10. Mackay (1992), Solomon et al. (1993), and Frank (1993) make a similar point.

11. HSDB 1997.

12. McMurray 1992, Henschler 1994.

13. Henschler 1994, Frank 1993.

14. Henschler 1994.

15. Howard (1990) notes that aliphatic organochlorines are generally resistant to biodegradation, and most aromatic organochlorines are as well (Webster 1990).

16. EPA 1986.

17. HSDB 1997.

18. Bonsor et al. 1988, Neilson et al. 1991.

19. A notable exception is the white-rot fungus *Phanerochaete chrusosporium,* which can degrade a number of chlorinated phenolics. Nevertheless, it is apparently ineffective at breaking down many other chlorinated phenolic compounds and the far more numerous group of chlorinated nonphenolic compounds (Gribble 1992).

20. Solomon et al. 1993, Ballschmitter 1996, Henschler 1994, Frank 1993. Solomon et al. (1993) explain that substituting a single chlorine atom for a hydrogen atom on an organic molecule typically increases the volume of the substance by 21 cubic centimeters per mole—enough to approximately double the volume of methane, for instance, with a single chlorine atom. For more on the relation between the size of molecules and their water solubility, see Stryer (1988).

21. Solomon 1996, Solomon et al. 1993, Mackay 1992.

22. Nebert and Gonzales 1987.
23. Webster 1990.
24. EPA 1994b.
25. For an excellent review of the nature and significance of natural organochlorines, see Stringer and Johnston (1995).
26. Gribble 1996a.
27. As Engvild (1986) points out, many identified plant organochlorines are chlorohydrins, which are extracted together with the corresponding epoxides. In the presence of acid used for extraction or cleanup in the laboratory, epoxides may incorporate chloride ions from the plant, the silica gel, or the solvent used in processing. In this way, a nonchlorinated epoxide may produce a chlorohydrin artifact that is mistaken for a natural compound.
28. One recent paper, for example, claimed to have established the natural production of trichloronorlichexanthones in lichens based simply on comparative chromatography of extracts of lichens collected in the wild in New Zealand and Europe, without any attempt to demonstrate that the compounds were actually synthesized by the lichen. It is possible that these organochlorines are natural, but it is also possible that they are industrial by-products, degradation products, or metabolites of man-made organochlorines that contaminate the environment in one place but not the other (Elix et al. 1990).
29. Gribble 1992, 1996a.
30. Luk et al. 1983.
31. Gribble 1992.
32. Gribble 1992, 1996a.
33. Gribble 1996a.
34. Singh et al. 1979, Graedel and Keene 1996, ATSDR 1989a. In addition to algae, fungi also produce chloromethane (Harper 1985), and an estimated 0.1 million tons is produced by natural combustion sources (Graedel and Keene 1996).
35. Lovelock 1975.
36. Neidleman and Geigert 1986, Gribble 1992, Fenical 1975, Hay and Fenical 1988.
37. Hay and Fenical 1988.
38. Hager 1982.
39. Gribble 1996a, Siuda and DeBernardis 1973.
40. Siuda and DeBernardis (1973) found that a colony of 50,000 ticks yielded 250 micrograms of 2,6-dichlorophenol, equivalent to 5 nanograms per tick. Dichlorophenol is 43.2 percent chlorine by weight, so each tick incorporated 2.16 nanograms of chlorine into 2,6-dichlorophenol. Human production of chlorine gas equals 40 million tons per year, or about 6.6 kilograms per person per year, based on a world population of 5 billion.
41. Stringer and Johnston 1995.
42. Neidleman and Geigert 1986.
43. Engvild 1995.
44. Asplund et al. 1989, Asplund and Grimvall 1991, Asplund 1995, Grimvall 1995.
45. Hjelm and Asplund 1995, Jonsson et al. 1995, Johansson et al. 1995a.
46. Hoekstra et al. 1995, Asplund 1995.
47. Lassen et al. 1995, Bollag and Dec 1995.
48. deJong et al. 1995.
49. Lassen et al. 1995.

50. Lassen et al. 1995. The specific rate of degradation depends on the precise chemical structure of the compounds and the environmental conditions in the soil (Lassen et al. 1995). Studies of groundwater show that AOX decreases at increasing depth, suggesting that chlorinated humic and fulvic acids are degraded with time (Gron 1995).

51. Bollag and Dec 1995.

52. deJong et al. 1995.

53. Bollag and Dec 1995.

54. Dolfing and Salomons 1995.

55. Asplund et al. 1989, Asplund and Grimvall 1991.

56. Asplund 1995. Stringer and Johnston 1995 provide additional criticisms of the use AOX for quantifying "natural" organochlorines.

57. Bumb et al. 1980.

58. Howlett 1995.

59. Rigo et al. 1995.

60. Brzuzy and Hites 1996b, EPA 1994b.

61. EPA 1998.

62. Brzuzy and Hites 1996a.

63. Reviewed in Schecter 1991. The original research reports are Ligon et al. (1989), Schecter et al. (1988), and Tong et al. 1990.

64. EPA 1994b.

65. Reviewed in Alcock and Jones (1996).

66. Czuczwa and Hites 1985, 1986; Czuczwa et al. 1984.

67. Kjeller and Rappe 1995.

68. Juttner et al. 1997.

69. Reviewed in Alcock and Jones (1996). Echoing these findings, EPA scientists, in a study of eleven lakes in remote parts of the United States, found that PCDD/F concentrations in pre-1930 sediments were at most one-tenth the levels in more recent layers (Cleverly et al. 1996).

70. Alcock and Jones (1996) provide a good discussion of problems with using sediment layers to infer that a pollutant was present before a certain date in the past.

71. Alcock et al. 1998. The ratio of dioxin levels in modern soil to those in nineteenth-century soil (3.28) in this and a related study (Kjeller et al. 1991) is lower than that in sediments. This difference may represent a peculiarity of the sample site (much industrial activity before the advent of chlorine or few chlorine-related sources in the modern era). It may also be due to the fact that soil concentrations integrate deposits from the air over a long period of time: much slower fluxes over a longer period of time will ultimately result in concentrations comparable to those caused by faster fluxes in a shorter period.

72. Brzuzy and Hites 1996b.

73. Brzuzy and Hites 1996b.

74. Czuczwa and Hites 1984. Additional data are reported in Czuczwa and Hites (1985, 1986).

75. Brzuzy and Hites 1996b.

76. For a useful introduction, see Stryer (1988). Lynn Margulis (1982) discusses the evolution of primary metabolism, and Morowitz (1968) gives the classic explanation of the organizational principles of natural biological and ecological chemistry.

77. Morowitz 1968.

78. This is, in fact, one of the core principles of the emerging field of green chemistry (Anastas and Williamson 1996).
79. The classic text is Ehrlich and Raven (1964); Rosenthal (1986) offers a general update.

Chapter 6

1. Ovid 1955.
2. Aftalion (1991), Trescott (1981), Taylor (1957), Ehrenfeld et al. (1993), Verbanic (1990), and Nader (1994) provide useful histories of the American and European chlorine industries.
3. Trescott 1981, Taylor 1957.
4. Trescott 1981; emphasis in the original.
5. Aftalion 1991, Trescott 1981, Taylor 1957.
6. Nader 1994.
7. Leder et al. 1994, Chlorine Institute 1991.
8. Ireland 1926, Aftalion 1991.
9. Aftalion 1991.
10. Aftalion 1991.
11. Chlorine Institute 1991, Leder et al. 1994. All subsequent figures on growth of the U.S. chlorine production are based on figures presented in these references.
12. Noble 1977.
13. Aftalion 1991.
14. CFCs were invented by Dupont chemist Thomas Midgley, whose biography reads like a tragic technological parable. In addition to the CFCs that would ultimately deplete the ozone layer, Midgley also invented tetraethyl lead, the gasoline additive that resulted in worldwide contamination of air, soil, and children's bodies with toxic lead compounds. At the peak of a successful scientific career that included his election as a fellow of the National Academy of Sciences and president of the American Chemical Society, Midgley was crippled by polio. Confined to his home, Midgley invented a pulley and harness system to help him get in and out of bed. In 1944, the harness system strangled him to death (Friedlander 1989).
15. Ehrenfeld et al. 1993.
16. Taylor 1957, Aftalion 1957.
17. Noble 1977.
18. Leder et al. 1994.
19. Carson 1962.
20. Lindqvist et al. 1991.
21. Lindqvist et al. 1991, Pacyna and Munch 1991.
22. Ayres 1997.
23. The data from Euro-Chlor, presented in Ayres (1997), are the most comprehensive available. They have the advantage of being based on a mass balance method, so that all mercury consumed is accounted for in way or another. My calculation of total mercury releases from the chlor-alkali industry uses this range and assumes 39 million tonnes global chlorine production—35.5 percent through the mercury process (Leder et al. 1994). The actual total may be higher, since many plants are not likely to be as well operated as those in Europe. Euro-Chlor's estimates of releases to water and air (0.2 and 1.9 grams of mercury per ton of chlo-

rine, respectively) are somewhat lower than estimates made by other parties: one review estimated mercury releases at 3 grams per ton of chlorine for a new chlor-alkali plant, and 10 grams per ton of chlorine for a well-operated existing facility (Schmittinger et al. 1986). Actual plants in Germany have been found to release 19 grams per ton (SRI 1993).

24. Harada 1995, Davies 1991.

25. Hill and Holman 1989.

26. Leder et al. 1994

27. Airey and Jones 1970, Johnston et al. 1993.

28. Maserti and Ferrara 1991.

29. Panda et al. 1990.

30. Leder et al. 1994.

31. Leder et al. 1994.

32. Energy requirements vary somewhat among the chlor-alkali cell types. The mercury cell requires 3,310 to 3,520 kilowatt-hours (kwh) per tons of chlorine, the diaphragm method 2,830 kwh per ton, and the membrane process 2,520 kwh per ton. Based on the proportion of each cell type in the world industry, the average energy requirement for the industry overall is slightly under 3,000 kwh per ton (SRI 1993).

33. SRI 1993.

34. Assuming an average global cost of 4.2 cents per kilowatt-hour for chlor-alkali customers (SRI 1993).

35. In the United States, 109 nuclear plants generated 673 billion kwh of electricity, for an average of about 6 billion kwh per plant per year (Energy Information Administration 1996).

36. Leder et al. 1994.

37. In Western Europe, the pattern of chlorine use is very similar, with slightly greater fractions used for propylene oxide, phosgene, hydrogen chloride, and hypochlorites, with smaller fractions in epoxy resins, pulp and paper, water treatment, and titanium dioxide (Leder et al 1994).

38. Leder et al. 1994.

39. Leder et al. 1994.

40. See Muir et al. (1993) for the first presentation of the idea of a chlorine use tree.

41. As figure 6.6 shows, growth is expected in a few much smaller applications, such as phosgene for polycarbonate and propylene chlorohydrin for propylene oxide, but the increases in these chlorine uses are less than one-tenth the growth expected in PVC (Mears 1995).

42. Leder et al. 1994, Shamel 1995.

43. Leder et al. 1994.

44. Leder et al. 1994, Mears 1995, Tullos 1995, Thayer 1990.

45. Leder et al. 1994.

46. Tullos 1995.

47. Endo 1990.

48. Mears 1995.

49. Leder et al. 1994.

50. Endo 1994.

51. Svalander 1996.

52. ECN 1992.
53. Endo 1990.
54. ECN 1992.
55. Shamel 1995.
56. Leder et al. 1994.
57. Tittle 1995.

Chapter 7

1. Scott 1815.
2. Mackay 1992.
3. Fischer 1994.
4. deOude (1993) reviews the principles of life cycle analysis.
5. There are some chlorination processes in the chemical industry that use hydrogen chloride rather than chlorine gas as a source of chlorine atoms, but the hydrogen chloride is a waste from other organochlorine synthesis processes in which elemental chlorine supplies the reactive chlorine atom. Because the elemental chlorine was produced by the chlor-alkali process, the hydrogen chloride too is implicated in the formation of by-products at the beginning of the chlorine life cycle.
6. Schmittinger et al. 1986.
7. HSDB 1997.
8. Concentrations of these compounds detected in chlorine gas range from 40 to 210 parts per billion (Hutzinger and Fiedler 1988). My calculation of annual loadings assumes world production of 39 million metric tons of chlorine each year (Leder 1994).
9. Rappe et al. (1991) report dioxins and furans in three samples of chlor-alkali electrode sludge, with total concentrations of PCDD/F of 641, 667 and 263 ppb, with TEQ values of 28, 28, and 13 ppb, respectively.
10. Kaminsky and Hites 1984.
11. Svensson et al. 1993.
12. Andersson et al. 1993.
13. This research is summarized in Versar (1996) and EPA (1998).
14. Environment Agency 1997. The quantity of dioxins discharged directly into the environment was estimated at 1.5 grams per year (TEQ).
15. Bonsor et al. 1988.
16. Suntio et al. 1988.
17. MacDonald et al. (1998) have found strong evidence in the sediment record that a pulp mill on Kamloops Lake, British Columbia, has been a major source of PCBs since the introduction of chlorine bleaching in the mid-1960s.
18. Bonsor et al. 1988.
19. Bonsor et al. 1988.
20. Solomon 1996. The quantity of organochlorines given is based on the fact that the mass of the organochlorines in bleaching by-products is typically about ten times the mass of the organically-bound chlorine alone (Bonsor et al. 1988).
21. Mantykoski et al. 1989.

22. See the discussion in chapter 8 and Solomon et al. (1993).
23. Bonsor et al. 1988.
24. Sodergren et al. 1993.
25. Stevens et al. 1990.
26. Beech and Diaz 1980, Biziuk et al. 1992.
27. Aggazzotti et al. 1990.
28. Amy et al. 1990, Ventresque et al. 1990, Jolley 1985. As these references note, the majority of the unidentified fraction are nonvolatile organochlorines, which are particularly likely to include persistent and bioaccumulative substances.
29. Rappe et al. 1989.
30. Onodera et al. 1989.
31. Mantykoski et al. 1989.
32. Oehme et al. 1989.
33. Oehme et al. 1989.
34. Hutzinger and Fiedler 1988.
35. Lahl 1993, 1994.
36. Lahl 1994.
37. Hutzinger and Fiedler 1988.
38. Hutzinger and Fiedler (1988) reported the presence of HCB and tetrachlorobenzene, while Rappe et al. (1990) found polychlorinated dibenzofurans at a concentration of 4.9 parts per quadrillion in a sample of hypochlorite-containing household bleach.
39. deLeer 1985.
40. Bertazzi and diDomenico (1994).
41. Rossberg et al. 1986.
42. Rossberg et al. 1986.
43. Only 77 percent of commercial-grade DDT is the intended substance p,p'-DDT; DDE, DDD, and o,p'-DDT account for 19.5 percent, and the remaining 3.5 percent of the product consists of unidentified contaminants (HSDB 1997).
44. HSDB 1997.
45. Rossberg et al. 1986.
46. This calculation assumes synthesis of EDC in an integrated chlorination and oxychlorination process followed by pyrolysis to VCM; it includes releases to air, water, heavy ends, and light ends, but it does not include nitrogen gas vented to the atmosphere or aqueous streams (Rossberg et al. 1986).
47. The listed compounds have been identified in the wastes from pyrolytic production of VCM from EDC and/or oxychlorination of ethylene to produce EDC, according to Dow (1990) and Environment Agency (1997); see additional discussion in chapter 8.
48. Rossberg et al. 1986.
49. All chlorobenzenes except monochlorobenzene and all chlorinated phenols (chlorinated benzenes with a hydroxyl group also attached to the ring) except dichlorophenol have been found to contain PCDDs or PCDFs, or both according to EPA (1994b).
50. Hutzinger and Fiedler 1988.
51. EPA 1998a.
52. EPA 1998, Esposito et al.1980, PTCN 1985.

53. EPA 1998, 1994b; Schecter et al. 1993; Zook and Rappe 1994.

54. Lexen and deWitt 1992.

55. Hutzinger and Fiedler 1988.

56. Heindl and Hutzinger 1987.

57. HCB is considered an indicator compound for dioxin formation because it is structurally related to dioxin, has been found in virtually all processes in which dioxin is known to be produced, and itself contains dioxins and furans at concentrations up to 200 parts per million (Oehme et al. 1989).

58. Rossberg et al. 1986, PTCN 1985.

59. HSDB 1997.

60. As I show in chapter 8, very large quantities of dioxins, furans, HCB, and PCBs are produced in the synthesis of chlorinated ethanes and ethylenes by oxychlorination, chlorinolysis, and pyrolysis from other aliphatic organochlorines.

61. Heindl and Hutzinger 1987.

62. Rappe and Marklund 1978.

63. Dioxin is generated when platinum catalysts used in the blending of high-octane gasolines. These catalysts accumulate organic compounds and are then regenerated at high temperature in the presence of methylene chloride, 1,1,1-trichlrooethane, and ethylene dichloride (Thompson et al. 1990, Versar 1996).

64. Marklund et al. 1987, 1990.

65. Kahlich et al. (1993) note that base-catalyzed dehydrochlorination of propylene chlorohydrin to yield propylene oxide generates such by-products as 1-chloro-2,3-epoxypropane, 3-chloropropane-1,2-diol, 1,2-propanediol, glycerol, acetone, and acetol. Wastes contain dichloropropane, 2,2-dichlorodiisopropyl ether, and epichlorohydrin, while wastewtaers contain AOX (total absorbable organically-bound halogen) at concentrations of 30 to 60 ppm.

66. The specific by-products depend on the isocyanate being produced. For instance, production of toluene diisocyanate from aromatic amines and phosgene will produce complex chlorinated aromatics, such as N-chloroformylchloroformamidines. Reacting free amines with phosgene results in allophanoyl chloride. (Ulrich 1989).

67. Muskopf and McCollister 1987.

68. EPA 1998, Esposito et al. 1980, PTCN 1985.

69. Hutzinger and Fiedler 1988.

70. TNO 1996.

71. Versar 1996. Soot from a fire in an office building with PCB-containing transformers in Binghamton, New York, was found to contain total PCDD/Fs in the parts per million range, with some samples in the parts per thousand range. Laboratory studies suggest that combustion may convert as much as 3 to 4 percent of PCB to PCDFs.

72. Brzuzy and Hites 1996b.

73. EPA (1998) estimates with "low confidence" that dioxin and furan emissions (TEQ) from landfill fires and open burning of trash are 1,000 grams per year each, while those from accidental vehicle fires are 10 grams per year. Data were inadequate to make a quantitative estimate with confidence for dioxin emissions from accidental fires in buildings. For more on this subject, see chapter 8.

74. Plimmer 1973, Catabeni et al. 1985.

75. Oberg et al. 1990, 1993; Schafer et al. 1993.

76. Burmaster and Harris 1982.

77. Of ninety-eight substances on EPA's list of pollutants reported at sites on the Superfund National Priorities List, forty are organochlorines. Of the top ten, five are organochlorines (Miller et al. 1991).

78. Silkworth 1994.

79. EPA has estimated that 37 percent of all wastes fed to U.S. incinerators are halogenated organics. Because organochlorines account for well over 90 percent of all organohalogen production, we can conclude that about 35 percent of these are organochlorine-containing wastes (Oppelt 1986, Dempsey and Oppelt 1993).

80. EPA 1990a.

81. EPA 1989.

82. UKDOE 1989. See also Hutzinger 1986.

83. Dellinger et al. 1988, EPA 1994b.

84. Gullett 1990.

85. EPA 1989.

86. Eklund et al. 1988. Similarly, combustion under well-controlled laboratory conditions of trichloroethylene produces a variety of persistent organochlorine PICs, including hexachloropentadiene, highly chlorinated benzenes and indenes, PCBs, and the dioxin-like chlorofulvalenes (Blankenship et al. 1994).

87. CARB 1990, Sakai et al. 1992, Green and Wagner 1993. Even under good combustion conditions—with high temperatures, adequate residence times, and low emissions of carbon monoxide and total volatile hydrocarbons—significant quantities of extremely hazardous PICs are emitted, including the dioxin-like compounds (Chang et al. 1993, Glasser et al. 1991).

88. Jay and Stieglitz 1995.

89. EPA 1990.

90. Markus et al. (1997) used a calibrated bioassay to quantify the activity of the cytochrome p4501A1 enzyme as a marker of the total dioxin-like toxicity of the fly ash, which exceeded that predicted by the quantity of dioxins, furans, and PCBs in the sample by a factor of two to five.

91. If all incinerators achieved 99.99 percent DRE, the incineration of 13 million tons of halogenated waste per year in the United States, as discussed above, would result in the emission to the air of 1,300 tons (2.6 million pounds) of unburned wastes per year.

92. EPA 1985c.

93. See, for instance, the 1986 analysis by U.S. EPA engineers (Staley et al. 1986), which concluded, "The trial burn data only indicate how well the incinerator was operating during the time that the data were being taken, typically only a period of a few days. No information is obtained on how the incinerator might respond if fuel, or especially waste, conditions change. Waste streams vary widely in composition and one incinerator may burn many different toxic substances over its useful life, resulting in unavoidable and frequent changes in waste feed conditions. It is difficult to generalize the results of a trial burn to predict how the composition of the incinerator exhaust will change under these varying conditions."

94. Licis and Mason 1989. See also the discussion of the implications of these findings in Costner and Thornton (1991).

95. Kramlich et al. 1989, Trenholm et al. 1984.

96. EPA 1994b. The source inventory has been revised and republished in EPA (1998).

97. Burnett 1994.

98. Rigo et al. 1995.

99. Goodman 1994.

100. Vinyl Institute 1996.

101. Versar 1996.

102. The EC report is Tukker et al. (1995); the others are presented in Weinberg and Finaldi (1997).

103. An extensive critique of the Rigo study can be found in Costner (1997).

104. Illustrating this point, even carbon monoxide, a widely accepted indicator of combustion conditions, is not consistently related to the emission of unburned wastes in full-scale incinerators (Staley 1986, EPA 1990). Further, although laboratory studies have found that the addition of HCl to a combustion mixture increases PCDD/F output (Wirts et al. 1998, Tiernan et al. 1993, Fangmark et al. 1993, Takeshita et al. 1992), studies at full-scale or pilot-scale incinerators have detected such a relationship only inconsistently (Ruuskanen et al. 1994, Manscher et al. 1990, Lenoir et al. 1991).

105. Rigo et al. (1995) analyzed HCl in stack gas for municipal and medical waste incinerators. For hazardous waste incinerators, Rigo's analysis was based on the percentage chlorine in the waste feed, a parameter that does not reflect the mass of chlorine either. If percentage chlorine stays the same and total waste feed is increased, then the mass of chlorine feed will increase but would not have been noted using Rigo's approach. Further, increasing the waste feed typically increases the stack gas flow rate, which will tend to reduce dioxin concentrations if the mass of dioxins emittedstays the same or increases less than the flow rate does.

106. Theisen 1991.

107. Halonen et al. 1995.

108. Bruce et al. (1991) found that the addition of potassium chloride, sodium chloride, or calcium chloride to a combustion reaction had no effect on the quantities of dioxins and furans formed and deposited in the fly ash. Addink et al. (1998) added sodium chloride to fly ash and found that it did not participate in the de novo formation of dioxins and furans. Lenoir et al. (1991) burned sodium chloride with polyethylene in a fluidized bed combustor and found no effect on the amount of dioxins and furans emitted.

109. Altwicker et al. 1993. In this study, increasing the feed of organically-bound chlorine resulted in a substantially higher ratio of chlorophenols to chlorobenzenes in the combustion products; chlorophenols are considered precursors for dioxin formation.

110. Kopponen et al. 1992.

111. Kolenda et al. 1994, Wilken 1994.

112. Mahle and Whiting 1980.

113. Liberti 1983.

114. Danish Environmental Protection Agency 1993.

115. Mattila et al. 1992, Ruuskanen et al. 1994, Frankenhaeuser et al. 1993.

116. Christmann et al. 1989b.

117. Vesterinen and Flyktmann 1996, Halonen et al. 1993, Huotari et al. 1996, Manninen et al. 1996. In all of these studies, dioxin levels in fly gas or flue gas increased with increasing feed of refuse-derived fuel to the burner, which was significantly higher in chlorine content than the organic matter used in comparison runs.

118. This study by Kanters et al. (1996) focused on emissions of chlorophenols as a surrogate for dioxin, due to the difficulty and expense of dioxin sampling and analysis.
119. Wagner and Green 1993. This study also measured emissions of chlorophenols as a dioxin surrogate.
120. Wikstrom et al. 1996.
121. Visalli 1987.
122. Mark 1994.
123. EPA 1988.
124. DTI 1995.

Chapter 8

1. Lucretius 1995.
2. Metcalf 1987. Pimentel and Levitan (1986) provide the calculations for the figure that only 0.1 percent of pesticides reach target pests.
3. In the United States alone, annual pesticide use on row crops is 450 to 500 million pounds per year (Pesek et al. 1989).
4. SRI 1993.
5. Scholz 1987.
6. 2,3,7,8-TCDD levels in Agent Orange were typically around 2 parts per million but in some cases may have been as high as 30 ppm (Schecter 1994).
7. EPA 1987a.
8. EPA 1998a.
9. EPA 1998a, Schecter et al. 1993. EPA concluded that 2,4-D contains dioxins and furans at about 700 parts per trillion (TEQ), resulting in a total annual discharge to the U.S. environment of 13 to 26 grams per year (TEQ), as of 1995.
10. EPA 1998a.
11. Kramarova et al. 1998.
12. Schecter 1991, Schecter 1994.
13. PAN 1997.
14. Carson 1962.
15. Jeyaratnam 1990.
16. Jeyeratnam 1990.
17. Pesek et al. 1989.
18. These data, from the U.S. Department of Agriculture, are presented in EWG (1995).
19. HSDB 1997.
20. HSDB 1997.
21. EWG 1995.
22. Of all the ozone-depleting organochlorines (CFCs, HCFCs, carbon tetrachloride, and 1,1,1-trichloroethane), about 50 percent were used as solvents, 20 percent as refrigerants, 15 percent for foam blowing, and the remainder for aerosol and miscellaneous uses (Makhijani and Gurney 1995). The other chlorinated solvents—dichloromethane, trichloroethylene,

and perchloroethylene—are used exclusively for coating, cleaning and degreasing, and stripping, and some uses within the chemical industry.

23. Rossberg et al. 1986.

24. Rossberg et al. 1986.

25. Rossberg et al. 1986.

26. Rossberg et al. 1986.

27. Heindl and Hutzinger 1987.

28. Environment Agency 1997.

29. EPA (1998a), based on EPA's midrange estimates. If EPA's new "order of magnitude" estimates (for sources on which data are very limited) are included, then backyard trash burning and landfill fires would also outrank this facility.

30. Of the total quantity of chlorinated solvents used in cleaning, degreasing, and coating operations, 80 to 90 percent escapes to the air (ATSDR 1995f, Wahlstrom and Lundquist 1993). In the pharmaceuticals industry, the fraction lost to the air is smaller, but the deposition of solvents in wastes destined for disposal is higher(Wahlstrom and Lundquist 1993).

31. Makhijani and Gurney 1995.

32. ATSDR 1995f, 1995g.

33. Oppelt 1986, Dempsey and Oppelt 1993.

34. EPA 1998b.

35. EPA (1998b) notes that at a typically sized dry cleaning shop, transfer machines may release to the air as much 501 gallons per year out of a total of 627 gallons consumed (80 percent). More modern dry-to-dry machines, depending on the equipment type, release a minimum of 51 gallons per year out of 178 gallons consumed (29 percent) to 434 gallons out of 561 gallons consumed (77 percent).

36. EPA 1998b. Based on air monitoring by OSHA at dry cleaning shops in 1997, EPA found that average exposures posed cancer risks of six per thousand and exceeded the reference concentration by a factor of 4.5. From monitoring data by the International Fabricare Institute in 1990, EPA calculated cancer risks of 1 per 100 or more and exposures that exceeded the reference concentration by a factor of 13.5 to 37.6 at shops using dry-to-dry and transfer machines, respectively.

37. Schreiber et al. 1993. Additional studies in New York, San Francisco, Germany, and the Netherlands, reviewed in EPA (1998b), confirm these findings.

38. Kleijn 1997.

39. EPA 1998b. Calculated cancer risks to colocated residents range from 1 in 100 to 10 per million, depending on the type of machines used and the monitoring program from which exposure data were obtained.

40. Miller and Uhler 1988.

41. Tichenor et al. 1990.

42. EPA 1998b.

43. SRI 1993.

44. This calculation extrapolates from Norwegian government estimates (SFT 1993), which were derived from industry data on releases. In the production of 1 ton of EDC, an estimated 7 grams of EDC are released to water. In the production of one ton of VCM from EDC, SFT estimated that one gram each of EDC and VCM are discharged to water, and 500 and 1,000 grams of EDC are emitted to air. In the polymerization of 1 ton of PVC, SFT estimated that 5,100 grams of VCM are released into the air. Extrapolations assume 1990 production rates

of 29,137 kilotons EDC per year, 18,495 kilotons of VCM per year, and 18,135 kilotons of PVC per year (SRI 1993).

45. SFT 1993.

46. ATSDR 1993a, 1995h.

47. ATSDR 1993a, 1995h.

48. Bowermaster 1993.

49. Curry et al. 1996.

50. My calculations assume that 15 million metric tons per year of EDC are produced by oxychlorination (half of world production [SRI 1993], assuming integrated oxychlorination and direct chlorination process in 1:1 molar ratios). Heavy and light ends are assumed to be produced at the rate of 2 kilograms each per ton, based on the fact that production of 168,796 tons of EDC in Sweden per year results in the generation of 335 and 333 tons per year of heavy and light ends, respectively (TNO 1996). This figure is slightly lower than that of Rossberg et al. (1986), who estimate 2.3 and 2.9 kg heavy and light ends per ton of VCM produced, respectively.

51. Rossberg et al. 1986.

52. Papp 1996.

53. Dow 1990.

54. This calculation assumes global production of 30 million kilograms of EDC heavy ends, as discussed in note 50 above.

55. Johnston et al. 1993.

56. TNO 1996.

57. Costner et al. 1995.

58. In research at a chemical plant in Russia, Khizbullia et al. (1998) found substantial quantities of dioxins and furans in the wastewater and wastewater sludge from the pyrolytic production of VCM from EDC; very large amounts were present in the waste incinerator emissions and residues from the pyrolytic production of VCM from EDC.

59. ICI 1994.

60. Evers 1989. That report estimated that 419 grams of dioxin TEQ are formed per 100,000 tons of EDC produced. World production of EDC by oxychlorination is about 15 million tons per year (see note 50), so this rate would result in the formation of over 60,000 grams of dioxin (TEQ) per year. Although presumably not all the dioxins created would be released directly to the environment, this quantity is more than fifty times greater than the annual dioxin emissions to air from all trash incinerators in the United States, the largest known source of dioxin emissions in that country. It is also double the 25,000 grams of dioxin (TEQ) per year that EPA estimates are carried into the environment by contaminated pentachlorophenol (EPA 1998), the largest identified source of dioxin to any environmental medium.

61. See SFT (1993), which uses the Norsk-Hydro data to estimate that 0.1 gram of dioxin are released directly to the environment for each 100,000 tons of EDC. This quantity does not include the much larger quantities—about twenty-three to thirty-eight times greater—of dioxins generated in the process that are directed into tars or which recirculate in the synthesis machinery.

62. Lower Saxony 1994.

63. Environment Agency 1997. Total dioxin generation associated with EDC/VCM synthesis was estimated at 27 grams (TEQ) per 200,000 tons of VCM, for a dioxin generation rate

of 13.5 grams (TEQ) per 100,000 tons—substantially more than the Norwegian estimate but less than the Dutch figure.

64. Costner et al. 1995.

65. DTI 1995, Environment Agency 1997, SFT 1993.

66. Contamination in the United Kingdom is described by Environment Agency (1997), in Germany by Lower Saxony (1994), and in the United States by Curry et al. (1996).

67. Ramacci et al. 1998.

68. Evers et al. 1988.

69. Evers et al. 1993, 1996.

70. Curry et al. 1996. PPG makes solvents and other organochlorines at the same facility, and these processes are also likely to have contributed to the contamination.

71. In this study, two samples of pure PVC contained 0.86 to 8.69 ppt TEQ (SEPA 1994).

72. The UK Ministry for Agriculture, Fisheries, and Food found that PVC food packaging contained dioxins at levels ranging from 2.6 to 6.9 ppt TEQ (MAFF 1995).

73. Wagenaar et al. 1996, Carroll et al. 1996.

74. Tukker et al. 1995.

75. DTI 1995.

76. My calculations are extrapolated from the figures for Sweden, where the lead input into PVC equals 0.653 percent of total PVC production, and the phthalate input equals 22.6 percent (TNO 1996). I have also assumed 19.1 million tons of worldwide PVC production per year (DTI 1995).

77. Extrapolated from the relevant figures for Sweden, where phthalate emissions associated with PVC are estimated at 102 tons per year to water, 95 tons per year to land, and 28 tons per year to air, for 101.5 kiloton per year PVC production (DTI 1995).

78. DTI 1995.

79. DTI 1995.

80. Pearson and Trissel 1993, Goldspiel 1994.

81. Plonait et al. 1993.

82. The Consumer Product Safety Commission issued a particularly trenchant warning on the release of lead from PVC blinds (*Chicago Tribune* 1996).

83. DTI 1995. Lead is much more commonly used as a stabilizer in pipes in Europe than in the United States.

84. Warren (1998), Jackson 1998, and *New York Times* (1998).

85. Mayer 1998.

86. Wirts et al. 1998.

87. Wallace 1990.

88. TNO 1996.

89. Fiedler et al. 1993. The soot samples contained dioxins and furans at 25,251 nanograms (TEQ) per kg, which Fiedler et al. note is equivalent to about 15,000 nanograms (TEQ) per square meter. In contrast, the Hessen Ministry of the Interior's recommended maximum for wipe samples from surfaces in homes and offices is 10 nanograms (TEQ) per square meter.

90. In a report for U.S. EPA, Versar (1996) cites analytical results that found dioxin and furan production from the combustion of vehicles that contained PVC at the following rates: 0.044 mg TEQ per automobile, 2.6 mg TEQ per subway car, and 10.3 mg TEQ per rail car.

91. The U.S. Vinyl Institute (Carroll 1995) has calculated, based on a study of soot residues measured within a limited radius of a fire at a plastics facility, that accidental fires are probably a relatively small contributor to the total dioxin burden. But over 90 percent of the dioxins produced in a structural fire are in the gaseous phase and escape into the atmosphere (Versar 1996, EPA 1998), so calculations from soot alone underestimate total dioxin emissions from accidental fires by at least an order of magnitude. EPA (1998) has concluded that the available data are inadequate to estimate with confidence the contribution of accidental structural fires to national dioxin emissions.

92. UBA 1992, German Environment Ministers 1992.

93. I have extrapolated from figures for Sweden (TNO 1996), which indicate that the stock of PVC in use (2 million tons) equals 22.47 years of current PVC production. I have assumed a similar stock-to-production ratio worldwide and annual PVC production of 19.1 million tons per year (DTI 1995).

94. Carroll (1995) notes that an average house contains 14 to 367 kg PVC, depending on the size and date of construction or remodeling.

95. Schecter and Kessler 1996. Deutsch and Goldfarb (1988) also reported elevated dioxin levels in a university building after an interior fire in a lecture hall that contained PVC components.

96. Hamilton-Wentworth 1997, Socha et al. 1997. The latter reference notes that dioxin levels in tree leaves downwind from the fire were 7 to 100 times above normal. Apparently pollutants on the leaves were washed from the leaves into the general environment by rain, because levels on leaves declined significantly after the first postfire rainstorm.

97. Based on preliminary data, EPA (1998a) estimates that landfill fires may emit on the order of 1,000 grams of dioxins and furans (TEQ) to the air each year in the United States, second only to trash incinerators among U.S. dioxin sources to the air.

98. I have extrapolated from figures for Sweden (TNO 1996), which indicate that the stock of PVC in use (2 million tons) equals 22.47 years of current PVC production, and contains 15,000 tons of lead and 288 thousand tons of phthalates. I have assumed a similar stock-to-production ratio worldwide and annual PVC production of 19.1 million tons per year (DTI 1995).

99. TNO 1996.

100. DTI 1995.

101. Association of Post-consumer Plastics Recyclers 1998.

102. DTI 1995.

103. Danish EPA 1993, Ecocycle 1994, DTI 1995, TNO 1994.

104. Assuming U.S. municipal waste incinerator capacity of 48 million tons per year (Versar 1996), 80 percent capacity utilization, and PVC content of 0.5 to 0.8 percent.

105. *Chemical Week* 1995b.

106. According to two studies, 9.4 percent (Marrack 1988) and 15 percent (Hasselriis and Constantine 1993) of infectious ("red-bag") waste in the United States is PVC, and as much as 18 percent of noninfectious hospital wastes are PVC (Hasselriis and Constantine 1993). In Denmark, PVC accounts for about 5 percent of all medical waste (DTI 1995).

107. See the discussion in Green (1993).

108. Christmann et al. 1989a, Theisen et al. 1989, Theisen 1991.

109. Yasuhara and Morita 1988, Blankenship et al. 1994.

110. I have extrapolated from the relevant figures for Sweden, where 249 tons of PVC enter the waste stream each year (TNO 1996), assuming 19.1 million tons of annual PVC production worldwide (DTI 1995).

111. TNO 1996.

112. Lemieux 1997. PCDD/F emissions per kilogram of waste were 6.3 times higher in the burning of waste from an avid recycler, which contained a very high fraction of PVC, than when waste from a nonrecycler with much lower quantities of PVC were burned. There were only two runs for each type of trash, so conclusions about the role of PVC in dioxin emissions are tentative. My calculations assume, based on EPA's figures, that an exemplary municipal waste incinerator releases 0.0035 micrograms per kilogram of waste, that the incinerator burns 182,000 kilograms of waste per day, and the non-recycler household burns 4.9 kilograms per day.

113. Christmann 1989a, EPA 1994a, Versar 1996.

114. Christmann 1989b.

115. Schaum et al. 1993, EPA 1994b, Aittola et al. 1993.

116. Versar 1996.

117. Derry and Williams 1960.

118. Myreen 1994. Total pulp production is almost twice as much (Matussek et al. 1996).

119. Bonsor et al. 1988.

120. This calculation assumes formation of AOX at the rate of 5 to 8 kg AOX per ton of pulp, and production at an average mill of 600 to 1,000 tons of pulp per day (Kroesa 1991).

121. This calculation assumes the discharge of about 4 kg of AOX per ton of pulp from chlorine-bleaching pulp mills (Bonsor et al. 1988), 100 million tons of bleached pulp production per year (Myreen 1994), and a total mass of 10 kg organochlorines for each kg AOX (Bonsor et al. 1988).

122. Enell 1992, Environment Canada 1991.

123. Paasivirta 1991.

124. Assuming 100 million tons per year of worldwide bleached pulp production and chloroform formation at the rate 0.3 kg per ton (Kroesa 1991).

125. Mantykoski et al. 1989. Kopponen et al. (1994) have also detected dioxins and furans in the fly ash of chlorine-bleached pulp mill sludge incinerators in quantities ranging from 10 to 3,400 parts per trillion TEQ.

126. Mantykoski 1989.

127. Bonsor et al. 1988.

128. Rappe (1990) conducted analyses on various bleached paper products. Rappe et al. (1990) have found dioxin in these products in concentration as high as 447 parts per trillion TEQ.

129. Furst 1991.

130. Sodergren et al. 1993.

131. McMaster et al. 1996.

132. Hocking 1991.

133. Weegar 1994.

134. Hauserman 1997, Regan 1992.

135. AET 1997.

136. Solomon et al. 1993, Solomon 1996.

137. Dahlmann and Morck 1993.

138. Solomon et al. 1993.

139. Rantio 1995; Koistinen et al. 1994a, 1994b; Solomon et al. 1993.

140. Rosenberg et al. (1994) found dioxin concentrations in air at a Finnish pulp mill that had completely replaced chlorine gas with chlorine dioxide. Concentrations of all 2,3,7,8-substituted PCDD/Fs ranged from 0.04 to 1.9 picograms (pg) per cubic meter, with TEQ values ranging from 0.002 to 0.2 pg per cubic meter.

141. Solomon et al. 1990.

142. Environment Canada 1991, Bonsor et al. 1988.

143. Solomon et al. 1993.

144. This calculation assumes world production of 100 million tons of bleached paper per year (Myreen 1994); AOX discharge at a rate of 0.7 kg per ton of pulp (Solomon et al. 1993); and 2 kg organochlorines per kilogram of AOX in effluents from chlorine-dioxide bleached pulp (Solomon et al. 1993).

145. The calculated quantity of bioaccumulative organochlorines is estimated based on a 0.0114 ratio EOCl/AOX in effluent from mills using 100 percent chlorine dioxide, where EOCl is the hexane extractable (hydrophobic) fraction (Solomon et al. 1993). An additional 200 tons of chlorophenolic compounds would be discharged, based on production of chlorophenols at the rate of 2 g per ton of pulp (Dahlmann and Morck 1993).

146. This calculation, which includes only kraft pulp mills, is based on EPA's (1997) estimate that U.S. mills, once all have adopted 100 percent chlorine dioxide bleaching, will discharge 11,200 tons per year AOX, assuming 2 kg of organochlorines per kg of AOX (Solomon et al. 1993).

147. Clapp et al. (1996) found that ECF pulps made in North America contained AOX at levels ranging from 124.6 to 391.6 mg per kg, with a mean value of 220.7 mg per kg. The figure given assumes that the mass of organochlorines is twice the mass of the organically-bound halide (Solomon et al. 1993) and that world bleached pulp production is 100 million tons per year.

148. Sodergren et al. 1993.

149. Cates et al. 1995, Lovblad and Maimstrom 1994.

150. McMaster et al. 1996.

151. The use of this argument by representatives of the Canadian Government is discussed in *Paper* (1993).

152. This history is based on the sworn congressional testimony of several EPA Region V employees and from EPA documents submitted as exhibits in the same congressional hearings (U.S. House 1983).

153. This story is told in compelling detail in a report by Paul Merrell and Carol Van Strum (1987), the report that first revealed the secret EPA-industry dioxin study to the public based on leaked documents.

154. The EPA-industry agreement is Farrar, Gellman, and McBride (1986); it is summarized, particularly in relation to Boise Cascade, in an EPA memo (Whittington 1986).

155. In a memo from a high-ranking EPA official (McBride 1987), EPA reiterated its promise not to release information to the public without prior arrangement with API.

156. Burnett 1994.

157. Vinyl Institute 1998.

158. Vinyl Institute 1998. Preliminary results were reported in Carroll et al. (1996).

159. Beekwilder 1989.

160. Scheck 1990.

Chapter 9

1. IJC 1996.

2. SDI's costs and flaws were summarized in Brand et al. (1987) and Nimroody (1988). My comparison of the philosophy of SDI to that of chemical-by-chemical pollution control was inspired by Bob Pollack (1999), who made a very different analogy to SDI.

3. Leinhardt 1994.

4. Baden 1994.

5. Dow 1995.

6. APHA 1994.

7. International Joint Commission, 1992, 1994, 1996.

8. At a full meeting in Paris of the Oslo and Paris Commissions on September 21–122, 1992, the environmental ministers of fifteen European nations agreed, "As a matter of principle for the whole Convention area, discharges and emissions of substances which are toxic, persistent and liable to bioaccumulate, in particular organohalogen substances, and which could enter the marine environment, should, regardless of their anthropogenic source, be reduced by the year 2000, to levels that are not harmful to man or nature with the aim of their elimination; to this end to implement substantial reductions in those discharges and emissions and where appropriate to supplement reduction measures with programmes to phase out the use of such substances."

9. At a full meeting in Turkey of the Barcelona Convention, October 15, 1993, the environmental ministers of the 18 Mediterranean nations agreed "to recommend that the contracting parties reduce and phase-out by the year 2005 inputs to the marine environment of toxic, persistent, and bioaccumulative substances in particular organohalogen compounds having those characteristics. In this framework, high priority is to be given to both diffuse sources and industrial sectors which are sources of organohalogen inputs."

10. Commoner 1987, 1990. In Commoner's analysis of air pollution trends between 1975 and 1987, emissions of particulates, sulfur dioxide, carbon monoxide, nitrogen oxides, and volatile organic compounds decreased by an average of just 18 percent. During this period, emissions of nitrogen oxides (a major cause of smog) increased by 2 percent, while most of the others declined by 15 to 30 percent. In contrast, Commoner shows that emissions of lead declined by more than 90 percent following restrictions on its use in gasoline, and PCBs and DDT show similar trends.

11. Commoner 1990.

12. For example, leaded gasoline, DDT, and the mercury chlor-alkali process continue to be used in many nations, and many other uses of lead and mercury have not been restricted. CFCs were phased out slowly and were replaced with other ozone-depleting compounds (the HCFCs), so ozone depletion is expected to continue for decades. Even after production of PCBs ended, large stocks have remained in many operating and discarded transformers. These examples illustrate the importance of global action to address all, rather than just some, sources of persistent pollutants.

13. Herman Daly (see Daly and Cobb 1989, for example) and others working in the field of ecological economics have challenged the assumption that economic growth must continue

forever in a capitalist economy. Daly argues that unchecked growth will inevitably lead to environmental damage, as well as socially unacceptable inequalities in the distribution of resources. Daly advocates economic "development," a continual improvement rather than expansion of society's economic activities to meet the economic, social, and environmental needs of the public better. This book's focus on Clean Production is the technological face of Daly's vision of a sustainable economy.

14. In a large ecosystem, small quantities of acids and bases can be diluted, so that the resulting change in pH is not of physiological or ecological significance. Further, most natural systems are buffered to some extent, so a small input of an acid or base results in no change in pH.

15. I have developed this argument from Brian Wynne's (1993) excellent discussion.

16. In his incisive analysis of what he calls the "Risk Society," the German sociologist Ulrich Beck (1992, 1995) has explored the sociological differences between risks and the novel kinds of hazards posed by current technologies.

17. Boehmer-Christansen (1994), Dethlefsen et al. (1993), and Freestone and Hey (1996) all provide useful histories of the precautionary principle.

18. Cameron and Abouchar 1996, Cameron 1994.

19. Reproduced in Dethlefsen et al. 1993.

20. Academic critiques include Bodansky (1994), Fleming (1996), and Nollkaemper (1996). Industry-funded criticisms include the reasoned position of Jacob (1995) and the strident version of Malkin and Fumento (1996). The latter document, prepared by an industry-funded think tank called the Competitive Enterprise Institute, is distributed by the Chlorine Chemistry Council.

21. PR Watch (1996).

22. Mongoven, Biscoe and Duchin 1994.

23. O'Riordan and Cameron 1994, von Moltke 1996.

24. Jacob 1994. Other weak versions are found in the definitions of Bodansky 1994, Haigh 1994, and Warner 1994, as well as that adopted by the European Union, in which precaution simply represents the "best available technology not entailing excessive costs" (Boehmer-Christansen 1994).

25. Jacob 1994. Gray (1996), Warner (1994), and Bodansky (1994) have made a similar argument.

26. Bodansky 1994.

27. See the discussion in Boehmer-Christansen (1994). Haigh (1994) has argued that "the precautionary principle is often taken to be synonymous with the setting of emissions standards from industrial plants based on what is technically achievable (best available technology). If this is so, then it can be argued that the precautionary principle has existed in Britain for over 100 years for aspects of pollution control."

28. Bodanksy 1994.

29. Boehmer-Christansen 1994. The responsibility to future generations is discussed in Kiss (1996).

30. The International Joint Commission first discussed the principle of reverse onus in its 1990 report. *Reverse onus* was originally a legal term that refers to the shifting of the burden of proof onto an accused party after the accuser has already prepared a compelling case. My usage, like the IJC's, departs from this narrow definition in that the reversal of the onus from the public to the proponent of a substance takes place not in the evaluation of each compound but is applied *a priori* to all organochlorines.

31. International Joint Commission 1990, 1992, 1994, 1996. The other international agreements include the Paris Commission and the Barcelona Convention, discussed earlier in this chapter. Although these agreements did not use the phrase *Zero Discharge,* they call for the elimination or phaseout of discharges of persistent bioaccumulative substances rather than their reduction to acceptable levels.

32. For reasons similar to my own, the International Joint Commission (1992) recommended that persistent substances be defined to include all chemicals with a half-life in any environmental medium of greater than eight weeks and all chemicals that bioaccumulate. This definition would be even more useful if the first criterion were expanded to include degradation products, since some synthetic chemicals are quickly broken down into persistent substances.

33. See, for example, the language of the Paris Commission and the Barcelona Convention, discussed earlier in this chapter. The International Joint Commission, in contrast, takes a position like mine.

34. International Joint Commission 1992.

35. Jackson 1993a.

36. Jackson 1993b, Hirschorn et al. 1993.

37. See the discussion of perc alternatives in EPA (1998b) and in chapter 10 of this book.

38. Geiser 1993.

39. Wahlstrom and Lundquist 1993, PAN 1996, Swedish Ministry of the Environment 1998..

40. Geiser 1993.

41. Peter Montague, editor of *RACHEL's Health and Environment Weekly,* has been a forceful advocate of reforming the corporate charter in order to limit corporate liberty and ensure that business activities do not damage society's well-being. My discussion draws on the argument he has developed.

42. APHA 1994.

43. International Joint Commission 1992. A detailed discussion of transition planning for a chlorine sunset is provided in GLU 1995.

44. This definition is intended to disqualify the argument that economic benefits to some individuals or organizations "trickle down" and, because prosperity is supposedly essential to health, serve compelling social needs. Such an argument, if accepted, would forbid society from ever choosing a course of action that imposes any economic costs on any individual or organization. Further, this argument, often repeated by industries facing regulations of all types, is based on a tenuous connection between private economic benefits and public welfare. There is no evidence that investments in health and safety have had negative effects on public health.

45. Winner 1986.

46. Environmentalists (Great Lakes United 1995) and the Oil Chemical and Atomic Workers (Miller and Lewis 1994) have put forth similar proposals for chemical taxes to fund transition planning and worker protection efforts.

47. See, for instance, Nollkaemper 1996.

48. IJCSAB 1989.

Chapter 10

1. Dubos 1965.

2. Leinhardt 1994.

3. Vinyl Institute 1995.
4. Malkin and Fumento 1996.
5. The only sector of moderate size in which the use of the best chlorine-free alternative provides questionable environmental benefit is the manufacture of titanium dioxide. There are two ways to separate titanium from the impure ore in which it is contained: one uses chlorine gas and the other uses sulfuric acid. The latter process produces very large quantities of acid wastes; these wastes can neutralized and/or recycled, but they can pose a significant environmental hazard if they are not handled properly. In fact, the sulfuric acid process has caused significant local environmental damage in European ecosystems. A Clean Production approach should seek to minimize unnecessary uses of white pigment (as in the ubiquitous white plastic shopping bag), to develop cleaner methods for the production of titanium dioxide, and to develop alternative pigments that can be produced through cleaner processes. A useful discussion of the chlorine and sulfuric acid processes can be found in a report by Environment Canada (CIS 1997).
6. Bleifus 1995.
7. CRA 1993.
8. In CRA's report, over three-fourths of the cost of phasing-out chlorine comes from pharmaceuticals and pesticides (which account for just over 2 percent of chlorine consumption), based on rather alarming scenarios of increased disease rates and reduced food supply. Another potentially expensive sector to replace, not addressed directly by CRA, is choline hydrochloride, a livestock feed additive for which an alternative (choline bitartrate) is available but at substantially higher cost (CIS 1997).
9. Plimke et al. 1994. In its estimate of the costs of phasing out chlorine, Environment Canada (CIS 1997) makes a similar mistake, calculating the gross capital costs of equipment changes rather than the net costs of a phase-in harmonized with existing investment cycles.
10. EPA 1998b. Machine wet cleaning is the dominant method in this new industry, although the techniques of multiprocess wet cleaning supplement machines at many shops.
11. EPA 1997d.
12. Jehassi 1993.
13. EPA 1998b. See also TURI (1997) for an evaluation of performance at a shop using machine wet cleaning.
14. Ward 1997, EPA 1998b.
15. Kostick 1989.
16. CRA (1993) estimates an annual cost of $2.51 billion (U.S.) to supply 10 million short tons of caustic from trona, equal to $276 per metric ton. Environment Canada estimates a cost of $130 million (Canadian) per year to supply 1.12 million metric tons of caustic, or $81 (U.S.) per ton (CIS 1997), assuming that $1 (Canadian) equals $0.70 (U.S.).
17. CIS 1997, Christaens 1990, Pinoir 1992.
18. CIS 1997, Christaens 1990, Pinoir 1992.
19. Wood 1992.
20. Bar and Mani 1992.
21. Pimentel et al. 1991.
22. Pesek et al. 1989.
23. Pretty 1995.
24. Discussed in Pretty (1995).
25. Hewitt and Smith (1995) discuss the example of the Gallo Winery in California, which reportedly converted over half its land to strictly organic production. At first, costs increased

and yields declined, but after a few years, yields returned to previous levels, and the company now reports a net savings. In Alabama, Georgia, and Virginia, growers of cotton and apples have reduced their pesticide use substantially with no reduction in yield (Hewitt and Smith 1995, Mullen et al. 1997). And a review of forty-nine individual studies by economists at Virginia Polytechnic University found that pesticide reduction programs in the United States had increased average yields and economic returns in seven major groups of crops: cotton, soybeans, vegetables, fruit, peanuts, tobacco, and alfalfa (discussed in Jaenicke 1997).

26. Pimentel et al. (1991) report that the U.S. government spent $26 billion on agricultural price supports in 1989.

27. CRA 1993.

28. Antle and Pingali 1994.

29. Pesek et al. 1989.

30. Pretty 1995.

31. Hewitt and Smith (1995) also point out that pesticide contamination is a major factor in fish kills and contamination of marine food chains, which provide more than 40 percent of annual protein for over half of the world's population.

32. Pimentel et al. 1991.

33. Pimentel et al. 1991.

34. Reported in Hewitt and Smith 1995.

35. Pretty 1995.

36. Stone 1992.

37. OECD 1995.

38. Pretty 1995.

39. World Wildlife Fund 1998.

40. Commoner et al. (1996) describe TCF bleaching and pulping processes.

41. Commoner et al. 1996.

42. Commoner et al. 1996, Moldenius 1995a, CIS 1997. In some TCF processes with some types of wood pulp, there can be a small loss of pulp strength, but the change does not have a significant impact on the pulp's applied performance (Moldenius 1995a).

43. Commoner et al. 1996, CIS 1997.

44. Kutney 1995.

45. Lee 1993.

46. Commoner et al. 1996.

47. Albert 1993, Commoner et al. 1996.

48. Grant 1996.

49. Commoner et al. 1996.

50. In contrast to the findings of CBNS, Radian, and the EDF task force, Environment Canada (CIS 1997) concluded that TCF bleaching may cost as much as $50 per ton more than chlorine bleaching. This estimate, however, was based on an outdated 1992 study of the costs of TCF bleaching; it also failed to take account of the substantial savings offered by effluent reduction and recycling in an advanced TCF mill. U.S. EPA (1997a) predicted that TCF bleaching would cost $19 to $34 more per ton than chlorine dioxide bleaching, but this estimate also failed to consider the most cost-efficient technologies in use at European mills and did not consider the savings associated with effluent reduction and recycling.

51. According to Helge Eklund (1995), chief executive of Sodra Cell, one of Sweden's largest pulp manufacturers, "Sodra Cell has proven that you can make money with TCF."

52. Johansson et al. 1995b. The most advanced of the chlorine dioxide effluent recycling systems is currently the Bleached Filtrate Recycle system of Champion International, which has been implemented at the company's mill in Canton, NC. As of December 1996, the mill (which uses the bleaching sequence oxygen/chlorine dioxide/extraction/chlorine dioxide) had reduced its effluent from the first three stages by about eighty percent, with the remainder (and all the effluent from the final chlorine dioxide stage) being discharged as effluent. The "recycled" portion is filtered, and the sludge is burned in the facility's furnace (Stratton and Ferguson 1998).

53. Moldenius 1995b, Lovblad 1996, Meadows 1995, Ferguson and Finchem 1997. The new Metsa-Rauma mill in Finland has the goal of total closure by the end of the century (Grant 1996).

54. Department of Energy 1998.

55. Union Camp 1998.

56. Albert 1993.

57. In 1990, the independent contractor Jacobs Engineering (Jacobs 1990a) evaluated the total potential for source reduction of chlorinated solvents in the United States for the Metropolitan Water District of Southern California and the Environmental Defense Fund. The resulting report concluded that up to 65 percent of solvent use could be eliminated in favor of ecologically superior alternatives that were technically and economically feasible at the time. This estimate assumed that alternatives with greater cost would not be used and that nothing could be done to overcome other constraints, including lack of capital at small firms, resistance to change, lack of information and training, and so on. With a concerted effort, Jacobs concluded, a much larger portion of solvent use could be eliminated. Jacobs' method turned out to be so conservative that during the period when the report was in preparation (1988–1991), most industries reduced their use of chlorinated solvents by much more than the report predicted was possible.

58. Sherman et al. (1998) and Makhijani and Gurney (1995) review alternatives to organochlorine solvents. Morganstern et al. (1996) and Donahue et al. (1996) discuss specific applications at length.

59. Jacobs Engineering 1990a.

60. Sherman et al. 1998, Makhijani and Gurney 1995, Morganstern et al. 1996.

61. Sherman et al. 1998, Makhijani and Gurney 1995.

62. Donahue et al. 1996.

63. Sherman et al. 1998, Makhijani and Gurney 1995.

64. Morganstern et al. 1996.

65. Makhijani and Gurney 1995, Rossi et al. 1991, APHA 1994.

66. Wahlstrom and Lundquist 1993.

67. Baas (1996).

68. CRA 1993.

69. CIS 1997, Makhijani and Gurney 1995.

70. Makhijani and Gurney 1995.

71. Makhijani and Gurney 1995.

72. CIS 1997.

73. Tellus Institute 1992. This study ranked common packaging materials for the pollution they created during manufacture and disposal, including energy consumption. Vinyl was by far the worst, with total environmental costs outranking aluminum, PET, and other plastics

by factors of 2.7, 4.8, and 8.8, respectively. When it first was released, Tellus's study reported an even greater difference between vinyl and other materials. The study was revised to reflect criticisms by the Vinyl Institute, and it is the updated figures that are reported here.

74. Christaensen et al. 1990. Under certain criteria (such as energy use and accident potential during manufacture), PVC was not judged demonstrably inferior to polyurethane, acrylonitrile-butadiene-styrene, and aluminum.

75. Ecotec 1995.

76. CIS 1997; see also Cray (1997), Belazzi (1994).

77. Wagener et al. 1993, MCEA 1997.

78. Ecotec 1995, Klausbruckner 1993, Cray 1997, Greenpeace 1996.

79. APHA 1996.

80. Wagener et al. 1993, Cray 1997.

81. Enquete Kommission 1994. In its study for the state of Hessen, Prognos found that 95 percent of all PVC uses could be easily substituted using available alternatives. Although some alternatives were more expensive, the majority of PVC uses could be substituted cost-effectively (with no or little increase in costs) and with a net increase in employment of 4 percent (Plimke et al. 1994).

82. CIS 1997. Similarly, when the Board of Health of Toronto recommended that the city stop using PVC pipe for environmental reasons, it estimated that a phaseout of most uses of PVC pipe would cost up to $100,000 per year, not excessive for a major PVC use in a very large city (City of Toronto 1996).

83. Hickling 1993. The report was based on the current market price of alternatives. It does not include installation costs, the capital costs to produce larger quantities of alternatives, or reductions in cost as substitutes achieve economies of scale. The figures presented are in U.S. dollars, converted from the original text at a rate of 70 cents per Canadian dollar.

84. CIS 1997.

85. CIS 1997, Roberts 1994, CRA 1993.

86. Plimke et al. 1994.

87. Roberts 1994; Illman 1993, 1994.

88. Komiya et al. 1996, McGhee et al. 1996.

89. Rivetti et al. 1996.

90. Muskopf and McCollister 1987.

91. Plimke et al. 1994. The alternative process, which produces epichlorohydrin from allyl alcohol rather than allyl chloride, is applied on a commercial scale by the Japanese firm Showa-Denko (Nagato 1991).

92. CRA 1993.

93. Anastas and Williamson 1996, Borman 1992.

94. Rivetti et al. 1996.

95. CRA 1993.

96. Jacobs 1990b.

97. Jacobs 1990b.

98. *Chemical Week* 1992.

99. Tundo et al. 1996.

100. For an introduction to green chemistry, see Anastas and Williamson (1996).

101. CIS 1997, CRA 1993.

102. Bellanca and Bailey 1977, Wray 1993, CIS 1997. In controlled tests, Szal et al. (1991) have determined that chlorinated wastewater is significantly toxic to fish at distances more than 500 meters from the outfall; unchlorinated wastewater was not toxic to fish. Scott et al. (1990) have documented a range of acute and sublethal effects among invertebrate species exposed to chlorinated wastewater.

103. Linden 1998, Blatchley et al. 1996, Whitby et al. 1984, CIS 1997, Mann and Cramer 1992, Fahey 1990, Trojan 1991. Ultimately the best way to protect rivers and lakes from sewage plant discharges is to stop putting feces in the water. Modern dry composting and dessicating toilets create a carefully controlled environment that degrades the waste without odor, producing a small residue of dry, sanitary humus that is removed once a year or so and serves as a safe garden fertilizer (Riggle 1996). Dry toilets of this type are already on the market and in use in communities around the world, from the United States to Sweden to Australia (see, for example, Barsheid 1991 and Design News 1990). Composting toilets have long been used in Vietnam and have recently been successfully tested as an environmentally sound sanitation method in Micronesia (Rapaport 1996). Dry toilets are not likely to gain rapid cultural and psychological acceptance in some industrialized nations, so UV-based sewage disinfection remains the best current option for effective wastewater disinfection.

104. Linden 1998, Valenti 1997.

105. Civil Engineering 1995. In Martinez, California, a new UV treatment system has saved the sanitary district about $100,000 per year in operating costs compared to the chlorine-based system it replaced (Billings 1996).

106. Huisman and Wood (1974) of the World Health Organization provide a comprehensive, though slightly dated, overview of all aspects of slow sand filtration. A more recent volume by the American Society of Civil Engineers includes papers on various aspects of the method, including Hendricks and Bellamy 1991, Sims and Slezak 1991, Haarhoff and Cleasby 1991, Letterman 1991, and Berg et al. 1991. Riesenberg et al. (1995) describe a positive experience in a small California community.

107. Lay and Allen 1992.

108. Huisman and Wood 1974.

109. Sims and Slezak 1991.

110. Huisman and Wood 1974, Berg et al. 1991.

111. In a 1990 study (Public Works 1990), the Idaho Division of Environmental Quality found that properly operated slow sand filters can adequately remove Giardia cysts. Fogel et al. (1993) show extremely efficient removal of Giardia but note that proper design standards must be followed for complete removal of Cryptosporidia. As Tanner (1997) has pointed out, U.S. EPA has determined that slow sand filtration can achieve at least 99 percent removal of Cryptosporidium cysts, making it "an acceptable technology to control Cryptosporidium in drinking water."

112. As of 1995, there were 100 new ozone systems in design or construction in the United States (Dimitriou 1994). Ozone has also been licensed as an alternative to chlorine in the food processing industry, where it can be used to disinfect animal products, fruit, vegetables, and other foods (Lamarre 1997).

113. Lamarre 1997, Glaze 1987, Ferguson et al. 1991, Singer 1990, Rice 1989, Bryant et al. 1992.

114. Meier (1988) provides a review. More recently, DeMarini et al. (1995) confirmed that ozone treatment caused no increase in mutagenicity by several measures and that treatment

with chlorine or with chloramine increased mutagenicity substantially. See Rice (1989), Glaze (1987), and Singer (1990).

115. DeMarini et al. 1995.

116. Parrotta and Bekdash 1998, Wolfe 1990, Kruithof and van der Leer 1990, Bryant et al. 1992.

117. Kruithof and van der Leer 1990.

118. Parrotta and Bekdash 1998.

119. Wolfe 1990.

120. LeChevallier et al. 1988, 1990, Herson et al. 1991.

121. In the United States, most slow sand facilities add a chlorine residual because it is required by law (see, for instance, Riesenberg et al. 1995, Sims and Slezak 1991).

122. Kruithof and van der Leer 1990, Patel 1992, Rice et al. 1991, Rice 1989.

123. CRA 1993.

124. In this study, Blatchley et al. (1996) studied the performance of UV and chlorination on the wastewater of West Lafayette, Indiana, and concluded that UV was more consistently effective and would cost 36 percent less to install and 32 percent less to operate than the chlorine system. At large plants, UV may be slightly more expensive than chlorine gas, which enjoys an economy of scale in capital costs that UV does not.

125. CIS 1997.

126. Valenti 1997.

127. Lamarre 1997.

128. Vore and Unhoch 1994, Sapers 1993.

129. NASA 1996, Herman 1993, Fountainhead 1994.

130. Fountainhead 1994.

131. Environment Canada (CIS 1997) estimates that each million dollars of revenue in the PVC sector produces 3.1 jobs, while substitutes for vinyl (weighted by the portion of vinyl they would replace) produce 5.8 jobs per million dollars of revenue. Given 1993 Canadian employment in the PVC industry of 6,378 persons, these figures imply a net increase of 5,555 jobs at constant revenue levels. Canadian PVC production capacity in 1993 was 450,000 tons per year, compared to global production of 19.1 million tons per year (DTI 1995). Assuming 90 percent capacity utilization and a similar ratio of capacity to employment in other nations, these figures imply a net of creation of over 240,000 new jobs worldwide.

132. The figure given is based on EPA's estimate of 158,000 jobs in the dry cleaning sector (Jehassi 1993).

133. CIS 1997.

134. According to the U.S. Department of Commerce, an investment of 37.7 dollars created one hour of labor as of 1991 (Renner 1991). My calculations assume a forty-hour week and forty-eight weeks of employment per year.

135. Kirschner 1993, 1994.

136. My figures are extrapolated from Environment Canada's estimate that there were 1,305 persons employed in chlor-alkali and EDC/VCM production in Canada in 1993 (CIS 1997). Of these, 250 were employed in the production of EDC/VCM, and the remainder were involved in the chlor-alkali process itself. Extrapolating from Canada's 3.6 percent share of global chlorine capacity gives an estimate of just under 30,000 chlor-alkali workers worldwide. In 1993, 37 percent of chlorine production in Canada went to production of EDC and VCM, roughly the same proportion as in the rest of the world. Using this same ex-

trapolation factor, there are an estimated 7,000 workers in the global production of EDC and VCM. In addition, Environment Canada estimates that there were 315 workers at PVC polymerization plants in Canada, with a combined capacity of 450,000 tons per year. Assuming annual global production of 19.1 million tons of PVC (DTI 1995) gives a total of less than 15,000 workers in this sector worldwide. As shown in chapter 6, consumption of chlorine for the synthesis of organochlorine products and intermediates other than PVC and its feedstocks are each about half that of PVC in the United States, and less than that internationally. Because these extrapolations are extremely imprecise, I have reported estimated ranges rather than specific figures.

137. UNEP 1998.

138. From a statement by William Reilly, EPA administrator at the time, reported in Bureau of National Affairs 1990.

139. Massachusetts Department of Environmental Protection 1998.

140. Pimentel et al. (1991) estimate that direct externalities of pesticide use cost the U.S. economy two to four billion dollars per year.

141. Russell et al. 1992.

142. This figure comes from a report to the International Joint Commission's Virtual Elimination Task Force (1993).

143. Malkin and Fumento 1996.

144. Dubos 1965, 1970. Langdon Winner (1986) has expanded on this theme in *The Whale and the Reactor.*

Chapter 11

1. Galton was an eminent scientist and advocate for the prestige and authority of science. He was also the founder of eugenics, the science and policy of directing human reproduction to achieve an increase in "desirable" traits in the population. I am grateful to Barnes (1985) for pointing out his words.

2. Dow 1995.

3. Willes et al. 1993.

4. Meeker 1994.

5. Mongoven, Biscoe and Duchin 1994.

6. Mongoven 1995.

7. The classic sociological exploration of the problem of defining science is Gieryn (1983); for a philosophical treatment, see Feyerabend (1993).

8. Karol 1995.

9. Popper 1953, 1992.

10. See Kamrin 1998 for a discussion.

11. As the sociologist Thomas Gieryn (1983) has shown, political considerations have frequently been the motive behind the efforts of scientists and scientific organizations to draw clear boundaries between scientific and nonscientific claims or approaches. As Gieryn describes it, the definition of science is quite flexible, and scientists for centuries have moved these boundaries in their public discourse to increase their influence, credibility, and resources.

12. Howlett 1995.

13. Karol 1995.

14. Some scientists have explicitly acknowledged this point; see, for example, Commoner (1990). Kelly Moore (1996) provides a useful sociological discussion of scientific organizations that have espoused this view.

15. Barnes 1985.

16. Habermas 1996. Barnes (1985) provides a useful discussion—but ultimately, in my view, a narrow one—of Habermas's views.

17. For a thorough discussion of the role of scientific experts in public policy, see Jasanoff (1990).

18. Lash et al. (1984) provide an excellent overview of the Gorsuch scandals. The first altered assessment was the report by EPA Region V on dioxin pollution in the Great Lakes, which I discussed in detail in chapter 8 (U.S. House 1983). The second was the establishment of a level of concern for dioxin clean-up at Times Beach, Missouri, which internal memos revealed had been established by EPA not on the basis of health protection but because it would "buy time" and provide "good press for Agency." Merrell and Van Strum (1987) discuss those events and present facsimiles of the memos.

19. Ruckleshaus 1983.

20. For a summary, see Jasanoff 1986.

21. Reilly 1990.

22. My thinking on this subject has been informed by Jeff Howard (1997), who addresses this issue in detail.

23. Ruckleshaus 1983.

24. Reilly 1990.

25. Fischli 1996.

26. Karstadt (1988) lists twenty specific steps in calculating risks, each of which requires a subjective choice by the assessor.

27. Albert 1989.

28. Cothern et al. 1986.

29. Shelia Jasanoff (1990, 1996), a scholar of science and law at Harvard University, has made this point forcefully, showing from several examples that scientific consensus is reached on potentially contentious issues only when there is a preexisting settlement of the moral and political issues that frame the application of the science. Jasanoff's insight suggests that we cannot separate science and policymaking as the risk assessment/management framework asserts that we can.

30. I have adapted this discussion from Weinberg and Thornton 1994.

31. The idea that scientific consensus requires an a priori political consensus comes from Jasanoff (1996).

32. These two opposite conclusions can be found in two reviews of breast cancer and organochlorines. First, a review by a number of prominent epidemiologists that was funded by the Chemical Manufacturers Association (Ahlborg et al. 1995) concluded that there was no evidence linking organochlorines to breast cancer risk. In contrast, a statement by twenty scientists concluded, "Credible science does, in fact, suggest that organochlorines may contribute to breast cancer risk, with profound implications for millions of women. Calls for action to protect women's health by preventing further contamination of the environment are well-grounded in science and in the fundamental principles of public health practice" (Arditti et al. 1993).

33. This position draws on a tradition of work over the last seventy years in the sociology of scientific knowledge. Barnes et al. (1996) provide a rigorous presentation of a position not

so different from mine. In much earlier works, now classic, Ludwik Fleck (1932) and Karl Mannheim (1952, 1991) had the same insights into the cultural character of knowledge, although Mannheim did not extend his analysis to science. My view differs, however, from the strict constructivist position first expounded by Latour and Woolgar (1986), who suggest that science is a purely social construction and that the empirical content of science is illusory. I agree with Latour and Woolgar's argument that the status of scientific knowledge as a set of purely objective "facts" is socially constructed, I disagree with their total dismissal of science's empirical content. My view also differs from that of Cole (1992, 1996), who argues that social factors affect the focus but not the result of scientific activity. In the case of organochlorines, the focus of scientific activity—which health effects are investigated, on local or global scales, with short or long time frames—determines conclusions like safe/unsafe or healthy/damaged, so Cole's dichotomy turns out to be a false one.

34. Sclove 1995. See also Winner 1986.

35. Epstein 1995.

36. Sclove 1995.

37. Paulsen 1993.

38. Beck 1992, 1995.

References

Adami HO, Bergstrom R, Sparen P, Baron J. Increasing cancer risk in younger birth cohorts in Sweden. *Lancet* 341:773–777, 1993.

Addink R, Espourteille F, Altwicker ER. Role of inorganic chlorine in the formation of polychlorinated dibenzo-p-dioxins/dibenzofurans from residual carbon on incinerator fly ash. *Environmental Science and Technology* 32:3356–3359, 1998.

Adlercreutz H. Phytogestrogens: Epidemiology and a possible role in cancer protection. *Environmental Health Perspectives* 103 (supp 7):103–112, 1995.

(AET) (Alliance for Environmental Technology). *Trends in World Bleached Chemical Pulp Production: 1990–1997*. Washington, DC, 1997.

Aftalion F. *A History of the International Chemical Industry.* Benfey OT, trans. Philadelphia: University of Pennsylvania Press, 1991.

Aggazzotti G, Fantuzzi G, Tartoni PL, Predieri G. Plasma chloroform concentrations in swimmers using indoor swimming pools. *Archives in Environmental Health* 45:175–180, 1990.

Ahlborg UG, Lipworth L, Titus-Ernstoff L, Hsieh CC, Hanberg A, Baron J, Trichopoulos D, Adami HO. Organochlorine compounds in relation to breast cancer, endometrial cancer and endometriosis: An assessment of the biological and epidemiological evidence. *Critical Reviews in Toxicology* 25:463–531, 1995.

Airey D, Jones PD. Mercury in the River Mersey, its estuary and tributaries during 1973 and 1974. *Water Research* 16:565–577, 1982.

Aittola JP, Paasivirta J, Vattulainen A. Measurements of organochloro compounds at a metal reclamation plant. *Chemosphere* 27:65–72, 1993.

Albert RE. Risk assessment for acid aerosols. *Environmental Health Perspectives* 79:201–202, 1989.

Albert RJ. Restrictive environmental regulations drive mills to operate effluent-free. *Pulp and Paper* 67(13):97–99, 1993.

Alcock RE, Jones DC. Dioxins in the environment: A review of trend data. *Environmental Science and Technology* 30:3133–3143, 1996.

Alcock RE, Behnisch PA, Jones KC, Hagenmaier H. Dioxin-like PCBs in the environment—human exposure and the significance of sources. *Chemosphere* 37:1457–1472, 1998a.

Alcock RE, McLachlan MS, Johnston AE, Jones KC. Evidence for the presence of PCDD/Fs in the environment prior to 1900 and further studies on their temporal trends. *Environmental Science and Technology* 32:1580–1587, 1998b.

Allan R, Peakall D, Cairns V, Cooley J, Fox G, Gilman A, Piekarz D, van Oostdam J, Villeneuve D, Whittle M. *Toxic Chemicals in the Great Lakes and Associated Effects,* Volume 1: *Contaminant Levels and Trends.* Ottawa: Environment Canada, 1991.

Allen JR, Barsotti DA, Lambrecht LK, Van Miller JP. Reproductive effects of halogenated aromatic hydrocarbons on nonhuman primates. *Annals of the New York Academy of Sciences* 320:419–425, 1979.

Alleva E, Brock J, Brouwer A, Colborn T, Fossi MC, Gray E, Guillette L, Hauser P, Leatherland J, MacLusky N, Mutti A, Palnza P, Parmigiani S, Porterfield S, Santti R, Stein SA, vom Saal F, Weiss B. Statement from the work session on environmental endocrine-disrupting chemicals: Neural, endocrine and behavioral effects. Erice, Italy: Ettore Majorana Centre for Scientific Culture, 1995.

Altwicker ER, Konduri RKNV, Lin C, Milligan MS. Formation of precursors to chlorinated dioxin/furans under heterogeneous conditions. *Combustion Science and Technology* 88: 349–368, 1993.

AMAP (Arctic Monitoring and Assessment Programme). *Arctic Pollution Issues: A State of the Arctic Environment Report.* Oslo: AMAP Directorate, 1997.

Amy G, Greenfield J, Cooper W. Correlations between measurements of organic halide and specific halogenated volatile organic compounds in contaminated groundwater. In: Jolley RL, Condie LW, Johnson JD, Katz S, Minear RA, Mattice JS, Jacobs VA, eds. *Water Chlorination: Chemistry, Environmental Impact and Health Effects,* Volume 6. Chelsea, MI: Lewis Publishers, 1990:691–702.

Anastas PT, Williamson TC. Green chemistry: An overview. In: Anastas PT, Williamson TC, eds. *Green Chemistry: Designing Chemistry for the Environment.* Washington, DC: American Chemical Society, 1996:1–19.

Anderson HA, Falk C, Hanrahan L, Olson J, Burse VW, Needham L, Paschal D, Patterson D, Hill RH, Great Lakes Consortium. Profiles of Great Lakes critical pollutants: A sentinel analysis of human blood and urine. *Environmental Health Perspectives* 106:279–289, 1998.

Anderson JN, Clark JH, Peck EJ. The relationship between nuclear receptor-estrogen binding and uterotrophic responses. *Biochemical and Biophysical Research Communications* 48:1460–1468, 1972.

Anderson JN, Peck EJ, Clark JH. Nuclear receptor-estrogen complex: Relationship between concentration and early uterotrophic response. *Endocrinology* 92:1488–1494, 1973.

Anderson JN, Peck EJ, Clark JH. Estrogen-induced uterine responses and growth: Relationship to receptor estrogen binding by uterine nuclei. *Endocrinology* 96:160–167, 1975.

Andersson P, Ljung K, Soderstrom G, Marklund S. *Analys av Polyklorerade Dibensofuraner och Polykloerarde Dibensodioxiner i Processprover fran Hydro Plast AB.* Umea, Sweden: Institute of Environmental Chemistry, University of Umea, 1993.

Antle JM, Pingali PL. Pesticides, productivity and farmer health: A Philippine case study. *American Journal of Agricultural Economics* 76:418–430, 1994.

APHA (American Public Health Association). Resolution 9304: Recognizing and addressing the environmental and occupational health problems posed by chlorinated organic chemicals. *American Journal of Public Health* 84:514–515, 1994.

APHA (American Public Health Association). Resolution: Prevention of dioxin generation from PVC plastic use by health care facilities. Adopted at APHA Annual Meeting, New York, November 1996.

Arditti R, Bertell R, Clapp R, Rpstein SS, Ginsburg R, Hall RH, Legator M, Montajue P, Navarro V, O'Brien M, Orris P, Ozonoff D, Pepper L, Post D, Schecter A, Sherman J, Soderstrom R, Toniolo P, Webster T, Wolff MS. Scientists endorsement In: Thornton J. *Chlorine, Health, and the Environment: The Breast Cancer Warning*. Washington DC: Greenpeace, 1993.

Arnetz BB, Raymond LW, Nicolich MJ, Vargo L. Mortality among petrochemical science and engineering employees. *Archives of Environmental Health* 46:237–247, 1991.

Arnold DL, Nera EA, Stapley R, Tolnai G, Claman P, Hayward S, Tryphonas H, Bryce F. Prevalence of endometriosis in rhesus (*Macaca mulatta*) monkeys ingesting PCB (Aroclor 1254): Review and evaluation. *Fundamental and Applied Toxicology* 31:42–55, 1996.

Arnold SF, Vonier PM, Collins BM, Klotz DM, Guillette LJ, McLachlan JA. In vitro synergistic interaction of alligator and human estrogen receptors with combinations of environmental chemicals. *Environmental Health Perspectives* 105 (supp 3):615–618, 1997.

Aschengrau A, Ozonoff D, Paulu C, Coogan P, Vezina R, Heeren T, Zhang Y. Cancer risk and tetrachloroethylene-contaminated drinking water in Massachusetts. *Archives of Environmental Health* 48:284–292, 1993.

Asplund G. Origin and occurrence of halogenated organic matter in soil. In: Grimvall A, deLeer EWB, eds. *Naturally-Produced Organohalogens*. Boston: Kluwer Academic, 1995: 35–48.

Asplund G, Grimvall A. Organohalogens: More widespread than previously assumed. *Environmental Science and Technology* 25:1347–1350, 1991.

Asplund G, Grimvall A, Petterson C. Naturally produced adsorbable organic halogens (AOX) in humic substances from soil and water. *Science of the Total Environment* 81: 239–248, 1989.

Association of Postconsumer Plastics Recyclers. APR takes a stand on PVC. Press release. Washington, DC, April 14, 1998.

ATSDR (Agency for Toxic Substances and Disease Registry). *Toxicological Profile for Chloromethane*. Washington, DC: U.S. Public Health Service, 1989a.

ATSDR (Agency for Toxic Substances and Disease Registry). *Toxicological Profile for 1,1,1-Trichloroethane*. Washington, DC: U.S. Public Health Service, October 1989b.

ATSDR (Agency for Toxic Substances and Disease Registry), *Toxicological Profile for Aldrin/Dieldrin*. Washington, DC: U.S. Public Health Service, 1992.

ATSDR (Agency for Toxic Substances and Disease Registry). *Toxicological Profile for 1,2-Dichloroethane*. Washington, DC: U.S. Public Health Service, 1993a.

ATSDR (Agency for Toxic Substances and Disease Registry), *Toxicological Profile for Methoxychlor*. Washington, DC: U.S. Public Health Service, 1993b.

ATSDR (Agency for Toxic Substances and Disease Registry), *Toxicological Profile for Pentachlorophenol (Update)*. Washington, DC: U.S. Public Health Service, 1993c.

ATSDR (Agency for Toxic Substances and Disease Registry). *Toxicological Profile for Trichloroethylene*. Washington, DC: U.S. Public Health Service, 1993d.

ATSDR (Agency for Toxic Substances and Disease Registry). *Toxicological Profile for Chloromethane*. Washington, DC: U.S. Public Health Service, 1994.

ATSDR (Agency for Toxic Substances and Disease Registry), *Toxicological Profile for Chlorfenvinphos*. Washington, DC: U.S. Public Health Service, 1995a.

ATSDR Agency for Toxic Substances and Disease Registry), *Toxicological Profile for Chlorpyrifos*. Washington, DC: U.S. Public Health Service, 1995b.

ATSDR (Agency for Toxic Substances and Disease Registry), *Toxicological Profile for Hexachlorobenzene (Update)*. Washington, DC: U.S. Public Health Service, 1995c.

ATSDR (Agency for Toxic Substances and Disease Registry), *Toxicological Profile for Mirex and Chlordecone.* Washington, DC: U.S. Public Health Service, 1995d.

ATSDR (Agency for Toxic Substances and Disease Registry), *Toxicological Profile for Polychlorinated Biphenyls.* Washington, DC: U.S. Public Health Service, 1995e.

ATSDR (Agency for Toxic Substances and Disease Registry), *Toxicological Profile for Tetrachloroethylene (Update)* Washington, DC: U.S. Public Health Service, 1995f.

ATSDR (Agency for Toxic Substances and Disease Registry). *Toxicological Profile for Trichloroethylene (Update).* Washington, DC: U.S. Public Health Service, 1995g.

ATSDR (Agency for Toxic Substances and Disease Registry), *Toxicological Profile for Vinyl Chloride (Update).* Washington, DC: U.S. Public Health Service, 1995h.

ATSDR (Agency for Toxic Substances and Disease Registry), *Toxicological Profile for alpha-, beta-, gamma-, and delta-Hexachlorocyclohexane (Update).* Washington, DC: U.S. Public Health Service, 1997a.

ATSDR (Agency for Toxic Substances and Disease Registry), *Case Studies in Environmental Medicine: Lead Toxicity.* Washington, DC: U.S. Public Health Service, 1997b.

Auger J, Kunstmann JM, Czyglik F, Jouannet P. Decline in semen quality among fertile men in Paris during the past 20 years. *New England Journal of Medicine* 332:281–285, 1995.

Axelson O, Hardell L. Letter. *Medical Journal of Australia* 144:612–613, 1986.

Axelson O, Davis DL, Forestiere F, Schneiderman M, Wagener D. Lung cancer not attributable to smoking. *Annals of the New York Academy of Sciences* 609:164–178, 1990.

Ayotte P, Dewailly E, Ryan JJ, Bruneau S, Lebel G. PCBs and dioxin-like compounds in plasma of adult Inuit living in Nunavik (Arctic Quebec). *Chemosphere* 34:1459–1468, 1997.

Ayres R. The life-cycle of chlorine, part I: Chlorine production and the chlorine-mercury connection. *Journal of Industrial Ecology* 1:81–94, 1997.

Baas LW. An integrated approach to cleaner production. In: Misra KB, ed. *Clean Production: Environmental and Economic Perspectives.* New York: Springer Verlag, 1996:211–230.

Babic-Gojmerac T, Kniewald Z, Kniewald J. Testosterone metabolism in neuroendocrine organs in male rats under atrazine and deethyl atrazine influence. *Journal of Steroid Biochemistry* 33:141–146, 1989.

Baden JA. The anti-chlorine chorus is hitting some bum notes. *Seattle Times,* May 4, 1994:B7 (available from Chlorine Chemistry Council at www.c3.org).

Bailar JC. Death from all cancers: Trends in sixteen countries. *Annals of the New York Academy of Sciences* 609:49–57, 990.

Bailar JC. How dangerous is dioxin. *New England Journal of Medicine* 324:260–262, 1991.

Ballschmitter K. Persistent, ecotoxic and bioaccumulative compounds and their possible environmental effects. *Pure and Applied Chemistry* 68:1771–1780, 1996.

Band PR, Le ND, Fang R, Threlfall WH, Atrakianakis G, Anderson JTL, Keefe A, Krewski D. Cohort mortality study of pulp and paper mill workers in British Columbia, Canada. *American Journal of Epidemiology* 146:186–194, 1997.

Baniahmad A, Tsai MJ, Burris TP. The nuclear hormone receptor superfamily. In: Tsai M-J, O'Malley BW, eds. *Mechanism of Steroid Hormone Regulation of Gene Transcription.* Austin: RG Landes, 1994:1–24.

Bar DH, Mani KN. *Bipolar Membrane Electrodialysis Technology for the Recycling of Waste Salts into Acids and Bases.* Warren NJ: Allied-Signal, Inc., 1992.

Barnes B. *About Science.* Oxford: Blackwell, 1985.

Barnes B, Bloor D, Henry J. *Scientific Knowledge: A Sociological Analysis*. Chicago: University of Chicago Press, 1996.

Barnett JB, Colborn T, Fourier M, Gierthy J, Grasman K, Kerkvliet N, Lahvis G, Luster M, McConnachi P, Myers JP, Osterhaus ADME, Repetto R, Rolland R, Rollins-Smith L, Smialowicz R, Smolen M, Walker S, Watkins D. Statement from the work session on chemically-induced alterations in the developing immune system: The wildlife/human connection. *Environmental Health Perspectives* 104 (supp 4):807–808, 1996.

Barnett JN, Rodgers KE. Pesticides. In: Dean JH, Luster MI, Munson AE, Kimber I, eds. *Immunotoxicology and Immunopharmacology*, 2nd ed. New York: Raven, 1994:191–211.

Barrie L, Macdonald R, Bidleman T, Diamond M, Gregor D, Semkin R, Strachan W, Alaee M, Backus S, Bewers M, Gobeil C, Halsall C, Hoff J, Li A, Lockhart L, Mackay D, Muir D, Pudykiewicz J, Reimer K, Smith J, Stern G, Schroeder W, Wagemann R, Wania F, Yunker M.: Sources, occurrence and pathways. In: Jensen J, Adare K, Shearer R, eds. *Canadian Arctic Contaminants Assessment Report*. Ottawa: Indian and Northern Affairs Canada, 1997.

Barshied RD. Composting toilets—is zero discharge possible. *Public Works* 122(9):45–47, September 1991.

Barsotti DA, Abrahamson LJ, Allen JR. Hormonal alterations in female rhesus monkeys fed a diet containing 2,3,7,8-tetrachlorodibenzo-p-dioxin. *Bulletin of Environmental Contamination and Toxicology* 21:463–469, 1979.

Baukloh V, Bohnet HG, Trapp M, Heeschen W, Feichtinger W, Kemeter P. Biocides in human follicular fluid. *Annals of the New York Academy of Sciences* 442:240–250, 1985.

Becher H, Flesch-Janys D. Dioxins and furans: Epidemiologica assessment of cancer risks and other human health effects. *Environmental Health Perspectives* 106 (supp 2):623–624, 1998.

Becher H, Flesch-Janys D, Kauppinen T, Kogevinas M, Steindorf K, Manz A, Wahrendorf J. Cancer mortality in German male workers exposed to phenoxy herbicides and dioxins. *Cancer Causes and Control* 7:3121–321, 1996.

Becher H, Steindorf K, Flesch-Janys D. Quantitative cancer risk assessment for dioxins using an occupational cohort. *Environmental Health Perspectives 106* (supp 2):663–670, 1998.

Beck U. *Risk Society: Towards a New Modernity*. Ritter M, trans. Thousand Oaks, CA: Sage, 1992.

Beck U. *Ecological Enlightenment: Essays on the Politics of the Risk Society*. Ritter MA, trans. Atlantic Highlands, New Jersey: Humanities Press, 1995.

Beech AJ, Diaz R. Nitrates, chlorates, and trihalomethanes in swimming pool water. *American Journal of Public Health* 70:79–81, 1980.

Beekwilder AHM. Memorandum to J. Postma and others: Research by Olie CS concerning PCDD/F formation in VCM production. Amsterdam: Akzo Engineering, November 9, 1989.

Behnke W, Zetsch C. Tropospheric photoxidation of perchloroethene and trichloroacetyl chloride and the role of aerosols. *Organohalogen Compounds* 14:273–276, 1993.

Beland P, DeGuise S, Girard C, Lagace A, Martineau D, Michaud R, Muir DCG, Norstrom RJ, Pelletier E, Ray S, Shugart LR. Toxic compounds and health and reproductive effects in St. Lawrence beluga whales. *Journal of Great Lakes Research* 19:766–775, 1993.

Belazzi T. *Alternatives to PVC Products*. Vienna: Greenpeace Austria, 1994.

Bellanca MA, Bailey DS. Effects of chlorinated effluents on the aquatic ecosystems of the lower James River. *Journal of the Water Pollution Control Federation* 49:639–45, 1977.

Benbrook C, Marqardt S, Groth EJ. *Pest Management at the Crossroads.* Yonkers, NY: Consumers Union, 1997.

Berg P, Tanner S, Shieh CY. Construction, operation, and maintenance costs. In: Logsdon GS, ed. *Slow Sand Filtration.* New York: American Society of Civil Engineers, 1991:165–90.

Bergmann A, Olsson M. Pathology of Baltic grey seal and ringed seal females with special reference to adrenocortical hyperplasia: Is environmental pollution the cause of a widely distributed disease syndrome? *Finnish Game Research* 44:47–62, 1985.

Bern HA. The fragile fetus. In: Colborn T, Clement C, eds. *Chemically-Induced Alterations in Sexual and Functional Development: The Wildlife/Human Connection,* Princeton: Princeton Scientific Publishing, 1992:1–8.

Berry M, Bove F. Birth weight reduction associated with residence near a hazardous waste landfill. *Environmental Health Perspectives* 105:856–861, 1997.

Bertazzi PA, diDomenico A. Chemical, environmental and health aspects of the Seveso, Italy, incident. In: Schecter A, ed. *Dioxins and Health.* New York: Plenum, 1994:587–632.

Bertazzi PA, Pesatori AC, Consonni D, Tironi A, Landi MT, Zocchetti C. Cancer incidence in a population accidentally exposed to 2,3,7,8-tetrachlorodibenzo-p-dioxin. *Epidemiology* 4:398–406, 1993.

Bertazzi PA, Zocchetti C, Guercilene S, Consonni D, Tironi A, Landi MT, Pesatori AC. Dioxin exposure and cancer risk: A 15-year mortality study after the Seveso accident. *Epidemiology* 8:646–652, 1997.

Bertazzi PA, Bernucci I, Brambilla G, Consonni D, Pesator AC. The Seveso studies on early and long-term effects of dioxin exposure: A review. *Environmental Health Perspectives* 106 (supp 2):625–633, 1998.

Billings CH. Evaluation of effectiveness of UV disinfection. *Public Works* 127(10):72–73, 1996.

Binder R, Lech J. Xenobiotics in gametes of Lake Michigan lake trout induce hepatic monooxygenase activity in their offspring. *Fundamental and Applied Toxicology* 4: 1042–1054, 1984.

Birnbaum LS. Workshop on perinatal exposure to dioxin-like compounds: V. Immunologic effects. *Environmental Health Perspectives* 103 (supp 2):157–159, 1995a.

Birnbaum LS. Developmental effects of dioxins. *Environmental Health Perspectives* 103 (supp 7):89–94, 1995b.

Bishop CA, Brooks RJ, Carey JH, Ng P, Norstrom RJ, Lean DS. The case for a cause-effect linkage between environmental contamination and development in eggs of the common snapping turtle (*Chelydra s. serpentina*) from Ontario, Canada. *Journal of Toxicology and Environmental Health* 33:521–548, 1991.

Biziuk M, Czerwinski J, Kozlowski E. Identification and determination of organohalogen compounds in swimming pool water. *International Journal of Environmental and Analytical Chemistry* 50:109–115, 1992.

Blair A, Zahm SH. Cancer among farmers. *Occupational Medicine State of the Art Reviews* 6:335–354, 1991.

Blair A, Stewart PA, Tolbert PE, Graumann D, Moran FX, Vaught J, Rayner J. Cancer and other causes of death among a cohort of dry cleaners. *British Journal of Industrial Medicine* 47:162–168, 1990.

Blair A, Linos A, Steward PA, Burmeister LF, Gibson R, Everett G, Schuman L, Cantor KP. Comments on occupational and environmental factors in the origin of non-Hodgkin's lymphoma. *Cancer Research* (supp) 52:5501s–5502s, 1992a.

Blair A, Zahm SH, Pearce NE, Heineman EF, Fraumeni JF. Clues to cancer etiology from studies of farmers. *Scandinavian Journal of Work and Environmental Health* 18:209–215, 1992b.

Blair A, Linos A, Stewart PA, Burmeister LF, Gibson R, Everett G, Schuman L, Cantor KP. Evaluation of risks for non-Hodgkin's lymphoma by occupation and industry exposures from a case-control study. *American Journal of Industrial Medicine* 23:301–312, 1993.

Blair A, Zahm SH. Agricultural exposures and cancer. *Environmental Health Perspectives* 103 (supp 8): 205–208, 1995.

Blair PB. Immunologic studies of women exposed in utero to diethylstilbestrol. In: Colborn T, Clement C, eds. *Chemically-Induced Alterations in Sexual and Functional Development: The Wildlife-Human Connection.* Princeton: Princeton Scientific Publishing, 1992:289–294.

Blair PB, Noller KL, Turiel J, Forghani B, Hagens S. Disease patterns and antibody responses to viral antigens in women exposed in utero to diethylstilbestrol. In: Colborn T, Clement C, eds. *Chemically-Induced Alterations in Sexual and Functional Development: The Wildlife-Human Connection.* Princeton: Princeton Scientific Publishing, 1992: 283–288.

Blankenship A, Chang DPY, Jones AD, Kelly PB, Kennedy IM, Matsumura F, Pasek R, Yang GS. Toxic combustion by-products from the incineration of chlorinated hydrocarbons and plastics. *Chemosphere* 28:183–196, 1994.

Blatchley ER, Bastian KC, Duggirala RK. Ultraviolet irradiation and chlorination/dechlorination for municipal wastewater disinfection: Assessment of performance limitations. *Water Environment Research* 68:194–204, 1996.

Bleifus J. Dioxin as a therapeutic agent and other PR tales. *In These Times,* March 20, 1995:12–13.

Bodansky D. The precautionary principle in U.S. environmental law. In: O'Riordan T, Cameron J, eds. *Interpreting the Precautionary Principle.* London: Earthscan Publications, 1994:203–228.

Boehmer-Christansen S. The precautionary principle in Germany—enabling government. In: O'Riordan T, Cameron J, eds. *Interpreting the Precautionary Principle.* London: Earthscan Publications, 1994:31–60.

Bollag J-M, Dec J. Incorporation of halogenated substances into humic material. In: Grimvall A, deLeer EWB, eds. *Naturally-Produced Organohalogens.* Boston: Kluwer Academic, 1995:161–169.

Bombick DW, Jankun J, Tullis K, Matsumura F. 2,3,7,8-tetrachlorodibenzo-p-dioxin causes increase in expression of c-erb-A and levels of protein-tyrosine kinases in selected tissues of responsive mouse strains. *Proceedings of the National Academy of Sciences USA* 85: 4128–4132, 1987.

Bond GG, Ott MG, Brenner FE, Cook RR. Medical and morbidity surveillance findings among employees potentially exposed to TCDD. *British Journal of Industrial Medicine* 40:318–324, 1983.

Bond GG, McLaren EA, Lipps TE, Cook RR. Update of mortality among chemical workers with potential exposure to the higher chlorinated dioxins. *Journal of Occupational Medicine* 31:121–123, 1989.

Bonsor N, McCubbin N, Sprague J. *Municipal Industrial Strategy for Abatement: Kraft Mill Effluents in Ontario.* Toronto: Environment Ontario, 1988.

Borman S. Aromatic amine route is environmentally safer. *Chemical and Engineering News* 70 (48):26–27, 1992.

Boslund MC. Hormonal factors in carcinogenesis of the prostate and testis in humans and in animal models. In: Huff J, Boyd J, Barrett JC, eds. *Cellular and Molecular Mechanism of Hormonal Carcinogenesis: Environmental Influences.* New York: Wiley-Liss, 1996:309–352.

Bove FJ, Fulcomer MC, Klotz JB, Esmart J, Dufficy EM, Savrin JE. Public drinking water contamination and birth outcomes. *American Journal of Epidemiology* 141:850–862, 1995.

Bowerman WW, Giesy JP, Best DA, Kramer VJ. A review of factors affecting productivity of bald eagles in the Great Lakes region: Implications for recovery. *Environmental Health Perspectives* 103 (supp 4):51–60, 1995.

Bowermaster J. A town called Morrisonville. *Audubon* 95(4):42–51, 1993.

Bowman RE, Schantz SL, Gross ML, Barsotti DA. Behavioral effects in monkeys exposed to 2,3,7,8-TCDD transmitted maternally during gestation and for four months of nursing. *Chemosphere* 18:235–242, 1989.

Boyd J. Estrogen as a carcinogen: The genetics and molecular biology of human endometrial carcinoma. In: Huff J, Boyd J, Barrett JC, eds. *Cellular and Molecular Mechanism of Hormonal Carcinogenesis: Environmental Influences.* New York: Wiley-Liss, 1996:151–173.

Boyle P, Maisonneve P, Napalkov P. Geographical and temporal patterns of incidence and mortality from prostate cancer. *Urology* 46:47–55, 1995.

Brand D, Thompson D, van Voorst D. Star Wars' hollow promise; funding cuts and technical delays beset SDI. *Time,* December 7, 1987:17–20.

Braungart M. *Halogenated Hydrocarbons: Principal Thoughts and Data About a Possible Ban and Substitution.* Hamburg: Hamburger Umwelt Institut, 1987.

Brouwer A, Reinjders PJH, Koeman JH. Polychlorinated biphenyl (PCB)-contaminated fish induces vitamin A and thyroid hormone deficiency in the common seal (*Phoca vitulina*). *Aquatic Toxicology* 15:99–106, 1989.

Brouwer A, Ahlborg UG, Van den Berg M, Birnbaum LS, Boersma ER, Bosveld B, Denison MS, Gray LE, Hagmar L, Holene E, Huisman M, Jacobson SW, Jacobson JL, Koopman-Esseboom C, Koppe JG, Kulig BM, Morse DC, Muckle G, Peterson RE, Sauer PJJ, Seegal RF, Smits-Van Prooije AE, Touwen BCL, Weisglas-Kuperus N, Winneke G. Functional aspects of developmental toxicity of polyhalogenated aromatic hydrocabons in experimental animals and human infants. *European Journal of Pharmacology* (Environmental Toxicology and Pharmacology Section) 293:1–40, 1995.

Brown LM, Pottern LM, Hoover RN, Devesa SS, Aselton P, Flannery JT. Testicular cancer in the United States: Trends in incidence and mortality. *International Journal of Epidemiology* 15:164–169, 1986.

Brown LM, Blair A, Gibson R, Everett GD, Cantor KP, Schuman LM, Burmeister LF, Van Lier SF, Dick F. Pesticide exposures and other agricultural risk factors for leukemia among men in Iowa and Minnesota. *Cancer Research* 50:6585–6591, 1990.

Brown NM, Manzolillo PA, Zhang JX, Wang J, Lamartiniere CA. Prenatal TCDD and predisposition to mammary cancer in the rat. *Carcinogenesis* 19:1623–1629, 1998.

Bruce KR, Beach LO, Gullett BK. The role of gas-phase Cl2 in the formation of PCDD/PCDF during waste combustion. *Waste Management* 11:97–102, 1991.

Brun GL, Howell GD, ONeill HJ. Spatial and temporal patterns of organic contaminants in wet precipitation in Atlantic Canada. *Environmental Science and Technology* 25:1249–1261, 1991.

Bryant EA, Fulton GP, Budd GC. *Disinfection Alternatives for Safe Drinking Water.* New York: Van Nostrand Reinhold, 1992.

Brzuzy LP, Hites RA. Global mass balance for polychlorinated dibenzo-p-dioxins and dibenzofurans. *Environmental Science and Technology* 30:1797–1804, 1996a.

Brzuzy LP, Hites RA. Response to comment on global mass balance for polychlorinated dibenzo-p-dioxins and dibenzofurans. *Environmental Science and Technology* 30:3647–3648, 1996b.

Buetow SA. *Epidemiology* of testicular cancer. *Epidemiologic Reviews* 17:433–449, 1995.

Bujan L, Mansat A, Pontonnier F, Miuesset R. Time series analysis of sperm concentration in fertile men in Toulouse, France, between 1977 and 1992. *British Medical Journal* 312:471–476, 1996.

Bumb RR, Crummett W, Artie S, Gledhill J, Hummel R, Kagel R. Trace chemistries of fire, source of chlorinated dioxins. *Science* 210:385–390, 1980.

Bureau of National Affairs. Reilly calls pollution technology too expensive. *BNA's Environment Watch* 1(27):1–2, October 15, 1990.

Burleson GR, Lebrec H, Yang YG, Ibanes JD, Pennington KN, Birnbaum LS. Effect of 2,3,7,8-tetrachlorodibenzo-p-dioxin (TCDD) on influenza host resistance in mice. *Fundamental and Applied Toxicology* 29:40–47, 1996.

Burmaster D, Harris R. Groundwater contamination: An emerging threat. *Technology Review* 85(5):50–61, 1982.

Burnett R. Memorandum and enclosures from Vinyl Institute executive director to VI executive board members and VI issues management committee members: Crisis management plans for the dioxin reassessment. Morristown, NJ: Vinyl Institute, September 6, 1994.

Bush B, Bennett AH, Snow JT. Polychlorinated biphenyl congeners, p,p′-DDE, and sperm function in humans. *Archives of Environmental Contamination and Toxicology* 15: 333–341, 1986.

Byers VS, Levin AS, Ozonoff DM, Baldwin RW. Association between clinical symptoms and lymphocyte abnormalities in a population with chronic domestic exposure to industrial solvent–contaminated domestic water supply and a high incidence of leukaemia. *Cancer Immunology Immunotherapy* 27:77–81, 1988.

Cameron J. The status of the precautionary principle in international law. In: O'Riordan T, Cameron J, eds. *Interpreting the Precautionary Principle.* London: Earthscan Publications, 1994:262–291.

Cameron J, Abouchar J. The status of the precautionary principle in international law. In: Freestone D, Hey E, eds. *The Precautionary Principle and International Law: The Challenge of Implementation.* Boston: Kluwer Law International, 1996:29–52.

Cantor KP. Bladder cancer, drinking water source and tap water consumption: A case-control study. *Journal of the National Cancer Institute* 79:1269–1279, 1987.

Cantor KP. Water chlorination, mutagenicity, and cancer epidemiology. *American Journal of Public Health* 84:1211–1213, 1994.

Cantor KP, Blair A, Everett G, Gibson R, Burmeister LF, Brown LM, Schuman L, Dick FR. Pesticides and other agricultural risk factors for non-Hodgkin's lymphoma among men in Iowa and Minnesota. *Cancer Research* 52:2447–2455, 1992.

Cantor KP, Shy CM, Chilvers C. Water pollution. In: Schottenfeld D, ed. *Cancer Epidemiology and Prevention.* Oxford: Oxford University Press, 1996:418–437.

Cantor KP, Lynch CF, Hildesheim ME, Dosemci M, Lubin J, Alavanja M, Craun C. Drinking water source and chlorination by-products: I: Risk of bladder cancer. *Epidemiology* 9:21–28, 1998.

CanTox, Inc. Interpretive review of the potential adverse effects of chlorinated organic chemicals on human health and the environment. *Regulatory Toxicology and Pharmacology* 20: 1994.

CARB (California Air Resources Board). *Proposed Dioxins Control Measure for Medical Waste Incinerators: Staff Report.* Sacramento: CARB Stationary Source Division, 1990.

Carlsen E, Giwercman A, Keiding N, Skakkebaek N. Evidence for decreasing quality of semen during past 50 years. *British Medical Journal* 305:609–613, 1992.

Carr BR, Griffin JE. Fertility control and its complications. In: Wilson JD, Foster DW, eds. *Textbook of Endocrinology,* Seventh Edition. Philadelphia: Saunders, 1985:452–475.

Carroll W. *Is PVC in House Fires the Great Unknown Source of Dioxin?* Washington,DC: Vinyl Institute, 1995.

Carroll WF, Borelli FE, Garrity PJ, Jacobs RA, Lewis JW, McCreedy RL, Weston AF. Characterization of dioxins and furans from ethylene dichloride, vinyl chloride and polyvinyl chloride facilities in the United States: Resin, treated wastewater and ethylene dichloride. *Organohalogen Compounds* 27:62–67, 1996.

Carson R. *Silent Spring.* Boston: Houghton Mifflin, 1962.

Castleman BI, Ziem GE. Toxic pollutants, science and corporate influence. *Archives of Environmental Health* 44:68, 1989.

Catabeni F, Cavallero A, Galli G. *Dioxin: Toxicological and Chemical Aspects.* New York: Spectrum, 1985.

Cates DH, Eggert C, Yang JL, Eriksson K-EL. Comparison of effluents from TCF and ECF bleaching of kraft pulps. *Tappi Journal* 78(12):93–98, 1995.

Catz DS, Fischer LM, Kelley DB. Androgen regulation of a laryngeal-specific myosin heavy chain mRNA isoform whose expression is sexually differentiated. *Developmental Biology* 171(2):448–57, 1995.

CCRIS (Chemical Carcinogenesis Research Information System) on-line database. Bethesda: National Library of Medicine, 1997.

Chang DPY, Glasser H, Hickman DC. Toxic products of medical waste incineration. In: Green AES, ed. *Medical Waste Incineration and Pollution Prevention.* New York: Van Nostrand Reinhold, 1993:73–96.

Chang, D, Richards M, Huffman G. Studies of POHC DE during simulated atomization failure in a turbulent flame reactor. In: *Land Disposal, Remedial Action, Incineration and Treatment of Hazardous Waste, Proceedings of the Fourteenth Annual Research Symposium.* Cincinnati: U.S. Environmental Protection Agency Hazardous Waste Engineering Laboratory (EPA 600/9-88/021), 1988.

Chappell PE, Lydon JP, Conneely OM, O'Malley BW, Levine JE. Endocrine defects in mice carrying a null mutation for the progesterone receptor gene. *Endocrinology* 138:4147–4152, 1997.

Cheek AO, Vonier PN, Oberdorster E, Burow BC, McLachlan JA. Environmental signaling: A biological context for endocrine disruption. *Environmental Health Perspectives* 106 (supp 1):5–10, 1998.

Cheek AO, Kow K, Chen J, McLachlan JA. Potential mechanisms of thyroid disruption in humans: interaction of organochorine compounds with thyroid receptor, transthyretin and thyroid-binding globulin. *Environmental Health Perspectives* 107:273–278, 1999.

Chemical Week. Arco, Sipsy in epoxides deal. *Chemical Week,* October 28, 1992:51.

Chemical Week. Vinyl chloride producers. *Chemical Week,* February 1, 1995a:9.

Chemical Week. Product focus: Polyvinyl chloride. *Chemical Week,* April 5, 1995b:63.

Chen Y-CJ, Yu M-LM, Rogan WJ, Gladen BC, Hsu C-C. A 6-year follow-up of behavior and activity disorders in the Taiwan Yu-cheng children. *American Journal of Public Health* 84:415–421, 1994.

Chiazze L, Nichols WE, Wong O. Mortality among employees of PVC fabricators. *Journal of Occupational Medicine* 19:623–628, 1977.

Chicago Tribune. Hazardous lead levels exist in plastic, vinyl miniblinds. July 8, 1996:A7.

Chlorine Chemistry Council and the Vinyl Institute. Statement on Greenpeace press briefing. Washington, DC: Chlorine Chemistry Council, 1997.

Chlorine Institute. *Production of Chlor-Alkali Chemicals.* Washington, DC: Chlorine Institute, 1991.

Chovanec A, Grath J, Schwaiger K, Nagy W, Schicho-Schreier I. Halogenated hydrocarbons in Austrian running waters—first results of a new water quality monitoring system. *Organohalogen Compounds* 14:85–88, 1993.

Christaens J. Satisfying the demand for alkali (presentation text). Second World Chlor-Alkali Symposium, Washington, DC, September 1990.

Christanesen K, Grove A, Hansen LE, Hoffman L, Hensen AA, Pommer K, Schmidt A. *Environmental Assessment of PVC and Selected Alternative Materials.* Copenhagen: Danish Ministry of the Environment, 1990.

Christmann W. Combustion of polyvinyl chloride—an important source for the formation of PCDD/PCDF. *Chemosphere* 19:387–392, 1989a.

Christmann W, Koppel KD, Partscht H, Rotard W. Determination of PCDDs/PCDFs in ambient air. *Chemosphere* 19:521–526, 1989b.

CIS (Chem-Info Services). *A Technical and Socio-Economic Comparison of Options to Products Derived from the Chlor-alkali Industry.* Burlington, Ontario: Environment Canada, 1997.

City of Toronto Public Health Department. Memorandum to Board of Health and City Services Committee: Health and environmental impacts associated with PVC and other pipe materials commonly used by the City of Toronto, Toronto, March 19, 1996.

Civil Engineering. Study finds ultraviolet effluent treatment efficient and cost-effective. *Civil Engineering* 65(9):24, 1995.

Clapp RT, Truemper CA, Aziz S, Reschke T. AOX content of paper manufactured with chlorine free pulps. *Tappi Journal* 79:111–113, 1996.

Clark JH, Schrader WT, O'Malley BW. Mechanisms of steroid hormone action. In: Wilson JD, Foster DW, eds. *Textbook of Endocrinology,* Seventh Edition. Philadelphia: Saunders, 1985:33–75.

Claus EB, Risch N, Thompson WD. Genetic analysis of breast cancer in the cancer and steroid hormone study. *American Journal of Human Genetics* 48:232–242, 1991.

Cleland GB, McElroy PH, Sonstegard RA. Immunomodulation in C57Bl/6 mice following the consumption of halogenatic aromatic hydrocarbon-contaminated coho salmon (*Onchorhynchus kisutch*) from Lake Ontario. *Journal of Toxicology and Environmental Health* 27:477–486, 1989.

Cleverly D, Monetti M, Phillips L, Cramer P, Heit M, McCarthy S, O'Rourke K, Stanley J, Winters D. A time-trends study of the occurrences and levels of CDDs, CDFs and dioxin-like PCBs in sediment cores from 11 geographically distributed lakes in the United States. *Organohalogen Compounds* 28:77–81, 1996.

Cohen L, Rose D, Wynder E. A rationale for dietary intervention in postmenopausal breast cancer patients. *Nutrition and Cancer* 10:1–10, 1993.

Cohen M, Commoner B, Eisl H, Bartlett P, Dickar A, Hill C, Quigley J, Rosenthal J. *Quantitative Estimation of the Entry of Dioxins, Furans, and Hexachlorobenzene into the Great Lakes from Airborne and Waterborne Sources.* Flushing, NY: Center for the Biology of Natural Systems, Queens College, City University of New York, 1995.

Cohn P, Klotz J, Bove F, Berkowitz M, Fagliano J. Drinking water contamination and the incidence of leukemia and non-Hodgkin's lymphoma. *Environmental Health Perspectives* 102:556–561, 1994.

Colborn T, Davidson A, Green S, Hodge R, Jackson C, Liroff R. *Great Lakes: Great Legacy?* Baltimore: Conservation Foundation and Institute for Research on Public Policy, 1991.

Colborn T, vom Saal FS, Soto AM. Developmental effects of endocrine-disrupting chemicals in wildlife and humans. *Environmental Health Perspectives* 101:378–384, 1993.

Colborn T, Dumanoski D, Myers JP. *Our Stolen Future.* New York: Dutton, 1996.

Colditz GA, Willett WC, Hunter DJ, Stampfer MJ, Manson JE, Hennekens CH, Rosner BA, Speizer FE. Family history, age and risk of breast cancer. *Journal of the American Medical Association* 270:338–343, 1994.

Cole S. *Making Science: Between Nature and Society.* Cambridge: Harvard University Press, 1992.

Cole S. Voodoo sociology: Recent developments in the sociology of science. *Annals of the New York Academy of Sciences* 775:274–287, 1996.

Commoner B. Failure of the environmental effort. *Environmental Law Reporter* 18:10195–10221, 1987.

Commoner B. *Making Peace with the Planet.* New York: Pantheon, 1990.

Commoner B, Cohen M, Bartlett PW, Dickar A, Eisl H, Hill C, Rosenthal J. *Zeroing Out Dioxin in the Great Lakes: Within Our Reach.* Flushing, NY: Center for the Biology of Natural Systems, Queens College, City University of New York, 1996.

Connor MS. Comparison of the carcinogenic risks from fish vs. groundwater contamination by organic compounds. *Environmental Science and Technology* 18:321–327, 1984.

Cortes DR, Basu I, Sweet CW, Brice KA, Hoff RM, Hites RA. Temporal trends in gas-phase concentrations of chlorinated pesticides measured at the shores of the Great Lakes. *Environmental Science and Technology* 32:1920–1927, 1998.

Costner P. *The Burning Question: Chlorine and Dioxin.* Washington DC: Greenpeace USA, 1997.

Costner P, Thornton J. *Playing with Fire: Hazardous Waste Incineration.* Washington, DC: Greenpeace USA, 1991.

Costner P, Cray C, Martin G, Rice B, Santillo D, Stringer R. *PVC: A Principal Contributor to the U.S. Dioxin Burden.* Washington, DC: Greenpeace USA, 1995.

Cothern C, Coniglio W, Marcus W. Estimating risk to human health: Trichloroethylene in drinking water is used as an example. *Environmental Science and Technology* 20(2): 111–116, 1986.

Council of Environmental Advisors (Der Rat von Sachverstandigen fur Umweltfragen). *Abfallwirtschaft Sondergutachten* (Waste management special report). Stuttgart: German Federal Environment Ministry, 1991.

Covello VT, Sandman PM, Slovic P. *Risk Communication, Risk Statistics and Risk Comparisons: A Manual for Plant Managers.* Washington, DC: Chemical Manufacturers Association, 1988.

CRA (Charles River Associates, Inc.) *Assessment of the Economic Benefits of Chlor-Alkali Chemicals to the United States and Canadian Economies.* Washington, DC: Chlorine Institute, 1993.

Crain DA, Guillette LJ, Rooney AA, Pickford DB. Alterations in steroidogenesis in alligators *(Alligator mississippiensis)* exposed naturally and experimentally to environmental contaminants. *Environmental Health Perspectives* 105:528–533, 1997.

Cranmer JM, Cranmer MF, Goad PT. Prenatal chlordane exposure: Effects on plasma corticosterone concentrations over the lifespan of mice. *Environmental Research* 35:204–210, 1984.

Cray C. Opportunities to phase out PVC. *New Solutions* 7(4): 17–29, 1997.

Cresanta JL. Epidemiology of cancer in the United States. *Primary Care* 19:419–441, 1992.

Crews D, Bergeron JM, McLachlan JA. The role of estrogen in turtle sex determination and the effect of PCBs. *Environmental Health Perspectives* 103 (supp 7):73–78, 1995.

Crisp TM, Clegg ED, Cooper RL, Wood WP, Anderson DG, Baetcke KP, Hoffman JL, Morrow MS, Rodier DJ, Schaeffer JE, Touart LW, Zeeman MG, Patel YM. Environmental endocrine disruption: An effects assessment and analysis. *Environmental Health Perspectives* 106 (supp 1):11–56, 1998.

Cummings AM, Gray LE. Antifertility effect of methoxychlor in female rats: Dose and time-dependent blockade of pregnancy. *Toxicology and Applied Pharmacology* 97:454–462, 1989.

Cummings AM, Metcalf JL, Birnbaum JL. Promotion of endometriosis by 2,3,7,8-tetrachlorodibenzo-p-dioxin in rat and mice: Time-dose dependence and species comparison. *Toxicology and Applied Pharmacology* 138:131–139, 1996.

Curry MS, Huguenin MT, Martin AJ, Lookingbill TR. *Contamination Extent Report and Preliminary Evaluation for the Calcasieu Estuary.* Silver Spring, MD: National Oceanic And Atmospheric Administration, 1996.

Czuczwa JM, Hites RA. Environmental fate of combustion-generated polychlorinated dioxins and furans. *Environmental Science and Technology* 186:444–449, 1984.

Czuczwa JM, Hites RA. Historical record of polychlorinated dioxins and furans in Lake Huron sediments. In: Keith LH, Rappe C, Choudhary G, eds. *Chlorinated Dioxins and Dibenzofurans in the Total Environment II.* Boston: Butterworth, 1985:59–63.

Czuczwa JM, Hites RA. Airborne dioxins and dibenzofurans: Sources and fates. *Environmental Science and Technology* 20:195–200, 1986.

Czucwa JM, Veety BD, Hites RA. Polychlorinated dibenzo-p-dioxins and dibenzofurans in sediments from Siskiwit Lake, Isle Royale. *Science* 226:568–569, 1984.

Dahlman O, Morck R. Chemical composition of the organic material in modern bleached kraft mill effluents. In: Sodergren A, ed. *Environmental Impact of Bleached Pulp Mill Effluents: Composition, Fate and Effects in the Baltic Sea.* Report of the Environment/Cellulose II Project. Solna: Swedish Environmental Protection Agency (Report 4047), 1993:135–150.

Dai LC, Phuong NTN, Thom LH, Thuy Tt, Van NT, Cam LH, Chi HtK, Thuy LB. A comparison of infant mortality rates between two Vietnamese villages sprayed by defoliants in wartime and one unsprayed village. *Chemosphere* 20:1005–1012, 1990.

Daly AK, Cholerton S, Armstrong M, Idle JR. Genotyping for polymorphisms in xenobiotic metabolism as a predictor of disease susceptibility: Mechanism-based predictions of interactions. *Environmental Health Perspectives* 102 (supp 9):55–62, 1994.

Daly HB. Reward reductions found more aversive by rats fed environmentally contaminated salmon. *Neurotoxicology and Teratology* 13:449–453, 1991.

Daly HB. Laboratory rat experiments show consumption of Lake Ontario salmon causes behavioral changes: Support for wildlife and human research results. *Journal of Great Lakes Research* 19:784–788, 1993.

Daly HB, Hertzler DR, Sargent DM. Ingestion of environmentally contaminated Lake Ontario salmon by laboratory rats increases avoidance of unpredictable aversive nonreward and mild electric shock. *Behavioral Neuroscience* 103:1356–1363, 1989.

Daly HB, Stewart PW, Lunkenheimer L, Sargent D. Maternal consumption of Lake Ontario salmon in rats produces behavioral changes in the offspring. *Toxicology and Industrial Health* 14:25–38, 1998.

Daly HE, Cobb JB. *For the Common Good: Redirecting the Economy Toward Community, the Environment, and a Sustainable Future.* Boston: Beacon Press, 1989.

Daniels JL, Olshan AF, Savitz DA. Pesticides and childhood cancers. *Environmental Health Perspectives* 105:1068–1077, 1997.

Danish Environmental Protection Agency. *PVC and Alternative Materials.* English trans. Copenhagen: Danish Environmental Protection Agency, 1993.

Danzo BJ. Environmental xenobiotics may disrupt normal endocrine function by interfering with the binding of physiological ligands to steroid receptors and binding proteins. *Environmental Health Perspectives* 105:294–301, 1997.

Dar E, Kanarek MS, Anderson HA, Sonzogaí WC. Fish consumption and reproductive outcomes in Green Bay, Wisconsin. *Environmental Research* 59:189–201, 1992.

Darnell J, Lodish H, Baltimore D. *Molecular Cell Biology,* Second Edition. New York: Freeman, 1990.

Davies D, Mes J. Comparison of the residue levels of some organochlorine compounds in breast milk of the general and indigenous Canadian populations. *Bulletin of Environmental Contamination and Toxicology* 39:743–749, 1987.

Davies FCW. Minimata disease: A 1989 update on the mercury poisoning epidemic in Japan. *Environmental Geochemistry and Health* 13:35–38, 1991.

Davis DL, Magee BH. Cancer and industrial chemical production. *Science* 206:1356–1358, 1979.

Davis DL, Hoel D, Fox J, Lopez AD. International trends in cancer mortality in France, West Germany, Italy, Japan, England, and Wales, and the United States. *Annals of the New York Academy of Sciences* 609:5–48, 1990.

Davis DL, Blair A, Hoel DG. Agricultural exposures and cancer trends in developed countries. *Environmental Health Perspectives* 100:39–44, 1992.

Davis DL, Bradlow HL, Wolff MS, Woodruff T, Hoel DG, Anton-Culver H. Medical hypothesis: Xeno-estrogens as preventable causes of breast cancer. *Environmental Health Perspectives* 101:372–376, 1993.

Davis DL, Dinse GE, Hoel DG. Decreasing cardiovascular disease and increasing cancer among whites in the United States from 1973 through 1987. *Journal of the American Medical Association* 271:431–437, 1994.

Davis DL, Telang NT, Osborne MP, Bradlow HL. Medical hypothesis: Bifunctional genetic-hormonal pathways to breast cancer. *Environmental Health Perspectives* 105 (supp 3): 571–576, 1997.

Dean AG, Imrey HH, Dusich K, Hall WN. Adjusting morbidity ratios in two communities using risk factor prevalence in cases. *American Journal of Epidemiology* 127:654–662, 1988.

Dearth M, Hites RA. Highly chlorinated dimethanofluorenes in technical chlordane and in human adipose tissue. *Journal of the American Society for Mass Spectrometry* 1:92–98, 1990.

deJong E, Field JA, Spinnler H-E, Cademier AE, deBont JAM. Significant fungal biogenesis of physiologically important chlorinated aromatics in natural environments. In: Grimvall A, deLeer EWB, eds. *Naturally-Produced Organohalogens.* Boston: Kluwer Academic, 1995: 251–259.

deLeer EWB. The identification of highly chlorinated ethers and diethers in river sediment near an epichlorohydrin plant. *Water Research* 19:1411–1419, 1985.

Dellinger B, Taylor P, Tiery D, Pan J, Lee CC. Pathways of PIC formation in hazardous waste incinerators. In: *Land Disposal, Remedial Action, Incineration and Treatment of hazardous Waste, Proceedings of the Fourteenth Annual Research Symposium.* Cincinnati: U.S. EPA Hazardous Waste Engineering Research Laboratory (EPA 600/9-88/021), 1988.

DeLorey D, Cronn D, Farmer J. Tropospheric latitudinal distributions of CF2Cl2, CFCl3, N20, CH3CCl3 and CCl4 over the remote Pacific Ocean. *Atmospheric Environment* 22: 1481–1494, 1988.

DeMarini DM, Abu-Shakra A, Felton CF, Patterson KS, Shelton ML. Mutation spectra in salmonella of chlorinated, chloraminated, or ozonated drinking water extracts: Comparison to MX. *Environmental and Molecular Mutagenesis* 26:270–285, 1995.

Dempsey C, Oppelt T. Incineration: A critical review update. *Air and Waste* 43:25–51, 1993.

deOude N. Product lifecycle assessment—development of a methodology. In: Jackson T, ed. *Clean Production Strategies: Developing Preventive Environmental management in the Industrial Economy.* Ann Arbor: Lewis, 1993:207–224.

Derry TK, Williams TI. *A Short History of Technology from the Earliest Times to* AD *1900.* New York: Dover, 1960.

Design News. Toilets flush without water. *Design News* 46(18):40, 1990.

de Swart RL, Ross PS, Vos JG, Osterhaus ADME. Impaired immunity in harbour seals (*Phoca vitulina*) exposed to bioaccumulated environmental contaminants: Review of a long-term feeding study. *Environmental Health Perspectives* 104 (supp 4):823–828, 1996.

Dethlefsen V, Jackson T, Taylor P. The precautionary principle—towards anticipatory environmental management. In: Jackson T, ed. *Clean Production Strategies: Developing Preventive Environmental Management in the Industrial Economy.* Boca Raton: Lewis, 1993:41–62.

Deutsch DG, Goldfarb TD. PCDD/PCDF contamination following a plastics fire in a university lecture hall building. *Chemosphere* 12:2423–2431, 1988.

Devesa SS, Blot WJ, Stone BJ, Miller BA, Tarone RE, Fraumeni JF. Recent cancer trends in the United States. *Journal of the National Cancer Institute* 87:175–182, 1995.

DeVito MG, Birnbaum LS. Toxicology of dioxins and related chemicals. In: Schecter A, ed. *Dioxins and Health.* New York: Plenum, 1994:139–162.

DeVito MJ, Birnbaum LS, Farland WH, Gasiewicz T. Comparisons of estimated human body burdens of dioxinlike chemicals and TCDD body burdens in experimentally exposed animals. *Environmental Health Perspectives* 103:820–831, 1995.

Dewailly E, Nantel A, Weber JP, Meyer F. High levels of PCBs in breast milk of Inuit women from Arctic Quebec. *Bulletin of Environmental Contamination and Toxicology* 43:641–646, 1989.

Dewailly E, Ayotte P, Bruneau S, Laliberte C, Muir DCG, Norstrom RJ. Inuit exposure to organochlorines through the food chain in arctic Quebec. *Environmental Health Perspectives* 101:618–620, 1993a.

Dewailly E, Brueneau S, Laliberte C, Belles-Iles M, Weber JP, Ayotte P, Roy R. Breast milk contamination by PCBs and PCDDs/PCDFs in arctic Quebec: Preliminary results on the immune status of Inuit infants. *Organohalogen Compounds* 13:403–406, 1993b.

Dewailly E, Ayotte P, Brueneau S, Laliberte C, Muir DCG, Norstrom RJ. Human exposure to polychlorinated biphenyls through the aquatic food chain in the Arctic. *Organohalogen Compounds* 14:173–176, 1993c.

Dewailly E, Ryan JJ, Lalibertie C, Bruneau S, Weber JP, Gingras S, Carrier G. Exposure of remote maritime populations to coplanar PCBs. *Environmental Health Perspectives* 102 (supp 1):205–209, 1994a.

Dewailly E, Dodin S, Verreault R, Ayotte P, Sauve L, Morin J. High organochlorine body burden in women with estrogen-receptor positive breast cancer. *Journal of the National Cancer Institute* 86:232–234, 1994b.

Dimitriou MA. Ozone treatment comes of age in water, wastewater applications. *Environmental Solutions,* December 1994:38.

Djordevic MV, Hoffman D, Fan J, Prokopczyk B, Citron ML, Stellman SD. Assessment of chlorinated pesticides and polychlorinated biphenyls in adipose breast tissue using a supercritical fluid extraction method. *Carcinogenesis* 15:2581–2585, 1994.

Dolfing J, Salomons W. Kinetic control of the biogeochemical formation of halogenated humic acids. In: Grimvall A, deLeer EWB, eds. *Naturally-Produced Organohalogens.* Boston: Kluwer Academic, 1995:221–223.

Donahue MD, Geiger JL, Kiamos AA, Nielsen KA. Reduction of volatile organic compound emissions during spray painting: A new process using supercritical carbon dioxide to replace traditional paint solvents. In: Anastas PT, Williamson TC, eds. *Green Chemistry: Designing Chemistry for the Environment.* Washington DC: American Chemical Society, 1996:152–167.

Donna A, Crosignani P, Robutti F, Betta PG, Bocca R, Mariana N, Ferrario F, Fissi R, Berrino F. Triazine herbicides and ovarian epithelial neoplasms. *Scandinavian Journal of Work and Environmental Health* 15:47–53, 1989.

Dorgan JF, Longcope C, Stephenson HE, Falk RT, Miller R, Franz C, Kahle L, Campbell WS, Tangrea JA, Schatzkin A. Serum sex hormone levels are related to breast cancer risk in postmenopausal women. *Environmental Health Perspectives* 105 (supp 3):583–586, 1997.

Doughtery R, Whitaker L, Smith L, Stalling D, Kuehl D. Negative ionization studies of human and food chain contamination with xenobiotic chemicals. *Environmental Health Perspectives* 35:103–118, 1980.

Dougherty RC, Whitaker MJ, Tang SY, Bottcher R, Keller M, Kuehl DW. Sperm density and toxic substances: A potential key to environmental health hazards. In: McKinney JD, et. *The Chemistry of Environmental Agents as Human Hazards.* Ann Arbor: Ann Arbor Science Publishers 1981:263–277.

Dow Chemical. Waste analysis sheet: Heavy ends from the distillation of ethylene dichlorine in ethylene dichloride production, Plaquemine, LA, February 21, 1990.

Dow Chemical. *Chlorine . . . A Building Block of Our World.* Midland, MI: Dow, 1995.

Doyle P, Roman E, Beral V, Brookes M. Spontaneous abortion in dry cleaning workers potentially exposed to perchloroethylene. *Occupational and Environmental Medicine* 54:848–853, 1997a.

Doyle TJ, Zheng W, Cerhan JR, Hong CP, Sellers TA, Kushi LH, Folsom AR. The association of drinking water source and chlorination by-products with cancer incidence among postmenopausal women in Iowa: A prospective cohort study. *American Journal of Public Health* 87:1168–1176, 1997b.

DTI (Danish Technical Institute). *Environmental Aspects of PVC.* Copenhagen: DTI, November 1995.

Dubos R. Science and man's nature. *Daedalus* 94:223–244, 1965.

Dubos R. *Reason Awake.* New York: Columbia University Press, 1970.

ECN (European Chemical News). ECN international project review, part 1. *European Chemical News,* March 1992, pp. 3–50.

Ecocycle Commission of the Government of Sweden. *PVC: A Plan to Prevent Environmental Impact.* Stockholm: Ecocycle Commission, 1994.

Ecotec Research and Consulting, IVAM Environmental Research, ZENIT GmbH. *New Clean and Low Waste Products Processes and Services, and Ways to Promote the Diffusion of Such Practices to Industry.* Brussels: Commission of the European Communities, 1995.

Edmonds LD, James LM. Temporal trends in the prevalence of congenital malformations at birth based on the birth defects monitoring program, United States, 1979–1987. *Morbidity and Mortality Weekly Report* 39(SS–4):19–23, 1990.

Egeland GM, Sweeney MH, Fingerhut MA, Wille KK, Schnorr TM, Halperin WE. Total serum testosterone and gonadotropins in workers exposed to dioxin. *American Journal of Epidemiology* 139:272–281, 1994.

Ehrenfeld J, Melhuish J, Bozdogan K, Maxwell J, Weiner S, Bucknall J, Rothenberg S, Vit W, Martin P, Griffith J, Field F, Najam A, Thomas C. *Dimensions of Managing Chlorine in the Environment: Report of the MIT/Norwegian Chlorine Policy Study.* Cambridge: Massachusetts Institute of Technology, Technology, Business and Environment Program, 1993.

Ehrlich PR, Raven PH. Butterflies and plants: A study in coevolution. *Evolution* 18:586–608, 1964.

Ekbom A, Trichopoulos D, Adami HO, Hsieh CC, Lan SH. Evidence of prenatal influence on breast cancer risk. *Lancet* 340:1015–1018, 1992.

Ekenat K, Weis KE, Katzenellenbogen JA, Katzenellenbogen BS. Different residues of the human estrogen receptor are involved in the recognition of structurally diverse estrogens and antiestrogens. *Journal of Biological Chemistry* 272, 5069–5075, 1997.

Eklund H. ECF vs TCF–A time to assess and a time to act. *Pulp and Paper* 69(5):83–85, 1995.

Eklund G, Pedersen J, Stromberg B. Methane, hydrogen chloride and oxygen form a wide range of chlorinated organic species in the temperature range 400°C–950°C. *Chemosphere* 173:575–586, 1988.

El-Bayoumi K. Environmental carcinogens that may be involved in human breast cancer etiology. *Chemical Research in Toxicology* 5:585–590, 1992.

Elix JA, Jian H, Wardlaw JH. A new synthesis of xanthones: 2,4,7-trichloronorlichexanthone and 4,5,6-trichloronorlichexanthone, two new lichen xanthones. *Australian Journal of Chemistry* 43:1745–1758, 1990.

Endo R. World demand for PVC (presentation materials). Second World Chlor-alkali Symposium, Washington, DC, September 1990.

Endometriosis Association. *What is Endometriosis?* Milwaukee: Endometreosis Association, 1993.

Enell M. AOX loadings on sea areas surrounding Sweden. In: *Environmental Fate and Effects of Bleached Pulp Mill Effluents.* Stockholm: Swedish Environmental Protection Agency (Report 4031), 1992.

Energy Information Administration, Office of Coal, Nuclear, Electric and Alternate Fuels. *Electric Power Annual 1995,* Volume 1. Washington, DC: U.S. Department of Energy (DOE/EIA-0348(95)/1, Distribution Category UC-950), 1996.

Engvild KC. Chlorine containing compounds in higher plants. *Phytochemistry* 25:781–791, 1986.

Engvild KC. The natural chlorinated plant hormone of pea, 4-chloroindole-3-acetic acid, an endogenous herbicide? In: Grimvall A, deLeer EWB, eds. *Naturally-Produced Organohalogens.* Boston: Kluwer Academic, 1995:227–234.

Enquete Kommission zum Schutz des Menschen und der Unmwelt (Outlook and crtiteria for evaluating environmentally-sound material cycles in industrialized society). Berlin: German Bundestag, 1994.

Enterline PE, Henderson V, Marsh G. Mortality of workers potentially exposed to epichlorohydrin. *British Journal of Industrial Medicine* 47:269–76, 1990.

Environment Agency. *Regulation of dioxin releases from the Runcorn Operations of ICI and EVC.* Warrington, UK: Environment Agency, 1997.

Environment Canada. *Canadian Environmental Protection Act: Priority Substances List Assessment Report No. 2: Effluents from Pulp Mills Using Bleaching.* Ottawa: Environment Canada, 1991.

EPA (U.S. Environmental Protection Agency). *Health Assessment Document for 2,3,7,8-tetrachlorodibenzo-p-dioxin.* Washington, DC: U.S. EPA Office of Health and Environmental Assessment (EPA/600-8-84/014f), 1985a.

EPA (U.S. Environmental Protection Agency). *Work/Quality Assurance Project Plan for the Bioaccumulation Study.* Washington, DC: U.S. EPA, Monitoring and Data Support Division, Office of Water Regulations and Standards, 1985b.

EPA (U.S. Environmental Protection Agency Science Advisory Board). *Report on the Incineration of Liquid Hazardous Wastes by the Environmental Effects, Transport and Fate Committee.* Washington, DC: U.S. EPA, 1985c.

EPA (U.S. Environmental Protection Agency). *Report to Congress on the Discharge of Hazardous Wastes to Publicly Owned Treatment Works.* Washington, DC: U.S. EPA Office of Water Regulations and Standards (EPA/530/SW-86-004), 1986.

EPA (U.S. Environmental Protection Agency). *National Dioxin Study: Report to Congress.* Washington, DC: U.S. EPA, Office of Air Quality Planning and Standards (EPA/530-SW-87-025), 1987a.

EPA (U.S. Environmental Protection Agency). *National Dioxin Study Tier 4—Combustion Sourcs: Enginering Analysis Report.* Washington, DC: U.S. EPA Office of Air Quality Planning and Standards (EPA/450-84-014h), 1987b.

EPA (U.S. Environmental Protection Agency). Specific comments on the Food and Drug Administration's evaluation of the environmental issues associated with the proposed rule on PVC. Washington, DC: U.S. EPA, Office of Federal Activities, May 23, 1988a.

EPA (U.S. Environmental Protection Agency). *Estimating Exposures to 2,3,7,8-TCDD. External Review Draft.* Washington, DC: U.S. EPA, Office of Research and Development (EPA/600/6-88/005a), 1988b.

EPA (U.S. Environmental Protection Agency). *A Cancer-Risk Specific Dose for 2,3,7,8-TCDD, External Review Draft.* Washington, DC: U.S. EPA, Office of Research and Development (EPA/600-6-99-007A), 1988c.

EPA (U.S. Environmental Protection Agency). *Background Document for the Development of PIC Regulations from Hazardous Waste Incinerators.* Washington, DC: U.S. Environmental Protection Agency, Office of Solid Waste, 1989.

EPA (U.S. Environmental Protection Agency). Standards for owners and operators of hazardous wastes incinerators and burning of hazardous wastes in boilers and industrial furnaces; proposed and supplemental proposed rule, technical corrections, and request for comments. *55 Federal Register* 82, April 27, 1990.

EPA (U.S. Environmental Protection Agency). *Health Assessment Document for* 2,3,7,8-*tetrachlorodibenzo*-p-*dioxin and Related Compounds, Volumes 1–3, Review Draft.* Washington, DC: U.S. EPA Office of Research and Development (EPA/600/BP-92-001), 1994a.

EPA (U.S. Environmental Protection Agency). *Estimating Exposures to Dioxin-like Compounds, Volumes 1–3, Review Draft.* Washington, DC: U.S. EPA Office of Research and Development (EPA/600/6-88-005), 1994b.

EPA (U.S. Environmental Protection Agency). *Region V. Risk Assessment for the Waste Technologies Industries (WTI) Hazardous Waste Incinerator Facility (East Liverpool, Ohio).* Chicago: U.S. EPA Region V, 1995.

EPA (U.S. Environmental Protection Agency). *Supplemental Technical Development Document for Effluent Limitations Guidelines and Standards for the Pulp, Paper, and Paperboard Category Subpart B (Bleached Papergrade Kraft and Soda) and Subpart E (Papergrade Sulfite).* Washington, DC: U.S. EPA Office of Science and Technology, Engineering and Analysis Division (DCN 14487), 1997a.

EPA (U.S. Environmental Protection Agency). National emission standards for hazardous air pollutants for source category: Pulp and paper production; effluent limitations guidelines, pretreatment standards, and new source performance standards: Pulp, paper and paperboard category. *63 Federal Register* 18504–18751, 1997b.

EPA (U.S. Environmental Protection Agency). *Making the Most of Your Cleaning Business.* Washington, DC: U.S. EPA Office of Pollution Prevention and Toxics, 1997c.

EPA (U.S. Environmental Protection Agency). *Wet Cleaning.* Washington, DC: U.S. EPA Office of Pollution Prevention and Toxics (EPA744-K-96-002), 1997d.

EPA (U.S. Environmental Protection Agency). *The Inventory of Sources of Dioxin in the United States (Review Draft).* Washington, DC: U.S. EPA Office of Research and Development (EPA/600/P-98-002a), 1998a.

EPA (U.S. Environmental Protection Agency). *Cleaner Technologies Substitutes Assessment for Professional Fabricare Processes.* Washington, DC: U.S. EPA Office of Pollution Prevention and Toxics (EPA/744-B-98-001), 1998b.

EPA (U.S. Environmental Protection Agency). *Chemical Testing and Information Gathering.* Washington, DC: U.S. EPA Office of Pollution Prevention and Toxics (available at www.epa.gov/opptintr/chemtest/index.htm), 1998c.

Epstein S. The construction of lay expertise: AIDS activism and the forging of credibility in the reform of clinical trials. *Science, Technology and Human Values* 20:408–437, 1995.

Epstein SS. Evaluation of the National Cancer Program and proposed reforms. *American Journal of Industrial Medicine* 24:109–133, 1993.

Eriksson M, Karlsson M. Occupational and other environmental factors and multiple myeloma: A population based case-control study. *British Journal of Industrial Medicine* 49:95–03, 1992.

Eriksson P, Fredriksson A. Developmental neurotoxicity of four ortho-substituted polychlorinated biphenyls in the neonatal mouse. *Environmental Toxicology and Pharmacology* 1:155–165, 1996.

Ernst M, Flesch-Janys D, Morgenstern I, Manz A. Immune cell functions in industrial workers after exposure to 2,3,7,8-tetrachlorodibenzo-p-dioxin: Dissocation of antigen-specific

T-cell response in cultures of diluted whole blood and of isolated peripheral blood mononuclear cells. *Environmental Health Perspectives* 106 (supp 2):701–705, 1998.

Esposito MP, Tiernan TO, Dryden FE. *Dioxins.* Washington, DC: U.S. Environmental Protection Agency (EPA/600/2-80-197), 1980.

European Commission. *European Workshop on the Impact of Endocrine Disrupters on Human Health and Wildlife.* Brussels: European Commission Environment and Climate Research Program (EUR 17549), 1996.

Evers E. *The Formation of Polychlorinated Dibenzofurans and Polychlorinated Dibenzo-p-dioxins and Related Compounds During Oxyhydrochlorination of ethylene.* Amsterdam: University of Amsterdam, Department of Environmental and Toxicological Chemistry, 1989.

Evers EHG, Ree K, Olie K. Spatial variations and correlations in the distribution of PCDDs, PCDFs and related compounds in sediments from the river Rhine. *Chemosphere* 17:2271–2288, 1988.

Evers EHG, Klamer H, Laane R, Govers H. Polychlorinated dibenzo-p-dioxin and dibenzofuran residues in estuarine and coastal North Sea sediments: Sources and distribution. *Environmental Toxicology and Chemistry* 12:1583–1598, 1993.

Evers EHG, Laane RWPM, Groeneveld GJJ, Olie K. Levels, temporal trends and risk of dioxins and related compounds in the Dutch aquatic environment. *Organohalogen Compounds* 28:117–122, 1996.

Exon JH, Koller LD. Effects of chlorinated phenols on immunity in rats. *International Journal of Immunopharmacology* 5:131–135, 1983.

Exon JH, Henningsen GM, Osborne CA, Koller LD. Toxicologic, pathologic, and immunotoxic effects of 2,4-dichlorophenol in rats. *Journal of Toxicology and Environmental Health* 14:723–730, 1984.

EWG (nvironmental Working Group and Physicians for Social Responsibility). *Weed Killers by the Glass.* Washington, DC: EWG, 1995.

Fabbri F. *Heavy Metals and Dioxins in Shellfish of the Venice Lagoon: Responsibility of the Petrochemical Plant at Porto Marghera.* Rome: Greenpeace Italy, 1996.

Facemire CF, Gross TS, Guillette LJ. Reproductive impairment in the Florida panther: Nature or nurture. *Environmental Health Perspectives* 103 (supp 4):87–92, 1995.

Fagliano J, Berry M, Bove F, Burke T. Drinking water contamination and the incidence of leukemia: An ecologic study. *American Journal of Public Health* 80:1209–1212, 1990.

Fahey RJ. The UV effect on wastewater. *Water Engineering and Management 137*:1–4, 1990.

Falck F, Ricci A, Wolff MS, Godbold J, Deckers P. Pesticides and polychlorinated biphenyl residues in human breast lipids and their relation to breast cancer. *Archives of Environmental Health* 47:143–146, 1992.

Fallon H, Tollerud D, Breslow N, Berlin J, Bolla K, Colditz G, Goetz C, Kaminski N, Kriebel D, Mottet K, Neugut A, Nicholson W, Olshan A, Rodgers K, Sprince N, Weisel C. *Veterans and Agent Orange: Health Effects of Herbicides Used in Vietnam.* Washington, DC: National Academy Press, 1993.

Fangmark I, Marklund S, Rappe C, Stromberg B, Berge N. Use of a synthetic refuse in a pilot combustion system for optimizing dioxin emission, part 2. *Chemosphere* 23:1233–1243, 1991.

Fangmark I, van Bavel B, Marklund S, Stromberg B, Berge N, Rappe C. Influence of combustion parameters on the formation of polychlorinated dibenzo-p-dioxins, dibenzofurans, benzenes, and biphenyls and polyaromatic hydrocarbons in a pilot incinerator. *Environmental Science and Technology* 27:1602–1610, 1993.

Farland W, Lorber M, Cleverly D. *WTI Screening Level Analysis.* Washington DC: U.S. Environmental Protection Agency Office of Research and Development, February 9, 1993.

Farrar MC, Gellman I, McBride AC. *USEPA/Paper Industry Cooperative Dioxin Screening Study.* Washington, DC: U.S. Environmental Protection Agency Office of Water, June 20, 1986.

Faustini A, Settimi L, Pacifici R, Fano V, Zuccaro P, Forastiere F. Immunological changes among farmers exposed to phenoxy herbicides: Preliminary observations. *Occupational and Environmental Medicine* 53:583–585, 1996.

Feichtinger W. Environmental factors and fertility. *Human Reproduction* 6:1170–1175, 1991.

Fein GG, Jacobson JL, Jacobson SW, Schwartz PM, Dowler JK. Prenatal exposure to polychlorinated biphenyls: Effects on birth size and gestational age. *Journal of Pediatrics* 105: 315–319, 1984.

Fenical W. Halogenation in the rhodophyta: A review. *Journal of Phycology* 11:245–259, 1975.

Ferencz C, Villasenor AC. Epidemiology of cardiovascular malformations: The state of the art. *Cardiology in the Young* 1:264–284, 1991.

Ferguson DW, Gramith JT, McGuire MJ. Applying ozone for organics control and disinfection: A utility perspective. *Journal of the American Water Works Association* 83(5):32–38, 1991.

Ferguson KH, Finchem KJ. Effluent minimization technologies move pulp mills closer to closure. *Pulp and Paper,* March 1997:55–65.

Ferroni C, Selis L, Mutti A, Folli D, Bergamaschi E, Franchini I. Neurobehavioral and neuroendocrine effects of occupational exposure to perchloroethylene. *Neurotoxicology* 13:243–248, 1992.

Feyerabend P. *Against Method,* Third Edition. London: Verson, 1993.

Fiedler H, Hutzinger O, Hosseinpour J. Analysis and remedial actions following an accidental fire in a kindergarten. *Organohalogen Compounds* 14:19–23, 1993.

Fingerhut MA, Halperin WE, Marlow DA, Piacitelli LA, Honchar PA, Sweeney MH, Greiff AL, Dill PA, Steenland K, Soroda AJ. Cancer mortality in workers exposed to 2,3,7,8-tetrachlorodibenzo-p-dioxin. *New England Journal of Medicine* 324:212–218, 1991.

Fisch H, Goluboff ET, Olson JH, Felshuh J, Broder SH, Barad DH. Semen analysis in 1283 men from the United States over a 25-year period: No decline in quality. *Fertility and Sterility* 65:1009–1014, 1996.

Fischer LJ. Chlorinated compounds: Research or ban. *Health and Environment Digest* 8: 9–11, 1994.

Fischer LM, Catz D, Kelley DB. Androgen-directed development of the *Xenopus laevis* larynx: Control of androgen receptor expression and tissue differentiation. *Developmental Biology* 170(1):115–126, 1995.

Fischer RC, Kramer W, Ballschmitter K. Hexachlorocyclohexane isomers as markers in the water flow of the Atlantic Ocean. *Chemosphere* 23:889–900, 1991.

Fischli AE. Conclusions. *Pure and Applied Chemistry* 68:1823–1824, 1996.

Fitchko J. *Literature Review of the Effects of Persistent Toxic Substances on Great Lakes Biota: Report of the Health of Aquatic Communities Task Force.* Windsor, ON: International Joint Commission, 1986.

Fitzhugh OG, Nelson AA, Quaife ML. Chronic oral toxicity of aldrin and dieldrin in rats and dogs. *Food and Cosmetic Toxicology* 2:551–562, 1964.

Fleck L. *Genesis and Development of a Scientific Fact.* Bradley F, Trenn TJ, trans. Chicago: University of Chicago Press, 1979.

Fleming D. The economics of taking care: An evaluation of the precautionary principle. In: Freestone D, Hey E, eds. *The Precautionary Principle and International Law: The Challenge of Implementation.* Boston: Kluwer Law International, 1996:147–170.

Flesch-Janys D, Berger J, Manz A, Nagel S, Ollroge I. Exposure to polychlorinated dibenzo-p-dioxins and furans and breast cancer mortality in a cohort of female workers of a herbicide producing plant in Hamburg, FRG. *Organohalogen Compounds* 13:381–384, 1993.

Flesch-Janys D, Berger J, Gurn P, Manz A, Nagel A, Waltsgott H, Dwyer JH. Exposure to polychlorinated dioxins and furans (PCDD/F) and mortality in a cohort of workers from a herbicide-producing plant in Hamburg, Federal Republic of Germany. *American Journal of Epidemiology* 142:1165–1175, 1995.

Flesch-Janys D, Steindorf K, Gurn P, Becher J. Estimation of the cumulated exposure to polychlorinated dibenzo-p-dioxins/furans and standardized mortality ratio analysis of cancer mortality by dose in an occupationally exposed cohort. *Environmental Health Perspectives* 106 (supp 2):655–662, 1998.

Fogel D, Isaac-Renton J, Guasparini R. Removing giardia and cryptosporidium by slow sand filtration. *American Water Works Association Journal* 85 (11):77–84, 1993.

Fountainhead Technologies. *How a Catalyst Purifier Works.* Providence, RI: Fountainhead Technologies, Inc., 1994.

Fox GA. Practical causal inference for ecoepidemiologists. *Journal of Toxicology and Environmental Health* 33:359–374, 1991.

Fox GA. Epidemiological and pathobiological evidence of contaminant-induced alterations in sexual development in free-living wildlife. In: Colborn T, Clement C, eds. *Chemically-Induced Alterations in Sexual and Functional Development: The Wildlife-Human Connection.* Princeton: Princeton Scientific Publishing, 1992:137–158.

Fox GA. What have biomarkers told us about the effects of contaminants on the health of fish-eating birds in the Great Lakes? The theory and a literature review. *Journal of Great Lakes Research* 19:722–736, 1993.

Frank H. Short-chain aliphatic hydrocarbons: Environmental levels and ecotoxicological properties. *Organohalogen Compounds* 14:267–270, 1993.

Frank H, Norokorpi Y, Scholl H, Renschen D. Trichloroacetate levels in the atmosphere and in conifer needles in Central and Northern Europe. *Organohalogen Compounds* 14:307–308, 1993a.

Frank H, Rether B, Scholl H, Stoll P. Phytotoxicity of haloacetic acids and of some derivatives. *Organohalogen Compounds* 14:311–312, 1993b.

Frank W, Frank H, Jans W, Funke W. Effects of some chlorocarbons on soil invertebrates. *Organohalogen Compounds* 14:313–314, 1993c.

Frankenhaeuser M, Manninen H, Kojo I, Ruuskanen J, Vartiainen T, Vesterinen R, Virkki J. Organic emissions from co-combustion of mixed plastics with coal in a bubbling fluidized bed boiler. *Chemosphere* 27:309–316, 1993.

Freestone D, Hey E. Origins and development of the precautionary principle. In: Freestone D, Hey E, eds. *The Precautionary Principle and International Law: The Challenge of Implementation.* Boston: Kluwer Law International, 1996:3–18.

Friedlander SK. Environmental issues: Implications for engineering design and education. In: Ausubel JH, Aladovich HE, eds. *Technology and Environment.* Washington, DC: National Academy Press, 1989:167–181.

Fry DM. Reproductive effects in birds exposed to pesticides and industrial chemicals. *Environmental Health Perspectives* 103 (supp 7):1165–172, 1995.

Fry D, Roone C, Speich S, Peard R. Sex-ratio skew and breeding patterns of gulls: Demographic and toxicological considerations. *Studies in Avian Biology* 10:26–43, 1987.

Fry DM, Toone CK. DDT-induced feminization of gull embryos. *Science* 213:922–924, 1981.

Fuhrer U, Ballschmitter K. Bromochloromethoxybenzenes in the marine troposphere of the Atlantic Ocean: A group of organohalogens with mixed biogenic and anthropogenic origin. *Environmental Science and Technology* 32:2208–2215, 1998.

Furst P, Wilmers K. Body burden with PCDD and PCDF from food. *Banbury Reports* 35:121–132, 1991.

Galton F. *English Men of Science: Their Nature and Nurture.* New York: D. Appleton, 1875.

Gammon MD, John EM. Recent etiologic hypotheses concerning breast cancer. *Epidemiologic Reviews* 15:163–168, 1993.

GAO (General Accounting Office). *Reproductive and Developmental Toxicants: Regulatory Actions Provide Uncertain Protection.* Washington, DC: General Accounting Office (GAO/PEMD-92-3), 1991.

Garry VF, Schreinmachers D, Jarkins ME, Griffith J. Pesticide appliers, biocides, and birth defects in rural Minnesota. *Environmental Health Perspectives* 104:394–399, 1996.

Geiser K. Material concerns: From pollution control to materials policy. In: Jackson T, ed. *Clean Production Strategies: Developing Preventive Environmental Management in the Industrial Economy.* Boca Raton, FL: Lewis, 1993:225–236.

Gellert RJ. Uterotrophic activity of polychlorinated biphenyls (PCBs) and induction of precocious reproductive aging in neonatally treated female rats. *Environmental Research* 16: 123–130, 1978.

Gerhard I. Reproductive risks of heavy metals and pesticides in women. In: Richardson M, ed. *Reproductive Toxicology.* New York: VCH, 1993:167–182.

Gerhard I, Derber M, Runnebaum B. Prolonged exposure to wood preservatives induces endocrine and immunologic disorders in women. *American Journal of Obstetrics and Gynecology* 165:487–488, 1991.

Gerhard I, Runnebaum B. Fertility disorders may result from heavy metal and pesticide contamination which limits effectiveness of hormone therapy. *Zentralblatt fur Gynakologie* 114:593–602, 1992.

Gerhard I, Eckrich W. Runnebaum B. Toxic substances and infertility: Organic solvents, pesticides. *Geburtsch u Frauenheilk* 53:147–160, 1993.

German Environment Ministers. *Impacts on the Environment from the Manufacture, Use and Disposal and Substitution of PVC.* Hamburg: German Joint Federal-State Committee on Environmental Chemicals for the German Environmental Ministers, 1992.

Germolec DR, Yang RSH, Ackermann MP, Rosenthal GJ, Boorman GA, Thompson M, Blair P, Luster MI. Toxicology studies of a chemical mixture of 25 groundwater contaminants: Immunosuppression in B6C3F mice. *Fundamental and Applied Toxicology* 13:377–387, 1991.

Geschwind SA, Stolwijk JAJ, Bracken M, Fitzgerald E, Stark A, Olsen C, Melius J. Risk of congenital malformations associated with proximity to hazardous waste sites. *American Journal of Epidemiology* 135:1197–1207, 1992.

Gieryn TF. Boundary work and the demarcation of science and non-science: Strains and interests in professional ideologies of scientists. *American Sociological Review* 48:781–795, 1983.

Giesy JP, Ludwig JP, Tillitt DJ. Deformities in birds of the Great Lakes region: Assigning causality. *Environmental Science and Technology* 28:128a–34a, 1994a.

Giesy JP, Ludwig JP, Tillitt DJ. Dioxins, dibenzofurans, PCBs and wildlife. In: Schecter A, ed. *Dioxins and Health*. New York: Plenum, 1994b:249–308.

Gilbertson MG, Kubiak T, Ludwig J, Fox G. Great Lakes embryo mortality, edema, and deformities syndrome (GLEMEDS) in colonial fish-eating birds: Similarity to chick edema disease. *Journal of Toxicology and Environmental Health* 33:455–520, 1991.

Gilman A, Dewailly E, Feeley M, Jerome V, Kuhnlein H, Kwavnick B, Neve S, Tracy B, Usher P, Van Oostdam J, Walker J, Wheatley B. Human health. In: Jensen J, Adare K, Shearer R, eds. *Canadian Arctic Contaminants Assessment Report*. Ottawa: Indian and Northern Affairs Canada, 1997.

Gladen BC, Rogan WJ. Effects of perinatal polychlorinated biphenyls and dichlorodiphenyl dichloroethene on later development. *Journal of Pediatrics* 119:58–63, 1991.

Gladen BC, Rogan WJ, Hardy P, Thullen J, Tingelstad J, Tully M. Development after exposure to polychlorinated biphenyls and dichlorodiphenyl dichloroethene transplacentally and through human milk. *Journal of Pediatrics* 113:991–995, 1988.

Glass AG, Hoover RN. The emerging epidemic of melanoma and squamous cell skin cancer. *Journal of the American Medical Association* 262:2097–2100, 1989.

Glass AG, Hoover RN. Rising incidence of breast cancer: Relationship to stage and receptor status. *Journal of the National Cancer Institute* 82:693–696, 1990.

Glasser H, Chang D, Hickman D. An analysis of biomedical waste incineration. *Journal of the Air and Waste Management Association* 41:1180–1188, 1991.

Glaze WH. Drinking water treatment with ozone. *Environmental Science and Technoloogy* 21:224–230, 1987.

Glotfelty DE, Seiber JN, Lilhedahl LA. Pesticides in fog. *Nature* 325:602–605, 1987.

GLU (Great Lakes United Clean Production Task Force). *Planning for the Sunset: A Case Study for Eliminating Dioxins by Phasing Out PVC Plastics*. Toronto: GLU, 1995.

GLWQB (Great Lakes Water Quality Board). *1987 Report on the Great Lakes Water Quality*. Windsor, ON: International Joint Commission, 1987.

Goh VG-H, Chia S-E, Ong C-N. Effects of chronic exposure to low doses of trichloroethylene on steroid hormone and insulin levels in normal men. *Environmental Health Perspectives* 106:41–44, 1998.

Goldey ES, Taylor DH. Developmental neurotoxicity following premating maternal exposure to hexachlorobenzene in rats. *Neurotoxicology and Teratology* 14:15–21, 1992.

Goldey ES, Kehn LS, Lau C, Rehnberg GL, Crofton KM. Developmental exposure to polychlorinated biphenyls (Aroclor 1254) reduces circulating thyroid hormone concentrations and causes hearing deficits in rats. *Toxicology and Applied Pharmacology* 135:77–88, 1995.

Goldman JM, Cooper RL, Edwards TL, Rehnberg GL, McElroy WK, Hein JF. Suppression of the luteinizing hormone surge by chlordimeform in ovariectomized, steroid-primed female rats. *Pharmacology and Toxicology* 68:131–136, 1991.

Goldspiel BR. Pharmaceutical issues: Preparation, administration, stability, and compatibility with other medications. *Annals of Pharmacotherapy* 28:S23–S26, 1994.

Gonzales FJ, Gelboin HV. Role of human cytochromes p450 in the metabolic activation of chemical carcinogens and toxins. *Drug Metabolism Reviews* 261:165–183, 1994.

Goodman D. Memorandum: Incineration task force. Washington, DC: Morrisville, NJ, August 22, 1994.

Goolsby D, Thurman EM, Pomes ML. Herbicides and their metabolites in rainfall: Origin, transport and deposition patterns across the midwestern and northeastern United States, 1990–1991. *Environmental Science and Technology* 31:1325–1333, 1997.

Gorski J, Hou Q. Embryonic estrogen receptors: Do they have a physiological function? *Environmental Health Perspectives* 103 (supp 7):69–72, 1995.

Graedel TE, Keene WC. Tropospheric budget of reactive chlorine. *Global Biogeochemical Cycles* 9:47–77, 1995.

Graedel TE, Keene WC. The budget and cycle of earth's natural chlorine. *Pure and Applied Chemistry* 68:1689–1697, 1996.

Graham S, Zielezny M, Marshall J, Priore R, Freudenheim J, Brasure J, Jaughey B, Nasca P, Zdeb M. Diet in the epidemiology of postmenopausal breast cancer in the New York State cohort. *American Journal of Epidemiology* 125:1327–1337, 1992.

Grant R. Rauma opens with the aim of closing up: Metsa-Rauma's totally chlorine-free greenfield mill. *Pulp and Paper International* 38(6):31–33, 1996.

Grasman KA, Fox GA, Scanlon PF, Ludwig JP. Organochlorine-associated immunosuppression in prefledgling caspian terns and herring gulls from the Great Lakes: An eco-epidemiological study. *Environmental Health Perspectives* 104 (supp 4):829–842, 1996.

Gray JS. Integrating precautionary scientific methods into decision-making. In: Freestone D, Hey E, eds. *The Precautionary Principle and Internatinal Law: The Challenge of Implementation.* Boston: Kluwer Law International, 1996: 133–146.

Gray LE, Ostby JS. In utero 2,3,7,8-tetrachlorodibenzo-p-dioxin (TCDD) alters reproductive morphology and function in female rat offspring. *Toxicology and Applied Pharmacology* 133:285–294, 1995.

Gray LE, Kelce WR, Monosson E, Ostby JS, Birnbaum LS. Exposure to TCDD during development permanently alters reproductive function in male Long Evans rats and hamsters: Reduced ejaculated and epididymal sperm numbers and sex accessory gland weights in offspring with normal androgenic status. *Toxicology and Applied Pharmacology* 131:108–118, 1995.

Gray LE, Ostby JS, Kelce WR. A dose-response analysis of the reproductive effects of a single gestational dose of 2,3,7,8-tetrachlorodibenzo-p-dioxin in male Long Evans hooded rat offspring. *Toxicology and Applied Pharmacology* 146:11–20, 1997a.

Gray LE, Wolf C, Mann P, Ostby JS. In utero exposure to low doses of 2,3,7,8-tetrachlorodibenzo-p-dioxin alters reproductive development of female Long Evans hooded rat offspring. *Toxicology and Applied Pharmacology* 146:237–244, 1997b.

Green AES. The future of medical waste incineration. In: Green AES, ed. *Medical Waste Incineration and Pollution Prevention.* New York: Van Nostrand Reinhold, 1993:170–207.

Green AES, Wagner JC. Toxic products of medical waste incineration. In: Green AES, ed. *Medical Waste Incineration and Pollution Prevention.* New York: Van Nostrand Reinhold, 1993:1–36.

Greenpeace. *What Is Clean Production?* Amsterdam: Greenpeace International, 1995.

Greenpeace. *PVC: The Poison Plastic.* Washington, DC: Greenpeace USA, 1996.

Gregor D, Gummer W. Evidence of atmospheric transport and deposition of organochlorine pesticides and plychlorinated biphenyls in Canadian arctic snow. *Environmental Science and Technology* 23:561–565, 1989.

Gregor DJ, Peters AJ, Teixeira C, Jones N, Spencer C. The historical residue trend of PCBs in the Agassiz Ice Cap, Ellesmere Island, Canada. *Science of the Total Environment* 160: 117–126, 1995.

Gregor D, Teixeira C, Rowsell R. Deposition of atmospherically transported polychlorinated biphenyls in the Canadian arctic. *Chemosphere* 33:227–244, 1996.

Greisemer RA, Eustis SL. Gender differences in animal bioassays for carcinogenicity. *Journal of Occupational Medicine* 36:855–859, 1994.

Gribble GW. Naturally occurring organohalogen compounds: A survey. *Journal of Natural Products* 55:1353–1395, 1992.

Gribble GW. The diversity of natural organochlorines in living organisms. *Pure and Applied Chemistry* 68:1699–1712, 1996a.

Gribble GW. *The Future of Chlorine.* Washington, DC: Heartland Institute, 1996b (available from Chlorine Chemistry Council at www.c3.org).

Griffith J, Duncan RC, Riggan W, Pellom AC. Cancer mortality in US counties with hazardous waste sites and groundwater pollution. *Archives of Environmental Health* 44:69–74, 1989.

Grimvall A. Evidence of naturally produced and man-made organohalogens in water and sediment. In: Grimvall A, deLeer EWB, eds. *Naturally-Produced Organohalogens.* Boston: Kluwer Academic, 1995:3–20.

Gron C. AOX in groundwater. In: Grimvall A, deLeer EWB, eds. *Naturally-Produced Organohalogens.* Boston: Kluwer Academic, 1995:49–64.

Gronemeyer H. Nuclear hormone receptors as transcriptional activators. In: Parker ML, ed. *Steroid Hormone Action.* Oxford: Oxford University Press: 1992:94–117.

Gronemeyer H, Laudet V. Transcription factors 3: Nuclear receptors. *Protein Profiles* 2: 1173–3351, 1995.

Grossman CJ. Regulation of the immune system by sex steroids. *Endocrine Reviews* 5: 435–455,1984.

Guillette JL, Gross TS, Masson GR, Matter JM, Percival JF, Woodward AR. Developmental abnormalities of the gonad and abnormal sex hormone concentrations in juvenile alligators from contaminated and control lakes in Florida. *Environmental Health Perspectives* 102:680–688, 1994.

Guillette LJ, Gross TS, Gross DA, Rooey AA, Percival HF. Gonadal steroidogenesis in vitro from juvenile alligators obtained from contaminated or control lakes. *Environmental Health Perspectives* 103 (supp 4):31–36, 1995.

Guimond R. Memorandum to C. Browner, Administrator: WTI Incinerator Issues, Washington, DC. U.S. Environmental Protection Agency Office of Solid Waste, January 22, 1993.

Gullett BK. The effect of metal catalysts on the formation of poychlorinated dibenzo-dioxin and polychlorinated dibenzofuran precursors. *Chemosphere* 20:1945–1952, 1990.

Guo YL, Lai TJ, Ju SH, Chen YC, Hsu CC. Sexual developments and biological findings in Yucheng children. *Organohalogen Compounds* 14:235–238, 1993.

Haarhoff J, Cleasby JL. Biological and physical mechanisms in slow sand filtration. In: Logsdon GS, ed. *Slow Sand Filtration.* New York: American Society of Civil Engineers, 1991: 19–68.

Habermas J. *The Habermas Reader.* Outhwaite W, ed. Cambridge, UK: Polity Press, 1996.

Hager LP. Mother nature likes some halogenated compounds. In: *Genetic Engineering of Microorganisms for Chemicals: Basic Life Sciences* 19:415–429. New York: Plenum, 1982.

Hagmar L, Bellander T, Englander V, Ranstam J, Attewell R, Skerfving S. Mortality and cancer morbidity among workers in a chemical factory. *Scandinavian Journal of Work and Environmental Health* 12:545–551, 1986.

Hagmar L, Halberg T, Leja M, Nilsson A, Schutz A. High consumption of fatty fish from the Baltic Sea is associated with changes in human lymphocyte subset levels. *Toxicology Letters* 77:335–342, 1995.

Hahn RA, Moolgavkar SH. Nulliparity, decade of first birth, and breast cancer in Connecticut cohorts, 1855 to 1945: An ecological study. *American Journal of Public Health* 79: 1503–1507, 1989.

Haigh N. The introduction of the precautionary principle into the UK. In: O'Riordan T, Cameron J, eds. *Interpreting the Precautionary Principle.* London: Earthscan Publications, 1994:229–251.

Hajek RA, Robertson AD, Johnston DA, Van NT, Tcholakian RK, Wagner LA, Conti CH, Meistrich ML, Contreras N, Edwards CL, Jones LA. During development, 17-alpha estradiol is a potent estrogen and carcinogen. *Environmental Health Perspectives* 105 (supp 3):577–582, 1997.

Hall NEL, Rosenman KD. Cancer by industry: Analysis of a population-based cancer registry with an emphasis on blue-collar workers. *American Journal of Industrial Medicine* 19:145–159, 1991.

Halonen I, Tarhanen J, Kopsa T, Palonen J, Vilokki H, Ruuskanen J. Formation of polychlorinated dioxins and dibenzofurans in incineration of refuse derived fuel and biosludge. *Chemosphere* 26:1869–1880, 1993.

Halonen I, Tarhanen J, Ruokojarvi, Tuppurainen K, Ruuskanen J. Effect of catalysts and chlorine source on the formation of organic chlorinated compounds. *Chemosphere* 30: 1261–1273, 1995.

Hamilton-Wentworth (Regional Municipality of of Hamilton-Wentworth, Ontario). Public Health Statement on Plasti-met Fire. Hamilton, Ontario, 1997 (available at http://www.hamilton-went.on.ca).

Hansen LG, Jansen HT. Letter. *Science* 266:528, 1994.

Harada M. Minimata disease: Methyl mercury poisoning in Japan caused by environmental pollution. *Critical Reviews in Toxicology* 25:1–24, 1995.

Hardell L, Eriksson M, Axelson O. On the misinterpretation of epidemiological evidence relating to dioxin-containing phenoxyacetic acids, chlorophenols and cancer effects. *New Solutions* 4(3):49–56, 1994a.

Hardell L, Eriksson O, Axelson O, Zahm SH. Cancer epidemiology. In: Schecter A, ed. *Dioxins and Health.* New York: Plenum, 1994b:525–542.

Hardell L, van Bavel B, Lindstrom G, Fredrikson M, Hagberg H, Liljegren G, Nordstrom M, Johansson B. Higher concentrations of specific polychlorinated biphenyl congeners in adipose tissue from non-Hodgkin's lymphoma patients compared with controls without a malignant disease. *International Journal of Oncology* 9:603–608, 1996.

Hardell L, Lundstrom G, van Bavel B, Fredrikson M, Liljegren. Some aspects of the etiology of non-Hodgkin's lymphoma. *Environmental Health Perspectives* 106 (supp 2):679–681, 1998.

Harper DB. Halomethane from halide ion: A highly efficient fungal conversion of environmental significance. *Nature* 315:55–57, 1985.

Harper PA, Prokipcak RD, Bush LE, Golas CL, Okey AB. Detection and characterization of the Ah receptor for 2,3,7,8-tetrachlorodibenzo-p-dioxin in the human colon adenocarcinoma cell line LS180. *Acta Biochemistry and Biophysics* 290:27–136, 1991.

Harris JR, Lippman ME, Veronesi U, Willett W. Breast cancer. *New England Journal of Medicine* 327:319–327, 1992.

Hartge P, Devesa SS, Fraumeni JF. Hodgkin's and non-Hodgkin's lymphomas. *Cancer Surveys* 19/20:423–453, 1994.

Hasselriis F, Constantine L. Characterization of today's medical waste. In: Green AES, ed. *Medical Waste Incineration and Pollution Prevention.* New York: Van Nostrand Reinhold, 1993:37–52.

Hattis D, Glowa J, Tilson H, Ulbrich B. Risk assessment for neurobehavioral toxicity: SGOMSEC joint report. *Environmental Health Perspectives* 104 (supp 2):217–226, 1996.

Hausserman J. Gulf may get river's waste. *St. Petersburg Times,* January 12, 1997:1B.

Havel V. Politics and conscience (1984). In: Havel V. *Open Letters: Selected Writings, 1965–1990.* New York: Random House, 1991:249–271.

Hay A, Silbergeld E. Assessing the risk of dioxin exposure. *Nature* 315:102–103, 1985.

Hay A, Silbergeld E. Dioxin exposure at Monsanto. *Nature* 320:569, 1986.

Hay ME, Fenical W. Marine plant-herbivore interactions: The ecology of chemical defense. *Annual Review of Ecology and Systematics* 19:111–145, 1988.

Hayes HM, Tarone RE, Casey HW, Huxsoll DL. Excess of seminomas observed in Vietnam service U.S. military working dogs. *Journal of the National Cancer Institute* 82:1042–1046, 1990.

Hayes HM, Tarone RE, Cantor KP, Jessen CR, McCurnin DM, Richardson RC. Case-control study of canine malignant lymphoma: Positive association with dog owner's use of 2,4-dichlorophenoxyacetic acid herbicides. *Journal of the National Cancer Institute* 82:1226–1231, 1991.

Hays SP. The role of values in science and policy: The case of lead. In: Needleman H, ed. *Human Lead Exposure.* Boca Raton, FL: CRC Press, 1992:277–299.

Heaton SN, Auerlich RJ, Bursian SJ, Giesy JP, Render JA, Tillitt DE, Kubiak TJ. Reproductive effects of feeding Saginaw Bay source fish to ranch mink (abstract). In: *Cause-Effect Linkages II Symposium Abstracts.* Traverse City, MI: Michigan Audubon Society, 1991: 24–25.

Heindl A, Hutzinger O. Search for industrial sources of PCDD/PCDF: III. Short-chain chlorinated hydrocarbons. *Chemosphere* 16:1949–1957, 1987.

Hendricks DW, Bellamy WD. Microorganism removals by slow sand filtration. In: Logsdon GS, ed. *Slow Sand Filtration.* New York: American Society of Civil Engineers, 1991:101–121.

Henriksen GL, Michalek JE, Swaby JA, Rahe AJ. Serum dioxin, testosterone, and gonadotropins in veterans of Operation Ranch Hand. *Epidemiology* 7:352–357, 1996.

Henschel DS, Martin JW, DeWitt JC. Brain asymmetry as a potential biomarker for developmental TCDD intoxication: A dose-response study. *Environmental Health Perspectives* 105:718–725, 1997.

Henschler D. Science, occupational exposure limits, and regulations: A case study on organochlorine solvents. *American Industrial Hygiene Association Journal* 51:523–530, 1990.

Henschler D. Toxicity of chlorinated organic compounds: Effects of the introduction of chlorine in organic molecules. *Angewandte Chemie International Edition (English),* 33:1920–1935, 1994.

Henschler D, Vamvakas S, Lammert M, Dekant W, Kraus B, Thomas B, Ulm K. Increased incidence of renal cell tumours in a cohort of cardboard workers exposed to trichloroethylene. *Archives of Toxicology* 70:131–133, 1995.

Herman E. A story about O. *Pool and Spa News,* May 17, 1993:38–40.

Herman-Giddens ME, Slora EJ, Wasserman RC, Bourdony CJ, Bhapkar MV, Koch GG, Hasemeier CM. Secondary sexual characteristics and menses in young girls seen in office practice: A study from the Pediatric Research in Office Settings Network. *Pediatrics* 99:505–512, 1997.

Herson DS, Marshall DR, Baker KH, Victoreen HT. Association of microorganisms with surfaces in distribution systems. *Journal of the American Water Works Association* 83(7):103–106, 1991.

Hewitt TI, Smith KR. *Intensive Agriculture and Environmental Quality: Examining the Newest Agricultural Myth.* Greenbelt, MD: Henry A. Wallace Institute for Alternative Agricutlure, 1995.

Hickling Corporation. *Economic Instruments for the Virtual Elimination of Persistent Toxic Substances in the Great Lakes Basin.* Windsor, ON: International Joint Commission, 1993.

Hickman D, Chang D, Glasser H. Cadmium and lead in bio-medical waste incinerators (presentation text). Eighty-second Annual Meeting of the Air and Waste Management Association, Anaheim, CA, June 1989.

Hildesheim ME, Cantor KP, Lynch CF, Dosemci M, Lubin J, Alavanja M, Craun C. Drinking water source and chlorination by-products: II. Risk of colon and rectal cancers. *Epidemiology* 9:29–35, 1998.

Hill G, Holman J. *Chemistry in Context,* Third Edition. London: Nelson Publishers, 1989:266–269.

Hilleman B, Long JR, Kirschner EM. Chlorine industry running flat out despite persistent health fears. *Chemical and Engineering News* 72(47):12–26, November 12, 1994.

Hirschorn J, Jackson T, Baas L. Towards prevention—the emerging environmental management paradigm. In: Jackson T, ed. *Clean Production Strategies: Developing Preventive Environmental Management in the Industrial Economy.* Boca Raton, FL: Lewis, 1993: 125–142.

Hjelm O, Asplund G. Chemical characterization of organohalogens in a coniferous forest soil. In: Grimvall A, deLeer EWB, eds. *Naturally-Produced Organohalogens.* Boston: Kluwer Academic, 1995:105–111.

Hocking J. Regulation of discharge of organochlorines from pulp mills in Canada. *Environmental Management* 15:195–204, 1991

Hoekstra EJ, Lassen P, van Leeuwen JGE, de Leer EWB, Carlsen L. Formation of organic chlorine compounds of low molecular weight in the chloroperoxidase-mediated reaction between chloride and humic materials. In: Grimvall A, de Leer EWB, eds. *Naturally Produced Organohalogens.* Boston: Kluwer Academic, 1995:149–158.

Hoel DG, Davis DL, Miller AB, Sondik EJ, Swerdlow AJ. Trends in cancer mortality in 15 industrialized countries, 1969–1986. *Journal of the National Cancer Institute* 84:313–320, 1992.

Holmberg PC, Nurminen M. Congenital defects of the central nervous system and occupational exposure during pregnancy: A case-referent study. *American Journal of Industrial Medicine* 1:167–176, 1980.

Howard J. *Chemical Pollution, Human Health, and Sustainability: Confronting the Limits and Politics of Scientific Expertise.* Thesis, Renssalaer Polytechnic Institute, 1997.

Howard P. *Handbook of Environmental Fate and Exposure Data for Organic Chemicals.* Volume 2: *Solvents.* Chelsea, MI: Lewis, 1990.

Howe GR, Hirohata T, Hislop TG, Iscovich JM, Yuan JM, Katsouyanni K, Lubin F, Marubbini E, Modan B, Rohan T, Toniolo P, Shunzhang U. Dietary factors and risk of breast cancer: Combined analysis of 12 case-control studies. *Journal of the National Cancer Institute* 82:561–569, 1990.

Howlett CT. Chlorine: The issue, the reality, and the solution (presentation text). American Chemical Society, Chicago, August 22, 1995.

Hoyer AP, Grandjean P, Jorgensen T, Brock JW, Hartvig HB. Organochlorine exposure and risk of breast cancer. *Lancet* 352:1816–1829, 1998.

Hrelia P, Maffei F, Vigagni F, Fimognari C, Flori P, Stanzani R, Cantelli Forti G. Interactive effects between trichloroethylene and pesticides at metabolic and genetic level in mice: Mechanism-based predictions of interactions. *Environmental Health Perspectives* 102 (supp 9): 31–34, 1994.

Hryb DJ, Khan S, Romas NA, Rosner W. Solubilization and partial characterization of the sex hormone–binding globulin receptor from human prostate. *Journal of Biological Chemistry* 264:5378–5383, 1989.

HSDB (Hazardous Substances Databank). Bethesda, MD: National Library of Medicine, 1995.

HSDB (Hazardous Substances Databank). Bethesda, MD: National Library of Medicine, 1997.

Hsieh CC, Lan SJ, Ekbom A, Petridou E, Adami HO, Trichopoulos D. Twin membership and breast cancer risk. *American Journal of Epidemiology* 136:1321–1326, 1992.

Hsu C-C, Yu M-LM, Chen Y-CJ, Guo Y-LL, Rogan WJ. The Yu-cheng rice oil poisoning incident. In: Schecter A, ed. *Dioxins and Health*. New York: Plenum, 1994:661–684.

Hudec T, Thean J, Kuehl D, Dougherty RC. Tris(dichloropropyl)phosphate, a mutagenic flame retardant: Frequent occurrence in human seminal plasma. *Science* 211:951–952, 1981.

Huestis SY, Servos MR, Dixon DG. Evaluation of temporal and age-related trends of chemically and biologically generated 2,3,7,8-tetrachlorodibenzo-p-dioxin equivalents in Lake Ontario trout, 1977–1993. *Environmental Toxicology and Chemistry* 16:154–164, 1997.

Huff J. Dioxins and mammalian carcinogenesis. In: Schecter A, ed. *Dioxins and Health*. New York: Plenum, 1994:389–408.

Huff J. Chemically induced cancers in hormonal organs of laboratory animals and of humans. In: Huff J, Boyd J, Barrett JC, eds. *Cellular and Molecular Mechanism of Hormonal Carcinogenesis: Environmental Influences*. New York: Wiley-Liss, 1996:77–101.

Huff J, Boyd B, Barrett JC. Hormonal carcinogenesis and environmental influences: Background and overview. In: Huff J, Boyd J, Barrett JC, eds. *Cellular and Molecular Mechanism of Hormonal Carcinogenesis: Environmental Influences*. New York: Wiley-Liss, 1996a: 3–23.

Huff J, Boyd J, Barrett JC. Preface to Huff J, Boyd J, Barrett JC, eds. *Cellular and Molecular Mechanism of Hormonal Carcinogenesis: Environmental Influences*. New York: Wiley-Liss, 1996b:xii–xvii.

Huff J, Weisburger E, Fung VA. Multicomponent criteria for predicting carcinogenicity: Database of 30 NTP chemicals. *Environmental Health Perspectives* 104 (supp 5):1105–1112, 1996c.

Huisman L, Wood WE. *Slow Sand Filtration*. Geneva: World Health Organization, 1974.

Huisman M, Koopman-Esseboom C, Fidler V, Hadders-Algra M, vander Paauw CG, Tuinstra LGMT, Weisglas-Kuperus N, Sauer PJJ, Touwen BCL, Boersma ER. Perinatal exposure to polychlorinated biphenyls and dioxins and its effect on neonatal neurological development. *Early Human Development* 41:111–27, 1995.

Hunter DJ, Willett WC. Diet, body size, and breast cancer. *Epidemiologic Reviews* 15:110–132, 1993.

Hunter DJ, Hankinson SE, Laden F, Colditz GA, Manson JE, Willett WC, Speizer FE, Wolff MS. Plasma organochlorine levels and the risk of breast cancer. *New England Journal of Medicine* 337:1253–1258, 1997.

Huong LD, Phuong NTN, Thuy TT, Hoan NTK. An estimate of the incidence of birth defects, hydatidiform mole and fetal death in utero between 1952 and 1985 at the obstetrical and gynecological hospital of Ho Chi Minh City, Republic of Vietnam. *Chemosphere* 18: 805–810, 1989.

Hutari J, Vesterinen R. PCDD/F emissions from co-combustion of RDF with peat, wood waste, and coal in FBC boilers. *Hazardous Waste and Hazardous Materials* 13:1–10, 1996.

Hutzinger O. Dioxin danger: incineration status report. *Analytical Chemistry* 586:633–642, 1986.

Hutzinger O, Fiedler H. *Formation of Dioxins and Related Compounds in Industrial Processes.* Pilot Study on International Information Exchange on Dioxins and Related Compounds no.173. Brussels: NATO Committee on the Challenges of Modern Society, 1988.

IARC (International Agency for Research on Cancer). Globocan: International Cancer Incidence and Mortality, 1990 (on-line database, available at www.iarc.fr). Lyon, 1998.

ICI Chemicals and Polymers, Ltd. *Report to the Chief Inspector HMIP Authorization AK6039, Improvement Condition, Part 8, Table 8.1, Item 2—Formation of Dioxins in Oxychlorination: Significance for Human Health ad Monitoring Proposals.* Runcorn, UK: ICI Chemicals and Polymers Safety and Environment Department, April 1994.

Illman DL. Green technology presents challenge to chemists. *Chemical and Engineering News* 71(36):26–31, 1993.

Illman DL. Environmentally benign chemistry aims for processes that don't pollute. *Chemical and Engineering News* 72 (36):22–27,1994.

IJCSAB (International Joint Commission Science Advisory Board). *1989 Report.* Windsor, ON: International Joint Commission, 1989.

IJCSAB (International Joint Commission Science Advisory Board). *1991 Report.* Windsor, ON: International Joint Commission, 1991.

IJCSAB (International Joint Commission Science Advisory Board). *1995 Report.* In: 1993–1995 *Priorities and Progress Under the Great Lakes Water Quality Agreement.* Windsor, ON: International Joint Commission, 1995.

International Joint Commission. *Fifth Biennial Report on Great Lakes Water Quality.* Windsor, Ontario, 1990.

International Joint Commission. *Sixth Biennial Report on Great Lakes Water Quality.* Windsor, Ontario, 1992.

International Joint Commission. Appendix D: The injury. In: *1993 Virtual Elimination Task Force Report.* Windsor, Ontario: 1993:87–105.

International Joint Commission. *Seventh Biennial Report on Great Lakes Water Quality.* Windsor, Ontario, 1994.

International Joint Commission. *Eighth Biennial Report on Great Lakes Water Quality.* Windsor, Ontario, 1996.

Ireland GW. *The Medical Department of the United States in the World War.* Volume 14: *Medical Aspects of* Gas *Warfare.* Washington, DC: Government Printing Office, 1926 (extracts available at http://www.ukans.edu/~kansite/ww_one/medical/gas.htm).

Irvine S, Cawood E, Richardson D, MacDonald E, Aitken J. Evidence of deteriorating semen quality in the United Kingdom: Birth cohort study in 577 men in Scotland over 11 years. *British Medical Journal* 312:467–471, 1996.

Jackson L. Children "put at risk from chemicals in toys." *London Sunday Telegraph,* November 9, 1997:11.

Jackson T. Preface. In: Jackson T, ed. *Clean Production Strategies: Developing Preventive Environmental Management in the Industrial Economy.* Boca Raton, FL: Lewis, 1993a.

Jackson T. Principles of clean production—an operational approach to the preventive paradigm. In: Jackson T, ed. *Clean Production Strategies: Developing Preventive Environmental Management in the Industrial Economy.* Boca Raton, FL: Lewis, 1993b:143–164.

Jacob TR. The precautionary principle—an industry perspective. *Risk Policy Report,* November 17, 1995:35–37.

Jacobs Engineering Group. *Potential for Source Reduction and Recycling of Halogenated Solvents: Summary Report.* Los Angeles: Metropolitan Water District and Environmental Defense Fund, 1990a.

Jacobs Engineering Group. *Source Reduction and Recycling of Halogenated Solvents in the Pharmaceutical Industry: Technical Support Document.* Los Angeles: Metropolitan Water District and Environmental Defense Fund, 1990b.

Jacobson JL, Jacobson SW. New methodologies for assessing the effects of prenatal toxic exposure on cognitive functioning in humans. In: Evans MS, ed. *Toxic Contaminants and Ecosystem Health: A Great Lakes Focus.* Ann Arbor: Wiley, 1988:374–388.

Jacobson JL, Jacobson SW. A 4-year followup study of children born to consumers of Lake Michigan fish. *Journal of Great Lakes Research* 19:776–783, 1993.

Jacobson JL, Jacobson SW. Intellectual impairment in children exposed to polychlorinated biphenyls in utero. *New England Journal of Medicine* 335:783–789, 1996.

Jacobson JL, Jacobson SW, Humphrey HEB. Effects of in utero exosure to polychlorinated biphenyls and related contaminants on cognitive functioning in young children. *Journal of Pediatrics* 116:38–45, 1990.

Jaenicke EC. *The Myths and Realities of Pesticide Reduction: A Reader's Guide to Understanding the Full Economic Impacts.* Greenbelt, MD: Henry A. Wallace Institute for Alternative Agriculture, 1997.

Jarman WM, Simòn M, Norstrom RJ, Burns SA, Bacon CA, Simonelt BRT, Risebrough RW. Global distribution of tris(4-chlorophenyl)methanol in high trophic level birds and mammals. *Environmental Science and Technology* 26:1770–1774, 1992.

Jasanoff S. *Risk Management and Political Culture.* New York: Russel Sage Foundation, 1986.

Jasanoff S. *The Fifth Branch: Science Advisers as Policymakers.* Cambridge: Harvard University Press, 1990.

Jasanoff S. Compelling knowledge in public decisions. In: Brooks A, Vandeveer SA, eds. *Saving the Seas.* College Park, MD: Maryland Sea Grant Press, 1996:229–251.

Jay K, Stieglitz L. Identification and quantification of volatile organic components in emissions of waste incineration plants. *Chemosphere* 30:1249–1260, 1995.

Jeffers P, Ward L, Woytowitch L, Wolfe N. Homogeneous hydrolysis rate constants for selected chlorinated methanes, ethanes, ethenes and propanes. *Environmental Science and Technology* 23:965–969, 1989.

Jehassi O. Multiprocess *Wet Cleaning: Cost and Performance Comparison of Conventional Dry Cleaning and an Alternative Process.* Washington, DC: U.S. Environmental Protection Agency Office of Pollution Prevention and Toxics (74-R-93-004), 1993.

Jenkins C. Memorandum to R. Loehr: Newly revealed fraud by Monsanto in an epidemiological study used by EPA to assess human health effects from dioxins. Washington, DC; U.S. Environmental Protection Agency, February 23, 1990.

Jennings AM, Wild G, Ward JD, Ward AM. Immunological abnormalities 17 years after accidental exposure to 2,3,7,8-tetrachlorodibenzo-p-dioxin. *British Journal of Industrial Medicine* 45:701–704, 1988.

Jensen AA. Chemical contaminants in human milk. *Residue Reviews* 89:1–128, 1983.

Jensen AA. Polychlorobiphenyls (PCBs), polychlorodibenzo-p-dioxins (PCDDs) and polychlorodibenzofurans (PCDFs) in human milk, blood and adipose tissue. *Science of the Total Environment* 64(3):259–293, 1987.

Jensen S, Kihlstrom JE, Olsson M, Lundberg C, Orberg J. Effects of PCB and DDT on mink (*Mustela vision*) during the reproductive season. *Ambio* 6:239, 1977.

Jeyeratnam J. Acute pesticide poisoning: A major global health problem. *World Health Statistics Quarterly* 43:139–143, 1990.

Johansson C, Boren H, Grimvall , Dahlman O, Morck R, Reimann A, Malcolm RL. Halogenated structural elements in naturally occurring organic matter. In: Grimvall A, deLeer EWB, eds. *Naturally-Produced Organohalogens.* Boston: Kluwer Academic, 1995:95–103.

Johansson N. PCDD/PCDF and PCB: The Scandinavian situation. *Organohalogen Compounds* 14:337–340, 1993.

Johansson NG, Clark FM, Fletcher DE. New technology development for the closed cycle bleach plant. International Non-Chlorine Bleaching Conference, Amelia Island, Florida, 1995.

John EM, Kelsey JL. Radiation and other environmental exposures and breast cancer. *Epidemiologic Reviews* 15:157–161, 1993.

Johnson J, Jolley R. Water chlorination: the challenge. In: Jolley RL, Condie LW, Johnson JD, Katz S, Minear RA, Mattice JS, Jacobs VA, eds. *Water Chlorination: Chemistry, Environmental Impact and Health Effects,* Volume 6. Chelsea, MI: Lewis Publishers, 1990:21–27.

Johnson KC, Rouleau J. Temporal trends in Canadian birth defects birth prevalences, 1979–1993. Canadian Journal of Public Health 88:169–176, 1997.

Johnson KL, Cummings AM, Birnbaum LS. Promotion of endometriosis in mice by polychlorinated dibenzo-p-dioxins, dibenzofurans, and biphenyls. *Environmental Health Perspectives* 105:750–755, 1997.

Johnston PA, Trondle S, Clayton R, Stringer R. *PVC—The Need for an Industrial Sector Approach to Environmental Regulation.* Exeter: University of Exeter Earth Resources Center (GERL Technical Note 04/93), 1993.

Jolley RL. Basic issues in water chlorination: A chemical perspective. In: Jolley RL, Bull RJ, Davis WP, Jatz S, Roberts MH, Jacobs VA, eds. *Water Chlorination: Chemistry, Environmental Impact, and Health Effects,* Volume 5. Chelsea, MI: Lewis, 1985:19–38.

Jonsson S, Pavasars I, Johansson C, Boren H, Gimvall A. Origin of organohalogens found in Baltic Sea sediments. In: Grimvall A, deLeer EWB, eds. *Naturally-Produced Organohalogens.* Boston: Kluwer Academic, 1995:339–352.

Juttner I, Henkelmann B, Schramm KW, Steinberg CEW, Winkler R, Kettrup A. Occurrence of PCDD/F in dated lake sediments of the Black Forest, southwestern Germany. *Environmental Science and Technology* 31:806–812, 1997.

Juuti S, Hirvonen A, Tarhanen J, Ruuskanen J. Trichloroacetic acid in conifer needles in polluted and clean environments. *Organohalogen Compounds* 14:281–284, 1993.

Kahlich D, Wiechern U, Lindner J. Propylene oxide. In: Elvers B, Hawkins S, Russey W, Schulz G, eds. *Ullman's Encyclopedia of Industrial Chemistry,* A22:243–247, 1993.

Kahn PC, Gochfeld M, Nygren M, Hansson M, Rappe C, Velez H, Ghent-Guenther T, Wilson WP. Dioxins and dibenzofurans in blood and adipose tissue of Agent Orange–exposed Vietnam veterans and matched controls. *Journal of the American Medical Association* 259:1661–1667, 1988.

Kaminsky R, Hites RA. Octachlorostyrene in Lake Ontario: Sources and fates. *Environmental Science and Technology* 18:275–279, 1984.

Kamrin MA. *Toxicology Primer.* Chelsea, MI: Lewis, 1988.

Kamrin MA. Environmental health, risk assessment, and democracy. *Environmental Health Perspectives* 106:216, 1998.

Kandel ER. A new intellectual framework for psychiatry. *American Journal of Psychiatry* 155:457–469, 1998.

Kang KS, Wilson MR, Hayashi T, Chang C-C, Trosko JE. Inhibition of gap junction intercellular communication in normal human breast epithelial cells after treatment with pesticides, PCBs, and PBBs, alone or in mixtures. *Environmental Health Perspectives* 104:192–200, 1996.

Kanitz S, Franco Y, Patrone V, Caltabellotta M, Raffo E, Rifggi C, Timitilli D, Ravera G. Association between drinking water disinfection and somatic parameters at birth. *Environmental Health Perspectives* 104:516–520, 1996.

Kanters MJ, Van Nispen R, Louw R, Mulder P. Chlorine input and chlorophenol emission in the lab-scale combustion of municipal solid waste. *Environmental Science and Technology* 30:2121–2126, 1996.

Karmaus W, Wolf N. Reduced birthweight and length in the offspring of females exposed to PCDFs, PCP and lindane. *Environmental Health Perspectives* 103:1120–1125, 1995.

Karol MH. Toxicological principles do not support the banning of chlorine. *Fundamental and Applied Toxicology* 24:1–2, 1995.

Karstadt M. Quantitative risk assessment: Qualms and questions. *Teratogenesis, Carcinogenesis, and Mutagenesis* 8:137–152, 1988.

Katzenellenbogen JA. The structural pervasiveness of estrogenic activity. *Environmental Health Perspectives* 103 (supp 7):99–102, 1995.

Kavlock RJ, Perreault SD. Multiple chemical exposures and risks of adverse reproductive function and outcome. In: Yang RSH, ed. *Toxicology of Chemical Mixtures.* New York: Academic Press, 1994:245–297.

Kelce WR, Monosson E. Gamcsik MP, Laws SC, Gray LE. Environmental hormone disruptors: Evidence that vinclozolin developmental toxicity is mediated by antiandrogenic metabolites. *Toxicology and Applied Pharmacology* 126:276–285, 1994.

Kelce WR, Stone CR, Laws SC, Gray LE, Kemppainen JA, Wilson EM. Persistent DDT metabolite p,p′-DDE is a potent androgen receptor antagonist. *Nature* 375:581–585, 1995.

Kelsey JL, Bernstein L. Epidemiology and prevention of breast cancer. *Annual Review of Public Health* 17:47–67, 1996.

Kelsey JL, Gammon MD. The epidemiology of breast cancer. *CA: A Cancer Journal for Clinicians* 41:146–165, 1991.

Kemner v. Monsanto Co. Civil No. 80-L-970, Circuit Court of St. Clair Country, IL, 1985.

Kerkvliet NI. Immunotoxicology of dioxins and related chemicals. In: Schecter A, ed. *Dioxins and Health.* New York: Plenum, 1994:199–217.

Kettles MA, Browning SR, Prince TS, Horstman SW. Triazine herbicide exposure and breast cancer incidence: An ecologic study of Kentucky counties. *Environmental Health Perspectives* 104:1222–1227, 1997.

Khizbullia F, Muslymova I, Khasanova I, Chernova L, Abdraschitov J. Evaluation of polychlorinated dibenzodioxins and dibenzofurans emissions from vinylchloride-monomer production. *Organohalogen Compounds* 26:225–229, 1998.

Kilburn KH, Warshaw RH. Prevalence of symptoms of systemic lupus erythematosus (SLE) and of fluorescent antinuclear antibodies associated with chronic exposure to trichloroethylene and other chemicals in well water. *Environmental Research* 67:1–9, 1992.

Kilburn KH, Warshaw RH. Chemical-induced autoimmunity. In: Dean JH, Luster MI, Munson AE, Kimber I, eds. *Immunotoxicology and Immunopharmacology.* New York: Raven Press, 1994:523–538.

King WD, Marrett LD. Case-control study of bladder cancer and chlorination by-products in treated water. *Cancer Causes and Control* 7:596–624, 1996.

Kirschner EM. Dow Chemical: Leading a quiet revolution. *Chemical Week,* September 28, 1993:36.

Kirschner EM. Environment, health concerns force shift in use of organic solvents. *Chemical and Engineering News* 72(25):13–20,1994.

Kiss A. The rights and interests of future generations and the precautionary principle. In: Freestone D, Hey E, eds. *The Precautionary Principle* and *International Law: The Challenge of Implementation.* Boston: Kluwer Law International, 1996:19–28.

Kjeller LO, Rappe C. Time trends in levels, patterns, and profiles for polychlorinated dibenzo-p-dioxins, dibenzofurans, and biphenyls in a sediment core from the Baltic proper. *Environmental Science and Technology* 29:346–355, 1995.

Kjeller LO, Jones KC, Johnston AE, Rappe C. Increases in the polychlorinated dibenzo-p-dioxin and -furan content of soils and vegetation since the 1840s. *Environmental Science and Technology* 25:1619–1627, 1991.

Kjeller LO, Jones KC, Johnston AE, Rappe C. Evidence for a decline in atmospheric emissions of PCDD/Fs in the UK. *Environmental Science and Technology* 30:1398–1403, 1996.

Klausbruckner B. Avoiding chlorinated substances in Vienna Hospital (presentation). Seventh Biennial Meeting of the International Joint Commission, Windsor, ON, 1993.

Kleijn R, Tukker A, van der Voet E. Chlorine in the Netherlands, part I, an overview. *Journal of Industrial Ecology* 1:95–116, 1997.

Klotz DM, McLachlan JA, Arnold SF. Identification of environmental chemicals with estrogenic activity using a combination of in vitro assays. *Environmental Health Perspectives* 104:1084–1089, 1996.

Klotz DM, Ladlie BL, Vonier PM, McLachlan JA, Arnold SF. o,p′-DDT and its metabolites inhibit progesterone-dependent responses in yeast and human cells. *Molecular and Cellular Endocrinology* 129:63–71, 1997.

Kocan RM, von Westernhagen H, Ladolt ML, Furstenburg G. Toxicity of sea surface microlayer: Effects of hexane extract on Baltic herring and Atlantic cod embryos. *Marine Environmental Research* 23:291–305, 1987.

Kogevinas M, Saracci R, Winkelmann R, Johnson ES, Bertazzi PA, Bueno de Mesquita BH, Kauppinen T, Littorin M, Lynge E, Neuberger M, Pearce N. Cancer incidence and mortality in women occupationally exposed to chlorophenoxy herbicides, chlorophenols, and dioxins. *Cancer Causes and Control* 4:547–553, 1993.

Kogevinas M, Kauppinen T, Winkelmann R, Becher H, Bertazzi PA, Bueno-de-Mesquita HB, Coggon D, Green L, Johnson E, Littorin M, Lynge E, Marlow DA, Matthews JD, Neuberger M, Benn T, Pannett B, Pearce N, Saracci R. Soft tissue sarcoma and non-Hodgkin's lymphoma in workers exposed to phenoxy herbicides, chlorophenols and dioxins: Two nested case-control studies. *Epidemiology* 6:396–402, 1995.

Kogevinas M, Becher H, Benn T, Bertazzi PA, Boffetta P, Bueno-de-Mesquita HB, Coggon D, Colin D, Flesch-Janus D, Fingerhut M, Green L, Kauppinen T, Littorin M, Lynge E, Matthews JD, Neuberger M, Pearce N, Saracci R. Cancer mortality in workers exposed to phenoxy herbicides, chlorophenols, and dioxins. *American Journal of Epidemiology* 145: 1061–1075, 1997.

Kohlmeier L, Rehm J, Hoffmeister H. Lifestyle and trends in worldwide breast cancer rates. *Annals of the New York Academy of Sciences* 609:259–268, 1990.

Kohn MC, Sewall CH, Lucier GW, Portier CJ. A mechanistic model of effects of dioxin on thyroid hormones in the rat. *Toxicology and Applied Pharmacology* 136:29–48, 1996.

Koistinen J, Paasivirta J, Nevalainen T, Lahtipera M. Chlorinated fluorenes and alkylfluorenes in bleached kraft pulp and pulp mill discharges. *Chemosphere* 28:2139–2150, 1994a.

Koistinen J, Paasivirta J, Nevalainen T, Lahtipera M. Chlorophenanthrenes, alkylchlorophenanthrenes and alkylchloronaphthalenes in kraft pulp mill products and discharges. *Chemosphere* 27:1261–1277, 1994b.

Koivusalo M, Vartiainen T. Drinking water chlorination by-products and cancer. *Reviews on Environmental Health* 12:81–90, 1997.

Koivusalo M, Jaakola JJK, Vartainen T, Hakulinen T, Karjalainen S, Pukkala E, Tuomisto J. Drinking water mutagenicity and gastrointestinal and urinary tract cancers: An ecological study in Finland. *American Journal of Public Health* 84:1223–1228, 1994.

Koivusalo M, Pukkala, Vartiainen T, Jaakola JJK, Hakulinen T. Drinking water chlorination and cancer—a historical cohort study in Finland. *Cancer Causes and Control* 8:192–200, 1997.

Kolata G. Chemicals that mimic hormones spark alarm and debate. *New York Times,* March 19, 1996:C1.

Kolenda J, Gass H, Wilken M, Jager J, Zeschmer-Lahl B. Determination and reduction of PCDD/F emissions from wood burning facilities. *Chemosphere* 29:1927–1938, 1994.

Komiya K, Fukoka S, Aminaka M, Hasegawa K, Hachiya H, Okamoto H, Watanabe T, Yoneda H, Fukawa I, Dozono T. New process for producing polycarbonate without phosgene and methylene chloride. In: Anastas PT, Williamson TC, eds. *Green Chemistry: Designing Chemistry for the Environment.* Washngton, DC: American Chemical Society, 1996:20–32.

Komulainen H, Kosma VM, Vaittinen SL, Vartiainen T, Kaliste-Korhonen E, Lotjonen S, Tuominen RK, Tuomisto J. Carcinogenicity of the drinking water mutagen 3-chloro-4-(dichloromethyl)-5-hydroxy-2(5H)-furanone in the rat. *Journal of the National Cancer Institute* 89:848–856, 1997.

Koninckx PR, Braet P, Kennedy SH, Barlow DH. Dioxin pollution and endometriosis in Belgium. *Human Reproduction* 9:1001–1002, 1994.

Koopman-Esseboom C, Morse DC, Weisglas-Kuperus N, Lutkeschipholt IJ, van der Paauw CG, Tuinstra LGMT, Brouwe A, Sauer PJJ. Effects of dioxins and polychlorinated biphenyls on thyroid hormone status of pregnant women and their infants. *Pediatric Research* 36:468–473, 1994.

Koppe JG. Dioxins and furans in the mother and possible effects on the fetus and newborn breast-fed baby. *Acta Paedeatrics Scandinavia* 360(Supp):146–153, 1989.

Koppe JG, Pluim E, Olie K. Breastmilk, PCBs, dioxins and vitamin K: A discussion paper. *Journal of the Royal Society of Medicine* 82:416–419, 1989.

Koppe JG, Pluim HJ, Olie K, van Wijnen J. Breast milk, dioxins and the possible effects on the health of newborn infants. *Science of the Total Environment* 106:33–41, 1991.

Kopponen P, Torronen R, Ruuskanen J, Tarhanen J, Vartiainen T, Karenlampi S. Comparison of cytochrome P4501A1 induction with the chemical composition of fly ash from combustion of chlorine containing material. *Chemosphere* 24:391–401, 1992.

Kopponen P, Valttila O, Talka E, Torronen R, Tarhanen J, Ruuskanen J, Karenlampi S. Chemical and biological 2,3,7,8-tetrachlorodibenzo-p-dioxin equivalents in fly ash from combustion of bleached kraft pulp mill sludge. *Environmental Toxicology and Chemistry* 13:143–148, 1994.

Kostick DS. Soda ash. In: *Soda Ash Minerals Yearbook*. Washington, DC: U.S. Department of the Interior, Bureua of Mines, 1989:951–979.

Kramarova E, Kogevinas M, Anh CT, Cau HD, Dai LC, Stellman SD, Parkin DM. Exposure to Agent Orange and occurrence of soft-tissue sarcomas or non-Hodgkin's lymphomas: An ongoing study in Vietnam. *Environmental Health Perspectives* 106 (supp 2):671–678, 1998.

Kramer MD, Lynch CF, Isaccson P, Hanson JW. The association of waterborne chloroform with intrauterine growth retardation. *Epidemiology* 3:407–413, 1992.

Kramlich JC, Poncelet EM, Charles RE, Seeker WR, Samuelsen GS, Cole JA. *Experimental Investigation Of Critical Fundamental Issues In Hazardous Waste Incineration.* Research Triangle, NC: U.S. Environmental Protection Agency Industrial Environmental Research Lab (600/2-89-048), 1989.

Krieger N, Wolff MS, Hiatt RA, Rivera M, Vogelman J, Orentreich N. Breast cancer and serum organochlorines: A prospective study among white, black and Asian women. *Journal of the National Cancer Institute* 86:589–599, 1994.

Krishnan K, Brodeur J. Toxic interactions among environmental pollutants: Corroborating laboratory observations with human experience. Mechanism-based predictions of interactions. *Environmental Health Perspectives* 102 (supp 9):11–17, 1994.

Kroesa R. *The Greenpeace Guide to Pulp*. Amsterdam: Greenpeace International, 1991.

Krotoszynski BK, O'Neill JH. Involuntary bioaccumulation of environmental pollutants in nonsmoking heterogeneous human population. *Journal of Environmental Science and Health A* 17:855–883, 1982.

Krotoszynski BK, Bruneau GM, O'Neil JM. Measurement of chemical inhalation exposure in urban population in the presence of endogenous effluents. *Journal of Analytical Toxicology* 3:225–234, 1979.

Kruithof JC, van der Leer RC. Practical experiences with UV-disinfection in the Netherlands. In: *Proceedings of the AWWA Seminar on Emerging Technologies in Practice.* Denver: American Water Works Association, 1990.

Kutney GW. Low-cost bleaching sequence changes yield low AOX pulp mill emissions. *Pulp and Paper* 69(1):85–86, 1995.

Kutz FW, Strassman SC, Sperling JF. Survey of selected organochlorine pesticides in the general population of the United States: Fiscal years 1970–1975. *Annals of the New York Academy of Science* 320:60–68, 1979.

Kutz FW, Wood PH, Bottimore DP. Organochlorine pesticides and polychlorinated biphenyls in human adipose tissue. *Reviews of Environmental Contamination and Toxicology* 120:1–82, 1991.

Kyyronen P, Taskinen H, Lindbohm ML, Hemminki K, Heinonen OP. Spontaneous abortions and congenital malformations among women exposed to tetrachloroethylene in dry cleaning. *Journal of Epidemiology and Community Health* 43:3346–351, 1989.

Lahl, U. Sintering plants of steel industry: The most important thermal PCDD/F source in industrial regions? *Organohalogen Compounds* 11:311–314, 1993.

Lahl U. Sintering plants of steel industry: PCDD emision status and perspectives. *Chemosphere* 29:1939–1945, 1994.

Lahvis GP, Wells RS, Kuejl DW, Stewart JL, Rhinehart HL, Via CS. Decreased in vitro lymphocyte responses in free-ranging bottlenose dolphins (*Tursios truncatas*) are associated with increased whole blood concentrations of polychlorinated biphenyls (PCBs) and p,p′-DDT and o,p′-DDE. *Environmental Health Perspectives* 103 (supp 4):67–72, 1995.

Lamarre L. A fresh look at ozone. *EPRI Journal* 22(4):6–15, July 1997.

Landrigan PJ. Commentary: environmental disease—a preventable epidemic. *American Journal of Public Health* 82:941–943, 1992.

Landrigan PJ, Graham DG, Anger WK, Barker J, Damstra T, Hattis D, Langston W, Lowndes HE, Marwah J, Morell P, Narahashi T, Nelson PP, Reiter LW, Rodier P, Rodricks J, Silbergeld EK, Spencer PS, Weiss B, Qyzga R, Mattison D, Emmergson JL, Thomas RD, Stratton KR, Paxton MB, Schneiderman MA, Pope AM, Grossblatt N, Sprague AM, Bennett GJ, Nisbet ICT, Tilson H, Leonard LV. *Environmental Neurotoxicology.* Washington, DC: National Academy Press, 1992.

Lane PA, Hathaway WE. Vitamin K in infancy. *Journal of Pediatrics* 106:351–355, 1985.

Lane RE, Hunter D, Malcolm D, Williams MK, Hudson TGF, Browne RC, McCallum RI, Thompson AR, deKretser AJ, Zielhuis RL, Cramer K, Barry PSI, Goldberg A, Beritic T, Vigliani EC, Truhaut R, Kehoe RA, King E. Diagnosis of inorganic lead poisoning: A statement. *British Medical Journal* 4:501, 1968.

Laseter JL, Dowty BJ. Association of biorefractories in drinking water and body burden in people. *Annals of the New York Academy of Sciences* 298:547–556, 1978.

Lash J, Gillman K, Sheridan D. *A Season of Spoils: The Reagan Administration's Attack on the Environment.* New York: Pantheon, 1984.

Lassen P, Randall A, Jorgensen O, Warwick P, Carlsen L. On the possible role of humic materials in the environmental organohalogen budget: The enzymatically mediated incorporation of 4-chlorophenol into humic acids. In: Grimvall A, deLeer EWB, eds. *Naturally-Produced Organohalogens.* Boston: Kluwer Academic, 1995:183–191.

Latour B, Woolgar S. *Laboratory Life: The Construction of Scientific Facts.* Princeton: Princeton University Press, 1986.

Lawrence DH. *Apocalypse.* New York: Viking, 1966.

Lay T, Allen MJ. Slow sand: Timeless technology for modern applications. *Journal of the American Water Works Association* 84(5):10, 1992.

Leatherland JF. Field observations on reproductive and developmental dysfunction in introduced and native salmonids from the Great Lakes. *Journal of Great Lakes Research* 19:737–751, 1993.

Leatherland JF. Changes in thyroid hormone economy following consumption of environmentally contaminated Great Lakes fish. *Toxicology and Industrial Health* 14:41–57, 1998.

Lebel G, Dodin S, Ayotte P, Marcoux S, Ferron LA, Dewailly E. Organochlorine exposure and the risk of endometriosis. *Fertility and Sterility* 69:221–228, 1998.

LeChevallier MW, Cawthon CD, Lee RG. Factors promoting survival of bacteria in chlorinated water supplies. *Applied and Environmental Microbiology* 54:649–654, 1988.

LeChevallier MW. *Treatment to Meet the Microbiological MCL in the Face of a Coliform Regrowth Problem.* Belleville, IL: American Water Works Service Company, 1990.

Leder A, Linak E, Holmes R, Sasano T. *CEH Marketing Research Report: Chlorine/Sodium Hydroxide.* Palo Alto: SRI International, 1994.

Lee A, Campbell B, Kelly W. *Dioxin and Furan Contamination in the Manufacture of Halogenated Organic Chemicals.* Cincinnati: U.S. Environmental Protection Agency Hazardous Waste Engineering Research Laboratory (EPA/600-S2-86/101), 1987.

Lee G. Government purchasers told to seek recycled products; Clinton executive order revises standards for paper. *Washington Post,* October 21, 1993:A29.

Leinhardt B. *Memorandum to Chlorine Outreach Champions: Tool Kit to Help Defeat Administration Chlorine Strategy.* Washington, DC: Chlorine Chemistry Council, February 8, 1994.

Leiss JK, Savitz DA. Home pesticide use and childhood cancer: A case-control study. *American Journal of Public Health* 85:249–252, 1995.

Lemieux PM. *Evaluation of Emissions from the Open Burning of Household Waste in Barrels:* Volume 1: *Technical Report.* Washington, DC: U.S. Environmental Protection Agency Office of Research and Development, Control Technology Center (EPA-600-R-97-134a), 1997.

Lenoir D, Kaune A, Hutzinger O, Mutzenrich G, Horch K. Influence of operating parameters and fuel type on PCDD/F emissions from a fluidized bed incinerator. *Chemosphere* 23:1491–500, 1991.

Leoni V, Fabiana L, Marinelli G, Puccetti G, Tarsitani GF, De Carolis A, Vescia N, Morini A, Aleandri V, Pozzi V, Cappa F, Barbati D. PCBs and other organochlorine compounds in blood of women with or without miscarriage: A hypothesis of correlation. *Ecotoxicology and Environmental Safety* 17:1–11, 1989.

Lerda D, Rizzi R. Study of reproductive function in persons occupationally exposed to 2,4-dichlorophenoxyacetic acid (2,4-D). *Mutation Research* 262:47–50, 1991.

Letterman RD. Operation and maintenance. In: Logsdon GS, ed. *Slow Sand Filtration.* New York: American Society of Civil Engineers, 1991:149–164.

Levin ED, Schantz SL, Bowman RE. Delayed spatial alternation deficits resulting from perinatal PCB exposure in monkeys. *Archives of Toxicology* 62:267–273, 1988.

Levine IN. *Physical Chemistry,* Fourth Edition. New York: McGraw-Hill, 1995.

Lewis-Michl EL, Melius JM, Kallenbach LR, Ju CL, Talbot TO, Orr MF, Lauridsen PE. Breast cancer risk and residence near industry or traffic in Nassau and Suffolk Counties, Long Island, New York. *Archives of Environmental Health* 51:255–265, 1996.

Lexen K, deWitt C. Polychlorinated dibenzo-p-dioxin and dibenzofuran levels and patterns in samples from different Swedish industries analyzed within the Swedish dioxin survey. *Organohalogen Compounds* 9:131–134, 1992.

Li FP, Fraumeni JF, Mantel N, Miller RW. Cancer mortality among chemists. *Journal of the National Cancer Institute* 43:1159–1164, 1989.

Li X, Johnson DC, Rozman KK. Effects of 2,3,7,8-tetrachlorodibenzo-p-dioxin (TCDD) on estrous cyclicity and ovulation in female Sprague-Dawley rats. *Toxicology Letters* 78: 219–222, 1995.

Liberti A, Goretti G, Russo MV. PCDD and PCDF formation in the combustion of vegetable wastes. *Chemosphere* 12:661–663, 1983.

Licis IJ, Mason HB. Boilers cofiring hazardous waste: Effects of hysteresis on performance measurements. *Waste Management* 9:101–108, 1989.

Ligon WV, Dorn SB, May RJ. Chlorodibenzofuran and chlorodibenzo-p-dioxin levels in Chilean mummies dated to about 2800 years before the present. *Environmental Science and Technology* 23:1286–1290, 1989.

Lijinsky W. Modulating effects of hormones on carcinogenesis. In: Huff J, Boyd J, Barrett JC, eds. *Cellular and Molecular Mechanisms of Hormonal Carcinogenesis: Environmental Influences*. New York: Wiley-Liss, 1996:57–76.

Linden KG. UV acceptance. *Civil Engineering* 68(3):58–61, 1998.

Lindqvist O, Johansson K, Aastrup M, Andersson A, Bringmark L, Hovsenius G, Hakanson L, Iverfeldt A, Meili M, Timm B. Mercury in the Swedish environment: Recent research on causes, consequences and corrective methods. *Water Air Soil Pollution* 55:1–32, 1991.

Lockey JE, Kelly CR, Cannon GW, Colby TV, Aldrich V, Livingston V. Progressive systemic sclerosis associated with exposure to trichloroethylene. *Journal of Occupational Medicine* 29:493–496, 1987.

Loganathan BG, Kannan K. Global organochlorine contamination trends: An overview. *Ambio* 23:187–191, 1994.

Lonky E, Reihman J, Darhill T, Mather J, Daly H. Neonatal behavioral assessment scale performance in humans influenced by maternal consumption of environmentally contaminated Lake Ontario fish. *Journal of Great Lakes Research* 22:198–212, 1996.

Lopez AD. Competing causes of death: A review of recent trends in mortality in industrialized countries with special reference to cancer. *Annals of the New York Academy of Sciences* 609:58–76, 1990.

Lopez-Carillo L, Blair A, Lopez-Cervantez M, Cebrian M, Rueda C, Reyes R, Mohar A, Bravo J. Dichlorodiphenyltrichloroethane serum levels and breast cancer risk: A case-control study from Mexico. *Cancer Research* 57:3728–3732, 1997.

Lovblad R. A strategy to minimize environmental impact from kraft pulp mill operation. GLOBE 96, Vancouver, 1996.

Lovblad R, Maimstrom J. Biological effects of kraft pulp mill effluents—a comparison between ECF and TCF pulp production. *Proceedings of the International Non–Chlorine Bleaching Conference,* Amelia Island, Florida, 1994.

Lovelock J. Natural halocarbons in the air and sea. *Nature* 256:193–194, 1975.

Lower Saxony Ministry of Environmental Affairs. Data report and press release, dioxin data from ICI facility, Wilhelmshaven, Germany, March 22, 1994.

Lu YF, Sun G, Wang X, Safe S. Inhibition of prolactin receptor gene expression by 2,3,7,8-tetrachlorodibenzo-o-dioxin in MCF-7 human breast cancer cells. *Archives of Biochemistry and Biophysics* 332:35–40, 1996.

Lucier GW, Schecter A. Human exposure assessment and the National Toxicology Program. *Environmental Health Perspectives* 106:623–627, 1998.

(Lucretius) Titus, Lucretius Carus. *De Rerum Naturum* (The Nature of Things). Esolen AM, trans. Baltimore: Johns Hopkins, 1995.

Ludwig JP, Giesy JP, Summer CL, Bowerman W, Auerlich R, Bursian S, Auman JH, Jones PD, Williams LL, Tillitt DE, Gilbertson M. A comparison of water quality criteria for the Great Lakes based on human and wildlife health. *Journal of Great Lakes Research* 19:789–810, 1993.

Ludwig JP, Kurita-Matsuba H, Auman HJ, Ludwig ME, Summer CL, Giesy JP, Tillitt DE, Jones PD. Deformities, PCBs and TCDD-equivalents in double-crested cormorants (*Phalacrocorax auritus*) and caspian terns (*Hydroprogne caspia*) of the upper Great Lakes 1986–1991: Testing a cause-effect hypothesis. *Journal of Great Lakes Research* 22:172–197, 1996.

Luk KC, Stern L, Weigele M, O'Brien RA, Spirt N. Isolation and identification of diazepam-like compounds from bovine urine. *Journal of Natural Products* 46:852–861, 1983.

Lundberg C. Effects of long-term exposure to DDT on the oestrus cycle and the frequency of implanted ova in the mouse. *Environmental Physiology and Biochemistry* 3:127–131, 1973.

Lundholm CE. The effects of DDE, PCB and chlordane on the binding of progesterone to its cytoplasmic receptor in the eggshell gland mucosa of birds and the endometrium of mammalian uterus. *Comparative Biochemistry and Physiology* 89C:361–368, 1988.

Luoma J. Scientists are unlocking secrets of dioxin's devastating power. *New York Times,* May 15, 1990:B7.

Mac MJ. Toxic substances and survival of Lake Michigan salmonids: Field and laboratory approaches. In: Evans M, ed. *Toxic Contaminants and Ecosystem Health: A Great Lakes Focus.* New York: Wiley, 1988:389–401.

Mac MJ, Edsall CC. Environmental contaminants and the reproductive success of lake trout in the Great Lakes: An epidemiological approach. *Journal of Toxicology and Environmental Health* 33:375–394, 1991.

Mac MJ, Edsall CC, Seelye J. Survival of lake trout eggs and fry reared in water from the upper Great Lakes. *Journal of Great Lakes Research* 11:520–529, 1985.

Mac MJ, Schwartz TR, Edsall CC, Frank AM. Polychlorinated biphenyls in Great Lakes lake trout and their eggs: Relations to survival and congener composition 1979–1988. *Journal of Great Lakes Research* 19:752–765, 1993.

MacDonald RW, Ikonomou MG, Paton DW. Historical inputs of PCDDs, PCDFs, and PCBs to a British Columbia interior lake: The effects of environmental controls on pulp mill emissions. *Environmental Science and Technology* 32:331–337, 1998.

Mackay D. Is chlorine really the evil element? *Environmental Science and Engineering,* November 1992:49–56.

Madigan MP, Ziegler RG, Benichou J, Byrne C, Hoover RN. Proportion of breast cancer cases in the United States explained by well-established risk factors. *Journal of the National Cancer Institute* 87:1681–1685, 1995.

MAFF. Dioxins in PVC food packaging. Food Surveillance Information Sheet Number 59. London: Ministry of Agriculture, Food, and Fisheries/Food Safety Directorate, 1995.

Mahle N, Whiting L. Formation of chlorodibenzodioxins by air oxidation and chlorination of bituminous coal. *Chemosphere* 9:693–699, 1980.

Makhijani A, Gurney KR. *Mending the Ozone Hole.* Cambridge: MIT Press, 1995.

Malkin M, Fumento M. *Rachel's Folly: The End of Chlorine.* Washington, DC: Competitive Enterprise Institute, 1996.

Malone KE. Diesthylstilbestrol (DES) and breast cancer. *Epidemiologic Reviews* 15: 108–109, 1993.

Mann MA, Cramer JA. Disinfecting with ultraviolet radiation. *Water Environment and Technology* 4(12):40–45, 1992.

Mannheim K. *Essays on the Sociology of Knowledge.* New York: Oxford University Press, 1952.

Mannheim, K. *Ideology and Utopia: An Introduction to the Sociology of Knowledge.* London: Routledge, 1991.

Manninen H, Perkio A, Vartiainen T, Ruuskanen J. Formation of PCDD/PCDF: Effect of fuel and fly ash composition on the formation of PCDD/PCDF in the co-combustion of refuse-derived and packaging-derived fuels. *Environmental Science and Pollution Research* 3:129–134, 1996.

Manscher OH, Heidam NZ, Vikelsoe J, Nielsen P, Blinksbjerg P, Madsen H, Pallesen L, Tiernan TO. The Danish incinerator dioxin study, part 1. *Chemosphere* 20:1779–1784, 1990.

Mantykoski K, Paasivirta J, Mannila E. Combustion products of biosludge from pulp mill. *Chemosphere* 191-6:413–416, 1989.

Manz A, Berger J, Dwyer JH, Flesch-Janys D, Nagel S, Waltsgott H. Cancer mortality among workers in chemical plant contaminated with dioxin. *Lancet* 338:959–964, 1991.

Margulis L. *Early Life.* Boston: Science Books International, 1982.

Mark F. *Energy Recovery Through Co-Combustion of Mixed Plastics Waste and Municipal Solid Waste.* Hamburg: Association of Plastics Manufacturers in Europe, 1994.

Marklund S, Rappe C. Tysklind M, Egeback K. Identification of polychlorinated dibenzofurans and dioxins in exhausts from cars run on leaded gasoline. *Chemosphere* 161:29–36, 1987.

Marklund S, Anderson R, Rappe C, Tysklind M, Egeback K, Bjorkman E, Grigoriadis V. Emissions of PCDDs and PCDFs in gasoline and diesel fueled cars. *Chemosphere* 20:553–561, 1990.

Markus T, Behnisch P, Hagenmaier H, Bock KW, Schrenk D. Dioxinlike components in incinerator fly ash: A comparison between chemical analysis data and results from a cell culture bioassay. *Environmental Health Perspectives* 105(12):475–481, 1997.

Marrack D. Hospital red bag waste: An assessment and management recommendations. *Journal of the Air Pollution Control Association* 38:1309–1311, 1988.

Marshall JR, Yinsheng Q, Junshi C, Parpia B, Campbell TC. Additional ecological evidence: Lipids and breast cancer mortality among women aged 55 and over in China. *European Journal of Cancer 28A*:1720–1727, 1992.

Maserti BE, Ferrara R. Mercury in plants, soil, and atmosphere near a chlor-alkali complex. *Water, Air, and Soil Pollution* 56:15–20, 1991.

Massachusetts Department of Environmental Protection. Toxic chemical use, waste by Massachusetts industries down dramatically (press release). Springfield, April 22, 1998.

Mattila H, Virtanen T, Vartiainen T, Ruuskanen J. Emissions of polychlorinated dibenzo-p-dioxins and dibenzofurans in flue gas from co-combustion of mixed plastics with coal and bark. *Chemosphere* 25:1599–1609, 1992.

Mattison DR. An overview on biological markers in reproductive and developmental toxicology: Concepts, definitions and use in risk assessment. *Biomedical and Environmental Sciences* 4:8–34, 1991.

Mattison DR, Ewing LL, Generoso WM, Paulsen CA, Robaire B, Sherins R, Wyrobek AJ, Hatch MC, Canfield RE, Finch C, Haney AF. Schwartz N, Miller RK, Erickson JD, Faulk

WP, Glasse SR, Longo LD, Needleman HL, Buelke-Sam JL, Silbergeld EK, Skalko RG, Spielberg SP, Eddy EM, Hecht N, Selevan S, Wilcox A. *Biological Markers in Reproductive Toxicology.* Washington, DC: National Academy Press, 1989.

Matussek H, Pappens RA, Kenny J. Unlucky 13th year for world production growth. *Pulp and Paper International* 38:22, 1996.

Mayani A, Barel S, Soback S, Almagor M. Dioxin concentrations in women with endometriosis. *Human Reproduction* 12:373–375, 1997.

Mayer CE. CPSC won't seek phthalate ban; agency asks that the chemicals not be used in some toys. *Washington Post,* December 2, 1998:C16.

McBride A. Letter to PK Hill, American Paper Industry. Washington, DC: U.S. Environmental Protection Agency Office of Water, January 13, 1987.

McConnachie PR. Zahalsky AC. Immunological consequences of exposure to pentachlorophenol. *Archives of Environmental Health* 46(4):249–253.

McConnachie PR, Zahalsky AC. Immune alterations in humans exposed to the termiticide technical chlordane. *Archives of Environmental Health* 47:295–301, 1992.

McDonald AD. Work and pregnancy. *British Journal of Industrial Medicine* 45:577–580, 1988.

McDowell M, Balarajan R. Testicular cancer and employment in agriculture. *Lancet* 1:510–511, 1984.

MCEA (Minnesota Center for Environmental Alternatives). PVC alternatives (fact sheet). Minneapolis, 1997.

McGhee WD, Paster M, Riley D, Ruettimann K, Solodar J, Waldman T. Generation of organic isocyanates from amines, carbon dioxide, and electrophilic dehydrating agents: Use of o-sulfobenzoic acid anhydride. In: Anastas PT, Williamson TC, eds. *Green Chemistry: Designing Chemistry for the Environment.* Washington, DC: American Chemical Society, 1996:49–58.

McGregor DB, Partensky C, Wilbourn J, Rice JM. An IARC evaluation of polychlorinated dibenzo-p-dioxins and polychlorinated dibenzofurans as risk factors in human carcinogenesis. *Environmental Health Perspectives* 106 (supp 2):755–760, 1998.

McKinney JD. Interactive hormonal activity of chemical mixtures. *Environmental Health Perspectives* 105:896–897, 1997.

McKinney JD, Pedersen LG. Do residue levels of polychlorinated biphenyls (PCBs) in human blood produce mild hypothyroidism? *Journal of Theoretical Biology* 129:231–241, 1987.

McLachlan JA. Functional toxicology: A new approach to detect biologically active xenobiotics *Environmental Health Perspectives* 101:386–387, 1993.

McLachlan JA, Newbold RR. Cellular and molecular mechanisms of cancers of the uterus in animals. In: Huff J, Boyd J, Barrett JC, eds. *Cellular and Molecular Mechanism of Hormonal Carcinogenesis: Environmental Influences.* New York: Wiley-Liss, 1996:175–182.

McMaster ME, Van Der Kraak GJ, Munkittrick KR. An epidemiological evaluation of the biochemical basis for steroid hormonal depressions in fish exposed to industrial wastes. *Journal of Great Lakes Research* 22:153–171, 1996.

McMurray J. *Organic Chemistry,* Third Edition. Belmont, CA: Brooks/Cole, 1992.

Meadows DG. Montseras going 100% TCF as it increases production and minimizes emissions. *Tappi Journal* 78:49–52, 1995.

Mears CL. Where will the chlorine industry be in ten years? (presentation materials) The Changing Chlorine Marketplace: Business Science and Regulations (conference), New Orleans, April 1995.

Meeker D. Summary of points—questions and discussions. *Journal of Vinyl Technology* 16:149–154, 1994.

Mehendale HM. Amplified interactive toxicity of chemicals at nontoxic levels: Mechanistic considerations and implications to public health: Mechanism-based predictions of interactions. *Environmental Health Perspectives* 102 (supp 9):139–150, 1994.

Meier JR. Genotoxic activity of organic chemicals in drinking water. *Mutation Research* 196:211–245, 1988.

Melius JM, Lewis-Michl EL, Kallenbach LR, Ju CL, Talbot TO, Orr MF, Lauridsen PE. *Residence Near Industries and High Traffic Areas and the Risk of Breast Cancer on Long Island.* Albany: New York State Department of Health, 1994.

Melnick RL, Boorman GA, Dellarco V. Water chlorination, 3-chloro-4-(dichloromethyl)-5-hydroxy-2(5H)-furanone (MX), and potential cancer risk. *Journal of the National Cancer Institute* 89:832–837, 1997.

Merrell P, Van Strum C. *No Margin of Safety: A Preliminary Report on Dioxin Pollution and the Need for Emergency Action in the Pulp and Paper Industry.* Toronto: Greenpeace, 1987.

Mes J, Marchand L. Comparison of some specific polychlorinated biphenyl isomers in human and monkey milk. *Bulletin of Environmental Contamination and Toxicology* 39:736–742, 1987.

Mes J, Weber D. Non-orthochlorine substituted coplanar polychlorinated biphenyl congeners in Canadian adipose tissue, breast milk and fatty foods. *Chemosphere* 19:1357–1365, 1989.

Mes J, Davies DJ, Turton D. Polychlorinated biphenyl and other chlorinated hydrocarbon residues in adipose tissue of Canadians. *Bulletin of Environmental Contamination and Toxicology* 28:97–104, 1982.

Mes J, Marchand L, Davies DJ. Organochlorine residues in adipose tissue of Canadians. *Bulletin of Environmental Contamination and Toxicology* 45:681–688, 1990.

Metcalf R. Benefit risk considerations in the use of insecticides. The Robert Van den Bosch Memorial Lectures, Berkeley, CA, Division of Biological Control and Department of Conservation and Resource Studies, University of California, November 1989.

Miles AK, Calkins DG, Coon NC. Toxic elements and organochlorines in Harbor Seals (*Phoca vitulina richardsi*), Kodiak, Alaska, USA. *Bulletin of Environmental Contamination and Toxicology* 48:727–732, 1992.

Miller AB, Bates D, Chalmers T, Coye MJ, Froines J, Hoel D, Schwartz J, Schulte P, Davis DL, Poore LM, Adams P. *Environmental Epidemiology: Public Health and Hazardous Wastes.* Washington, DC: National Academy Press, 1991.

Miller BA, Feuer EJ, Hankey BF. Recent incidence trends for breast cancer in women and the relevance of early detection: An update. *CA:A Cancer Journal for Clinicians* 43:27–41, 1993.

Miller LJ, Uhler AD. Volatile halocarbons in butter: Elevated tetrachloroethylene levels in samples obtained in close proximity to dry cleaning establishments. *Bulletin of Environmental Contamination and Toxicology* 41:469–474, 1988.

Miller R, Lewis SJ. Orderly transition required for any chemical sunsetting program. *New Solutions* 5(1):55–61, Fall 1994.

Mills PK, Newell GR, Johnson DE. Testicular cancer associated with employment in agricultural and oil and natural gas extraction. *Lancet* 1:207–210, 1984.

Moccia RD, Fox GA, Britton A. A quantitative assessment of thyroid histopathology of herring gulls (*Laurs argentatus*) from the Great Lakes and a hypothesis on the causal role of environmental contaminants. *Journal of Wildlife Diseases* 22:60–70, 1986.

Moldenius S. Pulp quality and economics of ECF vs. TCF. International Non-chlorine Bleaching conference, Amelia Island, Florida, 1995a.

Moldenius S. The introduction of zero chlorine pulp and paper. United Nations University World Congress on Zero Emissions, Tokyo, 1995b.

Mongoven Bioscoe and Duchin. *MBD Update and Analysis: For Chlorine Chemistry Council.* Washington, DC, May 18, 1994.

Mongoven JO. The precautionary principle. *Eco-logic,* March 1995:14.

Montague P. Bill Gaffey's work. *Rachel's Environment and Health Weekly 494,* May 16, 1996.

Montague P. The causes of lymph cancers. *Rachel's Environment and Health Weekly 562,* September 4, 1997.

Montzke SA, Butler JH, Elkins JW, Thompson TM, Clarke AD, Lock Lt. Present and future trends in the atmospheric burden of ozone-depleting halogens. *Nature* 398:690–694, 1999.

Moore K. Organizing integrity: American science and the creation of public interest organizations, 1955–1975. *American Journal of Sociology* 101:1592–1627, 1996.

Morganstern DA, Lelaacheur RM, Morita, DK, Borkowsky SL, Feng S, Brown GH, Luan L, Gross MF, Burk MJ, Tumas W. Supercritical carbon dioxide as a substitute solvent for chemical synthesis and catalysts. In: Anastas PT, Williamson TC, eds. *Green Chemistry: Designing Chemistry for the Environment.* Washington, DC: American Chemical Society, 1996: 132–151.

Morowitz HJ. *Energy Flow in Biology.* New York: Academic Press, 1968.

Morris RD. Drinking water and cancer. *Environmental Health Perspectives* 103(supp 8): 225–232, 1995.

Morris RD, Audet A-M, Angelillo IF, Chalmers TC, Mosteller F. Chlorination, chlorination by-products and cancer: A meta-analysis. *American Journal of Public Health* 82:955–963, 1992.

Morrison H, Saviz D, Semanciw R, Bulka B, Mao Y, Morison L, Wigle D. Farming and prostate cancer mortality. *American Journal of Epidemiology* 137:270–280, 1993.

Mosher WD, Bachrach CA. Understanding U.S. fertility: Continuity and change in the National Survey of Family Growth, 1988–1995. *Family Planning Perspectives* 28:4–12, 1996.

Moss AR, Osmond D, Bacchetti P, Torti FM, Gurgin V. Hormonal risk factors in testicular cancer: A case-control study. *American Journal of Epidemiology* 124:39–52,1986.

Mossing ML, Redetzke KA, Applegate HG. Organochlorine pesticides in blood of persons from El Paso, Texas. *Journal of Environmental Health* 47:312–313, 1985.

Moysich KB, Ambrosone CB, Vena JE, Shields PG, Mendola P, Kostnyiak P, Grezerstein H, Graham S, Marshall JR, Schisterman EF, Freudenheim JL. Environmental organochlorine exposure and postmenopausal breast cancer risk. *Cancer Epidemiology Biomarkers and Prevention* 7:181–188, 1998.

Mughal FH. Chlorination of drinking water and cancer: A review. Journal of Environmental Pathology, *Toxicology and Oncology* 11:287–292, 1992.

Muir CS, Black RJ. Time trends in hormone-dependent cancer. In: Huff J, Boyd J, Barrett JC, eds. *Cellular and Molecular Mechanisms of Hormonal Carcinogenesis: Environmental Influences.* New York: Wiley-Liss, 1996:25–40.

Muir D, Braune B, DeMarch B, Norstrom R, Wagermann R, Gamberg M, Poole K, Addison R, Bright D, Dodd M, Dischenko W, Eamer J, Evans M, Elkin B, Grundy S, Hargrave B, Hebert C, Johnstone R, Kidd K, Koenig B, Lockhart L, Payne J, Peddle J, Reimer K. Ecosystem uptake and effects. In: Jensen J, Adare K, Shearer R (eds.). *Canadian Arctic Contaminants Assessment Report.* Ottawa: Indian and Northern Affairs Canada, 1997.

Muir T, Eder T, Muldoon P, Lerner S. Case study: Application of a virtual elimination strategy to an industrial feedstock chemical–chlorine. In: *A Strategy for Virtual Elimination of Persistent Toxic Substances,* Volume 2. Windsor: International Joint Commission, 1993: 47–69.

Mullen JD, Norton GW, Reaves DW. Economic analysis of environmental benefits of integrated pest management. *Journal of Agricultural and Applied Economics* 29:243–253, 1997.

Mumtaz MM, DeRosa CT, Durkin PR. Approaches and challenges in risk assessments of chemical mixtures. In: Yang RSH, ed. *Toxicology of Chemical Mixtures.* New York: Academic Press, 1994:565–597.

Munger R, Isacson P, Hu S, Burns T, Hanson J, Lynch CF, Cherryholmes K, Van Dorpe P, Hausler WJ. Intrauterine growth retardation in Iowa communities with herbicide-contaminated drinking water supplies. *Environmental Health Perspectives* 105:308–314, 1997.

Munkittrick KR. Recent North American studies of bleached kraft mill impacts on wild fish. In: Sodergren A, ed. *Environmental Fate and Effects of Bleached Pulp Mill Effluents.* Stockholm: Swedish EPA (Report 4031), 1992.

Munkkittrick KR, van der Kraak GJ, McMaster ME, Porrt CB, van der Heuvel MR, Servos MR. Survey of receiving-water environmental impacts associated with discharges from pulp mills. *Environmental Toxicology* 13:1089–1101, 1994.

Murkies AL, Wilcox G, Davis SR. Clinical review 92: Phytoestrogens. *Journal of Clinical Endocrinology and Metabolism* 83:297–303, 1998.

Murphy R, Harvey C. Residues and metabolites of selected persistent halogenated hydrocarbons in blood specimens from a general population survey. *Environmental Health Perspectives* 60:115–120, 1985.

Muskopf JW, McCollister SB. Epoxy resins. In: Gerhartz W, ed. *Ullman's Encyclopedia of Industrial Chemistry.* New York: VCH Publishers, A9:547–564, 1987.

Mussalo-Rauhamaa H, Hasanen E, Pyysalo H, Kauppila AR, Pantzar P. Occurrence of beta-hexachlorocyclohexane in breast cancer patients. *Cancer* 66:2124–2128, 1990.

Mutti A, Franchini I. Toxicity of metabolites to dopaminergic systems and the behavioral effects of organic solvents. *British Journal of Industrial Medicine* 44:721–732, 1987.

Mutti A, Smargiassi A. Selective vulnerability of dopaminergic systems to industrial chemicals: Risk assessment of related neuroendocrine changes. *Toxicology and Industrial Health* 14(1–2):311–323, 1998.

Myreen B. Pulp and paper manufacture in transition. *Water Science and Technology* 29:1–9. 1994.

Nader F. Significance of chlorine chemistry to society. In: *Proceedings–Chlorinated Organic Chemicals: Their Effect on Human Health and the Environment.* Berlin: Toxicology Forum, 1994:87–117.

Nagato N. Allyl alcohols and derivatives. In: *Kirk-Othmer Encyclopedia of Chemical Technologies,* Fourth Edition. New York: Wiley, 2:144–160, 1991.

Najem GR, Greer TW. Female reproductive organs and breast cancer mortality in New Jersey counties and the relationship with certain environmental variables. *Preventive Medicine* 14:620–635, 1985.

Nakata H, Tanabe S, Tatsukawa R, Amano M, Miyazaki N, Petrov EA. Persistent organochlorines residues and their accumulation kinetics in Baikal seal (*Phoca sibrica*) from Lake Baikal, Russia. *Environmental Science and Technology* 29:2877–2885, 1995.

Nakhla AM, Khan MS, Romas NP, Rosner W. Estradiol causes the rapid accumulation of cAMP in human prostate. *Proceedings of the National Academy of Sciences* (USA) 91:5402–5405, 1994.

NASA. *Technology from Apollo Era Provides Fast, Efficient Water Purification.* Washington, DC: National Air and Space Administration, 1996 (available at http://www.jsc.nasa.gov/pao/spinoffs/waterpur.html).

Nebert DW, Gonzales FJ. P450 genes: Structure, evolution and regulation. *Annual Review of Biochemistry* 56:945–993, 1987.

Neidleman S, Geigert J. *Biohalogenation: Principles, Basic Roles, and Applications.* Chichester: Ellis Harwood, 1986.

Neilson AD, Allard AS, Hynning PA, Remberger M. Distribution, fate, and persistence of organochlorine compounds formed during production of bleached pulp. *Toxicological and Environmental Chemistry* 30:3–41, 1991.

Nesaretnam K, Corcoran D, Dils RR, Darbre P. 3,4,3′4-tetrachlorobiphenyl acts as an estrogen in vitro and in vivo. *Molecular Endocrinology* 10:923–926, 1996.

Neubert R, Jacob-Muller U, Helge H, Stahlmann R, Neubert D. Polyhalogenated dibenzo-p-dioxins and dibenzofurans and the immune system: 2. In vitro effects of 2,3,7,8-tetrachlorodibenzo-p-dioxin (TCDD) on lymphocytes of venous blood from man and a non-human primate (Callithrix jacchus). *Archives of Toxicology* 65:213–219, 1991.

Neubert R, Golor G, Stahlmann R, Helge H, Neubert D. Polyhalogenated dibenzo-p-dioxins and dibenzofurans and the immune system: 4. effects of multiple-dose treatment with 2,3,7,8-tetrachlorodibenzo-p-dioxin (TCDD) on peripheral lymphocyte subpopulations of a non-human primate. *Archives of Toxicology* 66:250–259, 1992.

New York Times. Toys R Us halting sale of some infant products. *New York Times,* November 14, 1998:C4.

Newbold R. Cellular and molecular effects of developmental exposure to diethylstilbestrol: Iimplications for other environmental estrogens. *Environmental Health Perspectives* 103 (supp 7):83—87, 1995.

Newbold RR, Hanson RB, Jefferson WN, Bullock BC, Haseman J, McLachlan JA. Increased tumors but uncompromised fertility in the female descendants of mice exposed developmentally to diethylstilbestrol. *Carcinogenesis* 19:1655–1663, 1998.

Nicholson WJ, Landrigan PJ. Human health effects of polychlorinated biphenyls. In: Schecter A, ed. *Dioxins and Health.* New York: Plenum, 1994:487–524.

Nimroody R. The S.D.I. drain: strategic defense initiative costs. *Nation,* January 16, 1988:41–42.

Nishikawa H, Katami, T, Takahara, Y, Sumida H, Yasuhara A. Emission of organic compounds by combustion of waste plastics involving vinyl chloride polymer. *Chemosphere* 25: 1954–1960, 1993.

Noble DF. *America by Design: Science, Technology, and the Rise of Corporate Capitalism.* New York: Oxford University Press, 1977.

Nollkaemper A. What you risk reveals what you value, and other dilemmas encountered in the legal assault on risks. In: Freestone D, Hey E, eds. *The Precautionary Principle and International Law: The Challenge of Implementation.* Boston: Kluwer Law International, 1996:73–96.

Nomura AMY, Kolonel LN. Prostate cancer: A current perspective. *American Journal of Epidemiology* 13:200–227, 1991.

Norheim G, Skaare JU, Wiig O. Some heavy metals, essential elements, and chlorinated hydrocarbons in polar bear (*Ursus maritimus*) at Svalbard. *Environmental Pollution* 77:51–57, 1992.

Norman JE, Robinette CD, Fraumeni JF. The mortality experience of army World War II chemical processing companies. *Journal of Occupational Medicine* 23:819–22, 1981.

Norstrom RJ, Muir DCG. Chlorinated hydrocarbon contaminants in arctic marine mammals. *Science of the Total Environment* 154:107–128, 1994.

Norstrom R, Gilman A, Hallet D. Total organically-bound chlorine and bromine in Lake Ontario herring gull eggs, 1977, by instrumental neutron activation and chromatographic methods. *Science of the Total Environment* 20:217–230, 1981.

Norstrom RJ, Simon M, Muir DCG. Polychlorinated dibenzo-p-dioxins and dibenzofurans in marine mammals of the Canadian North. *Environmental Pollution* 66:1–19, 1990.

NRC (National Research Council). *Toxicity Testing: Strategies to Determine Needs and Priorities.* Washington, DC: National Academy Press, 1984.

NRC (National Research Council). *Drinking Water and Health: Disinfectants and Disinfectant By-Products,* Volume 7. Washington, DC: National Academy Press, 1987.

O'Malley BW, Tsai-MJ. Overview of the steroid receptor superfamily of gene regulatory proteins. In: Parker ML, ed. *Steroid Hormone Action.* Oxford: IRL Press/Oxford University Press, 1992:45–63.

O'Riordan T, Cameron J. The history and contemporary significance of the precautionary principle. In: O'Riordan T, Cameron J, eds. *Interpreting the Precautionary Principle.* London: Earthscan Publications, 1994:12–30.

Oberg L. Peroxidase catalyzed oxidation of chlorophenols to polychlorinated dibenzo-p-dioxins and dibenzofurans. *Archives of Environmental Contamination and Toxicology* 19: 930–938, 1990

Oberg, LG, Wagman N, Andersson R, Rappe C. De novo formation of PCDD/F in compost and sewage sludge—a status report. *Organohalogen Compounds* 11:297–302, 1993.

OECD (Organisation for Economic Cooperation and Development). *Sustainable Agriculture: Concepts, Issues and Policies in OECD Countries.* Paris: OECD, 1995.

Oehme M. Dispersion and transport paths of toxic persistent organochlorines to the arctic—levels and consequences. *Science of the Total Environment* 106:43–53, 1991.

Oehme M, Furst P, Kruger C, Meemken H, Groebel W. Presence of polychlorinated dibenzo-p-dioxins, dibenzofurans and pesticides in arctic seal from Spitzbergen. *Chemosphere* 17: 1291–1300, 1988.

Oehme M, Mano S, Bjerke B. Formation of poly-chlorinated dibenzofurans and dibenzo-p-dioxins by production processes for magnesium and refined nickel. *Chemosphere* 18:1379–1389, 1989.

Oesch F, Oesch-Bartlomowicz, Arens J, Fahndrich F, Vogel E, Friedberg T, Glatt H. Mechanism-based predictions of interactions. *Environmental Health Perspectives* 102 (supp 9): 5–10, 1994.

Olin GR. The hazards of a chemical laboratory environment—a study of the mortality in two cohorts of Swedish chemists. *American Industrial Hygiene Association Journal* 39:557–562, 1978.

Olin GR, Ahlbom A. The cancer mortality among Swedish chemists graduated during three decades. *Environmental Research* 22:154–161, 1980.

Olsen GW, Bodner KM, Ramlow JM, Ross CE, Lipshultz LI. Have sperm counts been reduced 50 percent in 50 years? A statistical model revisited. *Fertility and Sterility* 63:887–893, 1995.

Olsson M, Karlsson B, Ahnland E. Diseases and environmental contaminants in seals from the Baltic and Swedish west coast. *Science of the Total Environment* 154:217–227, 1994.

Ono M, Kanna N, Wakimoto T, Tatsukawa R. Dibenzofurans: A greater global pollutant than dioxins? Evidence from analyses of open ocean killer whale. *Marine Pollution Bulletin* 18:640–643, 1987.

Onodera S, Ogawa M, Yamawaki C, Yamagishi K, Suzuki S. Production of polychlorinated phenoxyphenols predioxins by aqueous chlorination of organic compounds. *Chemosphere* 19:675–680, 1989.

Onstot J, Ayling R, Stanley J. *Characterization of HRGC/MS Unidentified Peaks from the Analysis of Human Adipose Tissue.* Volume 1: *Technical Approach.* Washington, DC: U.S. Environmental Protection Agency Office of Toxic Substances (560/6-87-002a), 1987.

Onstot J, Stanley J. *Identification of SARA Compounds in Adipose Tissues.* Washington, DC: U.S. Environmental Protection Agency Office of Toxic Substances (EPA-560/5-89-003), 1989.

Oppelt T. Incineration: A critical review. *Journal of the Air Pollution Control Association* 37:556–586, 1986.

Ott MG, Olson RA, Cook RR. Cohort mortality study of chemical workers with potential exposure to the higher chlorinated dioxins. *Journal of Occupational Medicine* 29:422–429, 1987.

Ovid. *Metamorphoses.* Innes M, trans. New York: Penguin, 1955.

Owen W. Basis for removing biologically active persistent toxic substances. In: Gilbertson M, Cole-Misch S, eds. *Applying Weight of Evidence: Issues and Practice.* Windsor: International Joint Commission, 1994.

Paasivirta J. *Chemical Ecotoxicology.* Chelsea, MI: Lewis, 1991.

Pacyna JM, Munch J. Anthropogenic mercury emission in Europe. *Water, Air, and Soil Pollution* 56:51–61, 1991.

Pajarinen J, Laippala P, Penttila A, Karhunen PJ. Incidence of disorders of spermatogenesis in middle aged Finnish men, 1981–91: Two necropsy series. *British Medical Journal* 314: 13–18, 1997.

PAN (Pesticide Action Network of North America). Denmark and Sweden move to further restrict pesticides. PANUPs Updates Service, March 18, 1996.

PAN (Pesticide Action Network of North America). Demise of the Dirty Dozen, 1995 Chart. San Francisco, 1997.

Panda KK, Lenka M, Panda BB. Monitoring and assessment of mercury pollution in the vicinity of a chlor-alkali plant. *Science of the Total Environment* 96:281–296, 1990.

Paper. Fewer fish with bigger livers; effects of chlorinated organic material in pulp mill effluent on aquatic life. *Paper* 218(7):14, July 1993.

Papke O. PCDD/F: Human background data for Germany, a 10-year experience. *Environmental Health Perspectives* 106 (supp 2):723–731, 1998.

Papp R. Organochlorine waste management. *Pure and Applied Chemistry* 68:1801–1808, 1996.

Paracelsus. *Selected Writings.* Jacobi J, ed. New York: Pantheon, 1951.

Parrotta MJ, Bekdash F. UV disinfection of small groundwater supplies. *American Water Works Association Journal* 90(2):71–81, 1998.

Patandin S, Dagnelie PC, Mulder PGH, de Coul EO, van der Veen JE, Weisglas-Kuperus N, Sauer PJJ. Dietary exposure to polychlorinated biphenyls and dioxins from infancy until adulthood: A comparison between breast-feeding, toddler, and long-term exposure. *Environmental Health Perspectives* 107:45–51, 1999.

Patel T. Water without the whiff of chlorine. *New Scientist,* November 7, 1992.

Patton GW, Walla MD, Bidleman TF, Barrie LA. Polycyclic aromatic and organochlorine compounds in the atmosphere of northern Ellesmere Island, Canada. *Journal of Geophysical Research* 96 (D6):10,867–10,877, 1991.

Paulozzi LJ. International trends in rates of hypospadias and cryptochidism. *Environmental Health Perspectives* 107:297–302, 1999.

Paulsen M. Adventures in chemical valley. *Detroit Metro-Times,* February 18, 1993:8–13.

Pearson SD, Trissel LA. Leaching of diethylhexyl phthalate from polyvinyl chloride containers by selected drugs and formulation components. *American Journal of Hospital Pharmacology* 50:1405–1409, 1993.

Pellizzari ED, Hartwell TD, Harris BS, Waddell RD, Whitaker DA, Erickson MD. Purgeable organic compounds in mother's milk. *Bulletin of Environmental Contamination and Toxicology* 28:322–328, 1982.

Pesek J, Brown S, Clancy KL, Coleman DC, Fluck RC, Goodman RM, Harwood R, Heffernan WD, Helmers GA, Hildebrand PE, Lockeretz W, Miller RH, Pimentel D, Qualset CO, Raun NS, Reynolds JT, Schiroth MN, Wiles R, Mason SE. *Alternative Agriculture: Report of the National Research Council Committee on the Role of Alternative Farming Methods in Modern Production Agriculture.* Washington, DC: National Academy Press, 1989.

Peterson RE, Moore RW, Mably TA, Bjerke DL, Goy RW. Male reproductive system ontogeny: Effects of perinatal exposure to 2,3,7,8-tetrachlorodibenzo-p-dioxin. In: Colborn T, Clement C, eds. *Chemically-Induced Alterations in Sexual and Functional Development: The Wildlife/Human Connection.* Princeton: Princeton Scientific Publishing, 1992:175–194.

Phuong NTN, Thuy TT, Phuong PK. An estimate of differences among women giving birth to deformed babies and those with hydatidiform mole seen at the Ob-Gyn Hospital of Ho Chi Minh City in the south of Vietnam. *Chemosphere* 19:801–803, 1989a.

Phuong NTN, Thuy TT, Phuong PK. An estimate of reproductive abnormalities in women inhabiting herbicide sprayed and non-herbicide sprayed areas in the South of Vietnam, 1952–1981, *Chemosphere* 18:843–846, 1989b.

Pimentel D, Levitan L. Pesticides: Amounts applied and amounts reaching pests. *BioScience* 36(2):86–91, 1986.

Pimentel D, McLaughlin L, Zepp A, Lagitan B, Kraus T, Kleinman P, Vancini F, Roach WJ, Graap E, Keeton WS, Selig G. Environmental and economic effects of reducing pesticide use. *BioScience* 41:402–409, 1991.

Pines A, Cucos S, Ever-Hadani P, Ron M. Some organochlorine insecticide and polychlorinated biphenyl blood residues in infertile males in the general Israeli population of the middle 1980s. *Archives of Environmental Contamination and Toxicology* 16:587–597, 1987.

Pinoir J. Caustic usage in alumina production (presentation materials). Third World Chlor-Alkali Symposium, Monte Carlo, 1992.

Pitot HC, Dragan YP. Facts and theories concerning the mechanisms of carcinogenesis. *FASEB Journal* 5:2280–2286, 1991.

Plimke E, Schessler R, Kaempf K. *Konversion Chlorchemie.* Wiesbaden: Prognos AG for Hessiches Umweltministerium, 1994.

Plimmer A. Photochemistry of dioxins. In: Blair E, ed. *Chlorodioxins: Origin and Fate. Advances in Chemistry* 120:50–56. Washington, DC: American Chemical Society, 1973.

Plonait SL, Nau H, Maier RF, Wittfoht W, Obladen M. Exposure of newborn infants to di-(2-ethylhexyl)-phthalate and 2-ethylhexanoic acid following exchange transfusion with polyvinyl chloride catheters. *Transfusion* 33:598–605, 1993.

Pluim HJ, de Vijlder JJM, Olie K, Kok JH, Vulsma T, van Tijn DA, van der Slikke JW, Koppe JG. Effects of pre-and postnatal exposure to chlorinated dioxins and furans on human neonatal thyroid hormone concentrations. *Environmental Health Perspectives* 101:504–508, 1993.

Plumacher J, Schroder P. Accumulation and fate of C2-chlorocarbons and trichloroacetic acid in spruce needles from an Austrian mountain site. *Organohalogen Compounds* 14:301–302, 1993.

Pohjanvirta R, Laitinen JT, Vakkuri O, Linden J, Kokkola T, Unkila M, Tuomisto J. Mechanism by which 2,3,7,8-tetrachlorodibenzo-p-dioxin (TCDD) reduces circulating melatonin levels in the rat. *Toxicology* 107:85–97, 1996.

Pollack R. *The Missing Moment: How the Unconscious Shapes Modern Science.* Boston: Houghton Mifflin, 1999.

Popper KR. The problem of induction (1953). In: Miller D, ed. *Popper Selections.* Princeton: Princeton University Press, 1985:101–117.

Popper KR. *The Logic of Scientific Discovery.* New York: Routledge, 1992.

Porter GE. Chlorine—an introduction. *Pure and Applied Chemistry* 68:1683–1687, 1996.

Porterfield SP. Vulnerability of the developing brain to thyroid abnormalities: Environmental insults to the thyroid system. *Environmental Health Perspectives* 102 (supp 2):125–130, 1994.

Portier CJ, Sherman CD. Potential effects of chemical mixtures on the carcinogenic process within the context of the mathematical multistage model. In: Yang RSH, ed. *Toxicology of Chemical Mixtures.* New York: Academic Press, 1994:665–687.

Portier CJ, Sherman CD, Kohn M, Edler L, Kopp-Schneider A, Maronpot RM, Lucier G. Modeling the number and size of hepatic focal lesions following exposure to 2,3,7,8-TCDD. *Toxicology and Applied Pharmacology* 138:20–30, 1996.

PR Watch. Women and children first: On the front line of the chlorine war. *PR Watch* 3(2):1–12, 1996.

Pretty JN. *Regenerating Agriculture: Policies and Practice for Sustainability and Self-Reliance.* London: Earthscan, 1995.

PTCN. Pesticides "possibly contaminated with dioxin" list compiled in OPP. *Pesticide and Toxic Chemical News* 1315:34–38, February 20, 1985.

Public Works. Evaluting slow sand filtration. *Public Works* 122(4):104–105, 1991.

Pujol P, Hilsenbeck SG, Chamness GC, Elledge RM. Rising levels of estrogen receptor in breast cancer over 2 decades. *Cancer* 74:1601–1606, 1994.

Rachootin P, Olsen J. The risk of infertility and delayed conception associated with exposures in the Danish workplace. *Journal of Occupational Medicine* 253:394–402, 1983.

Ramacci R, Ferrari G, Bonamin V. Sources of PCDDs/PCDFs from industrial and municipal waste water discharges and their spatial and temporal distribution in the sediments of the Venice lagoon. *Organohalogen Compounds* 39:91–95, 1998.

Rantio T. Chlorinated cymenes in effluents of two Finnish pulp mills in 1990–1993. *Chemosphere* 31:3413–3423, 1995.

Rapaport D. Aerobic composting toilets for tropical environments. *Biocycle* 37(7):77–82, 1996.

Rappe C. Environmentally stable chlorinated contaminants from the pulp and paper industry. In: Vaino H, Sorsa M, McMichael A, eds. *Complex Mixtures and Cancer Risk*. Lyon: International Agency for Research on Cancer, 1990:341–353.

Rappe C, Marklund S. Formation of polychorinated dibenzo-p-dioxins and dibenzofurans by burning or heating chlorophenates. *Chemosphere* 8:269–281, 1978.

Rappe C, Swanson S, Glas B, Kringstadt K, de Sousa P, Abe Z. Formation of PCDDs and PCDFs by the chlorination of water *Chemosphere* 19:1875–1880, 1989.

Rappe C, Andersson R, Lunstrom K, Wiberg K. Levels of polychlorinated dioxins and dibenzofurans in commercial detergents and related products. *Chemosphere* 21:43–50, 1990.

Rappe C, Kjeller LO, Kulp SE. Levels, profile and pattern of PCDDs and PCDFs in samples related to the production and use of chlorine. *Chemosphere* 23:1629–1636, 1991.

Raven PH, Johnson GB. *Biology,* Fourth Edition. Dubuque: William C. Brown, 1996.

Regan MB. A town's lifeblood—and polluter; pulp mill gave jobs to residents and chemicals to river. *Orlando Sentinel Tribune*. March 26, 1992:A1.

Reijnders PJH. Reproductive failure in common seals feeding on fish from polluted coastal waters. *Nature* 324:456–457, 1986.

Reilly W. Aiming before we shoot (speech). National Press Club, Washington, DC, September 26, 1990.

Renner M. *Jobs in A Sustainable Economy.* Washington, DC: Worldwatch Institute, 1991.

Repetto R, Baliga SS. *Pesticides and the Immune System: The Public Health Risks*. Washington, DC: World Resources Institute, 1996.

Rice DC. Behavioral impairment produced by low-level postnatal PCB exposure in monkeys. *Environmental Research* A80:S113–S121, 1999.

Rice DC, Evangelista de Duffard AM, Duffard R, Iregren A, Satoh H, Watanabe C. Lessons for neurotoxicology from selected model compounds: SGOMSEC joint report. *Environmental Health Perspectives* 104 (supp 2):205–215, 1996.

Rice EW, Scarpino PV, Reasoner DH, Logsdon GS, Wild DK. Correlation of coliform growth response with other water quality paramaters. *Journal of the American Water Works Association*. July 1991:98–102.

Rice P. Ozone oxidation productions—implications for drinking water treatment. In: Larson R, ed. *Biohazards of Drinking Water Treatment*. Chelsea, MI: Lewis Publishers, 1989:155–166.

Rickenbacher U, McKinney JD, Oatley SJ, Blake CC. Structurally specific binding of halogenated biphenyls to thyroxine transport protein. *Journal of Medical Chemistry* 29:641–648, 1986.

Rier SE, Martin DC, Bowman RE, Dmowski WP, Becker JL. Endometriosis in rhesus monkeys (*Macaca mulatta*) following chronic exposure to 2,3,7,8-tetrachlorodibenzo-p-dioxin. *Fundamental and Applied Toxicology* 21:433–441, 1993.

Rier SE, Martin DC, Bowman RE, Becker JL. Immunoresponsiveness in endometriosis: Implications of estrogenic toxicants. *Environmental Health Perspectives* 103 (supp 7):151–156, 1995.

Ries LAG, Kosary CL, Harkey BF, Miller BA, Harris, Edwards BK, eds. *SEER Cancer Statistics Review 1973–1994*. Bethesda: National Cancer Institute (NIH 97-2789), 1997.

Riesenberg F, Walters BB, Steele A. Slow sand filters for a small water system. *American Water Works Association Journal* 87(11): 48–56, 1995.

Rigel DS, Friedman RJ, Kopf AW. The incidence of malignant melanoma in the United States: Issues as we approach the 21st century. *Journal of the American Academy of Dermatology* 34:839–847, 1996.

Riggle D. Technology improves for composting toilets. *Biocycle* 37(4):39–43, 1996.

Rigo H, Chandler A, Lanier W . *The Relationship Between Chlorine in Waste Streams and Dioxin Emissions from Combustors.* New York: American Society of Mechanical Engineers, 1995.

Rivetti F, Romano U, Delledonne D. Dimethylcarbonate and its production technology. In: Anastas PT, Williamson TC, eds. *Green Chemistry: Designing Chemistry for the Environment.* Washngton, DC: American Chemical Society, 1996:70–80.

Robbins D, Johnston R. The role of cognitive and occupational differentiation in scientific controversies. In: Collins HM, ed. *Sociology of Scientific Knowledge: A Source Book.* Bath: University of Bath Press, 1982:219–238.

Roberts EA, Johnson KC, Dippold WG. Ah receptor mediating induction of cytochrome p4501A1 in a novel continuous human liver cell line (Mz-Hep-1). *Biochemical Pharmacology* 15:521–528, 1991.

Roberts G. Chlorine companies adopt defiant stand. *European Chemical News,* February 14, 1994:32–34.

Roe D, Pease W, Florini K, Silbergeld E, Leiserson K, Below C, Chang J, Abercrombie D. *Toxic Ignorance: The Continuing Absence of Basic Health Testing for Top-Selling Chemicals in the United States.* New York: Environmental Defense Fund, 1997.

Rogan WJ, Gladen BC, McKinney JD, Carreras N, Hardy P, Thullen J, Tingelstad J, Tully BA. Neonatal effects of transplacental exposure to PCBs and DDE. *Journal of Pediatrics* 109:335–341, 1986.

Rogan WJ, Gladen BC, Hung KL, Koong SL, Shih LY, Taylor JS, Wu YC, Yang D, Ragan B, Hsu CC. Congenital poisoning by polychlorinated biphenyls and their contaminants in Taiwan *Science* 241:334–336, 1988.

Rogan WJ, Gladen BC. PCBs, DDE, and child development at 18 and 24 months. *Annals of Epidemiology* 1:407–413, 1991.

Rogan WJ, Gladen BC. Neurotoxicology of PCBs and related compounds. *Neurotoxicology* 13:27–36, 1992.

Rohleder F. Dioxins and cancer mortality: Reanalysis of the BASF cohort (presentation text). Ninth International Symposium on Chlorinated Dioxins and Related Compounds, Toronto, September 17–22, 1989.

Rose DP, Boyar AP, Wynder EL. International comparisons of mortality rates for cancer of the breast, ovary, prostate and colon and per capita food consumption. *Cancer* 58: 2363–2371, 1986.

Rose G. Environmental factors and disease: The man made environment. *British Medical Journal* 294:963–965, 1987.

Rosenberg C, Kontsas H, Jappinen P, Tornaeus J, Hesso A, Vainio H. Airborne chlorinated dioxins and furans in a pulp and paper mill. *Chemosphere* 29:1971–1978, 1994.

Rosenthal GA. The chemical defenses of higher plants. *Scientific American* 254(1):94–99, 1986.

Rossberg M, Lendle W, Togel A, Dreher EL, Langer E, Rassaerts H, Kleinshmidt P, Strack H, Beck U, Lipper KA, Torkelson T, Loser E, Beutel KK. Chlorinated hydrocarbons. In: Gerhartz W, ed. *Ullman's Encyclopedia of Industrial Chemistry,* Fifth Edition. New York: VCH Publishers, A6:233–398, 1986.

Rossi M, Ellenbacker M, Geiser K. Techniques in toxics use reduction. *New Solutions* 2(2): 25–32, 1991.

Roth J, Grunfeld C. Mechanisms of action of peptide hormones and catecholamines. In: Wilson JD, Foster DW, eds. *Textbook of Endocrinology,* Seventh Edition. Philadelphia: Saunders, 1985:33–75.

Roth M. Effects of short chain aliphatic halocarbons on invertebrates of terrestrial and aquatic ecosystems. *Organohalogen Compounds* 14:285–286, 1993.

Rothman KJ, Greenland S. Causation and Causal Inference. In: Rothman K, Greenland S, eds. *Modern Epidemiology,* Second Edition. New York: Lippincott, 1998:7–28.

Rothmann N, Cantor KP, Blair A, Bush D, Brock JW, Helzlouer K, Zahm SH, Needham LL, Pearson GR, Hoover RN, Comstock GW, Strickland PT. A nested case-control study of non-Hodgkin's lymphoma and serum organochlorine residues. *Lancet* 350:240–244, 1997.

Ruckleshaus W. Science, risk and public policy. *Science* 221:1026–1028, 1983.

Rudell R. Predicting health effects of exposures to compounds with estrogenic activity: Methodological issues. *Environmental Health Perspectives* 105 (supp 3):655–664, 1997.

Ruder AM, Ward EM, Brown DP. Cancer mortality in female and male dry-cleaning workers. *Journal of Occupational Medicine* 36:867–874, 1994.

Russell DH, Buckley AR, Shah GN, Sipes IG, Blask DE, Benson B. Hypothalamic site of action of 2,3,7,8-tetrachlorodibenzo-p-dioxin (TCDD). *Toxicology and Applied Pharmacology* 94:496–502, 1988.

Russell M, Colglazier W, Tonn BE. The U.S. hazardous waste legacy. *Environment* 34(6): 13–39, 1992.

Ruuskanen J, Vartiainen T, Kojo L, Manninen H, Oksanen J, Frankenhauswer M. Formation of polychlorinated dibenzo-p-dioxins and dibenzofurans in co-combustion of mixed plastics with coal: Exploratory principal component analtysis. *Chemosphere* 28:1989–1999, 1994.

Rylander L, Hagmar L. Mortality and cancer incidence among women with a high consumption of fatty fish contaminated with persistent organochlorine compounds. *Scandinavian Journal of Work and Environmental Health* 21:419–426, 1995.

Rylander L, Stromberg U, Hagmar L. Decreased birthweight in infants born to women with a high dietary intake of fish contaminated with persistent organochlorine compounds. *Scandinavian Journal of Work and Environmental Health* 21:368–375, 1995.

Rylander L, Stromberg U, Hagmar L. Dietary intake of fish contaminated with persistent organochlorine compounds in relation to low birthweight. *Scandinavian Journal of Work and Environmental Health* 22:260–266, 1996.

Rylander L, Stromberg U, Dyremark E, Ostman C, Nilsson-Ehle P, Hagmar L. Polychlorinated biphenyls in blood plasma among Swedish fish consumers in relation to low birth weight. *American Journal of Epidemiology* 147:493–502, 1998.

Safe S, Safe L, Mullin M. Polychlorinated biphenyls: Congener-specific analysis of a commercial mixture and a human milk extract. *Journal of Agricultural and Food Chemistry* 33: 24–29, 1985.

Safe SH. Modulation of gene expression and endocrine response pathways by 2,3,7,8-tetrachlorodibenzo-p-dioxin and related compounds. *Pharmacology and Therapeutics* 67: 247–281, 1995.

Sager D, Girard D, Nelson D. Early postnatal exposure to PCBs: Sperm function in rats. *Environmental Toxicology and Chemistry* 10:737–746, 1991.

Sakai, S, Hiraoda M, Takeda N, Shiozaki K. Coplanar PCBs and PCDDs/PCDFs in municipal waste incineration. *Organohalogen Compounds* 9:215–219, 1992.

Sanders G, Eisenreich SJ, Jones KC. The rise and fall of PCBs: Time-trend data from temperate industrialized countries. *Organohalogen Compounds* 14:93–96, 1993.

Sapers J. Seeking a chemical-free pool. *New York Times,* July 15, 1993:C2.

Savage EP, Keefe TJ, Tessari JD, Wheeler HW, Applehans FM, Goes EA, Ford SA. National study of chlorinated hydrocarbon insecticide residues in human milk, USA: I. geographic distribution of dieldrin, heptachlor, heptachlor epoxide, chlordane, oxychlordane, and mirex. *American Journal of Epidemiology* 113(4):413–422, 1981.

Savitz DA. Re: Breast cancer and serum organochlorines: A prospective study among white, black and Asian women. *Journal of the National Cancer Institute* 86:1255, 1994.

Saxena MC, Siddiqui MKJ, Bhargava AK, Seth TD, Krishnamurti CR, Kutty D. Role of chlorinated hydrocarbon pesticides in abortions and premature labor. *Toxicology* 17:323–331, 1980.

Saxena MC, Siddiqui MKJ, Seth TD, Krishnamurti CR, Bhargava AK, Kutty D. Organochlorine pesticides in specimens from women undergoing spontanous abortion, premature or full-term delivery. *Journal of Analytical Toxicology* 5:6–9, 1981.

Schafer K, McLachlan MS, Reissinger M, Hutzinger O. An investigation of PCDD/F in a composting operation. *Organohalogen Compounds* 11:425–430, 1993.

Schantz SL, Laughlin N, van Valkenberg HC, Bowman RE. Maternal care by rhesus monkeys of infant monkeys exposed to either lead or 2,3,7,8-tetrachlorodibenzo-p-dioxin. *Neurotoxicology* 7:637–650, 1986.

Schantz SL, Levin ED, Bowman RE, Heironimus MP, Laughlin MP. Effects of perinatal PCB exposure on discrimination-reversal learning in monkeys. *Neurotoxicology and Teratology* 11:243–250, 1989.

Schantz SL, Levin ED, Bowman RE. Long-term neurobehavioral effects of perinatal polychlorinated biphenyl (PCB) exposure in monkeys. *Environmental Toxicology and Chemistry* 10:747–756, 1991.

Schapira DV, Kumar NB, Lyman GH. Obesity, body fat distribution, and sex hormones in breast cancer patients. *Cancer* 67:2215–2218, 1991.

Schaum J, Cleverly D, Lorber M, Phillips L, Schweer G. Sources of dioxin-like compounds and background exposure levels. *Organohalogen Compounds* 14:319–326, 1993.

Scheck MN. Catalytic oxychlorination process of aliphatic hydrocarbons as new industrial sources of PCDDs and PCDFs by Evers, Buring, Olie, Govers, University of Amsterdam. Letter from Assistant Director to HM Clayton, Norsk Hydro Polymers Limited, Durham, England. Wayne NJ: Vinyl Institute, January 26, 1990.

Schecter A. Dioxins in humans and the environment. In: *Biological Basis for Risk Assessment of Dioxins and Related Compounds. Banbury Report* 35:169–214, 1991.

Schecter A. Exposure assessment: Measurement of dioxins and related chemicals in human tissues. In: Schecter A, ed. *Dioxins and Health.* New York: Plenum, 1994:449–486.

Schecter A, Gasiewicz T. Health hazard assessment of chlorinated dioxins and dibenzofurans contained in human milk. *Chemosphere* 16:2147–2154, 1987.

Schecter A, Kessler H. Dioxin and dibenzofuran formation following a fire at a plastic storage warehouse in Binghamton, New York in 1995. *Organohalogen Compounds* 27:187–190, 1996.

Schecter A, Dekin A, Weerasinghe NCA, Arghestani S, Gross ML. Sources of dioxins in the environment: A study of PCDDs and PCDFs in ancient frozen eskimo tissue. *Chemosphere* 17:627–631, 1988.

Schecter A, Furst P, Kruger C, Meemken HA, Groebel W, Constable JD. Levels of polychlorinated dibenzofurans, dibenzodioxins, PCBs, DDT, DDE, hexachlorobenzene, dieldrin, hexachlorocyclohexanes and oxychlordane in human breast milk from the United States, Thailand, Vietnam, and Germany. *Chemosphere* 178:445–454, 1989a.

Schecter A, Mes J, Davies D. Polychlorinated biphenyl (PCB), DDT, DDE, and hexachlorobenzene (HCB) and PCDD/F isomer levels in various organs and autopsy tissue from North American patients. *Chemosphere* 18:811–818, 1989b.

Schecter A, Papke O, Ball M. Evidence for transplacental transfer of dioxins from mother to fetus: Chlorinated dioxin and dibenzofurans levels in the livers of stillborn infants. *Chemosphere* 21:1017–1022, 1990.

Schecter A, Papke O, Lis A, Ball M. Chlorinated dioxin and dibenzofuran content in 2,4-D amine salt from Ufa, Russia. *Organohalogen Compounds* 11:325–328, 1993.

Schecter A, Startin J, Wright C, Kelly M, Papke O, Lis A, Ball M, Olson JR. Congener-specific levels of dioxins and dibenzofurans in U.S. food and estimated daily dioxin toxic equivalent intake. *Environmental Health Perspectives* 102:962–966, 1994.

Schecter A, Toniolo P, Dai LC, Thuy TB, Wolff MS. Blood levels of DDT and breast cancer risk among women living in the north of Vietnam. *Archives of Environmental Contamination and Toxicology* 33:453–456, 1997.

Scherr PA, Hutchison GB, Neiman RS. Non-Hodgkin's lymphoma and occupational exposure. *Cancer Research* 52 (supp):5503s–5509s, 1992.

Schmidt JV, Bradfield CA. Ah receptor signaling pathways. *Annual Reviews in Cell and Development Biology* 12:55–89, 1996.

Schmittinger P, Curlin LC, Asawa T, Kotowski S, Beer HB, Greenberg AM, Zelfel E, Breitstadt R. Chlorine. In: W. Gerhartz, ed. *Ullman's Encyclopedia of Industrial Chemistry,* Fifth Edition. New York: VCH Publishers, A6:399–480, 1986.

Scholz B. Determination of polychlorinated dibenzo-p-dioxins and dibenzofurans in wastes of technical hexachlorocyclohexane. *Chemosphere* 16:1829–1834, 1987.

Schreiber JS, House S, Prohobic E, Smead G, Hudson C, Styk M, Lauber J. An investigation of indoor air contamination in residences above dry cleaners. *Risk Analysis* 13:335–344, 1993.

Schroder P. On the uptake and possible detoxification of short-chain aliphatic halocarbons in conifers. *Organohalogen Compounds* 14:277–280, 1993.

Schwartz J. Multinational trends in multiple myeloma. *Annals of the New York Academy of Sciences* 609:215–224, 1990.

Sclove RE. *Democracy and Technology.* New York: Guilford, 1995.

Scott GI, Davis WP, Marcus M, Ballous TG, Dahlin JA. Acute toxicity, sublethal effects, and bioconcentration of chlorinated products, viruses, and bacteria in edible shellfish. In: Jolley RL, Condie LW, Johnson JD, Katz S, Minear RA, Mattice JS, Jacobs VA, eds. *Water Chlorination: Chemistry, Environmental Impact, and Health Effects,* Volume 6. Chelsea, MI: Lewis, pp. 491–519, 1990.

Scott W. *The Lord of the Isles: A Poem.* New York: J. Eastburn and Company, 1818.

Scribner JD, Mottet NK. DDT acceleration of mammary gland tumors induced in the male Sprague-Dawley rat by 2-acetamidophenanthrene. *Carcinogenesis* 2:1236–1239, 1981.

Searle CE, Waterhouse JAH, Henman BA, Bartlett D, McCombie S. Epidemiological study of the mortality of British chemists. *British Journal of Cancer* 38:192–193, 1978.

Seegal RF, Brosch KO, Okoniewski RJ. Effects of in utero and lactational exposure of the laboratory rat to 2,4,2′,4′- and 3,4,3′,4′-tetrachlorobiphenyl on dopamine function. *Toxicology and Applied Pharmacology* 146:95–103, 1997.

Seegal RF, Schantz SL. Neurochemical and behavioral sequelae of exposure to dioxins and PCBs. In: Schecter A, ed. *Dioxins and Health*. New York: Plenum, 1994:409–448.

Semenza JC, Tolbert PE, Rubin CH, Guillette LJ, Jackson RJ. Reproductive toxins and alligator abnormalities at Lake Apopka, Florida. *Environmental Health Perspectives* 105: 1030–1032, 1997.

Seo BW, LI MH, Hansen LG, Moore RW, Peterson RE, Schantz SL. Effects of gestational and lactational exposure to coplanar polychlorinated biphenyl (PCB) congeners or 2,3,7,8-tetrachlorodibenzo-p-dioxin (TCDD) on thyroid hormone concentrations in weanling rats. *Toxicology Letters* 78:253–262, 1995.

SEPA (Swedish Environmental Protection Agency). *Sample reports 610s09 and 0610s017: PVC suspension/PVC plastic and PVC suspension used to make PVC.* Stockholm: Swedish Environmental Protection Agency, May 5, 1994.

SFT (Norwegian State Pollution Control Authority). *Input of Organohalogens to the Convention Area from the PVC Industry.* Submission to the Oslo and Paris Commissions. Oslo, November 1993.

Shamel RE. Where will the chlorine industry be in ten years (presentation materials). Changing Chlorine Marketplace: Business Science and Regulations (conference), New Orleans, April 1995.

Shane BS. Human reproductive hazards: Evaluation and chemical etiology. *Environmental Science and Technology* 23:1187–1195, 1989.

Sherins RJ. Are semen quality and male fertility changing? *New England Journal of Medicine* 332:327, 1995.

Sherman J, Chin B, Huibers PDT, Garcia-Valls R, Hatton TA. Solvent replacement for green processing. *Environmental Health Perspectives* 106 (supp 1):253–271, 1998.

Silbergeld EK. Developing formal risk assessment methods for neurotoxicants: An evaluation of the state of the art. In: Johnson BL, ed. *Advances in Neurobehavioral Toxicology: Applications in Environmental and Occupational Health.* Chelsea, MI: Lewis, 1990:133–149.

Silkworth JB. Love Canal: Development of the toxicologic evaluation of its complex chemical contamination. In: Yang RB, ed. *Toxicology of Chemical Mixtures.* New York: Academic Press, 1994:13–50.

Simic B, Kniewald Z, Davies JE, Kniewald J. Reversibility of the inhibitory effect of atrazine and lindane on cytosol 5-alpha dihydrotestosterone receptor complex formation in rat prostate. *Bulletin of Environmental Contamination and Toxicology* 46:92–99, 1991.

Simmons JE. Nephrotoxicity resulting from multiple chemical exposures and chemical interactions. In: Yang RSH, ed. *Toxicology of Chemical Mixtures.* New York: Academic Press, 1994:335–360.

Simonich SL, Hites RA. Global distribution of persistent organochlorine compounds. *Science* 269:1851–1854, 1995.

Sims RC, Slezak LA. Slow sand filtration: Present practice in the United States. In: Logsdon GS, ed. *Slow Sand Filtration.* New York: American Society of Civil Engineers, 1991:1–18.

Singer PC. Assessing ozonation research needs in water treatment. *Journal of the American Water Works Association* 82(10):78–88,1990.

Singh HB, Salas LH, Shigeishi H, Scriber E. Atmospheric halocarbons, hydrocarbons and sulfur hexafluoride: Global distributions, sources and sinks. *Science* 203:899–903, 1979.

Siuda JF, DeBernardis JF. Naturally occurring halogenated organic compounds. *Lloydia* 36: 107–143, 1973.

Snyder CA. Organic solvents. In: Dean JH, Luster MI, Munson AE, Kimber I, eds. *Immunotoxicology and Immunopharmacology.* New York: Raven Press, 1994:183–189.

Socha A, Abernethy S, Birmingham B, Bloxam R, Fleming S, McLaughlin D, Spry D. Plastimet Inc. Fire, Hamilton Ontario, July 9–12, 1997. Ottawa: Ontario Ministry of Environment and Energy, 1997.

Sodergren A. *Biological Effects of Bleached Pulp Mill Effluents: Final Report from the Environment/Cellulose I Project.* Solna, Sweden: National Swedish Environmental Protection Board, 1989.

Sodergren A, Larsson P, Knulst J, Bergqvist C. Transport of incinerated organochlorine compounds to air, water, microlayer and organisms. *Marine Pollution Bulletin* 2:18–24, 1990.

Sodergren A, Adolfsson-Erici M, Bengtsson BE, Jonsson P, Lagergren S, Rahm L, Wulff F. *Environmental Impact of Bleached Pulp Mill Effluents: Composition, Fate and Effects in the Baltic Sea: Report of the Environment/Cellulose II Project.* Solna: Swedish Environmental Protection Agency (Report 4047), 1993.

Solomon KR. Chlorine in the bleaching of pulp and paper. *Pure and Applied Chemistry* 68:1721–1730, 1996.

Solomon K, Bergman H, Huggett R, Mackay D, McKague B. *A Review and Assessment of the Ecological Risks Associated with the Use of Chlorine Dioxide for the Bleaching of Pulp.* Washington, DC: Alliance for Environmental Technlogy, 1993.

Soto A, Chyung K, Sonnenschein C. The pesticides endosulfan, toxaphene, and dieldrin have estrogenic effects on human estrogen-sensitive cells. *Environmental Health Perspectives* 102:378–383, 1994.

Soto AM, Sonnenschein C, Chung KL, Fernandez MF, Olea N, Serrano FO. The E-SCREEN assay as a tool to identify estrogens: An update on estrogenic environmental pollutants. *Environmental Health Perspectives* 103 (supp 7):113–122, 1995.

Spirta R, Steward PA, Lee JS. Retrospective cohort mortality study of workers at an aircraft maintenance facility. I. Epidemiological results. *British Journal of Industrial Medicine* 48:515–530, 1991.

SRI International. *The Global Chlor-Alkali Industry: Strategic Implications and Impacts.* Zurich, 1993.

Staley L, Richards M, Huffman G, Chang D. *Incinerator Operating Parameters Which Correlate with Performance.* Cincinnati: U.S. EPA Hazardous Waste Engineering Research Laboratory (EPA 600/2-86-091), 1986.

Stancel GM, Boettger-Tong HL, Chiapette C, Hyder SM, Kirkland HL, Murthy L, Loose-Mitchell DS. Toxicity of endogenous and environmental estrogens: What is the role of elemental interactions? *Environmental Health Perspectives* 103 (supp 7):29–34, 1995.

Stanley JS. *Broad Scan Analysis of the FY 1982 National Human Adipose Tissues Survey Specimens.* Volume 1: *Executive Summary.* Washington, DC: U.S Environmental Protection Agency Office of Toxic Substances (EPA 560-5-86-035), 1986a.

Stanley JS. *Broad Scan Analysis of the FY 1982 National Human Adipose Tissue Survey Specimens.* Volume 3: *Semivolatile Compounds.* Washington, DC: U. S. Environmental Protection Agency (EPA 560/5-86/037), 1986b.

Stanley JS, Ayling RE, Cramer PH, Thornburg KR, Remmers JC, Breen JJ, Schwemberger J, Kiang HK, Watanabe K. Polychlorinated dibenzo-p-dioxin and dibenzofuran concentration levels in human adipose tissue samples from the continental United States collected from 1971 to 1987. *Chemosphere* 20:895–892, 1990a.

Stanley JS, Cramer PH, Ayling RE, Thornburg KR, Remmers JC, Breen JJ, Schwemberger J. Determination of the prevalence of polychlorinated diphenyl ethers (PCDPEs) in human adipose tissue samples. *Chemosphere* 20:981–985, 1990b.

Stevens A, Moore L, Slocum C, Smith L, Seeger D, Ireland J. By-products of chlorination at ten operating utilities. In: Jolley RL, Condie LW, Johnson JD, Katz S, Minear RA, Mattice JS, Jacobs VA, eds. *Water Chlorination: Chemistry, Environmental Impact and Health Effects,* Volume 6. Chelsea, MI: Lewis Publishers, pp. 579–604, 1990.

Stone R. Researchers score victory over pesticides—and pests—in Asia. *Science* 256: 1272–1273, 1992.

Stratton SC, Ferguson M. Progress report on the BFRTM technology demonstration, December 1996. *Pulp and Paper Canada,* March 1998:45–47.

Stringer RL, Johnston PA. Occurrence and toxicology of natural and anthropogenic organohalogens and relevance to environmental protection. In: Grimvall A, deLeer EWB, eds. *Naturally-Produced Organohalogens.* Boston: Kluwer Academic, 1995:415–427.

Stryer L. *Biochemistry,* Third Edition. New York: Freeman, 1988.

Subramian A, Tanabe S, Tatsukawa R, Saito S and Miyazaki N. Reduction in testosterone levels by PCBs and DDE in Dall's porpoises of the northwestern North Pacific. *Marine Pollution Bulletin* 19:643–646, 1987.

Suntio LR, Shiu WY, Mackay D. A review of the nature and properties of chemicals present in pulp mill effluents. *Chemosphere* 17:1249–1290, 1988.

Suskind RR. Human health effects of 2,4,5-T and its toxic contaminants. *Journal of the American Medical Association* 251:2372–2380, 1984.

Svalander JR. Pursuing new horizons: Taking PVC beyond the millennium (presentation materials). PVC '96 New Perspectives Conference, Brighton, UK, April 1996.

Svensson BG, Barregard L, Sallsten G, Nilsson A, Hansson M, Rappe C. Exposure to polychlorinated dioxins and dibenzofurans from graphite electrodes in a chloralkali plant. *Chemosphere* 27:259–262, 1993.

Svensson BG, Hallberg T, Nilsson A, Schutz A, Hagmar L. Parameters of immunological competence in subjects with high consumption of fish contaminated with persistent organochlorine compounds. *International Archives of Occupational and Environmental Health* 65:351–358, 1994.

Svensson BG, Mikoczy Z, Stromberg U, Hagmar L. Mortality and cancer incidence among Swedish fishermen with a high dietary intake of persistent organochlorine compounds. *Scandinavian Journal of Work and Environmental Health* 21:106–115, 1995.

Swain WR. Human health consequences of consumption of fish contaminated with organochlorine compounds. *Aquatic Toxicology* 11(3):357–377, 1988.

Swain WR. Effects of organochlorine chemicals on the reproductive outcomes of humans who consumed contaminated Great Lakes fish: An epidemiologic consideration. *Journal of Toxicology and Environmental Health* 33:587–640, 1991.

Swan SH, Elkin EP, Fenster L. Have sperm densities declined? A reanalysis of global trend data. *Environmental Health Perspectives* 105:1228–1232, 1997.

Swedish Ministry of the Environment. Sweden's offensive policy on chemicals (Factsheet 6.1). Stockholm, 1998.

Sweeney A. Reproductive epidemiology of dioxins. In: Schecter A, ed. *Dioxins and Health*. New York: Plenum, 1994:549–586.

Szal GM, Nolan PM, Keenedy LE, Barr CP, Bilger MD. The toxicity of chlorinated wastewater: Instream and laboratory case studies. *Research Journal of the Water Pollution Control Federation* 63:910–920, 1991.

Takei GH, Kauahikaua SM, Leong GH. Analyses of human milk samples collected in Hawaii for residues of organochlorine pesticides and polychlorobiphenyls. *Bulletin of Environmental Contamination and Toxicology* 30:606–613, 1983.

Takeshita R, Akimoto Y, Nito S. Relationship between the formation of polychlorinated dibenzo-p-dioxins and dibenzofurans and the control of combustion, hydrogen chloride level in flue gas, and gas temperature in a municipal waste incinerator. *Chemosphere* 24: 589–598, 1992.

Talmage DW, Bice DE, Bloom JC, Koller LD, Lamm E, Luster MI, Meggs WJ, Munson AE, Rodgers KE, Rose NR, Schulte PA, Travis CC, Tucker ES, Vogt RF, Waldmann TA, Burleson GR, Thomas RD, Beliles RP, Schneiderman MA, Kelly K, Crossgrove RE, Corriveau D, Walz J. *Biological Markers in Immunotoxicology*. Washington, DC: National Academy Press, 1992.

Tanner SA. Slow sand filtration: Still a timeless technology under the new regs. *Journal of the American Water Works Association* 89(12):14, 1997.

Tarone RE, Hayes HM, Hoover RN, Rosenthal JF, Brown LM, Pottern LM, Javadapour N, O'Connell KJ, Stutzman RE. Service in Vietnam and risk of testicular cancer. *Journal of the National Cancer Institute* 83:1497–1499, 1991.

Taruski AG, Olney CE, Winn HE. Chlorinated hydrocarbons in cetaceans. *Journal of the Fish Research Board of Canada* 32:2205–2209, 1975.

Tatsukawa R, Tanabe S. Fate and bioaccumulation of persistent organochlorine compounds in the marine environment. In: Baumgartner DJ, Dudall IM, eds. *Oceanic Processes in Marine Pollution,* Volume 6. Malabar, FL: Krieger, 1990:39–55.

Taylor F. *A History of Industrial Chemistry*. New York: Abelard Schuman, 1957.

Tellus Institute. CSG/Tellus Packaging Study for the Council of State Governments, U.S. Environmental Protection Agency, and New Jersey Department of Environmental Protection and Energy. Boston, 1992.

Tezak Z, Simic B, Kniewald J. Effect of pesticides on oestradiol receptor complex formation in rat uterus cytosol. *Food Chemical Toxicology* 30:879–885, 1992.

Thayer A. Chlor-alkali industry focuses on transition. *Chemical and Engineering News,* October 8, 1990:16–17.

Theisen J. *Untersuchung der Moglichen Umweltgefahrdung Beim Brand Von Kunststoffen* (Investigation of Possible Environmental Dangers Caused by Burning Plastics). Berlin: German Umweltbundesamt (104-09-222), 1991.

Theisen J, Funcke W, Balfanz E, Konig J. Determination of PCDFs and PCDDs in fire accidents and laaboratory combustion tests involving PVC-containing materials. *Chemosphere* 19:423–428, 1989.

Theobald HM, Peterson RE. Developmental and reproductive toxicity of dioxins and other Ah receptor agonists. In: Schecter A, ed. *Dioxins and Health*. New York: Plenum, 1994: 309–346.

Theobald HM, Peterson RE. In utero and lactational exposure to 2,3,7,8-tetrachlorodibenzo-p-dioxin: effects on development of the male and female reproductive system of the mouse. *Toxicology and Applied Pharmacology* 145:124–135, 1997.

Thiess AM, Frentzel-Beyme R, Link R. Mortality study of persons exposed to dioxin in a trichlorophenol-process accident that occurred in the BASF AG on November 17, 1953. *American Journal of Industrial Medicine* 3:179–189, 1982.

Thomas L. *The Medusa and the Snail.* New York: Viking, 1979.

Thomas TL, Decoulle P. Mortality among workers employed in the pharmaceutical industry: A preliminary investigation. *Journal of Occupational Medicine* 21:619–623, 1979.

Thomas TL, Waxweiler RJ, Moure-Eraso R, Itaya S, Fraumeni JF. Mortality patterns among workers in three Texas oil refineries. *Journal of Occupational Medicine* 24:135–141, 1982.

Thomas V, Spiro C. An estimation of dioxin emissions in the United States. *Toxicology and Environmental Chemistry* 50:1–37, 1995.

Thompson T, Clement R, Thornton N, Luyt J. Formation and emission of PCDDs/PCDFs in the petroleum refining industry. *Chemosphere* 20 10:1525–1532, 1990.

Tiernan TO, Taylor ML, Garrett JH, VanNess GF, Solch JG, Deis DA, Wagel DJ. Chlorodibenzodioxins, chlorodibenzofurans and related compounds in the effluents from combustion processes. *Chemosphere* 12:595–606, 1993.

Thornton J, McCally M, Orris P, Weinberg J. Dioxin prevention and medical waste incinerators. *Public Health Reports* 111:298–313, 1996.

Thummel CS. From embryogenesis to metamorphosis: Regulation and function of Drosophila nuclear receptor superfamily members. *Cell* 83:871–877, 1995.

Tichenor BA, Sparks LE, Jackson MD, Guo Z, Mason M, Plyunket CM, Rasor SA. Emissions of perchloroethylene from dry cleaned fabrics. *Atmospheric Environment* 24A: 1219–1229, 1990.

Tillitt DE, Gale RW, Meadows JC, Zajicek JL, Peterman PH, Heaton SN, Jones PD, Bursian SJ, Kubiak TJ, Giesy JP, Aulerich RJ. Dietary exposure of mink to carp from Saginaw Bay. 3. Characterization of dietary exposure to planar halogenated hydrocarbons, dioxin equivalents, and biomagnification. *Environmental Science and Technology* 30:283–291, 1996.

Tittle WL. Where will the chlorine industry be in ten years (presentation materials). Changing Chlorine Marketplace: Business Science and Regulations (conference), New Orleans, April 1995.

TNO Institute of Environmental and Energy Technology. *Sources of Dioxin Emissions into the Air in Western Europe.* Brussels: Eurochlor, 1994.

TNO Centre for Technology and Policy Studies. *A PVC Substance Flow Analysis for Sweden: Report for Norsk-Hydro.* Apeldoorn, Netherlands: TNO, 1996.

Tomatis L, Aito A, Day NE, Heseltine E, Kaldor J, Miller AB, Parkin DM, Riboli E. *Cancer: Causes, Occurrence and Control.* Lyon: World Health Organization (IARC Scientific Publication 100), 1990.

Tong HY, Gross MLO, Schecter A, Monson SJ, Dekin A. Sources of dioxin in the environment: Second stage study of PCDD/Fs in ancient human tissue and environmental samples. *Chemosphere* 20:987–992, 1990.

Tonn T, Esser C, Schneider EM, Steinmann-Steiner-Haldenstatt W, Gleichmann E. Persistence of decreased T-helper cell function in industrial workers 20 years after exposure to 2,3,7,8-tetrachlorodibenzo-p-dioxin. *Environmental Health Perspectives* 104:422–426, 1996.

Toppari J, Larsen JC, Christiansen P, Giwercman A, Grandjean P, Guillette LJ, Jegou B, Jensen TK, Jouannet P, Keidig N, Leffers H, McLachlan JA, Meyer O, Muller J, Rajpert-DeMeyts E, Scheike T, Sharpe R, Sumpter J, Skakkebaek NE. Male reproductive health and environmental xenoestrogens. *Environmental Health Perspectives* 104 (supp 4):741–806, 1996.

Toren K, Persson B, Wingren G. Health effects of working in pulp and paper mills: Malignant diseases. *American Journal of Industrial Medicine* 29:123–130, 1996.

Trenholm A, Agorman P, Jungclaus G. *Performance Evaluation of Full-Scale Hazardous Waste Incinerators.* Volume 1: *Executive Summary.* Cincinnati: U.S. Environmental Protection Agency, Industrial Environmental Research Lab, 1984.

Trenholm A, Lee CC. Analysis of PIC and total mass emissions from an incinerator. In: *Land Disposal, Remedial Action, Incineration and Treatment of Hazardous Waste,* Proceedings of the Twelfth Annual Research Symposium. Cincinnati: U.S. Environmental Protection Agency Hazardous Waste Engineering Research Laboratory (EPA 600/9-86/022), 1986.

Trenholm A, Thurnau R. Total mass emissions from a hazardous waste incinerator. In: *Land Disposal, Remedial Action, Incineration, and Treatment of Hazardous Waste,* Proceedings of the Thirteenth Annual Research Symposium. Cincinnati: U.S. Environmental Protection Agency Hazardous Waste Engineering Laboratory (EPA/600/9-87/015), 1987.

Trescott MM. *The Rise of the American Electrochemicals Industry, 1880–1910.* Westport: Greenwood Press, 1981.

Tritscher AM, Clark GS, Lucier GW. Dose-response effects of dioxins: Species comparison and implications for risk assessment. In: Schecter A, *Dioxins and Health.* New York: Plenum, 1994:227–248.

Tritscher AM, Seacat AM, Yager JD, Groopman JD, Miller BD, Bell D, Sutter TR, Lucier GW. Increased oxidative DNA damage in livers of 2,3,7,8-tetraclorodibenzo-p-dioxin treated intact but not ovariectomatized rats. *Cancer Letters* 98:219–225, 1996.

Trojan Technologies. *The Facts on Disinfection Costs.* London, Ontario: Trojan Technologies, 1991.

Tryphonas H. Immunotoxicity of PCBs (Aroclors) in relation to Great Lakes. *Environmental Health Perspectives* 103 (supp 9):35–46, 1995.

Tschirley F. Dioxin: Concern that this material is harmful to health or the environment may be misplaced. *Scientific American* 254(2):29–35, 1986.

Tukker A, deGroot JLB, van de Hofstadt P. *PVC in Europe: Environmental Concerns, Measures and Markets.* Apeldoorn, Netherlands: TNO Policy Research Group, 1995.

Tullos BG. The caustic/chlorine relationship (presentation materials). Changing Chlorine Marketplace: Business Science and Regulations (conference), New Orleans, April 1995.

Tundo P, Selva P, Marques CA. Selective monomethylation of arylacetonitriles and methyl arylacetates by dimethylcarbonate: A process without production of waste. In: Anastas PT, Williamson TC, eds. *Green Chemistry: Designing Chemistry for the Environment.* Washington, DC: American Chemical Society, 1996:81–91.

TURI (Toxics Use Reduction Institute). *Garment Wet Cleaning: Cleaner Technology Demonstrations Site Case Study: Utopia Cleaners, Arlington, MA.* Lowell, MA: University of Lowell, 1997.

Tysklind M, Soderstrom G, Rappe C. PCDD and PCDF emissions from scrap metal melting processes at a steel mill. *Chemosphere* 19:705–710, 1989.

UNEP (United Nations Environmental Program). *Executive Summary of the Ozone Trends Panel.* New York, 1988.

UNEP (United Nations Environmental Program). *Environmental Effects of Ozone Depletion: 1994 Assessment.* New York, 1994.

UNEP (United Nations Environmental Program). *Environmental Effects of Ozone Depletion (Interim Report)*. New York, 1996.

UNEP (United Nations Environmental Program). Decisions Adopted by The Governing Council At Its Nineteenth Session: 13C. International Action to Protect Human Health and the Environment through Measures which Will Reduce and/or Eliminate Emissions and Discharges of Persistent Organic Pollutants, including the Development of an International Legally Binding Instrument. Geneva, 1997.

UNEP (United Nations Environmental Program). *International Declaration on Cleaner Production.* New York, 1998.

U.S. Department of Commerce. *Statistical Abstract of the United States.* Washington, DC: U.S. Government Printing Office, 1993.

U.S. Department of Energy. *Recycling Chlorine in the Pulp Manufacturing Process: NICE3: National Industrial Competitiveness through Energy, Environment, Economics.* Washington, DC: DOE Office of Industrial Technology, 1998 (available at http://www.oit.doe.gov/Access/nice3/4063-227.html).

U.S. District Court, Northern District of Ohio, Eastern Division. *Greenpeace, Inc., et al., vs. Waste Technologies Industries, et al.* Case no. 4:93CV083, March 5, 1993.

U.S. House of Representatives, Committee on Energy and Commerce, Subcommittee on Oversight and Investigations. *Environmental Protection Agency: Investigation of Superfund and Agency Abuses, Part 1: Alteration of EPA Scientific Studies; House Hearing Report Y4EN2/3 98-81, March 18, 1983.* Washington, DC: U.S. Government Printing Office, 1983.

U.S. House of Representatives, Committee on Government Operations. *The Agent Orange Coverup: A Case of Flawed Science and Political Manipulation; House Report HR 101-672.* Washington, DC: U.S. Government Printing Office, 1990.

UBA. German Federal Office of the Environment. *Environmental Damage by PVC: An Overview.* Berlin: Umweltbundesamt, 1992.

Uhler B. Atrazine. *Journal of Pesticide Reform* 11:4–9, 1991.

UKDOE (United Kingdom Department of the Environment). *Dioxins in the Environment.:* Pollution Paper 27. London, 1989.

Ulrich H. Isocyanates, organic. In: Elvers B, Hawkins S, Ravenscroft M, Schulz G, eds. *Ullman's Encyclopedia of Industrial Chemistry.* New York: VCH Publishers, A14:611–626, 1989.

Union Camp Corporation. *Ozone Bleaching Technology.* Franklin, VA: Union Camp, 1998 (available at http://www.unioncamp./com/technology/ozone.html).

Valenti M. Lighting the way to improved disinfection. *Mechanical Engineering* 119(7): 82–86, July 1997.

Van der Gulden JWJ, Zielhuis GA. Reproductive hazards related to perchloroethylene. *International Archives of Occupational and Environmental Health* 61:235–242, 1989.

Van Strum C, Merrell P. Dioxin human health damage studies: Damaged data? *Journal of Pesticide Reform* 10:8–13, 1990.

Van Waeleghem K, De Clercq N, Vermeulen L, Schoonjans F, Comhaire F. Deterioration of sperm quality in young healthy Belgian men. *Human Reproduction* 11:325–329, 1996.

van't Veer P, Lobbezo IE, Martin-Moreno JM, Guallar E, Gomez-Aracena J, Kardinal AFM, Kohlmeier L, Martin BC, Strain JJ, Thamm M, van Zoonen P, Baumann BA, Huttunen JH, Kok FH. DDT (dicophane) and postmenopausal breast cancer in Europe: Case-control study. *British Medical Journal* 315:81–85, 1997.

van Birgelen APJM, Fase KM, van derk Kolk J, Poiger J, Brouwe A, Seinen W, van den Berg M. Synergistic effect of 2,2′,4,4′,5,5′-hexachlorobiphenyl and 2,3,7,8-tetrachlorodibenzo-p-dioxin on hepatic porphyrin levels in the rat. *Environmental Health Perspectives* 104:550–557, 1996.

Vaughan TL, Stewart PA, Davis S, Thomas DB. Work in dry cleaning and the incidence of cancer of the oral cavity, larynx, and oesophagus. *Occupational and Environmental Medicine* 54:692–695, 1997.

Ventresque C, Bablon G, Legube B, Jadas-Hecart A, Dore D. Development of chlorine demand kinetics in a drinking water treatment plant. In: Jolley RL, Condie LW, Johnson JD, Katz S, Minear RA, Mattice JS, Jacobs VA, eds. *Water Chlorination: Chemistry, Environmental Impact and Health Effects,* Volume 6. Chelsea, MI: Lewis Publishers, 1990:715–728.

Verbanic CJ. Can chlorine and caustic recycle the good times? *Chemical Business,* September 1990:24–35.

Versar, Inc. *Formation and Sources of Dioxin-like Compounds: A Background Issue Paper for U.S. Environmental Protection Agency, National Center for Environmental Assessment.* Springfield, VA: Versar, 1996.

Vesterinen R, Flyktmann M. Organic emissions from co-combustion of RDF with wood chips and milled peat in a bubbling fluidized bed boiler. *Chemosphere* 32:681–689, 1996.

Villeneuve DC, Chu I, Rousseaux CG. Toxicology studies on mixtures of contaminants in the Great Lakes. In: Yang RSH, ed. *Toxicology of Chemical Mixtures.* New York: Academic Press, 1994:51–61.

Vineis P, D'Amore F. The role of occupational exposure and immunodeficiency in B-cell malignancies. *Epidemiology* 3:266–270, 1992.

Vinyl Institute. Proposed EPA Chlorine Study Declares War on Modern Society, Vinyl Institute Says (press release). Wayne, New Jersey, February 15, 1994.

Vinyl Institute. *Comments on EPA's Reassessment of Dioxin.* Morristown, NJ: Vinyl Institute, 1995.

Vinyl Institute. *The Vinyl Institute Dioxin Characterization Program Phase I Report.* Morristown, NJ: Vinyl Institute, 1998.

Visalli JR. A comparison of dioxin, furan and combustion gas data from test programs at three MSW incinerators. *Journal of the Air Pollution Control Association* 3712:1451–1463, 1987.

vom Saal FS, Montano MM, Wang MH. Sexual differentiation in mammals. In: Colborn T, Clement C, eds. *Chemically-Induced Alterations in Sexual and Functional Development: The Wildlife/Human Connection.* Princeton: Princeton Scientific Publishing, 1992:17–84.

vom Saal FS, Nagel SC, Palanza P, Boechler M, Parmigiani S, Welshons WV. Estrogenic pesticides: Binding relative to estradiol in MCF-7 cells and effects of exposure during fetal life on subsequent territorial behavior in male mice. *Toxicology Letters* 77:343–350, 1995.

vom Saal FS, Timms BG, Montano MM, Palanza P, Thayer KA, Nagel SC, Dhar MD, Ganjam VK, Parmigiani S, Welshons WV. Prostate enlargement in mice due to fetal exposure to low doses of estradiol or diethylstilbestrol and opposite effects at high doses. *Proceedings of the National Academy of Sciences* USA 94:2056–2061, 1997.

Von Moltke K. The relationship between policy, science, technology, economics, and law in the implementation of the precautionary principle. In: Freestone D, Hey E, eds. *The Precautionary Principle and International Law: The Challenge of Implementation.* Boston: Kluwer Law International, 1996:97–108.

Vonier PM, Crain DA, McLachlan JA, Guillette LJ, Arnold SF. Interaction of environmental chemicals with the estrogen and progesterone receptors from the oviduct of the American alligator. *Environmental Health Perspectives* 104:1318–1322, 1996.

Vore RD, Unhoch MJ. The use of PHMB as a sanitizer in domestic Spas. Abstracts from the 1st Annual NSPI Sanitizer Chemistry Symposium. Alexandria, VA: National Spa and Pool Institute, 1994.

Wagenaar H, Langeland K, Hardman R, Sergeant Y, Brenner K, Sandra P, Rappe C, Tiernan T. Analyses of PCDDs and PCDFs in virgin suspension PVC resin. *Organohalogen Compounds* 27:72–74, 1996.

Wagener KB, Batich CD, Green AES. Polymer substitutes for medical grade polyvinyl chloride. In: Green AES, ed. *Medical Waste Incineration and Pollution Prevention.* New York: Van Nostrand Reinhold, 1993:155–169.

Wagner J, Green A. Correlation of chlorinated organic compound emissions from incineration with chlorinated organic input. *Chemosphere* 26:2039–2054, 1993.

Wahlstrom B, Lundqvist B. Risk reduction and chemicals control—lessons from the Swedish chemicals action programme. In: Jackson T, ed. *Clean Production Strategies: Developing Preventive Environmental Management in the Industrial Economy.* Ann Arbor: Lewis, 1993:237–260.

Walker MK, Peterson RE. Aquatic toxicity of dioxins and related chemicals. In: Schecter A, ed. *Dioxins and Health.* New York: Plenum, 1994:347–388.

Walker SE, Keisler LW, Caldwell CW, Kier AB, vom Saal FS. Effects of altered prenatal hormonal environment on expression of autoimmune disease in NZB/NZW mice. *Environmental Health Perspectives* 104 (supp 4):815–822, 1996.

Wallace D. *In the Mouth of the Dragon.* Yonkers, NY: Consumers Union, 1990.

Wallace LA, Pellizzari E, Hartwell T, Rosenzweig M, Erickson M, Sparacino C, Zelon H. Personal exposure to volatile organic compounds. I. Direct measurements in breathing-zone air, drinking water, food, and exhaled breath. *Environmental Research* 35:293–319, 1984.

Wallace LA, Pellizzari ED, Hartwell TD, Whitmore R, Zelon H, Perritt R, Sheldon L. The California TEAM Study: Breath concentrations and personal exposures to 26 volatile compounds in air and drinking water of 188 residents of Los Angeles, Antioch, and Pittsburg, CA. *Atmospheric Environment* 22:2141–2163, 1988.

Waller CL, Juma BW, Gray LE, Kelce WR. Three-dimensional quantitative structure-activity relationships for androgen receptor ligands. *Toxicology and Appplied Pharmacology* 137: 219–227, 1996.

Waller K, Swan SH, DeLorenze G, Hopkins B. Trihalomethanes in drinking water and spontaneous abortion. *Epidemiology* 9:134–140, 1998.

Walrath J, Li FP, Hoar SK, Mead MW, Fraumeni JF. Causes of death among female chemists. *American Journal of Public Health* 75:883–884, 1985.

Waltermire T. The vinyl plastics industry (presentation materials). Goldman, Sachs Fourth Annual Chemical Investor Forum, New York, May 1996.

Walters LM, Rourke AW, Eroschenko VP. Purified methoxychlor stimulates the reproductive tract in immature female mice. *Reproductive Toxicology* 7:599–606, 1993.

Ward M. Greener dry cleaners. *Technology Review* 100(8):22, November 21, 1997.

Warner F. What if? versus if it ain't broke don't fix it. In: O'Riordan T, Cameron J, eds. *Interpreting the Precautionary Principle.* London: Earthscan Publications, 1994:102–109.

Warren S. Toy makers say bye-bye to "plasticizers." *Wall Street Journal,* November 12, 1998.

Webster T. Why dioxins and other halogenated hydrocarbons are bad news. *Journal of Pesticide Reform* 9:32–35, 1990.

Webster T, commoner B. Overview: The dioxin debate. In: Schecter A, ed. *Dioxins and Health.* New York: Plenum, 1994:1–50.

Weegar AK. More contamination, less concern: Public apathy about poisoned wildlife slows the search for answers. *Maine Times,* July 15, 1994:2.

Weinberg J, Finaldi L. Foreword to Costner P. *The Burning Question: Chlorine and Dioxin.* Amsterdam: Greenpeace International 1997.

Weinberg J, Thornton J. Precautionary inference. In: Gilbertson M, Cole-Misch S, eds. *Applying Weight of Evidence: Issues and Practice.* Windsor: International Joint Commission, 1994.

Weisel CP, Jo WK. Ingestion, inhalation, and dermal exposures to chloroform and trichloroethene from tap water. *Environmental Health Perspectives* 104:48–51, 1996.

Weisenburger DD. Epidemiology of non-Hodgkin's lymphoma: Recent findings regarding an emerging epidemic. *Annals of Oncology* 5 (supp 1):s19–s24, 1994.

Weiss B. Risk assessment: The insidious nature of neurotoxicity and the aging brain. *Neurotoxicology* 11:305–314, 1990.

Weiss NS. Cancer in relation to occupational exposure to perchloroethylene. *Cancer Causes and Control* 6:257–266, 1995.

Welch LS, Paul ME. *Reproductive and Developmental Hazards: Case Studies in Environmental Medicine 29.* Washington, DC: Agency for Toxic Substances and Disease Registry, 1993.

Wester PG, DeBoer J. Determination of polychlorinated terphenyls in biota and sediments with gas chromatography/mass spectrometry. *Organohalogen Compounds* 14:121–124, 1993.

Westin JB. Carcinogens in Israeli milk: A study in regulatory failure. *International Journal of Health Services* 23:497–517, 1993.

Whitby GE, Palmateer G, Cook WG, Maarschalkerweerd, Huber D, Flood K. Ultraviolet disinfection of secondary effluent. *Journal of the Water Pollution Control Federation* 56: 844–850, 1984.

White E. Projected changes in breast cancer incidence due to the trend toward delayed childbearing. *American Journal of Public Health* 77:495–497, 1987.

Whitlock JP. The aromatic hydrocarbon receptor, dioxin action, and endocrine homeostasis. *Trends in Endocrinology and Metabolism* 5:183–188, 1994.

Whitten PL. Chemical revolution to sexual revolution: Historical changes in human reproductive development. In: Colborn T, Clement C, eds. *Chemically-Induced Alterations in Sexual and Functional Development: The Wildlife-Human Connection.* Princeton: Princeton Scientific Publishing, 1992:311–334.

Whittington WA. Memorandum: EPA/Paper Industry Dioxin Investigation. Washington, DC: U.S. Environmental Protection Agency Office of Water, July 10, 1986.

Whorton D, Krauss RM, Marshall S, Milby TH. Infertility in male pesticide workers. *Lancet* 2(8051):1259–1261, 1977.

Whorton D. Milby TH, Krauss RM. Stubbs HA. Testicular function in DBCP exposed pesticide workers. *Journal of Occupational Medicine* 21:161–166, 1979.

Wienecke J, Kruse H, Huckfeldt U, Eickhoff W, Wasserman O. Organic compounds in the flue gas of a hazardous waste incinerator. *Chemosphere* 30:907–913, 1995.

Wiklund K, Dich J, Holm LE. Testicular cancer among agricultural workers and licensed pesticide applicators in Sweden. *Scandinavian Journal of Work and Environmental Health* 12: 630–631, 1986.

Wikstrom E, Lofvenius G, Rappe C. Influence of level and form of chlorine on the formation of chlorinated dioxins, dibenzofurans, and benzenes during combustion of an artificial fuel in a laboratory reactor. *Environmental Science and Technology* 305:1637–1644, 1996.

Wilken M. *Dioxin Emissions from Furnaces, Particularly from Wood Furnaces.* Berlin: Engineering Group for Technological Environmental Protection for the German Environmental Protection Agency, February 1994.

Willes, RF, Nestmann ER, Miller PA, Orr JC, Munro IC. Scientific principles for evaluating the potential for adverse effects from chlorinated organic chemicals in the environment. *Regulatory Toxicology and Pharmacology* 18:313–356, 1993.

Willett KL, Ulrich EM, Hites RA. Differential toxicity and environmental fates of hexachlorocyclohexane isomers. *Environmental Science and Technology* 32:2197–2207, 1998.

Willett WC, Hunter DJ, Stampfer MJ, Colditz G, Manson JE, Spiegelman D, Rosner B, Hennekens CH, Speizer FE. Dietary fat and fiber in relation to risk of breast cancer: A 8-year follow-up. *Journal of the American Medical Association* 268:2037–2044, 1992.

Williams DT, LeBel GL. Polychlorinated biphenyl congener residues in human adipose tissue samples from five Ontario municipalities. *Chemosphere* 20:33–42, 1990.

Williams MK. Lead pollution on trial. *New Scientist* 51:578–580, 1971.

Wingren G, Persson B, Thoren K, Axelson O. Mortality pattern among pulp and paper mill workers in Sweden: A case-referent study. *American Journal of Industrial Medicine* 20: 769–774, 1991.

Winner L. *The Whale and the Reactor: A Search for Limits in an Age of High Technology.* Chicago: University of Chicago Press, 1986.

Wirts M, Lorenz W, Bahadir M. Does co-combustion of PVC and other plastics lead to enhanced formation of PCDD/F? *Chemosphere* 37:1489–1500, 1998.

WMO/UNEP (World Meteorological Organization and United Nations Environmental Programme). *Report of the Third Meeting of the Ozone Research Managers.* Geneva: World Meteorological Organization, 1996.

Wolff MS, Toniolo PG, Lee EW, Rivera M, Dubin N. Blood levels of organochlorine residues and risk of breast cancer. *Journal of the National Cancer Institute* 85:648–652, 1993.

Wolfe RL. Ultraviolet disinfection of potable water. *Environmental Science and Technology* 24:768–773, 1990.

Wolff MS, Toniolo PG, Lee EW, Rivera M, Dubin N. Blood levels of organochlorine residues and risk of breast cancer. *Journal of the National Cancer Institute* 85:648–652, 1993.

Wolff MS, Collman GW, Barrett JC, Huff JC. Breast cancer and environmental risk factors: Epidemiological and experimental findings. *Annual Review of Pharmacology and Toxicology* 36:573–596, 1996.

Wood A. Dow move adds a twist to lime. *Chemical Week,* January 15, 1992:22.

World Wildlife Fund. *Resolving the DDT Dilemma.* Washington, DC: World Wildlife Fund, 1998.

Wray TK. Chlorine: A deadly disinfectant. *Hazmat World,* July 1993:84–85.

Wren CD. Cause-effect linkages between chemicals and populations of mink (*Mustela vision*) and otter (*Lutra canadensis*) in the Great Lakes Basin. *Journal of Toxicology and Environmental Health* 33:549–586, 1991.

Wright S. *Evolution and the Genetics of Populations.* Volume 2: *The Theory of Gene Frequencies.* Chicago: University of Chicago Press, 1969.

Wurtz JM, Bourguet W, Renaud JP, Vivat V, Chambon P, Mora D, Gronemeyer H. A canonical structure for the ligand-binding domain of nuclear receptors. *Nature Structural Biology* 3:87–94, 1996.

Wynder EL, Rose DP, Cohen LA. Diet and breast cancer in causation and therapy. *Cancer* 58:1804–1813, 1986.

Wynne B. Uncertainty and environmental learning. In: Jackson T, ed. *Clean Production Strategies: Developing Preventive Environmental Management in the Industrial Economy.* Boca Raton: Lewis, 1993:63–84.

Yadav AK, Singh TP. Pesticide-induced changes in peripheral thyroid hormone levels during different reproductive phases in *Heteropneustes fossilis. Ecotoxicology and Environmental Safety* 13:97–103, 1987.

Yanchinski S. New analysis links dioxin to cancer. *New Scientist* 124(1688):24, 1989.

Yang RSH. Introduction to the toxicology of chemical mixtures. In: Yang RSH, ed. *Toxicology of Chemical Mixtures.* New York: Academic Press, 1994a:1–10.

Yang RSH. Toxicology of chemical mixtures derived from hazardous waste sites or application of pesticides and fertilizers. In: Yang RSH, ed. *Toxicology of Chemical Mixtures.* New York: Academic Press, 1994b:99–117.

Yang YG, Lebrec H, Burleson GR. Effect of 2,3,7,8-tetrachlorodibenzo-p-dioxin on pulmonary influenza virus titer and natural killer activity in rats. *Fundamental and Applied Toxicology* 23:125–131, 1994.

Yasuhara A, Morita M. Formation of chlorinated aromatic hydrocarbons by thermal decomposition of vinylidene chloride polymer. *Environmental Science and Technology* 22: 646–650, 1988.

Zack JA, Gaffey WR. The mortality experience of workers employed at the Monsanto company plant in Nitro, West Virginia. *Environmental Science Research* 26:575–591, 1983.

Zack JA, Suskind RR. The mortality experience of workers exposed to tetrachlorodibenzodioxin in a trichlorophenol process accident. *Journal of Occupational Medicine* 22:11–14, 1980.

Zahm SH, Blair A. Pesticides and non-Hodgkin's lymphoma. *Cancer Research* (supp): 5485s–5488s, 1992.

Zahm SH, Blair A. Cancer among migrant and seasonal farmworkers: An epidemiologic review and research agenda. *American Journal of Industrial Medicine* 24:753–766, 1993.

Zahm SH, Devesa SS. Childhood cancer: Overview of incidence trends and environmental carcinogens. *Environmental Health Perspectives* 103 (supp 6):177–184, 1995.

Zava DT, Bleu M, Duwe G. Estrogenic activity of natural and synthetic estrogens in human breast cancer cells in culture. *Environmental Health Perspectives* 105 (supp 3):637–645, 1997.

Zelikoff JT. Fish immunotoxicology. In: Dean JH, Luster MI, Munson AE, Kimber I, eds. *Immunotoxicology and Immunopharmacology.* New York: Raven Press, 1994:71–95.

Zhou YX, Waxman DJ. Activation of peroxisome proliferator-activated receptors by chlorinated hydrocarbons and endogenous steroids. *Environmental Health Perspectives* 106 (supp 4):983–988, 1998.

Zile MH. Vitamin A homeostasis endangered by environmental pollutants. *Proceedings of the Society for Experimental Biology and Medicine* 201:141–153, 1992.

Zober AM, Ott MG, Messerer P. Morbidity follow up study of BASF employees exposed to 2,3,7,8-tetrachlorodibenzo-p-dioxin (TCDD) after a 1953 chemical reactor incident. *Occupational and Environmental Medicine* 51:479–486, 1994.

Zook DR, Rappe C. Environmental sources, distribution and fate. In: Schecter A, ed. *Dioxins and Health.* New York: Plenum, 1994:79–113.

Index